ASTRONOMICAL DATA ANALYSIS
SOFTWARE AND SYSTEMS IV

A SERIES OF BOOKS ON RECENT DEVELOPMENTS IN ASTRONOMY AND ASTROPHYSICS

A.S.P. CONFERENCE SERIES PUBLICATIONS COMMITTEE

Dr. Sallie L. Baliunas, Chair
Dr. John P. Huchra
Dr. Roberta M. Humphreys
Dr. Catherine A. Pilachowski

© Copyright 1995 Astronomical Society of the Pacific
390 Ashton Avenue, San Francisco, California 94112

All rights reserved

Printed by BookCrafters, Inc.

First published 1995

Library of Congress Catalog Card Number: 95-79535
ISBN 0-937707-96-1   ISSN 1080-7926

D. Harold McNamara, Managing Editor of Conference Series
408 ESC Brigham Young University
Provo, UT 84602
801-378-2298

A SERIES OF BOOKS ON RECENT DEVELOPMENTS IN ASTRONOMY AND ASTROPHYSICS

Vol. 1-Progress and Opportunities in Southern Hemisphere Optical Astronomy: The CTIO 25th Anniversary Symposium
ed. V. M. Blanco and M. M. Phillips   ISBN 0-937707-18-X

Vol. 2-Proceedings of a Workshop on Optical Surveys for Quasars
ed. P. S. Osmer, A. C. Porter, R. F. Green, and C. B. Foltz   ISBN 0-937707-19-8

Vol. 3-Fiber Optics in Astronomy
ed. S. C. Barden   ISBN 0-937707-20-1

Vol. 4-The Extragalactic Distance Scale: Proceedings of the ASP 100th Anniversary Symposium
ed. S. van den Bergh and C. J. Pritchet   ISBN 0-937707-21-X

Vol. 5-The Minnesota Lectures on Clusters of Galaxies and Large-Scale Structure
ed. J. M. Dickey   ISBN 0-937707-22-8

Vol. 6-Synthesis Imaging in Radio Astronomy: A Collection of Lectures from the Third NRAO Synthesis Imaging Summer School
ed. R. A. Perley, F. R. Schwab, and A. H. Bridle   ISBN 0-937707-23-6

Vol. 7-Properties of Hot Luminous Stars: Boulder-Munich Workshop
ed. C. D. Garmany   ISBN 0-937707-24-4

Vol. 8-CCDs in Astronomy
ed. G. H. Jacoby   ISBN 0-937707-25-2

Vol. 9-Cool Stars, Stellar Systems, and the Sun. Sixth Cambridge Workshop
ed. G. Wallerstein   ISBN 0-937707-27-9

Vol. 10-The Evolution of the Universe of Galaxies. The Edwin Hubble Centennial Symposium
ed. R. G. Kron   ISBN 0-937707-28-7

Vol. 11-Confrontation Between Stellar Pulsation and Evolution
ed. C. Cacciari and G. Clementini   ISBN 0-937707-30-9

Vol. 12-The Evolution of the Interstellar Medium
ed. L. Blitz   ISBN 0-937707-31-7

Vol. 13-The Formation and Evolution of Star Clusters
ed. K. Janes   ISBN 0-937707-32-5

Vol. 14-Astrophysics with Infrared Arrays
ed. R. Elston   ISBN 0-937707-33-3

Vol. 15-Large-Scale Structures and Peculiar Motions in the Universe
ed. D. W. Latham and L. A. N. da Costa   ISBN 0-937707-34-1

Vol. 16-Atoms, Ions and Molecules: New Results in Spectral Line Astrophysics
ed. A. D. Haschick and P. T. P. Ho   ISBN 0-937707-35-X

Vol. 17-Light Pollution, Radio Interference, and Space Debris
ed. D. L. Crawford   ISBN 0-937707-36-8

Vol. 18-The Interpretation of Modern Synthesis Observations of Spiral Galaxies
ed. N. Duric and P. C. Crane   ISBN 0-937707-37-6

Vol. 19-Radio Interferometry: Theory, Techniques, and Application, IAU Colloquium 131
ed. T. J. Cornwell and R. A. Perley   ISBN 0-937707-38-4

Vol. 20-Frontiers of Stellar Evolution, celebrating the 50th Anniversary of McDonald Observatory
ed. D. L. Lambert   ISBN 0-937707-39-2

Vol. 21-The Space Distribution of Quasars
ed. D. Crampton
ISBN 0-937707-40-6

Vol. 22-Nonisotropic and Variable Outflows from Stars
ed. L. Drissen, C. Leitherer, and A. Nota
ISBN 0-937707-41-4

Vol. 23-Astronomical CCD Observing and Reduction Techniques
ed. S. B. Howell
ISBN 0-937707-42-4

Vol. 24-Cosmology and Large-Scale Structure in the Universe
ed. R. R. de Carvalho
ISBN 0-937707-43-0

Vol. 25-Astronomical Data Analysis Software and Systems I
ed. D. M. Worrall, C. Biemesderfer, and J. Barnes
ISBN 0-937707-44-9

Vol. 26-Cool Stars, Stellar Systems, and the Sun, Seventh Cambridge Workshop
ed. M. S. Giampapa and J. A. Bookbinder
ISBN 0-937707-45-7

Vol. 27-The Solar Cycle
ed. K. L. Harvey
ISBN 0-937707-46-5

Vol. 28-Automated Telescopes for Photometry and Imaging
ed. S. J. Adelman, R. J. Dukes, Jr., and C. J. Adelman
ISBN 0-937707-47-3

Vol. 29-Workshop on Cataclysmic Variable Stars
ed. N. Vogt
ISBN 0-937707-48-1

Vol. 30-Variable Stars and Galaxies, in honor of M. S. Feast on his retirement
ed. B. Warner
ISBN 0-937707-49-X

Vol. 31-Relationships Between Active Galactic Nuclei and Starburst Galaxies
ed. A. V. Filippenko
ISBN 0-937707-50-3

Vol. 32-Complementary Approaches to Double and Multiple Star Research, IAU Collouquium 135
ed. H. A. McAlister and W. I. Hartkopf
ISBN 0-937707-51-1

Vol. 33-Research Amateur Astronomy
ed. S. J. Edberg
ISBN 0-937707-52-X

Vol. 34-Robotic Telescopes in the 1990s
ed. A. V. Filippenko
ISBN 0-937707-53-8

Vol. 35-Massive Stars: Their Lives in the Interstellar Medium
ed. J. P. Cassinelli and E. B. Churchwell
ISBN 0-937707-54-6

Vol. 36-Planets and Pulsars
ed. J. A. Phillips, S. E. Thorsett, and S. R. Kulkarni
ISBN 0-937707-55-4

Vol. 37-Fiber Optics in Astronomy II
ed. P. M. Gray
ISBN 0-937707-56-2

Vol. 38-New Frontiers in Binary Star Research
ed. K. C. Leung and I. S. Nha
ISBN 0-937707-57-0

Vol. 39-The Minnesota Lectures on the Structure and Dynamics of the Milky Way
ed. Roberta M. Humphreys
ISBN 0-937707-58-9

Vol. 40-Inside the Stars, IAU Colloquium 137
ed. Werner W. Weiss and Annie Baglin
ISBN 0-937707-59-7

Vol. 41-Astronomical Infrared Spectroscopy: Future Observational Directions
ed. Sun Kwok
ISBN 0-937707-60-0

Vol. 42-GONG 1992: Seismic Investigation of the Sun and Stars
ed. Timothy M. Brown
ISBN 0-937707-61-9

Vol. 43-Sky Surveys: Protostars to Protogalaxies
ed. B. T. Soifer   ISBN 0-937707-62-7

Vol. 44-Peculiar Versus Normal Phenomena in A-Type and Related Stars
ed. M. M. Dworetsky, F. Castelli, and R. Faraggiana   ISBN 0-937707-63-5

Vol. 45-Luminous High-Latitude Stars
ed. D. D. Sasselov   ISBN 0-937707-64-3

Vol. 46-The Magnetic and Velocity Fields of Solar Active Regions, IAU Colloquium 141
ed. H. Zirin, G. Ai, and H. Wang   ISBN 0-937707-65-1

Vol. 47-Third Decinnial US-USSR Conference on SETI
ed. G. Seth Shostak   ISBN 0-937707-66-X

Vol. 48-The Globular Cluster-Galaxy Connection
ed. Graeme H. Smith and Jean P. Brodie   ISBN 0-937707-67-8

Vol. 49-Galaxy Evolution: The Milky Way Perspective
ed. Steven R. Majewski   ISBN 0-937707-68-6

Vol. 50-Structure and Dynamics of Globular Clusters
ed. S. G. Djorgovski and G. Meylan   ISBN 0-937707-69-4

Vol. 51-Observational Cosmology
ed. G. Chincarini, A. Iovino, T. Maccacaro, and D. Maccagni   ISBN 0-937707-70-8

Vol. 52-Astronomical Data Analysis Software and Systems II
ed. R. J. Hanisch, J. V. Brissenden, and Jeannette Barnes   ISBN 0-937707-71-6

Vol. 53-Blue Stragglers
ed. Rex A. Saffer   ISBN 0-937707-72-4

Vol. 54-The First Stromlo Symposium: The Physics of Active Galaxies
ed. Geoffrey V. Bicknell, Michael A. Dopita, and Peter J. Quinn   ISBN 0-937707-73-2

Vol. 55-Optical Astronomy from the Earth and Moon
ed. Diane M. Pyper and Ronald J. Angione   ISBN 0-937707-74-0

Vol. 56-Interacting Binary Stars
ed. Allen W. Shafter   ISBN 0-937707-75-9

Vol. 57-Stellar and Circumstellar Astrophysics
ed. George Wallerstein and Alberto Noriega-Crespo   ISBN 0-937707-76-7

Vol. 58-The First Symposium on the Infrared Cirrus and Diffuse Interstellar Clouds
ed. Roc M. Cutri and William B. Latter   ISBN 0-937707-77-5

Vol. 59-Astronomy with Millimeter and Submillimeter Wave Interferometry
ed. M. Ishiguro and Wm. J. Welch   ISBN 0-937707-78-3

Vol. 60-The MK Process at 50 Years: A Powerful Tool for Astrophysical Insight
ed. C. J. Corbally, R. O. Gray, and R. F. Garrison   ISBN 0-937707-79-1

Vol. 61-Astronomical Data Analysis Software and Systems III
ed. Dennis R. Crabtree, R. J. Hanisch, and Jeannette Barnes   ISBN 0-937707-80-5

Vol. 62-The Nature and Evolutionary Status of Herbig Ae / Be Stars
ed. P. S. Thé, M. R. Pérez, and E. P. J. van den Heuvel   ISBN 0-937707-81-3

Vol. 63-Seventy-Five Years of Hirayama Asteroid Families: The role of Collisions in the Solar System History
ed. R. Binzel, Y. Kozai, and T. Hirayama   ISBN 0-937707-82-1

Vol. 64-Cool Stars, Stellar Systems, and the Sun, Eighth Cambridge Workshop
ed. Jean-Pierre Caillault   ISBN 0-937707-83-X

Vol. 65-Clouds, Cores, and Low Mass Stars
ed. Dan P. Clemens and Richard Barvainis   ISBN 0-937707-84-8

Vol. 66- Physics of the Gaseous and Stellar Disks of the Galaxy
ed. Ivan R. King   ISBN 0-937707-85-6

Vol. 67-Unveiling Large-Scale Structures Behind the Milky Way
ed. C. Balkowski and R. C. Kraan-Korteweg   ISBN 0-937707-86-4

Vol. 68-Solar Active Region Evolution: Comparing Models with Observations
ed. K. S. Balasubramaniam and George W. Simon   ISBN 0-937707-87-2

Vol. 69-Reverberation Mapping of the Broad-Line Region in Active Galactic Nuclei
ed. P. M. Gondhalekar, K. Horne, and B. M. Peterson   ISBN 0-937707-88-0

Vol. 70-Groups of Galaxies
ed. Otto G. Richter and Kirk Borne   ISBN 0-937707-89-9

Vol. 71-Tridimensional Optical Spectroscopic Methods in Astrophysics
ed. G. Comte and M. Marcelin   ISBN 0-937707-90-2

Vol. 72-Millisecond Pulsars—A Decade of Surprise, ed. A. A. Fruchter
M. Tavani, and D. C. Backer   ISBN 0-937707-91-0

Vol. 73-Airborne Astronomy Symposium on the Galactic Ecosystem: From Gas to Stars to Dust
ed. M. R. Haas, J. A. Davidson, and E. F. Erickson   ISBN 0-937707-92-9

Vol. 74-Progress in the Search for Extraterrestrial Life,
ed. G. Seth Shostak   ISBN 0-937707-93-7

Vol. 75-Multi-Feed Systems for Radio Telescopes
ed. D. T. Emerson and J. M. Payne   ISBN 0-937707-94-5

Vol. 76-GONG '94: Helio- and Astero-Seismology from the Earth and Space
ed. Roger K. Ulrich, Edward J. Rhodes, Jr., and Werner Däppen   ISBN 0-937707-95-3

Vol. 77-Astronomical Data Analysis Software and Systems IV
ed. R. A. Shaw, H. E. Payne, and J. J. E. Hayes   ISBN 0-937707-96-1

Inquiries concerning these volumes should be directed to the:
 Astronomical Society of the Pacific
 CONFERENCE SERIES
 390 Ashton Avenue
 San Francisco, CA 94112-1722
 415-337-1100
 e-mail asp @ stars.sfsu.edu

ASTRONOMICAL SOCIETY OF THE PACIFIC
CONFERENCE SERIES

Volume 77

ASTRONOMICAL DATA ANALYSIS
SOFTWARE AND SYSTEMS IV

Meeting held at Baltimore, Maryland
25-28 September 1994

Edited by
R. A. Shaw, H . E. Payne,
and J. J. E. Hayes

Contents

Preface xix

Conference participants xxi

Conference photograph xxxiv

Part 1. Education and Public Policy

The Electronic PictureBook and Astronomy's Education Initiative (invited talk) 3
    R. A. Brown, J. Ishee, and C. Lallo

The View From NASA Headquarters: Trends And Changes in Mission Operations And Data Analysis Programs (invited talk) 8
    G. Riegler

Part 2. Network Information Systems

Modelling Astrophysical Turbulent Convection (invited talk) 15
    N. Brummell and J. Toomre

Distributed Software for Observations in the Near Infrared 25
    V. Gavryusev, C. Baffa, and E. Giani

The New Astrophysics Data System 28
    G. Eichhorn, S. S. Murray, M. J. Kurtz, A. Accomazzi, and C. S. Grant

Development of an ADS Data Dictionary Standard 32
    C. S. Grant, A. Accomazzi, G. Eichhorn, M. J. Kurtz, and S. S. Murray

ADS Abstract Service Enhancements 36
    A. Accomazzi, C. S. Grant, G. Eichhorn, M. J. Kurtz, and S. S. Murray

WWW as a Support for the Long Term LBT Archive 40
    L. Fini

The World Wide Web: Cornerstone of the EUVE Science Archive 44
    K. McDonald, B. Stroozas, B. Antia, B. Roberts, K. Chen, N. Craig, and C. Christian

The EINSTEIN On-Line Service 48
    D. E. Harris, C. S. Grant, and H. Andernach

AstroWeb – Internet Resources for Astronomers 52
    R. E. Jackson, H.-M. Adorf, D. Egret, A. Heck, A. Koekemoer, F. Murtagh, and D. Wells

Indexing and Searching Distributed Astronomical Data Archives ... 55
    R. E. Jackson

An Information System for Proposal Submission and Handling 58
  A. M. Chavan and M. A. Albrecht
Design of a Remote Proposal Submission System 62
  A. Richmond, M. Duesterhaus, Song Yom, E. Schlegel, A. Smale, and N. White
Electronic Submission of HST Phase I Observing Proposals 65
  H. E. Payne and D. J. Asson
A Generalized Mosaic-to-SQL Interface with Extensions to Distributed Archives 68
  F. Pasian and R. Smareglia
WDB—A Web Interface to Sybase 72
  B. F. Rasmussen
Two Small Astronomical Catalogs Available In Hypertext 76
  K. M. Strom
User Interfaces to HST Data: StarView and the World Wide Web 80
  J. J. Travisano
ISSA-PS, The Postage Stamp Server for IRAS Imaging Data 84
  D. Van Buren, R. Ebert, and D. Egret

Part 3. Graphical User Interfaces and Visualization

A Portable GUI Development System—The IRAF Widget Server (invited talk) 89
  D. Tody
The AstroVR Collaboratory, An On-line Multi-User Environment for Research in Astrophysics (invited talk) 99
  D. Van Buren, P. Curtis, and D. A. Nichols
A Graphical User Interface for a Development Environment (GUIDE) .. 109
  A. Bhatnagar
Tcl/Tk with DRAMA - A Natural for Building User Interfaces to Instrumentation Systems? 113
  T. J. Farrell, J. A. Bailey, and K. Shortridge
A Method for Visualizing Time Variability in X-Ray Images 117
  P. Giommi, N. E. White, and L. Angelini
ASpect: A Multi-Wavelength Spectrum Analysis Package for IRAF 121
  S. J. Hulbert, J. D. Eisenhamer, Z. G. Levay, and R. A. Shaw
Applying Public Access Programming Techniques To SAOimage 125
  E. Mandel and D. Tody
WIP – An Interactive Graphics Software Package 129
  J. A. Morgan

Providing a Common GUI to Image Processing Tasks under PCIPS .. 133
    O. M. Smirnov and N. E. Piskunov
A Graphical Planning and Scheduling Toolkit for Astronomical Spacecraft 136
    S. C. Kleiner
A Graphical Front End for the WIYN Telescope Engineering Data System 140
    J. W. Percival
Space and the Spaceball 144
    R. Gooch
Visualization of GBT Geometry 148
    D. C. Wells

Part 4. Archives and Databases

What Happened to the Results? 155
    S. G. Ansari and A. Micol
The Hubble Space Telescope Data Archive 158
    K. D. Borne, S. A. Baum, A. Fruchter, and K. S. Long
The Hubble Data Archive: Opening the Treasures of the HST to the Community 162
    J. A. Pollizzi, III
Designing an Open E-Mail Interface to a Data Archive 166
    J. Richon
An Object-Oriented Approach to Astronomical Databases 169
    R. J. Brunner, K. Ramaiyer, A. Szalay, A. J. Connolly, and R. H. Lupton
Data Archive System for Kiso Observatory and Okayama Astrophysical Observatory 173
    S. Ichikawa, S. Yoshida, M. Yoshida, T. Horaguchi, and M. Hamabe
Exploring Interactive Archive Data Presentation at the COSSC ... 176
    J. M. Jordan, M. Cresitello-Dittmar, and J. S. Allen
Accessing the Digitized Sky Survey 179
    J. E. Morrison
GUIDARES: Reading the Guide Star Catalog in Very Many Ways ... 182
    O. Yu. Malkov and O. M. Smirnov
Storing and Distributing GONG Data 185
    M. Trueblood, W. Erdwurm, and J. A. Pintar

Part 5. Data Models and Formats

What is an Astronomical Data Model? 191
    A. Farris and R. J. Allen

Propagating Uncertainties and Units in Data Structures 195
    J. McDowell and M. Elvis

An Abstract Data Interface 199
    D. J. Allan

Proposed FITS Keywords and Column Headers for ALEXIS Mission Data
    Files 203
    J. Bloch and J. Theiler

ASC Data Structures and Model 207
    M. Conroy, R. Simon, J. McDowell, and K. Barry

Reformatting the Ginga Database to FITS and the Creation of a Data Products Archive 211
    R. H. D. Corbet, C. Larkin, J. A. Butcher, J. P. Osborne, and
    J. A. Nousek

Source-searching in Photon-event Lists without Imaging 215
    C. G. Page

The OGIP FITS Working Group 219
    M. F. Corcoran, L. Angelini, I. George, T. McGlynn, K. Mukai,
    W. Pence, and A. Rots

Organizing Observational Data at the Telescope 221
    M. Peron, D. Baade, M. A. Albrecht, and P. Grosbøl

IMPORT/EXPORT: Image Conversion Tools for IRAF 225
    M. Fitzpatrick

Convert: Bridging the Scientific Data Format Chasm 229
    D. G. Jennings, W. D. Pence, and M. Folk

Representations of Celestial Coordinates in FITS 233
    E. W. Greisen and M. Calabretta

A Generic Data Exchange Scheme Between FITS Format and C Structures 237
    W. Peng and T. Nicinski

A Proposed Convention for Writing FITS Data Tapes: DRAFT 0 241
    ROSAT/ASCA/XTE Development Team

FITSIO Subroutine Library Update 245
    W. D. Pence

FITS Checksum Verification in the NOAO Archive 247
    R. Seaman

Part 6. Object Detection and Classification

Automated Classification of a Large Database of Stellar Spectra 253
    R. K. Gulati, R. Gupta, P. Gothoskar, and S. Khobragade

Classification of Objects in the Guide Star Catalog 257
    O. Yu. Malkov and O. M. Smirnov

Object Detection Using Multi-Resolution Analysis 260
    F. Murtagh, W. Zeilinger, J.-L. Starck, and A. Bijaoui

Unsupervised Catalog Classification 264
    F. Murtagh

Astronomical Image Compression Using the Pyramidal Median Transform 268
    J.-L. Starck, F. Murtagh, and M. Louys

Clustering Analysis Algorithms and Their Applications to Digital POSS-II
    Catalogs 272
    R. R. de Carvalho, S. G. Djorgovski, N. Weir, U. Fayyad, K. Cherkauer, J. Roden, and A. Gray

Part 7. Image Restoration and Analysis

Multiresolution and Astronomical Image Processing (invited talk) 279
    J.-L. Starck, F. Murtagh, and A. Bijaoui

Improvements in Filter Design for Removing Galactic "Cirrus" from IRAS
    Images 289
    J. P. Basart, Lun X. He, P. N. Appleton, and J. A. Pedelty

Star Finding and PSF Determination using Image Restoration 293
    R. N. Hook and L. B. Lucy

Registering, PSF-Matching and Intensity-Matching Images in IRAF . . . 297
    A. C. Phillips and L. E. Davis

Restoration of HST WFPC2 Images in Gyro-Hold Mode 301
    J. Mo and R. J. Hanisch

MEM Task for Image Restoration in IRAF 305
    N. Wu

Part 8. Statistical Analysis

Statistical Consulting Center for Astronomy 311
    E. D. Feigelson, M. G. Akritas, and J. L. Rosenberger

Stochastic Relaxation as a Tool for Bayesian Modeling of Astronomical Images 315
    I. C. Busko

Spatial Models and Spatial Statistics for Astronomical Data 319
    L. Pásztor and L. V. Tóth

Spatial Structure of NGC 6822: An Example for Statistical Modeling of Astronomical Data 323
    L. Pásztor, C. Gallart, A. Aparicio, and J. M. Vílchez

Cheating Poisson: A Biased Method for Detecting Faint Sources in All-Sky Survey Data 327
    J. W. Lewis

Bias-Free Parameter Estimation with Few Counts, by Iterative Chi-Squared Minimization 331
    K. Kearns, F. Primini, and D. Alexander

A Method for Minimizing Background Offset Errors When Creating Mosaics 335
    M. W. Regan and R. A. Gruendl

Part 9. Simulation

CCDs at ESO: A Systematic Testing Program 341
    T .M. C. Abbott and R. H. Warmels

Modeling Scattered Light in the HST Faint Object Spectrograph 345
    H. Bushouse, M. Rosa, and Th. Mueller

Simulation of HST PSFs using Tiny Tim 349
    J. Krist

QPSIM: An IRAF/PROS Tool for Source Simulation 353
    K. R. Manning, J. DePonte, and F. Primini

The SAO AXAF Simulation System 357
    D. Jerius, M. Freeman, T. Gaetz, J. P. Hughes, and W. Podgorski

Simulations of Pinhole Imaging for AXAF: Distributed Processing Using the MPI Standard 361
    D. Nguyen and B. Hillberg

Part 10. Software Systems

FTOOLS: A FITS Data Processing and Analysis Software Package 367
    J. K. Blackburn
Migrating the Starlink Network from VMS to Unix 371
    C. Clayton
Porting CGS4DR to Unix 375
    P. N. Daly
The Array Limited Infrared Control Environment 379
    P. N. Daly, A. Bridger, D. A. Pickup, and M. J. Paterson
The SAX-LEGSPC Data Reduction and Analysis System: An Example of a Minimalist Approach 383
    F. Favata, A. N. Parmar, U. Lammers, G. Vacanti, M. Busetta, J. J. Mathieu, and P. Isherwood
The Data Analysis System For The PDS Detector On-board the SAX Satellite 387
    D. Dal Fiume, F. Frontera, L. Nicastro, M. Orlandini, and M. Trifoglio
Use of Inheritance Techniques in Real-Time Systems under DRAMA ... 391
    T. J. Farrell and K. Shortridge
Interfacing the Tk Toolkit to ADAM 395
    D. L. Terrett
The Stellar Dynamics Toolbox NEMO 398
    P. Teuben
Data Reduction Software for SWAS 402
    Z. Wang
A New PROS Task for Calculating HRI Source Intensities or Upper Limits 406
    J. C. Chen, M. A. Conroy, J. DePonte, and F. A. Primini
The PROS Big Picture: A High-Level Representation of a Software System 410
    J. DePonte, J. Chen, K.R. Manning, D. Schmidt, and D. Van Stone
An IRAF-Independent Interface for Spatial-Region Descriptors 414
    D. Schmidt
Recreating Einstein Level One Processing Exposure Masks and Background Maps in IRAF 418
    D. Van Stone, M. Garcia, and J. McDowell
The ASC Pipeline: Concept to Prototype 422
    A. Mistry, D. Plummer, and R. Zacher
Evolution of EUVE Guest Observer Data Reduction 425
    E. C. Olson
The OPUS Pipeline: A Partially Object-Oriented Pipeline System 429
    J. Rose, R. Akella, S. Binegar, T. H. Choo, C. Heller-Boyer, T. Hester, P. Hyde, R. Perrine, M. A. Rose, and K. Steuerman

A Retrospective View of Miriad 433
    R. J. Sault, P. J. Teuben, and M. C. H. Wright
The IDL Astronomy User's Library 437
    W. B. Landsman

Part 11. Data Modeling and Analysis

Spectroscopic reduction and analysis programs at the DAO (invited talk)   443
    G. Hill
Interactive Fitting of EUVE Emission Line Spectra 453
    M. Abbott
On the Combination of Undersampled Multiframes 456
    H.-M. Adorf
Interpolation of Irregularly Sampled Data Series—A Survey 460
    H.-M. Adorf
Calculating the Position and Velocity Components of HST 464
    T. B. Ake
Robust Data Analysis Methods for Spectroscopy 468
    P. Ballester
Application of the Linear Quadtree to Astronomical Databases 472
    P. Barrett
Wavelength Calibration at Moderately High Resolution 476
    J.-P. De Cuyper and H. Hensberge
An Approach for Obtaining Polarization Information from COBE-DMR . 480
    P. B. Keegstra, C. L. Bennett, and G. F. Smoot
Discrimination of Point-like Objects in Astronomical Images using Surface
    Curvature 484
    A. Llebaria, P. Lamy, and P. Malburet
Automated Globular Cluster Photometry with DASHA 488
    O. M. Smirnov and A. P. Ipatov
MACSQIID: A package for the reduction of data from the SQIID Infrared
    Camera 492
    J. W. MacKenty
EMSAO: Radial Velocities from Emission Lines in Spectra 496
    D. J. Mink and W. F. Wyatt
A Technique for Determining Proper Motions from Schmidt Plate Scans at
    ST ScI 500
    R. L. Williamson II, D. J. MacConnell, and W. J. Roberts
Towards a General Definition for Spectroscopic Resolution 503
    A. W. Jones, J. Bland-Hawthorn, and P. L. Shopbell

A Test for Weak Cosmic Ray Events on CCD Exposures 507
    J. Bland-Hawthorn, S. Serjeant, and P. L. Shopbell

Automated Spectral Reduction in the IRAF Fabry-Perot Package 510
    P. L. Shopbell and J. Bland-Hawthorn

Robust Estimation and Image Combining 514
    C. Y. Zhang

ADASS '94 - A Summary And A Look To The Future 518
    G. H. Jacoby

Author index 523

Index 527

Preface

This volume contains the papers presented at the fourth annual conference on Astronomical Data Analysis Software and Systems (ADASS IV) which was held at the Omni Inner Harbor Hotel in Baltimore, Maryland, from the 26th through the 28th of October 1994. There were 279 registered participants at the meeting with 61 people representing 14 countries outside the United State and Canada. The large attendance, and the number of participants from outside North America, is testimony to the vitality of this field and underscores the importance of ADASS as the world's premier meeting on astronomical software.

In a way, this strong interest in astronomical software is not surprising, given the flood of new, high quality data being collected daily, and the rapid pace of advancements in the computer hardware and software industries. (Who had heard of the World Wide Web or NCSA Mosaic two years ago?) To help keep up with new trends in software as they relate to advances in astronomy, the ADASS conferences are organized around certain topics of current interest. The special topics for ADASS IV were Astronomical Data Modelling and Analysis, Design and Development of Graphical User Interfaces (GUIs), Network Information Systems, and Parallel and Distributed Processing. This volume is organized into sections and subsections which reflect these categories, as well as sections on Archives and Databases, Image Restoration, Statistical Analysis, and Systems Software, for which there were many contributed papers.

There were 39 oral papers presented, as well as 93 poster papers and 17 computer demonstrations which were available throughout the meeting. This year's conference also included a number of BoF (Birds of a Feather) sessions on special topics including GUIs, FITS, IRAF Site Management, and IDL. In addition there was an IRAF-STSDAS-PROS-EUVE status report and discussion on the afternoon of the last day of the meeting. This year's ADASS was followed by two related, tag-along meetings: an Electronic Preprint Distribution Systems, organized by Bob Hanisch, and an IRAF Developers' Workshop (held at Space Telescope Science Institute), organized by Dick Shaw.

The conference was sponsored by the Space Telescope Science Institute, the National Optical Astronomy Observatories, the Smithsonian Astrophysical Observatory, the National Radio Astronomy Observatory, the National Aeronautics and Space Administration, the National Research Council of Canada, and the National Science Foundation. Corporate sponsors for the conference included Hughes STX, Sun Microsystems, Inc., Resource One, Sybase, Inc., Silicon Graphics, Inc., Open Concepts, Inc., Research Systems, Inc., and Digital Equipment Corporation. We are very grateful to both the sponsoring institutions and the corporate sponsors for their generous support.

A conference of this size cannot possibly succeed without the efforts and dedication of a large number of people; we are indebted to them all. The conference Program Organizing Committee was comprised of: Rudi Albrecht (Space Telescope–European Coordinating Facility), Roger Brissenden (Smithsonian Astrophysical Observatory), Carol Christian (Center for EUV Astronomy, U.C. Berkeley), Tim Cornwell (National Radio Astronomy Observatory), Dennis Crabtree (Canadian Astronomy Data Centre, Dominion Astrophysical Observatory), Daniel Durand (Canadian Astronomy Data Centre, Dominion Astrophysical Observatory), Bob Hanisch (Space Telescope Science Institute), F. Rick Harnden, Jr. (Smithsonian Astrophysical Observatory), George Jacoby

(National Optical Astronomy Observatories), Barry Madore (Infrared Processing and Analysis Center), Dick Shaw (Space Telescope Science Institute), Karen Strom (U. Massachusetts), and Doug Tody (National Optical Astronomy Observatories).

The Local Organizing Committee was chaired by Betty Stobie, who along with Bob Hanisch labored tirelessly to make ADASS IV a success. The LOC also benefitted from the eager and able support of: Angie Clarke, Jonathan Eisenhamer, Jeff Hayes, Phil Hodge, Zolt Levay, Harry Payne, Mary Alice Rose, Krista Rudloff, Dick Shaw, and Nelson Zarate from the Space Telescope Science Institute. The LOC was also joined by Eric Smith of NASA/Goddard Space Flight Center, and Peter Teuben of U. Maryland. Special thanks are extended to Tom DiGiacinto, Phil Grant, Jamie Lipinski, and Otto Wassenius for their able assistance in arranging and setting up the large number of computer workstations and network connections for the software demonstrations that have become such an integral part of ADASS.

The proceedings of this conference are once again being made available on the World Wide Web thanks to the permission of the Editors of the Astronomical Society of the Pacific. This experiment in electronic publishing has proven to be successful, in that a large number of last year's papers have been viewed and/or retrieved via the Web. Indeed, it would seem that the age of electronic publishing is fully upon us, for every paper, and all but eight illustrations, were submitted to us in electronic form. We eagerly await the (not too distant) time when all other science research publications will be "on the Web" as well.

Richard A. Shaw
Space Telescope Science Institute

Harry E. Payne
Space Telescope Science Institute

Jeffrey J. E. Hayes
Space Telescope Science Institute

March 1995

Cover Illustration: Two B−band images of the Phoenix dwarf galaxy at different epochs (*top*), and the difference of the two using new software to register and scale the images (*bottom left*) and to match their PSFs (*bottom right*). In the fully matched case about a dozen Cepheid variables are readily apparent, as is one high proper-motion star just below and left of center. Figures provided by Andrew Phillips and Lindsey Davis (see their paper, page 297).

Participant List

Mark Abbott, CEA/UC Berkeley, 2150 Kittredge St., Berkeley, CA 94704 USA (mabbott@cea.berkeley.edu)

Alberto Accomazzi, SAO, 60 Garden Street, Cambridge, MA 02138 USA (alberto@cfa.harvard.edu)

Hans-Martin Adorf, ST-ECF/ESO, Karl-Schwarzschild-Str. 2, Garching b. Muenchen, D-85748 Germany (hmadorf@eso.org)

Tom Ake, CSC/GSFC, Code 681/CSC, Greenbelt, MD 20771 USA (hrsake@hrs.gsfc.nasa.gov)

Miguel A. Albrecht, European Southern Obs., K-Schwarzschild-Str 2, Garching, D-85748 Germany (archeso@ac3.hq.eso.org)

Rudolf Albrecht, Space Tel. Europ. Coord. Fac, Karl Schwarzschild Str.2, Garching, Bavaria D-85748 Germany (ralbrech@eso.org)

David Allan, University of Birmingham, Dept. of Physics and Space Research, Edgbaston, Birmingham, West Midlands B15 2TT United Kingdom (dja@star.sr.bham.ac.uk)

Jesse Allen, Hughes-STX, 14201 Mac Farlane Green Ct., #4104, Upper Marlboro, MD 20772-5972 USA (allen@prufrock.gsfc.nasa.gov)

Marsha Allen, Johns Hopkins University, Dept. of Physics and Astronomy, 34th and Charles St., Baltimore, MD 21218 USA (allen@pha.jhu.edu)

Ron Allen, ST ScI, 3700 San Martin Drive, Baltimore, MD 21218 USA (rjallen@stsci.edu)

Gaetano Andreoni, European Southern Obs., Casilla 19001, Santiago, Chile (gandreon@eso.org)

Drew Asson, ST ScI, 3700 San Martin Drive, Baltimore, MD 21218 USA (asson@stsci.edu)

Karin Loya Babst, CSC/GSFC, 7926 Helmart Drive, Laurel, MD 20723 USA (kbabst@sssp.gsfc.nasa.gov)

Pascal Ballester, European Southern Obs., Karl-Schwarzschildstr. 2, Garching, D-85748 Germany (pballest@eso.org)

Klaus Banse, European Southern Obs., Karl-Schwarzschild-Str. 2, Garching, 85748 Germany (kbanse@eso.org)

Jeannette Barnes, NOAO/IRAF, P.O. Box 26732, Tucson, AZ 85726 USA (jbarnes@noao.edu)

James Barrett, SUNY at Stony Brook, Astronomy Program, SUNY, Stony Brook, NY 11794-2100 USA (jbarrett@astro.sunysb.edu)

Paul Barrett, NASA/GSFC & USRA, Code 668.1, Greenbelt, MD 20771 USA (barrett@piglet.gsfc.nasa.gov)

John Basart, Iowa State University, 333 Durham Center, Ames, IA 50011 USA (jpbasart@iastate.edu)

Stefi Baum, ST ScI, 3700 San Martin Drive, Baltimore, MD 21204 USA (sbaum@stsci..edu)

David Bazell, GSC, 6100 Chevy Chase Drive, Laurel, MD USA (dbazell@ame.gsfc.nasa.gov)

Participant List

Terrence Beck, ACC/GSFC, 3445 Andrews Ct #202, Laurel, MD 20724 USA (beck@cassini.gsfc.nasa.gov)
Aron Benett, NRAO, P.O. Box 2, Green Bank, WV 24944-0002 USA (abenett@gb.nrao.edu)
Piero Benvenuti, ESA/ESO, K.-Schwarzschild-Str.2, Garching, Germany (pbenvenu@eso.org)
Louis Bergeron, ST ScI, 18 Kings Crossing Ct., Cockeysville, MD 21030 USA (bergeron@stsci.edu)
Avnish Bhatnagar, The Aerospace Corporation, P.O. Box 92957 - M2/259, Los Angeles, CA 90009-2957 USA (bhatnagar@dirac2.span.nasa.gov)
Chris Biemesderfer, ferberts associates, P.O. Box 1180, Oracle, AZ 85623 USA (cbiemes@pulsar.ferberts.noao.edu)
James Kent Blackburn, NASA/GSFC:Hughes STX, NASA/Goddard Space Flight Center, Code 664.0, Greenbelt, MD 20771 USA (kent@kentaurus.gsfc.nasa.gov)
James Blackwell, CSC/GHRS/GSFC, Code 681.0, Greenbelt, MD 20771 USA (hrsblackwell@stars.gsfc.nasa.gov)
Jeffrey Bloch, Los Alamos National Lab., Mail Stop D436, Group NIS-2, Los Alamos, NM 87545 USA (jbloch@lanl.gov)
Richard Bochonko, University of Manitoba, Astronomy, Winnipeg, MB R3T 2M8 Canada (richard_bochonko@umanitoba.ca)
Bruce Bohannan, KPNO, P.O. Box 26732, Tucson, AZ 85726 USA (bruce@noao.edu)
Kirk Borne, ST ScI, 3700 San Martin Drive, Baltimore, MD 21218 USA (borne@stsci.edu)
Peter B. Boyce, AAS, 2000 Florida Ave NW, Suite 400, Washington, DC 20009 USA (pboyce@blackhole.aas.org)
Christine Boyer, ST ScI, 3700 San Martin Drive, Baltimore, MD 21218 USA (heller@stsci.edu)
Roger Brissenden, SAO, 60 Garden Street, Cambridge, MA 02138 USA (rjb@koala.harvard.edu)
Bob Brown, ST ScI, 3700 San Martin Drive, Baltimore, MD 21218 USA (rbrown@stsci.edu)
Robert Brunner, The Johns Hopkins University, 3400 N. Charles Street, Baltimore, MD 21218 USA (rbrunner@skysrv.pha.jhu.edu)
Bob Burns, NRAO, 520 Edgemont Road, Charlottesville, VA 22903-2475 USA (bburns@nrao.edu)
Howard Bushouse, ST ScI, 3700 San Martin Drive, Baltimore, MD 21218 USA (bushouse@stsci.edu)
Ivo Busko, INPE, Astrophysics Division, C.P. 515, CEP 12.201.970, Sao Jose doe Campos - SP, Brazil (busko@das.inpe.br)
Joan Centrella, Drexel University, Dept. of Physics & Atmospheric Science, Philadelphia, PA 19104 USA (joan@sparrow.drexel.edu)
Alberto Maurizio Chavan, European Southern Obs., Karl-Schwarzschild-Str 2, Garching, D-85748 Germany (amchavan@eso.org)

Judy Chen, SAO, 60 Garden Street, MS 12, Cambridge, MA 02138 USA (chen@cfa263.harvard.edu)

Cynthia Cheung, NASA/GSFC, Code 631, Greenbelt, MD 20771 USA (ccheung@nssdca.gsfc.nasa.gov)

Carol Christian, CEA/UC Berkeley, 2150 Kittredge St., Berkeley, CA 94707 USA (carolc@cea.berkeley.edu)

Chris Clayton, Rutherford Appleton Lab, Chilton, Didcot, Oxon, OX11 OQX U.K. (cac@star.rl.ac.uk)

Andree Coffre, Observatoire de NANCAY, Observatoire de NANCAY, Neuvy Sur Barangeon, 18330 France (andree@nanrh1.obs-nancay.fr)

Bruce Cogan, Mt. Stromlo Observatory, Private Bag, Weston Creek, A.C.T., 2606 Australia (cogan@mso.anu.edu.au)

Maureen Conroy, SAO, 60 Garden Street, Cambridge, MA 02138 USA (mo@cfa.harvard.edu)

Sharon Conroy, Hughes STX, 7701 Greenbelt Rd., Suite 400, Greenbelt, MD 20770 USA (conroy@nssdca.gsfc.nasa.gov)

Robin Corbet, GSFC, NASA/GSFC, Code 666, Greenbelt, MD 20771 USA (corbet@astro.psu.edu)

Michael Corcoran, GSFC-USRA, Code 668, Greenbelt, MD 20771 USA (corcoran@barnegat.gsfc.nasa.gov)

Mark Cornell, McDonald Observatory, University of Texas, RLM 15.308, Austin, TX 78712 USA (cornell@puck.as.utexas.edu)

Tim Cornwell, NRAO, P.O. Box O, Socorro, NM 87801 USA (tcornwel@nrao.edu)

Dennis Crabtree, DAO/CADC, 5071 W. Saanich Rd., Victoria, BC V8X 4M6 Canada (crabtree@dao.nrc.ca)

Mark Cresitello-Dittmar, Hughes STX, Code 668.1, Greenbelt, MD 20771 USA (mdittmar@enemy.gsfc.nasa.gov)

Heather Dalterio, AAS, 1651 Lamont St. NW #24, Washington, DC 20010 USA (dalterio@blackhole.aas.org)

Philip N. Daly, Joint Astronomy Center, 660 N. Aohoku Place, University Park, Hilo, HI 96720 USA (pnd@jach.hawaii.edu)

Julian Daniels, Johns Hopkins University, 206B Bloomberg, 3400 N. Charles St., Baltimore, MD 21218-2695 USA (daniels@msx3.pha.jhu.edu)

Lindsey Davis, NOAO/IRAF Group, 950 North Cherry Avenue, Tucson, AZ 85726-6732 USA (davis@noao.edu)

Reinaldo de Carvalho, Caltech, Astronomy Dept. 105-24, Pasadena, CA 91125 USA (reinaldo@astro.caltech.edu)

Jean-Pierre De Cuyper, Royal Observatory of Belgium, Astrophysics Department, Ringlaan 3, Ukkel, B1180 Belgium (jeanpier@astro.oma.be)

Michele De La Pena, CSC/IUE Observatory, 10000-A Aerospace Road, GT 1 Bldg, Lanham-Seabrook, MD 20706 USA (delapena@iuegtc.gsfc.nasa.gov)

Cor de Vries, SRON Leiden, Niels Bohrweg 2, Leiden, 2333 AL Netherlands (devries@rulrol.leidenuniv.nl)

Janet DePonte, SAO, 60 Garden Street, Cambridge, MA 02138 USA (janet@cfa.harvard.edu)

Susana Delgado, I.A.C, C/ Via Lactea s/n, La Laguna, 38200 Spain (sdm@iac.es)

Laurent Denis, C.N.R.S. Obs. de NANCAY, Observatoire de NANCAY, Neuvy Sur Barangeon, 18330 France (denis@nanda2.obs-nancay.fr)

Ralf J. Dettmar, ESA/ST ScI/Ruhr University, 3700 San Martin Drive, Baltimore, MD 21218 USA (dettmar@stsci.edu)

Erik Deul, Sterrewacht Leiden, P.O. Box 9513, Leiden, 2300 RA Netherlands (erik.deul@strw.leidenuniv.nl)

Jesse Doggett, ST ScI, 3700 San Martin Dr., Baltimore, MD 21208 USA (doggett@stsci.edu)

Daniel Durand, CADC/DAO, 5071 West Saanich Road, Victoria, BC V8X 4M6 Canada (durand@dao.nrc.ca)

Guenther Eichhorn, SAO, 60 Garden Street, Cambridge, MA 02138 USA (gei@head-cfa.harvard.edu)

Jonathan Eisenhamer, ST ScI, 3700 San Martin Drive, Baltimore, MD 21218 USA (eisenhamer@stsci.edu)

Giuseppina Fabbiano, AXAF Science Center/CfA, 60 Garden Street, Cambridge, MA 02138 USA (pepi@cfa.harvard.edu)

Tony Farrell, Anglo-Australian Observatory, P.O. Box 296, Epping, N.S.W. 2121 Australia (tjf@aaoepp.aao.gov.au)

Allen Farris, ST ScI, 3700 San Martin Drive, Baltimore, MD 21218 USA (farris@stsci.edu)

Fabio Favata, European Space Agency, P.O. Box 299/SA, Noordwijk, 2200 AG The Netherlands (fabio.favata@astro.estec.esa.nl)

Keith Feggans, Advanced Computer Concepts, 300 69th Place, Seat Pleasant, MD 20743 USA (hrsfeggans@hrs.gsfc.nasa.gov)

Eric Feigelson, Penn State University, 525 Davey Laboratory, Dept. of Astro. & Astrophysics, University Park, PA 16802 USA (edf@astro.psu.edu)

Luca Fini, Osservatorio Di Arcetri, L.GO E. Fermi, 5, Firenze, 50125 Italy (lfini@arcetri.astro.it)

Mike Fitzpatrick, NOAO/IRAF Group, P.O. Box 26732, Tucson, AZ 85726-6732 USA (fitz@noao.edu)

Marty Fredrickson, Sybase Federal, Herndon, VA 22070 USA (martyf@sybase.com)

Immanuel Freedman, The Experts, 7529 Greenbelt Road #910, Greenbelt, MD 20770 USA (gauixf@fnma.com)

Marian Frueh, McDonald Obs./U Texas, P.O. Box 1337, Fort Davis, TX 79734 USA (frueh@astro.as.utexas.edu)

Tom Garrard, Caltech, 220-47 Downs, Pasadena, CA 91125 USA (tlg@citsrl.srl.caltech.edu)

Bob Garwood, NRAO, 2786 Timber Lane, Charlottesville, VA 22901 USA (bgarwood@nrao.edu)

Severin Gaudet, CADC/DAO, 5071 West Saanich Road, Victoria, BC V8X 4M6 Canada (gaudet@dao.nrc.ca)

Vladimir Gavryusev, Oss. Astro.di Arcetzi, Largo E. Fermi, 5, Firenze, 50125 Italy (vladimiz@arcetri.astro.it)

Paolo Giommi, ESA/ESRIN, Via G. Gallilei, C.P. 64, Frascati, Rome 00044 Italy (giommi@mail.esrin.esa.it)

John Glaspey, CFHT, P.O. Box 1597, Kamuela, HI 96743 USA (glaspey@cfht.hawaii.edu)

Brian Glendenning, NRAO, 520 Edgemont Rd., Charlottesville, VA 22903 USA (bglenden@nrao.edu)

Richard Gooch, CSIRO, P.O. Box 76, Epping, N.S.W., 2121, Australia (rgooch@atnf.csiro.au)

Maia Good, Hughes STX, Code 631, GSFC, Greenbelt, MD 20771 USA (good@rsdps.gsfc.nasa.gov)

Carolyn Stern Grant, SAO, 60 Garden St., MS 70, Cambridge, MA 02138 USA (stern@cfa.harvard.edu)

Emily A. Greene, NASA/GSFC/Hughes STX, Code 664, Greenbelt, MD 20771 USA (greene@legacy.gsfc.nasa.gov)

Eric W. Greisen, NRAO, 520 Edgemont Road, Charlottesville, VA 22903-2475 USA (egreisen@nrao.edu)

Ted Groner, SAO, 2251 Calle Comodo, Tucson, AZ 85705 USA (ted@sol2.sao.arizona.edu)

Robert Gruendl, University of Maryland, Department of Astronomy, College Park, MD 20742 USA (gruendl@astro.umd.edu)

Bernt Grundseth, CFH Telescope Corporation, P.O. Box 1597, Kamuela, Hawaii 96743 USA (bernt@cfht.hawaii.edu)

Ravi Kumar Gulati, IUCAA, Post Bag 4, Ganeshkhind, 1, Pune, Maharashtra 411007 India (gulati@iucaa.ernet.in)

Peter Hallett, Research Systems, Inc., 2995 Wilderness Place, Suite 203, Boulder, CO 80301 USA (peter@rsinc.com)

Nancy Hamilton, Dept. of Defense, 930 Penobscot Harbour, Pasadena, MD 21122 USA

Robert Hanisch, ST ScI, 3700 San Martin Drive, Baltimore, MD 21218 USA (hanisch@stsci.edu)

F. Rick Harnden, SAO, 60 Garden Street, Cambridge, MA 02138 USA (frh@cfa.harvard.edu)

Dan Harris, CfA, 60 Garden Street, Cambridge, MA 02138 USA (harris@cfa.harvard.edu)

R. Lee Hawkins, Wellesley College/KNAL, Whitin Observatory, Wellesley, MA 02181 USA (lhawkins@annie.wellesley.edu)

Jeffrey Hayes, ST ScI, 3700 San Martin Drive, Baltimore, MD 21218 USA (hayes@stsci.edu)

Andre Heck, Strasbourg Astronomical Obs., 11, rue de l'Universite, Strasbourg, F-67000 France (heck@cdsxb6.u-strasbg.fr)

Graham Hill, DAO, 5071 West Saanich Road, Victoria, BC V8X 4M6 Canada (hill@dao.nrc.ca)

Phil Hodge, ST ScI, 3700 San Martin Drive, Baltimore, MD 21218 USA (hodge@stsci.edu)

Participant List

Charles Holmes, Johns Hopkins University, 3400 N. Charles St., 135 Bloomberg Center, Baltimore, MD 21218 USA (cholmes@pha.jhu.edu)

Richard Hook, ST-ECF/ESO, Karl Schwarzchild Str-2, Garching, D-85748 Germany (rhook@eso.org)

Jim Horstkotte, NRAO, 520 Edgemont Rd., Charlottesville, VA 22903-2475 USA (jhorstko@nrao.edu)

Sethanne Howard, JPL/NASA HQ, Mail Code SZE, NASA Headquarters, Washington, DC 20546-0001 USA (showard@smtpgmgw.ossa.hq.nasa.gov)

Jin-chung Hsu, ST ScI, 3700 San Martin Drive, Baltimore, MD 21218 USA (hsu@stsci.edu)

Stephen Hulbert, ST ScI, 3700 San Martin Drive, Baltimore, MD 21218 USA (hulbert@stsci.edu)

Gareth Hunt, NRAO, 520 Edgemont Road, Charlottesville, VA 22903-2475 USA (ghunt@nrao.edu)

Eric Huygen, K.U.Leuven, Heuvenstraat 100 B, B-3520 Zonhoven, Belgium (rik@sron.rug.nl)

Peter Hyde, ST ScI, 3700 San Martin Drive, Baltimore, MD 21218 USA (phyde@stsci.edu)

Shin-ichi Ichikawa, Nat'l Astro. Obs. of Japan, 2-21-1 Osawa, Mitaka, Tokyo, 181 Japan (ichikawa@cl.mtk.nao.ac.jp)

Yasuhide Ishihara, Fujitsu Limited, Earth Sciences Systems Dept., 9-3, Nakase 1-Chome, Mihama-ku, Chiba, Chiba 261 Japan (ishi@sp.se.fujitsu.co.jp)

Kevin Jackson, Research Systems, Inc., 4041 Powder Mill Rd, Suite 300, Calverton, MD 20705 USA (kevinj@rsinc.com)

Robert Jackson, Computer Sciences Corp., 3700 San Martin Dr., Baltimore, MD 21218 USA (jackson@stsci.edu)

George Jacoby, KPNO/NOAO, P.O. Box 26732, Tucson, AZ 85712 USA (gjacoby@noao.edu)

Rieks Jager, Space Research Organization, Sorbonnelaan 2, 3584 CA, Utrecht, The Netherlands (rieksj@sron.ruu.nl)

Donald Jennings, ADF/NASA/GSFC/Hughes STX, Code 631, Building 28, Greenbelt, MD 20771 USA (jennings@tcumsh.gsfc.nasa.gov)

Diab Jerius, SAO, 60 Garden Street, Cambridge, MA 02138 USA (jerius@cfa.harvard.edu)

Mark Johnston, ST ScI, 3700 San Martin Drive, Baltimore, MD 21218 USA (johnston@stsci.edu)

James Jordan, Hughes STX, NASA/GSFC, Code 668.1, Greenbelt, MD 20771 USA (jmj@enemy.gsfc.nasa.gov)

Mary Beth Kaiser, Johns Hopkins University, Dept. of Physics and Astronomy, Baltimore, MD 21218 USA (kaiser@pha.jhu.edu)

Monica Kane Stewart, NASA Science Internet, Moffett Field, CA 20414 USA (monica@nsipo.arc.nasa.gov)

Kristin Kearns, 60 Garden Street, MS 70, Cambridge, MA 02138 USA (kkearns@cfa.harvard.edu)

Phillip Keegstra, Hughes STX/COBE, 7601 OraGlen Dr., Suite 100, Greenbelt, MD 20770 USA (keegstra@csdr2.gsfc.nasa.gov)

Ajit Kembhavi, IUCAA, Post Bag 4, Ganeshkhind, Pune 411 007, India (akk@iucaa.ernet.in)

Steven Kleiner, Harvard-Smithsonian CfA, 60 Garden Street, Mail Stop 66, Cambridge, MA 02138 USA (kleiner@cfa.harvard.edu)

John Kressel, Hughes STX, Code 668.1, Greenbelt, MD 20771 USA (kressel@enemy.gsfc.nasa.gov)

John Krist, ST ScI, 3700 San Martin Drive, Baltimore, MD 21218 USA (krist@stsci.edu)

Mark Kyprianou, ST ScI, 3700 San Martin Drive, Baltimore, MD 21218 USA (kyp@stsci.edu)

Wayne Landsman, Hughes STX, Code 681, NASA/GSFC, Greenbelt, MD 20771 USA (landsman@stars.gsfc.nasa.gov)

Zoltan Levay, ST ScI, 3700 San Martin Drive, Baltimore, MD 21218 USA (levay@stsci.edu)

Cristian Levin, European Southern Obs., Casilla 2584, Central Casillas, Santiago, Chile (clevin@eso.org)

Jim Lewis, CEA/UC Berkeley, 2150 Kittredge St., Berkeley, CA 94704 USA (jwl@cea.berkeley.edu)

Shian Lin, ST ScI, 3700 San Martin Drive, Baltimore, MD 21218 USA (lin@stsci.edu)

Don Lindler, Advanced Computer Concepts, 11518 Gainsborough Road, Potomac, MD 20854 USA (hrslindler@hrs.gsfc.nasa.gov)

Antoine Llebaria, LAS (CNRS), Traverse du Siphon, Marseille, 13012 France (llebaria@astrsp-mrs.fr)

Knox S. Long, ST ScI, 3700 San Martin Drive, Baltimore, MD 21218 USA (long@stsci.edu)

Rick Lyon, Hughes STX, Code 934, Building 28, Greenbelt, MD 20771 USA (lyon@phase.gsfc.nasa.gov)

Dyer Lytle, NOAO - IRAF, 950 N. Cherry Ave., Tucson, AZ 85719 USA (lytle@noao.edu)

John MacKenty, ST ScI, 3700 San Martin Drive, Baltimore, MD 21218 USA (mackenty@stsci.edu)

Barry Madore, NASA/IPAC/NED, Caltech 100-22, Pasadena, CA 91125 USA (barry@ipac.caltech.edu)

Patricia Mailhot, SAO, 160 Concord Ave, Cambridge, MA 02138 USA (mailhot@sma2.harvard.edu)

Oleg Malkov, Inst. of Astro./Russ.AS, 48 Pyatnitskaya Str, Moscow 109017, Russia (omalkov@airas.msk.su)

Eric Mandel, SAO, 60 Garden Street, Cambridge, MA 02138 USA (eric@cfa.harvard.edu)

Kathleen Manning, SAO, 60 Garden Street, MS 70, Cambridge, MA 02138 USA (manning@cfa.harvard.edu)

Ralph Martin, Royal Greenwich Observatory, Madingley Road, Cambridge, CB3 0EZ United Kingdom (ralf@mail.ast.cam.ac.uk)

Roel Martinez, NRAO/UNAM IA, Mexico, 520 Edgemont Rd., Charlottesville, VA 22903-2475 USA (rmartinez@nrao.edu)

John Mattox, NASA/GSFC, Code 668.1, Greenbelt, MD 20771 USA (john_mattox@gsfc.nasa.gov)

Kelley McDonald, Center for EUV Astrophysics, 2150 Kittredge St., Berkeley, CA 94704 USA (kelley@cea.berkeley.edu)

Jonathan McDowell, SAO, 60 Garden Street, Cambridge, MA 02138 USA (jcm@urania.harvard.edu)

Thomas McGlynn, GSFC/USRA, Code 668.1, Greenbelt, MD 20771 USA (mcglynn@grossc.gsfc.nasa.gov)

Suzanne McKibbin, ST ScI, 3700 San Martin Drive, Baltimore, MD 21218 USA (mckibbin@stsci.edu)

Glenn Miller, ST ScI, 3700 San Martin Drive, Baltimore, MD 21218 USA (miller@stsci.edu)

Doug Mink, Center for Astrophysics, 60 Garden Street, Cambridge, MA 02138 USA (mink@cfa.harvard.edu)

Atul Mistry, SAO/TRW, 60 Garden St. MS4, Cambridge, MA 02138 USA (atm@atomic.harvard.edu)

Jinger Mo, ST ScI, 3700 San Martin Drive, Baltimore, MD 21218 USA (jinger@stsci.edu)

James Morgan, University of Maryland, Astronomy Program, College Park, MD 20742 USA (morgan@astro.umd.edu)

Stephen Morris, NRCC / HIA / DAO / CADC, 5071 West Saanich Rd., Victoria, B.C. V8X 4M6 Canada (morris@dao.nrc.ca)

Jane Morrison, ST ScI, 3700 San Martin Drive, Baltimore, MD 21218 USA (morrison@stsci.edu)

Stephen Murray, SAO, 60 Garden Street, MS 2, Cambridge, MA 02138 USA (ssm@cfa.harvard.edu)

Fionn Murtagh, ST-ECF, Karl-Schwarzschild-Str. 2, Garching, D-85748 Germany (fmurtagh@eso.org)

Jayant Murthy, The John Hopkins University, Baltimore, MD 21218 USA (murthy@pha.jhu.edu)

Dan Nguyen, SAO, 60 Garden St, MS-81, Cambridge, MA 02138 USA (dtn@dumbo.harvard.edu)

Luciano Nicastro, TESRE/CNR, Via Gobetti 101, Bologna, 40129 Italy (nicastro@botesa.tesre.bo.cnr.it)

Jan E. Noordam, NFRA, P.O. Box 2, Dwingeloo, 7990 AA The Netherlands (jnoordam@nfra.nl)

Chris O'Dea, ST ScI, 3700 San Martin Drive, Baltimore, MD 21218 USA (odea@stsci.edu)

Earl J. O'Neil Jr., KPNO, P.O. Box 26732, Tucson, AZ 85726 USA (oneil@noao.edu)

Eric Olson, CEA/UC Berkeley, 2150 Kittredge Street, Berkeley, CA 94704 USA (ericco@cea.berkeley.edu)

Mauro Orlandini, TESRE/CNR Bologna, Via Gobetti 101, Bologna, 40129 Italy (orlandini@botes1.tesre.bo.cnr.it)

Clive Page, Leicester University, X-ray Astronomy Group, Dept. of Physics & Astronomy, Leicester, LE1 7RH UK (cgp@star.le.ac.uk)

Fabio Pasian, Oss. Astr. Trieste, Via G B Tiepolo 11, Trieste, I 34131 Italy (pasian@ts.astro.it)

Scott Paswaters, Interferometrics/NRL, 4555 Overlook Ave SW, Code 7660, Washington, DC 20375 USA (scott@argus.nrl.navy.mil)

Laszlo Pasztor, MTA TAKI, 1022 BP., Herman Otto UT 15, Hungary (h2295pas@huella.bitnet)

Hallie Patterson, NASA Science Internet, Moffett Field, CA 20414 USA (hallie@nsipo.arc.nasa.gov)

Thomas Pauls, Naval Research Laboratory, Code 7220, Washington, DC 20375-5351 USA (pauls@atlas.nrl.navy.mil)

Harry Payne, ST ScI, 3700 San Martin Drive, Baltimore, MD 21218 USA (payne@stsci.edu)

Robert Payne, NRAO, P.O. Box 2, Green Bank, WV 24944 USA (rpayne@nrao.edu)

Jeffrey Pedelty, NASA/GSFC, Code 934, Greenbelt, MD 20771 USA (pedelty@janksy.gsfc.nasa.gov)

William Pence, NASA/GSFC, Code 668, Greenbelt, MD 20771 USA (pence@tetra.gsfc.nasa.gov)

Wei Peng, Fermilab, Computing Division, MS-234, P.O. Box 500, Batavia, IL 60510 USA (wpeng@fndaut.fnal.gov)

Jeffrey W. Percival, Space Astro Lab-U. of Wisc., 1150 University Ave, Madison, WI 53706 USA (jwp@sal.wisc.edu)

Mario Perez, Applied Research Corporation, 8201 Corporate Drive, Suite 1120, Landover, Maryland 20785 USA (perez@indy.arclch.com)

Michele Peron, European Southern Obs., K-Schwarzschild-Str 2, Garching, D-85748 Germany (mperon@eso.org)

Richard Perrine, CSC, 3700 San Martin Drive, Baltimore, MD 21218 USA (perrine@stsci.edu)

Andrew C. Phillips, Lick Observatory, University of California, Santa Cruz, CA 95064 USA (phillips@lick.ucsc.edu)

Benoit Pirenne, ST-ECF, Karl Schwarzschild-Str. 2, Garching, D-85748 Germany (bpirenne@eso.org)

Joseph Pollizzi, ST ScI, 3700 San Martin Drive, Baltimore, MD 21218 USA (pollizzi@stsci.edu)

Karen Presley, TRW/SAO, 60 Garden Street MS21, Cambridge, MA 02138 USA (kbarry@cfatrw1.harvard.edu)

Mauro Pucillo, Astro. Obs. of Trieste, Via Tiepolo, 11 - P.O. Box Succ. 5, Trieste, I-34131 Italy (pucillo@oat.ts.astro.it)

Deborah Puku, NASA Science Internet, Moffett Field, CA 20414 USA (puku@nispo.arc.nasa.gov)

Guy Purcell, University of Alabama, 206 Gallalee Hall, Dept. of Physics & Astronomy, Tuscaloosa, AL 35487-0324 USA (purcell@crux.astr.ua.edu)

Bo Frese Rasmussen, ST-ECF, Karl Schwarzschild-Strasse 2, Garching, D-85748 Germany (bfrasmus@eso.org)

Mike Regan, University of Maryland, Department of Astronomy, College Park, MD 20742 USA (mregan@astro.umd.edu)

Alan Richmond, GSFC/NASA, Greenbelt, MD 20771 USA (richmond@guirar.gsfc.nasa)

Joel Richon, ST ScI, 3700 San Martin Drive, Baltimore, MD 21218 USA (richon@stsci.edu)

Guenter Riegler, NASA HQ-Astrophysics, Code SZ - 300 E St. SW, Washington, DC 20546 USA (griegler@gm.ossa.hq.nasa.gov)

Timothy Roberts, NRAO, 520 Edgemont Rd., Charlottesville, VA 22903-2454 USA (troberts@nrao.edu)

David Robinson, Institute of Astronomy, Madingley Road, Cambridge, CB3 0HA UK (drtr@mail.ast.cam.ac.uk)

Jim Rose, ST ScI, 3700 San Martin Drive, Baltimore, MD 21218 USA (rose@stsci.edu)

Mary Alice Rose, ST ScI, 3700 San Martin Drive, Baltimore, MD 21218 USA (marose@stsci.edu)

Arnold Rots, USRA-GSFC/NASA, Code 668, Greenbelt, MD 20771 USA (arots@xebec.gsfc.nasa.gov)

Lee Rottler, UCO Lick Observatory, 1156 High Street, Santa Cruz, CA 95064 USA (rottler@lick.ucsc.edu)

Krista Rudloff, ST ScI, 3700 San Martin Drive, Baltimore, MD 21218 USA (rudloff@stsci.edu)

Bert Rust, NIST, Building 101, Room A-238, Gaithersburg, MD 20899 USA (bwr@cam.nist.gov)

Darrell Schiebel, NRAO, 520 Edgemont Rd., Charlottesville, VA 22903-2475 USA (dschieb@nrao.edu)

Marcel Schlapfer, NASA Science Internet, Moffett Field, CA 20414 USA (marcel@nsipo.arc.nasa.gov)

Barry Schlesinger, Hughes STX/FITS Office, 7701 Greenbelt Road, Suite 400, Greenbelt, MD 20770 USA (bschlesinger@nssdca.gsfc.nasa.gov)

Dennis Schmidt, SAO, MS 83, 60 Garden St., Cambridge, MA 02138 USA (dennis@head-cfa.harvard.edu)

Ethan J. Schreier, ST ScI, 3700 San Martin Drive, Baltimore, MD 21218 USA (schreier@stsci.edu)

Joseph Schwarz, ESO, Karl-Schwarzschild Str-2, Garching, Germany (jschwarz@eso.org)

Keith Scollick, GSFC/CSC, Code 668.1, Greenbelt, MD 20771 USA (scollick@skview.gsfc.nasa.gov)

Rob Seaman, NOAO/IRAF, 950 North Cherry Avenue, Tucson, AZ 85719 USA (seaman@noao.edu)

William L. Sebok, University of Maryland, Department of Astronomy, College Park, MD 20742 USA (wls@astro.umd.edu)

Dick Shaw, ST ScI, 3700 San Martin Drive, Baltimore, MD 21218 USA (shaw@stsci.edu)

Eliot Shepard, AXAF Science Center-SAO, MS 21 SAO, 60 Garden St., Cambridge, MA 02138 USA (shepard@cfa.harvard.edu)

Patrick Shopbell, Rice University, P.O. Box 1892, Houston, TX 77251-1892 USA (pls@pegasus.rice.edu)

Bernard Simon, ST ScI, 3700 San Martin Drive, Baltimore, MD 21239 USA (bsimon@stsci.edu)

Richard Simon, NRAO, 520 Edgemont Rd., Charlottesville, VA 22903-2475 USA (rsimon@nrao.edu)

Rameshwar Sinha, Tata Institute, Post Bag 3, Poona University Campus, Ganeshkhind, Pune 411007 India (rsinha@iucaa.ernet.in)

Oleg Smirnov, Inst. of Astro./Russ.AS, 48 Pyatnitskaya Str, Moscow 109017, Russia (oms@astro.free.net)

Eric Smith, LASP/GSFC, Code 681, Greenbelt, MD 20771 USA (esmith@hubble.gsfc.nasa.gov)

Jean-Luc Starck, CEA, DSM DAPNIA, F-91191, Gif-sur-Yvette Cedex France Gif-sur-Yvette, F-91191 France (starck@ariane.saclay.cea.fr)

Alan Steiner, NASA Science Internet, NASA/Ames Research Center, MS 204-14, Mountain View, CA 94035 USA (alan@nsipo.arc.nasa.gov)

David M. Stern, Research Systems, Inc., 2995 Wilderness Place, Suite 203, Boulder, CO 80301 USA (stern@rsinc.com)

Malcolm Stewart, Royal Observatory Edinburgh, Blackford Hill, Edinburgh, EH9 3HJ UK (m.stewart@roe.ac.uk)

Elizabeth Stobie, ST ScI, 3700 San Martin Drive, Baltimore, MD 21218 USA (stobie@stsci.edu)

Karen Strom, Five College Astronomy Dept., GRC 519B, University of Massachusetts, Amherst, MA 01002 USA (kstrom@hanksville.phast.umass.edu)

Sridharan Sudarsan, Hughes STX, 7701 Greenbelt Road, Suite 400, Greenbelt, MD 20770 USA (sudarsan@killians.gsfc.nasa.gov)

Ralph Swick, The X Consortium, One Memorial Drive, Cambridge, MA 02142 USA (swick@x.org)

Edward Swing, Dept. of Defense, 9800 Savage Rd, Ft. Meade, MD 20755-6000 USA

Philip Taylor, Royal Greenwich Observatory, Madingley Road, Cambridge, CB3 0EZ UK (pbt@mail.ast.cam.ac.uk)

Sandra Terranova, SAO, 60 Garden Street, Cambridge, MA 02138 USA (terranova@cfa.harvard.edu)

David Terrett, Rutherford Appleton Lab, Chilton, Didcot, Oxfordshire, OX11 0QX UK (d.terrett@rl.ac.uk)

Peter Teuben, Astronomy Dept. - U. of Maryland, College Park, MD 20742 USA (teuben@astro.umd.edu)

Vadakkanthara Thaila, NODC/NOAA, 1825 Connecticut Ave NW, Room 403, Washington, DC 20235 USA (thaila@nodc.noaa.gov)

James Theiler, Los Alamos Nat'l Laboratory, MS D436, Los Alamos, NM 87545 USA (jt@lanl.gov)

Doug Tody, NOAO/IRAF, P.O. Box 26732, Tucson, AZ 85749 USA (tody@noao.edu)

Susan Tokarz, SAO, 60 Garden Street, Cambridge, MA 02138 USA (tokarz@cfa.harvard.edu)

Juri Toomre, JILA, Boulder, CO 08309-0440 USA (jtoomre@solarz.colorado.edu)

Jay Travisano, CSC/ST ScI, 3700 San Martin Drive, Baltimore, MD 21218 USA (jay@stsci.edu)

Ralph Tremmel, Max-Planck-Inst. for Astro., Konigstuhl 17, Heidelberg, 69117 Germany (tremmel@mpis-hd.mpg.de)

Francesco Tribioli, Oss. Astr. di Arcetri, E. Fermi 5, Firenze, 50127 Italy (tribioli@arcetri.astro.it)

Massimo Trifoglio, C.N.R./Itesre, Via Gobetti/101, Bologna, 40129 Italy (massimo@botes2.tesre.bo.cnr.it)

Mark Trueblood, National Solar Observatory, P.O. Box 26732, Tucson, AZ 85726-6732 USA (trublood@noao.edu)

Luc Turbide, Universite de Montreal, C.P. 6128, Succ. Centre-ville, Montreal, QC H3C-3J7 Canada (turbide@astro.umontreal.ca)

Ben Turgeon, York University, 4700 Keele St, Dept. Physics & Astronomy, North York, Ontario M3J 3P1 Canada (turgeon@nereid.sal.ists.ca)

Frank Valdes, NOAO/IRAF, P.O. Box 26732, Tucson, AZ 85726 USA (fvaldes@noao.edu)

Dave Van Buren, IPAC, MS 100-22 Caltech, Pasadena, CA 91125 USA (dave@ipac.caltech.edu)

David Van Stone, Harvard Smithsonian-CfA, 60 Garden Street, Cambridge, MA 02138 USA (vanstone@cfa.harvard.edu)

Tony Villasenor, NASA Headquarters, Washington, DC 20546-0001 USA

Stephen A. Voels, Hughes STX/GSFC, Code 631, Greenbelt, MD 20771 USA (voels@mystry.gsfc.nasa.gov)

Martin Vogelaar, Kapteyn Astronomical Inst., Landleven 12, Groningen, 9747 AD The Netherlands (vogelaar@astro.rug.nl)

Patrick Wallace, Starlink, Rutherford Appleton Lab, Chilton, DRAL, OXON UK (p.wallace@rutherford.ac.uk)

Dennis Wang, NRL/Interferometrics, 4555 Overlook Ave SW, Code 7660, Washington, DC 20375-5000 USA (wang@cedar.nrl.navy.mil)

Zhong Wang, SAO, 60 Garden Street, Cambridge, MA 02138 USA (zwang@cfa.harvard.edu)

Michael Ward, McDonald Observatory, University of Texas, P.O. Box 1337, Fort Davis, TX 79734 USA (ward@astro.as.utexas.edu)

Rein H. Warmels, ESO, Karl Schwarzschild-Str 2, Garching, D-85748 Germany (rwarmels@eso.org)

Archibald Warnock, 6652 Hawkeye Run, Columbia, MD 21044 USA (warnock@clark.net)

Joyce Watson, SAO/SIMBAD, 60 Garden Street, Cambridge, MA 02138 USA (watson@cfa.harvard.edu)

Scott Weiss, Sybase, Inc., 1870 Embarcadero Rd., Palo Alto, CA 94303 USA (sweiss@sybase.com)

Don Wells, NRAO, 520 Edgemont Road, Charlottesville, VA 22903-2475 USA (dwells@nrao.edu)

Nicholas White, LHEA/GSFC, Code 668 GSFC, Greenbelt, MD 20771 USA (white@adhoc.gsfc.nasa.gov)

Richard White, NASA/GSFC, Code 932, Greenbelt, MD 20771 USA (rwhite@amarna.gsfc.nasa.gov)

Andreas J. Wicenec, Astronomisches Institut, Waldhaeuserstr. 64, Tuebingen, 72076 Germany (wicenec@ait.physik.uni-tuebingen.de)

Ramon Williamson, ST ScI, 3700 San Martin Drive, Baltimore, MD 21218 USA (ramon@stsci.edu)

George Wolf, SW Missouri State University, Department of Physics and Astronomy, Springfield, MO 65804 USA (gww836f@vma.smsu.edu)

Nailong Wu, ST ScI, 3700 San Martin Drive, Baltimore, MD 21218 USA (nailong@stsci.edu)

Nelson Zarate, ST ScI, 3700 San Martin Drive, Baltimore, MD 21218 USA (zarate@stsci.edu)

Cheng-Yue Zhang, ST ScI, 3700 San Martin Drive, Baltimore, MD 21218 USA (zhang@stsci.edu)

Fourth Annual Conference on Astronomical Data Analysis Software and Systems
1994 September 26–28 Baltimore, MD

Conference Photograph

1. O. Malkov
2. I. Busko
3. R. de Carvalho
4. M. Trueblood
5. J. McDowell
6. B. Jackson
7. T. Farrell
8. J. Doggett
9. D. J. Asson
10. A. Rots
11. K. Blackburn
12. A. Wicenec
13. V. Gavryusev
14. L. Davis
15. G. Eichhorn
16. O. Smirnov
17. R. R. Harnden
18. J. Barnes
19. N. Hamilton
20. R. Sinha
21. L. E. Bergeron
22. R. Martin
23. G. Hill
24. J. Morrisson
25. L. Rottler
26. S. S. Murray
27. B. Schlesinger
28. D. Tody
29. C. A. Christian
30. F. Valdes
31. F. Tribioli
32. G. Fabbiano
33. K. Kearns
34. J.-C. Hsu
35. J. W. Percival
36. E. Greisen
37. R. Simon
38. J. Eisenhamer
39. R. Hanisch
40. E. Stobie
41. Z. Levay
42. P. Hodge
43. J. Watson
44. J. Hayes
45. J. Toomre
46. H. Payne
47. R. Brissenden
48. T. Ake
49. W. Peng
50. W. Pence
51. K. L. Babst
52. L. Nicastro
53. J. R. Lewis
54. R. Corbet
55. C. Clayton
56. D. Wang
57.
58. A. Heck
59. S. Hulbert
60. D. Wells
61. D. Robinson
62. R. Hook
63. N. Zarate
64. E. Swing
65. D. Crabtree
66. E. Schreier
67.
68. M. Pucillo
69. M. Conroy
70. R. K. Gulati
71. J. DePonte
72. M. Vogelaar
73. J. Chen
74. A. Kembhavi
75. L. Fini
76. S. Delgado
77. R. Shaw
78. K. Rudloff

Conference Photograph

79. D. Jerius
80. B. Sebok
81. P. Taylor
82. K. Long
83. P. Daly
84. F. Pasian
85. K. Presley
86. M. Albrecht
87. R. Albrecht
88. B. F. Rasmussen
89. S. Gaudet
90. D. Van Stone
91. L. Turbide
92. K. Manning
93. P. Benvenuti
94. A. Mistry
95. A. Accomazzi
96.
97. M. Ward
98. M. Frueh
99. B. Grundseth
100. M. De La Peña
101. A. Llebaria
102. C. de Vries
103. J. Morgan
104.
105.
106. D. Durand
107. G. Wolf
108. J. Mo
109. J. Krist
110. K. Scollick
111. J. Blackwell
112. H. C. Minh
113. T. McGlynn
114. H. Bushouse
115. E. Huygen
116. S. Ichikawa
117. J. Glaspey
118. T. Groner
119. Y. Ishihara
120. N. Wu
121. R. Simon
122. R. Tremmel
123. C. Stern Grant
124. R. Payne
125. R. Gooch
126. A. Farris
127. D. Harris
128. D. Schmidt
129. S. Paswaters
130. J. Theiler
131. R. Jager
132. M. Orlandini
133. M. Corcoran
134. M. Trifoglio
135. T. Beck
136. M. B. Kaiser
137. R. Garwood
138. P. Hyde
139. J. Pedelty
140. L. Pásztor
141. J. Pollizzi
142. J. Basart
143. P. Keegstra
144. D. Lindler
145. D. Schiebel
146. R. Martinez
147. J. Rose
148.
149. R. Williamson
150. G. Jacoby
151. B. F. Madore
152. C. Heller-Boyer
153. M. A. Rose
154. S. McKibbin
155. G. Hunt
156. C. Levin
157. B. Bohannan
158. J. Horstkotte
159. T. Roberts
160. G. Andreoni
161. J. Richon
162.
163. D. Van Buren
164. S. Morris
165. M. Cresitello-Dittmar
166. A. Benett
167. B. Turgeon
168. B. Cogan
169. C. O'dea
170. C.-Y. Zhang
171. M. Stewart
172. J.-P. De Cuyper
173. B. Glendenning
174. J. Allen
175. J. Travisano
176. P. Teuben
177. P. Shopbell
178. P. Ballester
179. M. Fitzpatrick
180. C. Biemesderfer
181. E. Smith
182. F. Murtagh
183. A. M. Chavan
184. F. Favata
185.
186. M. Abbott
187. J. Noordam
188. T. Cornwell
189. E. C. Olson
190. E. Deul
191. E. J. O'Neil, Jr.
192. Z. Wang
193. S. Tokarz
194. R. Bochonko
195. P. Wallace
196. J. Bloch
197. D. Terrett
198. M. Peron
199. M. Cornell
200. D. Lytle
201. E. Feigelson
202. K. Banse
203. K. Strom
204. R. Warmels
205. K. Borne
206. D. Mink
207. J.-L. Starck
208.
209. A. Phillips

Part 1. Education and Public Policy

The Electronic PictureBook and Astronomy's Education Initiative

R. A. Brown, J. Ishee, and C. Lallo

Space Telescope Science Institute[1], 3700 San Martin Dr., Baltimore, MD 21218

Abstract. The Exploration in Education (ExInEd) program at the Space Telescope Science Institute seeks to promote basic teaching and learning by exploiting the connections between astronomy research and education. Our primary product category consists of Electronic Picture-Books (EPBs): educational show-and-tell software allowing astronomers to communicate their ideas and results to a broad audience.

In 1990, the Bahcall committee called for an "educational initiative" associated with its decadal plan for astronomy research (Bahcall 1991). It urged that new resources and innovative partnerships be developed to foster beneficial connections between the exploration of the universe and basic teaching and learning and, in particular, to address the deficit our society faces in the math, science, and technology skills of the next generations.

A study and report entitled An Education Initiative in Astronomy (EIA) (Brown 1990) is an informal appendix to the Bahcall committee report. This work is the product of seventeen professionals in education or astronomy, who deliberated the question of strategies and approaches for connecting astronomy research and basic education. The connection is not obvious, after all: the former activity is the acquisition of advanced knowledge about the universe, which may be understandable in detail by only a few; the latter is aimed at helping all inquirers learn "the basics," that is, information and concepts often long known by past—even ancient—scholars.

The EIA report affirmed manifold opportunities for astronomy research to contribute to education, noting that the astronomy enterprise has many relevant assets, including a broad domain of inquiry, role models, advanced technology, cadres of semi-professionals and amateurs, and interdisciplinary connections. Two prime objectives promoted by EIA were sparking pre-college interest in science and improving the accuracy of astronomical information presented to students and the public. One strategy recommended by EIA was using innovative technologies to disseminate science results, and in particular, using advanced communications networks to facilitate delivery of electronic information to computers in homes and schools.

Today, many new programs are seeking to exploit the connections between astronomy research and education. The Exploration in Education (ExInEd) program at the Space Telescope Science Institute (ST ScI) is one such program. The focus of ExInEd is to enable individual authors in the community to communicate the ideas and results of their activities in the form of computer multimedia software, which is distributed over Internet, via on-line services, and on durable media.

The Electronic PictureBook (EPB) is our primary product category. It is educational software of the show-and-tell variety, consisting of a HyperCard stack that runs on the Macintosh platform. The basic elements of an EPB are picture-caption pairs that are augmented with a structural framework (indices, navigational buttons, overlays) and additional content (an introduction, a glossary, reading lists, credits). The EPB uses an introduction to orient the user in the topic at hand, and provides navigation tools that enable him or her to explore the pictures and captions in either a hierarchical way (via indices and sub indices) or sequentially, as in a slide show.

EPBs have a serious—but unstructured—educational purpose. They are designed to: (1) capture student interest, (2) provide quality information, and (3) construct understanding and skills. We hope that by browsing through their exciting images and informative captions, children and adults will receive the message that science is fun, that learning is a process of exploration and discovery, and that these activities can be enjoyable and rewarding. We are encouraged, based on voluminous, unsolicited correspondence and favorable reviews, that EPBs are being well received by the press and public, by students, and by school teachers in particular.

From the author's standpoint, ExInEd provides a simple and inviting interface. An EPB author has simply to provide ExInEd with pictures and captions along with a brief introduction, and ExInEd does the rest. We have the technical capabilities to accept images and text in a wide variety of forms and formats, which we transform or translate appropriately for the EPB. Editorial responsibilities are shared between the author and ExInEd staff. Once all the edited materials are in-hand, production of an EPB is generally completed in a period of a few weeks.

At the current time, ExInEd has completed and is distributing ten EPB titles over the Internet, through on-line services and bulletin boards, and on diskette and CD-ROM:

- *Gems of Hubble*
  Stephen P. Maran   Published 1994. Version 2.0.
  Gems of Hubble is a collection of recent images obtained by NASA's *Hubble Space Telescope (HST)*. Topics include the *HST*'s launch and first servicing mission, stars and star clusters, galaxies and black holes, solar system objects, nebulae, and cosmology.

- *The Planetary System*
  Dr. David Morrison with The Astronomical Society of the Pacific
                                             Published 1994. Version 1.0.1.
  The Planetary System is a pictorial survey of our solar system that includes all nine planets, the Moon, and small bodies. The images it contains were selected for their beauty and for how well they illustrate the wide range of physical and geological features present in the solar system.

- *Endeavour Views the Earth*
  Dr. Jay Apt   Published 1993. Version 1.0.
  Endeavour Views the Earth is a collection of Space Shuttle images of Earth's northern hemisphere. Image captions by NASA astronaut Jay Apt touch upon timely issues in geography, agronomy, geology, meteorology, and environmental science. Linked to The World Factbook.

- *The World Factbook '92*
  Central Intelligence Agency   Published 1993. Version 1.0.
  The World Factbook is a survey of the people, culture, geography, government, economy, political conditions, communications, and defense forces of 261 countries and regions, as well as information on the United Nations and other international organizations. Each of the 261 entries includes a map of that country or region with major population centers highlighted. Can be linked to *Endeavour Views the Earth*.

- *Scientific Results from the Goddard High Resolution Spectrograph*
  James Blackwell, Jennifer Sandoval, Steven Maran, and Steven Shore
  Published 1993. Version 1.2.2.
  Scientific Results from the Goddard High Resolution Spectrograph is a collection of images and the spectra obtained with the Goddard High Resolution Spectrograph (GHRS) during the first years of operation of the *Hubble Space Telescope*. Captions were authored with the assistance of the GHRS science team. Topics include the northern aurora of Jupiter, Beta Pictoris, Chi Lupi, Alpha Tauri, AU Microscopii, AR Lacertae, Nova Cygni 1992, Zeta Ophiuchi, 3C273, R31, and 30 Doradus.

- *Volcanic Features of Hawaii and Other Worlds*
  Peter Mouginis-Mark   Published 1993. Version 1.0.
  This EPB describes and compares Hawaiian volcanoes and volcanic features with features believed to have a similar origin on the Moon, Mars, Venus, and the Jovian moon, Io. Accompanying text highlights these comparisons, explains the features, and describes past, current, and planned research by volcanologists. Specific topics include volcanic structures, lava flows, eruption types and resultant land forms, rocks, boulder fields, and radar studies. The glossary is linked to the text.

- *Images of Mars*
  The Planetary Society   Published 1993. Version 1.1.4.
  Images of Mars is a collection of re-processed photos of the Martian surface taken during the *Viking* mission. In addition to being an update on U.S. and Russian interest in Mars, the introduction provides a brief look at the broad history of human interest, touching on the early observations of Tycho Brahe, Johann Kepler, Christiaan Huygens, William Herschel, Asaph Hall, and Giovanni Schiaparelli; the imaginative theories of Percival Lowell; and the popular views of science fiction writers Edgar Rice Burroughs, H. G. Wells, and Ray Bradbury.

- *Impact Catastrophe That Ended the Mesozoic Era*
  Dr. William K. Hartmann   Published 1993. Version 1.0.
  This series of seven original paintings by artist and planetary scientist William K. Hartmann depicts the impact and resulting effect of a 10 km asteroid believed to have hit the Yucatan Peninsula in Mexico about 65 million years ago. The climatic changes set in motion by the impact are thought to have had a significant role in the extinction of many species at the time, including the dinosaurs.

- *Apollo 11 at Twenty-Five*
  Dr. Roger D. Launius   Published 1994. Version 1.0.

This EPB was published on the occasion of NASA's 25th celebration of the moment when humans first walked on the Moon. It contains an historical survey of the Apollo program for the period from May 1961, with its announcement by President John F. Kennedy, to December 1972, when the program ended with the flight of Apollo 17. Special emphasis on the Apollo 11 mission.

- *Magellan Highlights of Venus*
  Dr. Steve Saunders and Dr. Ellen Stofan   Published 1993. Version 1.0. This EPB contains images from the *Magellan* spacecraft's first 243-day mapping cycle of Venus. An introduction summarizes the most important science findings. Image captions describe both planetary features—Venusian impact craters, mountains, tessera terrain, coronae, arachnoids, volcanoes, channels, and dunes—and the methods scientists have used to analyze them.

ExInEd has also created its first CD-ROM, entitled Space Science Library of Electronic PictureBooks. It includes all current EPBs plus GIF-format images of 330 individual images in 640x480 pixel format.

In the coming months, we propose to expand our collaborations with at least five new EPB titles:

- *Comparing Earth and Its Planetary Neighbors.* Patricia Barnes-Svarney, an established science writer, has collected 50 images of corresponding features amongst the planets and composed captions for them.

- *The New Solar System.* David Morrison has agreed to select a recent set of planetary images for a new tour of the solar system based on *HST*, Galileo, Clementine, and ground-based observations. He will write the captions, and we will produce the EPB. This will be a valuable update of our current title, The Planetary System.

- *Other Worlds From Earth.* Bill Hartmann will prepare a series of paintings that take perspectives on the solar system from distances 1 AU, 10 AU, 100 AU... out to a planetary system around another star. The purposes of this "powers of ten" sequence are to conceptually connect our solar system with possible others while clarifying the distance scales involved.

- *Comet Strikes Jupiter!* Steve Maran, who edited the images and wrote the captions for Gems of Hubble, has agreed to select SL-9/Jupiter material for an EPB based on the spectacular events of July '94 and to compose an introduction and captions.

- *PlanetQuest.* Daniel Glover, a NASA engineer based at Lewis Research Center, is compiling a history of NASA's planetary spacecraft.

ExInEd plans to produce additional multimedia products for both the Macintosh and the Windows PC platforms. As noted, EPBs are currently based on the HyperCard engine and are thus available only for the Macintosh computer. In the future, we will be using the Apple Media Tool (AMT) authoring system in addition to HyperCard, and will be producing runtime executable software for Windows also. The content will be derived from EPBs, which we will continue to produce. This will expand ten-fold the installed base of computers that can use our products.

ExInEd is also exploring new products specifically for Internet distribution. The Electronic InfoCapsule (EIC), for example, will be small cross-platform interactive computer presentation incorporating images, sound, text, and video. Authored with Apple Media Tool, EICs will run identically on the Macintosh and Windows PC platforms.

All ExInEd titles currently available can be obtained by downloading them from any one of several electronic sources, including SSO's electronic bulletin board system (bbs) at 410-516-4880 using user ID "guest" and password "guest", then searching under the heading "Conferences"; via anonymous ftp over the Internet at host address *stsci.edu* in directory */ExInEd*; through America Online using the search term "ExInEd" in the Macintosh software library; and from ExInEd's Mosaic/WWW Home Page[1].

ExInEd EPB's can also be purchased on durable media, diskette or CD-ROM, by contacting The Astronomical Society of the Pacific (415-337-2624) between 9 a.m. and 3 p.m. Pacific Time or writing to: The Astronomical Society of the Pacific, 390 Ashton Ave., San Francisco, California 94112. They may also be purchased from NASA's Central Operation of Resources for Educators (CORE) by calling 216-774-1051, ext. 293 or writing to NASA CORE, Lorain County Joint Vocational School, 15181 Route 58 South, Oberlin, OH 44074.

To receive a copy of An Education Initiative in Astronomy, to obtain more information on the ExInEd program, or to provide comments or suggestions on the EPB concept, please write to: ExInEd, Space Telescope Science Institute, 3700 San Martin Drive, Baltimore, MD 21218. FAX: 410-516-7450. Internet: *ExInEd@stsci.edu*.

Acknowledgments. Support for this research was provided by NASA Grant NAGW–3048 through the Space Telescope Science Institute, which is operated by the Association of Universities for Research in Astronomy.

References

Bahcall, J. N., ed. 1991, The Decade of Discovery in Astronomy and Astrophysics (Washington, D.C., National Academy Press)

Brown, R. A., ed. 1990, An Education Initiative in Astronomy (Baltimore, Space Telescope Science Institute)

---

[1] http://www.stsci.edu/exined-html/exined.home.html

The View From NASA Headquarters: Trends And Changes in Mission Operations And Data Analysis Programs

G. Riegler

NASA Headquarters, Astrophysics Division, Washington DC

Abstract. The Astrophysics Program at NASA is mission oriented. With seven flying missions and twenty additional (international collaborations) planned, NASA provides an ever increasing amount of exciting astrophysical data, and supports a large portion of astrophysics data analysis. There have been significant efforts during the past seven years that support the development of astronomical data analysis software and systems. The future contains many changes that offer us significant challenges. I describe here some of those changes, and NASA's responses to them.

1. Introduction

NASA supports seven astrophysics missions at present; they are: the *International Ultraviolet Explorer* (*IUE*), the *Hubble Space Telescope* (*HST*), the *Röntgen Satellite* (*ROSAT*), the *Compton Gamma Ray Observatory* (*CGRO*), the *Extreme Ultraviolet Explorer* (*EUVE*), and the *Advanced Spacecraft for Cosmology and Astrophysics* (*ASCA*). Figure 1 shows the Astrophysics Mission Plan Chart which lists all current NASA astrophysics missions, planned and in progress, as of January 1995. (All launch dates are subject to change.) While the chart indicates the broad scope of the NASA astrophysics missions, it is important to realize that NASA supports not only the mission, but also the associated mission science centers and/or mission data archives. Many mission-related software and archive developments were outgrowths of the 1987 Astrophysics Data System (ADS) Workshops. In response to these workshops, discipline archive centers were established in two of three Astrophysics disciplines: the High Energy Archive (HEASARC) and the Infrared Archive (IPAC). The Astronomical Data Center at the NASA Space Science Data Center (NSSDC) continues to support the archival needs of the broader astronomy community. In 1989 NASA also funded the Astrophysics Software and Research Aids Program (AS&RA) and the development of various software facilities through the Astrophysics Data Program (ADP).

Many developments related to the support of astronomical software coincided with a time of growth and mission expansion. NASA maintains some of those software tools by continuing to fund, for example, FITS, the HEASARC BROWSE, *HST* image reconstruction, CASA/*IUE* access, the IDL astronomy library, and SkyView.

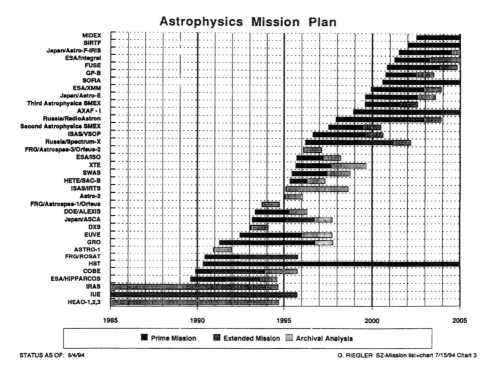

Figure 1. NASA Astrophysics Mission Plan as of January 1995.

2. Changes for MO & DA

The environment continues to change. The directives from the NASA Administrator and the White House are clear: They want smaller, faster, better, and cheaper missions. They want to see a revolution in concepts, not an evolution of old concepts. In addition we now need to show that new projects represent the diverse population of scientists. In a broader sense, perhaps the most significant change is that the previous pre-eminence of science has been replaced by drivers of technology, economics, education, and public outreach.

When funding increased year to year, NASA could continue programs so long as they produced valuable data. Now that funding is level and is expected to decrease, we ask the science community to use a new paradigm to help us restructure old programs and select new programs. Programs must maintain the highest likely science return, improve our technological leadership, and yet remain cost-effective.

2.1. Changes to the ADS and Mission Programs

As an example of the restructuring, consider the Astrophysics Data System which was designed to enable remote, coherent access to distributed astrophysics data holdings. The program was characterized by very high science return with minimal oversight from NASA Headquarters. ADS achieved an impressive growth, especially with its abstract service. But as information technology evolved, the World-Wide Web, the Mosaic interface, and other Internet tools became very popular for remote data access. NASA responded to customer concerns and preferences, and, guided by cost efficiency, redirected the ADS project. The ADS project will reduce the classical remote access to distributed data holdings and concentrate on the abstract service.

Other restructuring plans affect the data analysis plans for several flying missions. *IUE*, *ROSAT*, *EUVE*, and *GRO* will complete their missions within the next few years. A Senior Review Panel recently held a comparative review of their science merits to determine if their requests for mission extension are valid. Acceptable requests for mission extension now must show credible plans for mission continuation past the prime phase at one-third to one-half of the prime phase funding level. In the future NASA will cap post-launch costs. A project may stretch its observing time if it can find ways to cut operating costs so the cap is maintained. This reduced scope and reduced cost come with increased risk that varies with size of mission. However, an Explorer-class mission, or a "Small Explorer," after completion of its primary science objective, can accept more risks than an Observatory-class mission early in its intended lifetime.

2.2. Changes to the Software Funding Environment

NASA supports many projects related to the development of software and data analysis tools. Some support is provided directly to institutions such as the HEASARC, IPAC, and ADC, and some projects are supported through the Astrophysics Data Program (ADP). The ADP began in 1986 as the "Space Astrophysics Data Analysis Program." The purpose of the ADP is to optimize the scientific return from space astrophysics missions, and to enable broad scientific investigations requiring analysis of data from one or several space-based data sets.

Previous funding cycles for the ADP considered three types of proposals. Type 1 proposals were for research involving space astrophysics data sets, Type 2 proposals were for applied research to improve and enhance space-based observing and data analysis, and Type 3 proposals were for applied research to improve access to, and management of, space-based astronomical data. Software projects faired well in the ADP: during each of the last three cycles (FY92, FY93, and FY94), ADP projects of types 2 and 3 received 45–49% of the total funding. Contrary to rumors, this funding percentage did not drop in the latter years.

During the FY88–92 cycles, NASA selected software-related proposals on the basis of their *general* usefulness to science. To verify whether or not the objectives were met, NASA conducted a survey of 13 ADP software projects funded during this period. Of the 13 projects surveyed, only 10 ever produced a product. Of the 13 projects, 8 sent their product to at least one person outside their own institution. Except for one or two high-usage products, the others, on average, were used by only two or three other persons. The basic objective—providing tools for the science community—was often not met!

In response to the changing environment and to correct the problem, NASA will improve the selection process for software development projects. NASA support for these projects will continue within the ADP, but the previous—somewhat strained—distinction between Type 2 and Type 3 proposals will be dropped: The old Type 2 and 3 Proposals are combined into a new Type 2. ADP will solicit proposals for:

- **Science research** (previously Type 1). The Type 1 proposals are those whose dominant emphasis is the analysis and interpretation of data from the above-named space astrophysics missions.

- **Software tools** (previously Types 2 and 3). The Type 2 proposals may include the writing of algorithms for analyzing data, planning space-based astronomical observations, correlating or displaying space-based astronomical data; deconvolving sources in crowded fields, absolute radiance calibration techniques, statistically modeling the appearance of the sky in specific wavelength regions, providing tools for the NASA/IPAC Extragalactic Database (NED); improving or adding reduction or analysis packages to any of the Astrophysics Science and Data Centers or to Astronomical Data Reduction and Analysis Systems; or other similar software activities. In addition, Type 2 proposals may include research to improve access to, and management of, space-based astronomical data.

A more significant change to the program concerns the peer review of proposals. In the past the Type 1 proposals were considered in competition only with each other for funding. Similarly for Type 2 and Type 3. This new cycle will put all proposals, regardless of type, into the same competitive base. This new procedure will ensure that all proposals share and are judged on the common goal of improving the scientific output from NASA Astrophysics mission data. Instead of artificially separating software development proposals from other ADP proposals, all proposals will be reviewed in the science-topical panel where they would generate the most direct benefits. Each proposer will designate the science-topical panel where the proposed software development effort will have greatest impact. The research areas are: Solar System; Star Formation and Pre-Main Sequence Stars; Main Sequence Stars; Post-Main Sequence Stars and Collapsed Objects; Binary systems; Interstellar Medium and Galactic Structure; Galaxies; Large-Scale Cosmic Structures. In addition to their usefulness for science, each proposal's cost effectiveness will be considered as selection criterion.

With the help of the science and technical communities NASA can meet the new challenges of producing first class and cost-effective science We solicit your best efforts and cooperation to continue providing the best science and data analysis tools.

Part 2. Network Information Systems

Modelling Astrophysical Turbulent Convection

N. Brummell and J. Toomre

Joint Institute for Laboratory Astrophysics, Department of Astrophysical, Planetary and Atmospheric Sciences, University of Colorado, Boulder, CO 80309-0440

Abstract. Numerical simulations of highly turbulent flows in astrophysics benefit greatly from recent advances in high performance computing. Yet such three-dimensional modelling raises major problems in capturing, moving, and analyzing the resulting massive data sets necessary to sample the evolving intricate dynamics. Thus archiving, networking, and visualization are as essential as the actual act of computing when it comes to real scientific progress.

1. Introduction

An interdisciplinary team of researchers at several institutions (Universities of Colorado, Chicago, and Minnesota, Michigan State University, and National Center for Atmospheric Research) is working jointly on problems in geophysical and astrophysical fluid dynamics (GAFD) to utilize massively-parallel architectures to increase the spatial resolution in three-dimensional simulations of GAFD turbulence. Within our "grand challenge" team efforts, we are employing variously pseudo-spectral, finite-difference, multi-grid, and piecewise-parabolic method (PPM) approaches in studying the intense turbulence encountered in both planetary and stellar settings. The scale of these simulations requires corresponding progress in the computational sciences, both in order to develop and optimize software for massively-parallel computers and to capture and visualize the resulting massive data sets.

Our research related to astrophysics has so far concentrated largely on turbulent compressible convection within stars. In particular, the vigorous turbulence that results from convective instability within rotating stars not only serves to transport heat but also to redistribute angular momentum and chemical species, and can yield magnetic dynamo action. A hallmark of such turbulence constrained by rotation and stratification is that *large-scale coherent structures and strong mean flows can coexist with the intense smaller-scale turbulence.* Our work with three-dimensional direct simulations of turbulent compressible convection influenced by rotation is motivated by trying to understand how stars and giant planets rotate differentially, and in particular to examine the redistribution of angular momentum that leads to differential rotation within a star like the sun. As one example, these studies seek to resolve a major challenge raised recently by helioseismology, wherein acoustic oscillations of the sun are used to probe the structure and dynamics of the solar convection zone. Preliminary results from helioseismology (cf. Gough & Toomre 1991) suggest that the sun is rotating differentially with depth and latitude in a manner very different from predictions based on earlier numerical simulations. Those spherical shell simulations of rotating anelastic convection (e.g., Gilman & Miller 1986; Glatzmaier 1987) dealt with effectively laminar convective flows, whereas our recent work suggests that *fully turbulent* compressible flows appear to yield very different transport properties for angular momentum. This holds out the hope that in due course simulations of GAFD turbulence carried out in appropriate spherical geometries and with the inclusion of adequate stellar physics (realistic equations of state and opacities, ionization effects) may help to resolve the basic dynamical puzzle now being raised by helioseismology. However, understanding such nonlinear dynamics at a fundamental level raises formidable challenges because of the broad range of physical scales that must be resolved. High-performance computing offers the opportunity to make substantial inroads for studying the properties of such astrophysical turbulence.

2. Profile of Turbulence Simulations

The basic difficulty with trying to model turbulence in most natural settings, and certainly so in a stellar convection zone, is that the active dynamical scales of convection encompass a formidable range. Turning to the sun, we expect that the largest-scale turbulent flows deep within the convection zone should possess scales comparable to the overall depth of that zone (occupying the outer 30 percent by radius, or about 200,000 km in depth). Yet the density scale height near the top of the zone is only about 100 km, and one readily observes there intense turbulent convection in the form of granulation with horizontal scales of order 1,000 km. Detectable there, too, are supergranulation flow patterns with scales of order 30,000 km. However, we can estimate that viscous dissipation is operating at scales of order 0.1 km or smaller, suggesting that the active dynamical scales of turbulence in such a zone range at least from 10^5 to 10^{-1} km, or thus over six decades in each physical dimension. The most intricate three-dimensional turbulence simulations to date have just begun dealing with order 1024^3 spatial modes, capturing three decades in each dimension. As we will discuss below, such 10^9 spatial zone calculations to study the evolution of turbulence for typically 10^5 time steps make rather interesting demands on processors, memory, communications, and data capture. Though we may have to wait for quite some time before 10^{18} zone simulations become tractable (if ever), we can take some comfort in realizing that the hints of structured turbulence first realized with 128^3 simulations are now seen in fascinating detail with the 1024^3 "hero" calculations of the day. Another three decades of resolution (and degrees of freedom) in each physical dimension may well herald other surprises in the resulting turbulent dynamics. The optimism of theoretical physics encourages great extrapolations to be made, and in that spirit we hope that the turbulence calculations of 4096^3-class likely by the end of this decade and century may provide substantial clues about the inner workings of a real star like the sun. Yet we recognize that all such modelling is of the large-eddy simulation (LES) variety, requiring some manner of sub-grid-scale (SGS) representation of the unresolved scales.

2.1. Memory Requirements and Operation Counts

The nature of our computational challenges are revealed by considering some of the attributes of a 1024^3-mode turbulence simulation. For fully compressible convection, we typically use as working variables the three components of velocity (u, v, w), pressure p, and temperature T. This implies that the memory requirements are 160 GB, based on 10^9 spatial zones × 5 variables × factor 4 of working space, for a total of about 20 Gwords. On a 1024-node massively parallel processor (mpp) system, we effectively need 160 MB of memory per node. Turning to estimates of operation counts for these simulations, the floating point operation (flop) count per zone per time step range from about 200 for pseudo-spectral codes to about 4000 for PPM codes. Since the intricate dynamics needs to be evolved for about 100,000 times steps, the total operation count for a given simulation are 2×10^{16} flop using pseudo-spectral codes and 4×10^{17} flop using PPM codes. Once a real Tflop s^{-1} machine (1000 Gflop s^{-1}) is realized, such a pseudo-spectral run could be accomplished in 2×10^4 s (about 6 hours), whereas the PPM run requires 4×10^5 s (about 110 hours). At the sustained 20 Gflop s^{-1} that is about the best performance now, these numbers translate into 10^6 s (280 hours) and 2×10^7 s (5500 hours).

2.2. Data Flow and Data Capture

The choice of computational fluid dynamics algorithms to be favored on a given mpp architecture is considerably influenced by the effective internal communication speed between processors. The use of fully explicit and spatially local methods, such as PPM, places the least demands on such communication. With PPM, our 3–D simulations on regular grids can be subdivided into many subvolumes, with only values of variables on boundary zones a few layers deep having to be exchanged after each time step between processors working on adjacent subvolumes. If the operation count is high, this is a relatively infrequent event. In contrast, pseudo-spectral codes assign successive physical planes to the processors, and carry out fast transforms on these planes within processors, followed by data transposes when calculations are required in the final dimension. Alternatively, the transforms may be carried out in parallel across processors, with either approach requiring fairly intensive data movement. Since the operation counts with pseudo-spectral codes are typically far more modest than with PPM, communication currently places major constraints upon the actual performance realized with the principal mpp contenders.

Of greater concern is how to extract and capture sufficient data to be used for subsequent visualization and analysis of these intricate turbulent flows. A single data snapshot of five fields and two tracers at 1024^3-resolution yields 56 GB per dump. If we sample the time evolution fairly sparsely after every 100 time steps (or 10^3 samples per run), then the full data set is about 56 TB per run. Not only is that number rather large given most current mass storage facilities, but the data flow rate will also be problematic as the computing machines get faster. On current 20 Gflop s^{-1} machines, the required data outflow for the PPM code is a tractable 2.8 MB s^{-1}, and for the pseudo-spectral code (with about 20-fold fewer operations per time step) a more difficult 56 MB s^{-1}. Present multi-TB mass stores have sustained transfer rates of 3 to 20 MB s^{-1}. Data output from a Tflop s^{-1} machine would in turn be 0.14 GB s^{-1} (for PPM) and 2.8 GB s^{-1} (for pseudo-spectral), or thus at worst a factor 10^3 greater than current mass

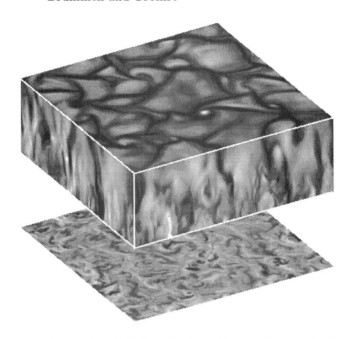

Figure 1. Perspective full domain view of vertical velocity at one instant in time in 3-D turbulent convection simulation ($256^2 \times 130$). Here $R_a = 10^7, T_a = 2 \times 10^7, P_r = 0.1$, with the rotating f-plane domain at latitude $45°$. The rendering has upward-flowing, hot fluid as bright, with downward-flowing, cold fluid as darker tones. Note the curvaceous network at the upper surface and the smaller-scale turbulence at the bottom.

store rates. Substantial improvements in rates of data output and capture are anticipated, but it is likely that such data movement places the severest restraints on efficient simulations of turbulence. The options to help survive this looming data crunch are to sample full dumps less often, or to take compressed dumps, or used weighted averages of the data, or to do smaller problems. More acceptable is to recognize that investments must be made for data movement and capture that may turn out to be comparable in cost to that of the mpp itself.

3. Computational Approach

Our goals have been to first study the dynamics of increasingly turbulent states in simplified "local area" geometries with the influence of rotation incorporated via f-plane modelling, and then to proceed toward a more physical "global" geometry encompassing an entire rotating spherical shell of fluid. Our local-area models consist of a planar layer of fluid positioned tangentially to a rotating sphere at a latitude ϕ. The model configuration for the fully compressible treatments is essentially that of Cattaneo et al. (1991). The physics of the model is simplified by employing a perfect gas, and fluid properties such as the specific

Figure 2. Perspective 3–D view of enstrophy (vorticity squared) at one instant in time as a companion to Figure 1. High enstrophy is bright and opaque whereas low enstrophy is dark and translucent. Distinctive enstrophy structures are present, with evolving cellular flows near the top of the layer yielding sheets and tubes of vorticity, whereas in the intense turbulence at greater depths the vorticity is often organized into randomly aligned tubes. Large coherent tube-like structures intermingle with small-scale random vortex tubing in the interior.

heats, shear viscosity, and thermal conductivity are also assumed constant. The flow is confined between two horizontal, impenetrable, stress-free boundaries, and the fields are assumed to be periodic in the horizontal. The upper boundary is held at a given temperature, whereas the heat flux is prescribed on the lower boundary. We then proceed to solve as an initial value problem the partial differential equations for the conservation of mass, momentum, and energy, along with the equation of state. The simulations are characterized by several nondimensional parameters: the Prandtl number P_r, the ratio of viscosity to thermal conductivity, the Taylor number T_a, the ratio of Coriolis to diffusive terms, the Rayleigh number R_a, the ratio of buoyancy forcing to viscous and thermal dissipation, and the Rossby number R_o, the ratio of nonlinear advective to Coriolis forces.

Our main computations shown here involve the solution of the fully compressible equations using a hybrid finite-difference and pseudo-spectral code. The vertical structure is treated by fourth-order finite differences in the interior. The horizontal components are treated by a pseudo-spectral method that calculates all linear operations and derivatives in spectral space (k_x, k_y, z) and performs the nonlinear multiplications in configuration space (x, y, z), with the transform between spaces achieved by fast transform methods. The time discretization is based on an explicit two-level Adams-Bashforth scheme, and the thermal conduction terms are treated implicitly with a Crank-Nicholson method.

4. Representative Results from 3–D Simulations

We have studied the basic features of fully-developed rotating compressible turbulence, with special regard to the types of flow structures, topologies, mean-flow generation, and energy balances that may be achieved. The simulations typically involve a density contrast across the layer of $\chi = 11$, spanning roughly 5 pressure scale heights. Other parameters fixed are the ratio of the specific heats, $\gamma = 5/3$, and the aspect ratio of the box, at $4:4:1$. A series of runs has been made sparsely populating the four-dimensional parameter space (R_a, T_a, P_r, ϕ) of the model but revealing a rich set of results. We will briefly sample some of the characteristics of such turbulent convection.

Let us first comment on the effects of rotation on the structure and topology of the turbulent convective flows. We find that convective flows in the upper thermal boundary layer consist of a nearly laminar, cellular network, atop a fully turbulent interior, punctuated by vertically-coherent structures emanating from the upper surface (e.g., Brummell, Hurlburt, & Toomre 1993, 1995a,b). There is pronounced asymmetry between the strong, narrow downdrafts and the weak, broad upflows. This is evident in Figure 1, which shows the local-area computational domain with the vertical velocity on the visible surfaces shaded according to its strength (see also Figure 5). A plane close to the lower surface of the domain has been removed to show clearly the difference in scales between the fully-developed turbulence there and the laminar upper surface. The smoothness of the upflow regions near the surface is achieved by rapid horizontal expansion of the ascending fluid elements which will tend to smooth out any small-scale fluctuations. The cell packing near the top is irregular, and cells are born and obliterated in an intricate time-dependent fashion. The Coriolis forces of rotation have changed the characteristics of both the surface network and the turbulent interior from that of earlier nonrotating models (Cattaneo et al. 1991). The surface network has become more curvaceous, less connected and more time-dependent with the addition of rotational constraints. These changes are due to inertial oscillations of surface flows induced by the underlying rotation which mobilizes the network, and to the enhancement of a mechanism for cell destruction and creation by the background pool of vorticity. In the latter, vorticity concentrated at the interstices of the downflow network lanes is amplified by the rotational influence to produce vortices which are strong enough to evacuate their interiors, thus causing a reversal of the vertical flow and the eventual destabilization of the vortex itself. These "spinners" break apart the network at the junctions and allow new upflowing cells to form there.

We find that the nature of the flows in the turbulent interior are subtly altered by rotation. The Coriolis forces provide a linear mechanism for coupling the vertical and horizontal motions, and thus with rotation the isotropy of energy and enstrophy is enhanced. This manifests itself as the concentration of vorticity into distinctive tube-like structures with a very random orientation in the interior of the flow. These vortex tubes are illustrated in Figure 2, which displays the enstrophy (vorticity squared) field in a volume rendering. Near the surface, the laminar cellular flow is dominated by vortex tubes where the coherent downflows form, and by weaker sheets of vorticity generated by the

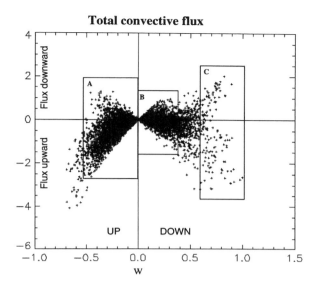

Figure 3. Phase diagram of pointwise convective flux F_c as function of the vertical velocity in a horizontal plane near midlayer for turbulent convection with $R_a = 5 \times 10^5$, $T_a = 10^5$ and $P_r = 0.1$, with $R_o = 7$. The boxes A, B and C delineate horizontal sites of upflow, downflow, and strong downflow within the horizontal sampling plane.

downflow lanes. At greater depths, the small-scale turbulence is characterized by intricately interwoven small intense vortex tubes. Larger structures with vertical coherence are always in existence, and their presence in conjunction with Coriolis effects can lead to enhanced Reynolds stresses and thus a generation of mean flows (cf. Brummell, Xie, Toomre, & Baillie 1995). The vortex interactions of the small-scale turbulent tubes and the larger-scale vertically-coherent strong downflowing structures provide a horizontal mixing mechanism which retards the vertical mixing by the convection, yielding a mean thermal stratification in the vertical which is substantially nonadiabatic over much of the interior. This has the important and surprising consequence that with rotation, buoyancy driving of such turbulence will be experienced throughout the layer depth, and not just dominantly in the thermal boundary layers near the top and bottom of the layer.

There is a distinctive transition from laminar to turbulent flows at greater depths when the Prandtl number P_r in the simulations was decreased below a value of about one-third. The flows are vigorously time dependent and turbulent, yet they exhibit coherent flow patterns recognizable over several scale heights, contrary to the assumptions of mixing-length approaches. Further, the surface appearance of the convection, always being cellular and irregular, does not belie whether the deeper structures are laminar or turbulent. The transition to turbulence also results in remarkable changes in how the convection transports the energy. Namely, in the laminar cases the strong downflows serve to both carry most of the enthalpy flux upward and a significant kinetic energy

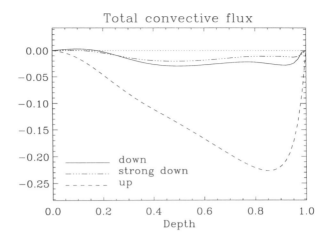

Figure 4. Horizontally-averaged convective flux F_c as function of depth for the case sampled in Figure 3, revealing that the near cancellation of enthalpy and kinetic fluxes within the downflows leads to the upflows dominantly carrying the convective flux.

flux downward: the upflow sites make only a modest contribution to both fluxes. However, once there is transition to turbulence at the greater depths, there is a prominent change in the role of the downflows, in which the (upward) enthalpy and (downward) kinetic flux contributions now nearly cancel each other, leaving the disorganized turbulent motions outside of those coherent structures to effect most of the overall energy transport. Thus whereas the larger scales control the convective patterns and their evolution, the transport is left to the smaller turbulent scales. This is a striking example of how transports achieved by convection can be radically modified by transitions to turbulence. Further, there are major fluctuations in the total convective flux (sum of enthalpy and kinetic fluxes) from site to site across any given horizontal plane, as the phase diagram of vertical velocity w and convective flux F_c in Figure 3 reveals. In the upflows (A), the pointwise F_c values are sufficiently skewed that there is a net upward flux (negative) when averaged over all points, whereas the distribution is more nearly symmetrical for the moderate downflows (B) and so too for the strong downflows (C), with only a small value for the net transport emerging after averaging. This is indeed seen in Figure 4 of the horizontally-averaged F_c with depth as contributed in turn by the upflows and the moderate and strong downflow regions. The contributions from the upflows dominate in the convective transport.

The results shown in Figure 4 are extraordinary, implying that throughout the turbulent layer nearly all the convective flux is carried by the upflowing material and not the downflows, despite the fact that the downflows are much more coherent in space and time compared to the random fluctuations of the smaller scale upflows. The near-zero horizontal average value in the downflow at any depth is due to a cancellation of the two components of the convective flux, i.e., the enthalpy flux and the kinetic flux. Cattaneo et al. (1991) proposed a theory

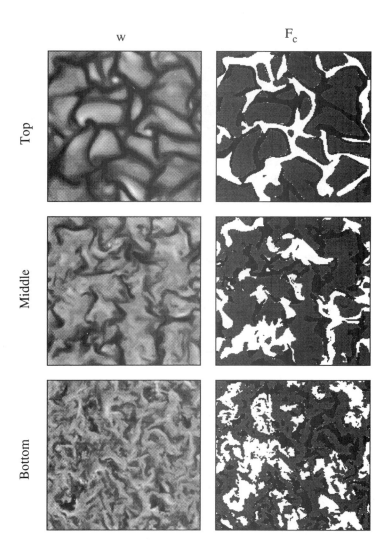

Figure 5. Vertical velocity w (left) and convective flux F_c (right) shown in three horizontal slices near the upper surface, middle and lower surface in single snapshot of vigorous turbulent convection for $R_a = 5 \times 10^6$, $T_a = 10^7$, $P_r = 0.1$, and $R_o = 2.24$. Upflows are lighter tones, downflows are darker tones. The change from smooth, nearly laminar flows at the top to disordered, turbulent flows deeper down is clearly evident. For F_c, regions of upflow are blanked out with neutral grey, and downflow regions encoded either as white or black according to the sense of F_c. Figure 4 indicates that horizontal averaging of these spatially intermittent fluxes in downflows leads to near cancellation.

as to why that might be, suggesting that cancellation of the constituent fluxes at each point in the downflows is due to a Bernoulli-like balance achieved in the dynamics, provided that the motion is nearly steady, the mean stratification is nearly adiabatic, and viscous and thermal conduction are negligible. These conditions are reasonably well met in the turbulent simulations. Figure 5, however, points out that this theory may be overly simplistic. Shown is the vertical velocity and measures of the convective flux F_c for three depths at a typical time in the simulation. The convective flux is shown only in the downflows (the upflows are obscured with a grey mask) with black marking regions of positive convective flux and white marking regions of negative flux. This rendering reveals that the cancellation is neither pointwise nor even local. If it conformed to the theory, then the whole downflow regions would be colored neutrally gray, or at least would have a very high frequency speckling between small black and white values. Instead, we see interleaved patches of black and white where the values of F_c are actually quite large and definitely non-zero. Surprisingly, however, the horizontal average (i.e., the sum of all the black and white patches) is nearly zero. The cancellation of the convective flux appears to be more a phenomenon related to major intermittency in the flux production, both in time and space, rather than by pointwise cancellation as was first thought. These are rather subtle matters, emphasizing that highly turbulent flows can possess transport properties very different from those of laminar flows.

This work was supported in part by NSF through grant ECS–9217394 and by NASA through grants NAG5–2218 and NAG5–2256. The simulations were carried out at the Pittsburgh Supercomputer Center under grant MCA93S005P.

References

Brummell, N. H., Hurlburt, N. E., & Toomre, J., 1993, in GONG 1992: Seismic Investigation of the Sun and Stars, ASP Conf. Ser., Vol. 42, ed. T. M. Brown (San Francisco, ASP), p. 61

Brummell, N. H., Hurlburt, N. E., & Toomre, J., 1995a,b, ApJ, submitted

Brummell, N. H., Xie, X., Toomre, J., & Baillie, C. 1995, in Proc. GONG'94, Helio- and Asteroseismology, ASP Conf. Ser., ed. R. K. Ulrich (San Francisco, ASP), in press

Cattaneo, F., Brummell, N. H., Toomre, J., Malagoli, A., & Hurlburt, N. E. 1991, ApJ, 370, 282

Gilman, P. A., & Miller, J. 1986, ApJ, 61, 585

Glatzmaier, G. A. 1987, in The Internal Solar Angular Velocity, eds. B. R. Durney & S. Sofia (Dordrecht, Reidel), p. 263

Gough, D. O., & Toomre, J. 1991, ARA&A, 29, 627

Distributed Software for Observations in the Near Infrared

V. Gavryusev[1] and C. Baffa

Osservatorio Astrofisico di Arcetri, Largo E. Fermi, 5, Firenze, 50125 Italy

E. Giani

Dipartamento di Astronomia, Università di Firenze

Abstract. We have developed an integrated system that performs astronomical observations in Near Infrared bands operating two-dimensional instruments at the Italian National Infrared Facility's ARNICA[2] and LONGSP[3]. This software consists of several communicating processes, generally executed across a network, as well as on a single computer. The user interface is organized as widget-based X11 client. The interprocess communication is provided by sockets and uses TCP/IP.

The processes denoted for control of hardware (telescope and other instruments) should be executed currently on a PC dedicated for this task under DESQview/X, while all other components (user interface, tools for the data analysis, etc.) can also work under UNIX. The hardware independent part of software is based on the Athena Widget Set and is compiled by GNU C to provide maximum portability.

1. Design

The design of a data acquisition package is, as always, a long struggle between many different and contrasting needs. Also, we have to bear in mind the fundamental consideration that software which gives the user fast operation and an immediate feeling for data quality raises the total (system + operator) efficiency. This has the same effect of a bigger telescope or a more sensitive instrument.

For effective support of the observations of the Arcetri two-dimensional infrared instruments, we need to put together cheap hardware, an easy and intuitive user interface, the fastest data acquisition, the best quick look possible, complete data documentation, infinite and safe data storage, code portability, and the possibility of shifting to a more powerful platform.

As usual, we were forced to choose a middle path between these conflicting requirements. As a start, we used an AT-Bus based machine for hardware

---

[1] Visiting Astronomer, Nuclear Physics Institute of Moscow State University, Moscow, Russia

[2] http://helios.arcetri.astro.it:/home/idefix/Mosaic/instr/arnica/arnica.html

[3] http://helios.arcetri.astro.it:/home/idefix/Mosaic/instr/longsp/longsp.html

construction. However, we found that neither DOS or Windows will do the work (Baffa 1991). This is due to a number of reasons; DOS is short of memory and lacks multi-tasking, while Windows is slow and completely lacks the code portability we wanted. We finally chose to use DESQview/X and the GNU C compiler as a reasonable compromise between our different needs (Gavryusev et al. 1993; Di Giacomo et al. 1993).

From DESQview/X, we get an X server, so we have excellent code portability, a native DOS Extender to have fewer memory problems, and the ability to run in a DOS-like exclusive task mode, which gives us a workable time control over our data acquisition (DESQview 1993). We can compile and run most of our modules on different platforms—PC's and workstations, under DESQview/X and UNIX. We can spread our processes over a network, so we can get data at a telescope, control the instrument from Firenze in Italy, and display the data on a computer in Baltimore.

2. Structure

It is always the best choice to solve independent tasks by independent software. Independently executed parts can be compiled by different compilers (if necessary), debugged separately, and their subsequent versions not influence other pieces of the software (if interface demands are followed). The situation is the same as with the use of DLL libraries. Moreover, if the interprocess communication interface includes support for networking, the software immediately becomes network distributed software. Such organization of the program can, in principle, cause some complexity from the point of view of the user, but this problem can be covered by providing well chosen default options and the permission to easily change them only for experienced users.

In our case we have to differentiate between a number of tasks: (1) a graphical user interface, (2) the display of images, (3) the hardware interface, and (4) data transfer between computers. These tasks are solved by different processes, generally more then one for each task when it is possible. The interprocess communication is provided by sockets and uses TCP/IP if the processes are executed on different computers.

It is important to remember that we plan to use the same control program (without recompilation) for the two different infrared instruments: the camera ARNICA (Lisi et al. 1993) and the long slit spectrograph LONGSP (Gennari et al. 1993). This should be possible because the instrument dependent parts of the software can start or finish their work dynamically, or depending on their contents, an easily editable resource file for the current session.

Figure 1 shows the typical view of the screen during an observation. There is the window with main menu (marked *xnir*), the control/information window during the acquisition process (marked *ACQUISITION*) with an open display of the current image, obtained from ARNICA during the "Single Frame" measurements (marked *Image (acquisition)*). The menu for starting the frame viewer (marked *Display Frame*) is shown with a submenu that has a default list of the hosts where the display can be sent. In addition the view-window (here, SAOimage) is open as well.

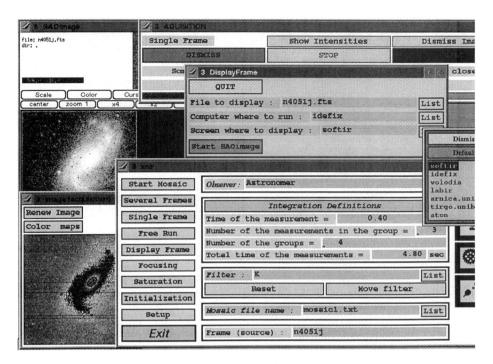

Figure 1. The typical view of screen during the observations.

Acknowledgments. We are grateful to R. Stanga and F. Lisi for useful discussions, to A. Di Giacomo for some of the preparation work, and to Professor F. Pacini who invited one of us (VG) to Arcetri.

References

Baffa, C. 1991, Arcetri Technical Report
DESQview/X User Guide 1993, Quarterdeck
Di Giacomo, A., Giani, E., & Baffa, C. 1993, Arcetri Technical Report
Gavryusev, V., Giani, E., & Baffa, C. 1993, Arcetri Technical Report
Gennari, S., & Vanzi, L. 1993, UCLA Conference
Lisi, F., Baffa, C., & Hunt, L. K. 1993, SPIES International Symposium on optical Engineering and Photonics in Aerospace and Remote Sensing (Orlando, SPIE)

The New Astrophysics Data System

G. Eichhorn, S. S. Murray, M. J. Kurtz, A. Accomazzi, C. S. Grant

Smithsonian Astrophysical Observatory, 60 Garden St., Cambridge, MA 02138

Abstract. The Astrophysics Data System (ADS) is moving towards public domain access software for its data holdings. Several ADS services are already accessible through the World Wide Web (WWW) (e.g., abstract service, catalogs, *Einstein* archive) and more are planned.

1. Introduction

The Astrophysics Data System has been restructured in the recent past. We now concentrate on making data available through the World Wide Web (WWW). The main emphasis of the restructured ADS Project will be the operation and development of the ADS Astrophysics Science Information and Abstract Service (ASIAS) (formerly the ADS Abstract Service). We also will consolidate the data assets already created by the ADS Project so that they will be administered within the restructured project and continue to be made available.

This paper introduces some of the new capabilities that are already available and then describes some of the improvements that are in development.

2. Astrophysics Science Information and Abstract Service (ASIAS)

The ASIAS has been very successful in providing the astronomy researcher the capability to search the astronomical literature. It currently provides access to over 160,000 astronomical abstracts with a sophisticated search engine. The WWW interface to the ASIAS was made available in 1994 February. It is accessible through the ADS Abstract Service[1]. It uses the same search engine as the original ADS abstract service. The data are returned as hypertext documents.

Figure 1 shows the results of two ASIAS queries. On the left is the result of a query which returned three abstracts. The links at the bibliographic codes point directly to the abstracts. This allows the user to retrieve one abstract at a time. Several abstracts can be retrieved at the same time by selecting the check boxes at the requested abstracts and pushing the "Retrieve" button. On the right is a full abstract. This particular abstract has a link to data tables at the Centre de Donnees in Strasbourg, France. The CDS provides on-line access to data tables for some publications. These data are automatically cross-referenced with the ASIAS abstracts. The link is displayed whenever a particular abstract has data associated with it.

---

[1] http://adsabs.harvard.edu/abstract_service.html

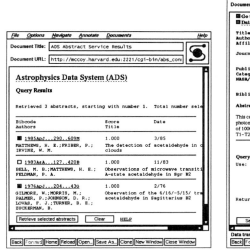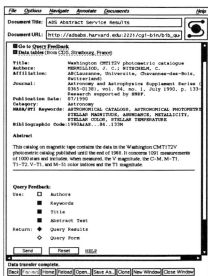

Figure 1. ASIAS query result. Short results with three references (left) and full abstract (right).

The query feedback feature allows the user to automatically build and execute a new query from the current abstract. This allows the user to easily find more information about a given subject.

3. Catalog Data

The ADS now provides access to most of the old ADS catalogs through the WWW. It is accessible through the ADS Catalog Service[2] This WWW access tool provides access to catalogs at SAO, at CASA (University of Colorado), the Center for EUV Astronomy at Berkeley, and at the University of Minnesota. For some catalogs, like the plate scan data at the University of Minnesota, the ADS is the primary means of data access. It is therefore important that we maintain this catalog access.

Figure 2 shows the query form from the Catalog Service for one of the plate scan catalogs at the University of Minnesota. In this form the user can select which fields of the catalog to retrieve, what selection criteria to apply for different fields, and what output format to use for the individual fields. Currently the data can be retrieved as ASCII tables, ADS tables, FITS tables, and, for the plate scan catalogs, finder charts. The user can select to either display the data or to store the file directly on disk.

---

[2] http://adscat.harvard.edu/catalog_service.html

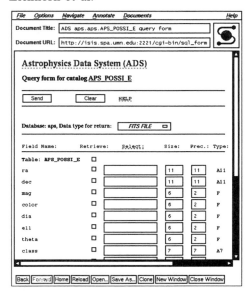

Figure 2. ADS Catalog Service query form for an APS catalog.

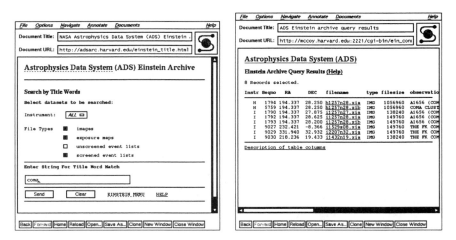

Figure 3. ADS *Einstein* Archive Service query form (left) and query results (right).

4. Archive Data

We are currently providing access to the *Einstein* Observatory data set maintained at SAO. This WWW interface again provides the same functionality as the original ADS interface. Figure 3 shows one of the query forms for the *Einstein* Archive Service (left) and the results of a query (right). The form on the left provides the capability to query the *Einstein* archive by words in the description of data files. Figure 3 (right) shows the result of such a query with the word COMA in the title. It returns information about the data files that fit the query. The filenames are hyperlinks that allow the user to directly download these files. The *Einstein* Archive Service is available through the ADS *Einstein* Archive Service[3]

5. Future Plans

We plan to greatly enhance the utility of the ASIAS in the next year by expanding its data base, providing access to full articles and other data associated with the articles. Accomazzi et. al. (1995) describe the future plans of the ASIAS in more detail.

The central location catalog service provided by the ADS helps to deal with the general problem of how the user community can know about the existence of valuable data. We plan to continue to provide this important (and unique) function through our WWW Catalog Server. We encourage other data holders to make their catalogs available on the WWW. We have a set of programs that provides WWW access to catalog data. They currently work with several databases and can easily be adapted to other databases. Instructions on how to get and install the catalog server are at Catalog Server Installation[4]

In order to facilitate the access to archive data, we will work with our current data providers to help in the transition from ADS servers using our proprietary protocol to servers based on WWW protocols. We will also include links within the ADS WWW Services HomePage[5] to other sources of astronomical data so that users can have a single starting point for finding these resources.

The efforts described above will extend the scope of the abstract service and expand it into a wide ranging digital library service with greatly enhanced utility for the astronomical community

Acknowledgments. This project is funded by the NASA Astrophysics program under grant NCCW-0024.

References

Accomazzi, A., Grant, C. S., Eichhorn, G., Kurtz, M. J., & Murray, S. S. 1995, this volume, p. 36

---

[3] http://adsarc.harvard.edu/einstein_service.html

[4] http://adsdoc.harvard.edu/install.html

[5] http://adswww.harvard.edu/ads_services.html

Development of an ADS Data Dictionary Standard

C. S. Grant, A. Accomazzi, G. Eichhorn, M. J. Kurtz, and S. S. Murray
Smithsonian Astrophysical Observatory, 60 Garden St., Cambridge, MA 02138

Abstract. We present a proposed standard for data dictionaries associated with catalogs being accessed through the Astrophysics Data System (ADS) Mosaic Catalog Service. Each catalog made available for searching must have a data dictionary, which consists of a set of specified keywords describing format, content and location of data. The data dictionary provides the ability to describe catalogued data accurately, and can be used to facilitate examining data from different sources in a common environment.

1. Introduction

The ADS Mosaic Catalog Service[1] currently provides access to about 150 data catalogs located at various nodes across the country. This service was developed to provide a simple means of making on-line catalogs available for searching. Sites with catalogs already in a relational database need only install a server which we provide and create a data dictionary with tools we will provide.

In the current implementation of the service, users may search on any field contained within the catalog, either by typing in a Standard Query Language (SQL) request, or by filling in a table template (QBT). In order to provide the capability of searching catalogs by position, our software needs to be able to identify what kinds of positions are contained in which columns. Because we do not maintain the data, we have no control over what coordinate types and units are in the catalogs. We therefore propose a standard method of identifying column contents with an associated data dictionary file. The data dictionary can also be extended beyond positional column identification and used to key on any astronomically-interesting field (such as magnitude or class).

2. Data Dictionaries

We propose having one file per database which lists information about each table in the database (**Database Data Dictionary**). In addition, we propose having one file per table which lists information about each column in the table (**Table Data Dictionary**). Files will be in FITS-like format, with *keyword=value* pairs. We will provide software which creates these files from the catalog doc-

---

[1]http://adscat.harvard.edu/catalog_service.html

Table 1.   Database Data Dictionary

Keyword	Description
DB_NODE	Node name
NAMEnnn	Database name
TBLEnnn	Table name
DESCnnn	Description of table
COLSnnn	Number of columns in table
ROWSnnn	Number of rows in table
SUBJnnn	Colon separated list of subjects (for lists)
KWDnnn	Colon separated list of keywords (for searching)
URLnnn	URL for description file

umentation file. Nodes will maintain the files so that updates to the databases can be handled without requiring ADS intervention.

Both the descriptions of the catalogs and the keywords assigned to the catalogs are WAIS-indexed. This allows users to search the catalog content for information about the catalogs (such as which catalogs contain redshift).

2.1. Database Data Dictionary

There is one **Database Data Dictionary** file per database at a node. This file is used by the software to construct lists of tables for selection ("by node" or "by subject"). The field specifications are detailed in Table 1. The subjects are used for grouping similar catalogs together. The keywords are used for WAIS searches to enable users to find a catalog about a particular subject. Both of these are to be taken from word sets already in use by the ADS project. The software provided to help generate the data dictionary files will facilitate the selection of appropriate subjects and keywords.

2.2. Table Data Dictionary

There is one **Table Data Dictionary** per table at a node. The counter, nnn, specifies the column number in the table. Additional keywords (such as NAXIS1 and NAXIS2) are added when a FITS table is written out containing results of a query. The field specifications are detailed in Table 2.

The keywords which begin with "DB_" correspond to keywords in the associated Database Data Dictionary. These are essentially a header to the data dictionary, giving information about the table (as opposed to information about the columns). TTYPE and TFORM are the only two required keywords for each column. Unused keywords may be omitted.

3.   Data and Auxiliary Flags

There are two keywords in the Table Data Dictionary which indicate to the server software that the corresponding column contains special attributes. The data flag, **TDFLGnnn**, describes the type of attribute (such as position or image); the auxiliary flag, **TAFLGnnn**, describes additional information associated with the data flag (such as coordinate specifications or a directory pathname). Table 3 lists examples of data flags.

Table 2.   Table Data Dictionary

Keyword	Description
DB_NODE	Node name
DB_NAME	Database name
DB_TBLE	Table name
DB_DESC	Description of table
DB_COLS	Total number of columns in table
DB_ROWS	Total number of rows in table
DB_SUBJ	Colon separated list of subjects (for lists)
DB_KWD	Colon separated list of keywords (for searching)
DB_URL	URL for description file
TTYPEnnn	Column name
TFORMnnn	Format of data in database
TUNITnnn	IAU/NSSDC units code
TDESCnnn	Description of column
TEXMPnnn	Example of data
TFMATnnn	Desired output format
TWDTHnnn	Width of output column
TJUSTnnn	Justification of output column
TDFLGnnn	Data flag
TAFLGnnn	Auxiliary data flag
TACCYnnn	Accuracy of data in database
TLMINnnn	Lower limit of valid data values
TLMAXnnn	Upper limit of valid data values
TDMINnnn	Minimum value of column in database
TDMAXnnn	Maximum value of column in database
TSIZEnnn	Size in bytes of data referenced in columns

Table 3.   Data Flags

TDFLG	Description
p	positional primary search field
s	positional secondary search field (s1, s2, etc.)
t	path name of ASCII text file
i	path name of binary data file
m	astronomical magnitude
e	ellipticity
z	size in arcmin
o	orientation in degrees
c	object classification

3.1. Positional Flags

Positional data flags indicate to the server software that the corresponding column contains coordinates. The primary set of positions to be used in searching (determined by the data providers) is flagged with a data flag "p". There can be only one set of primary positions, but there can be multiple secondary position data flags. The positional auxiliary flags (see Table 4 for examples) describe what kind of coordinates are represented in a colon-separated list, including: which coordinate (x or y), the coordinate units (DD, RAD, sex1, sex2, etc.), the coordinate name where appropriate (some sexagesimal representations), the coordinate system (EQ, GAL, or ECL), and the coordinate epoch where appropriate (B1950 or J2000).

Table 4. Positional Auxiliary Flags

TAFLG	Example
x:sex1:EQ:B1950	122430.0
y:sex1:EQ:B1950	+261630.0
x:sex2:RAH:EQ:J2000	12
x:sex2:RAM:EQ:J2000	27
x:sex2:RAS:EQ:J2000	00.1
y:sex2:DECD:EQ:J2000	+25
y:sex2:DECM:EQ:J2000	59
y:sex2:DECS:EQ:J2000	54.2
x:DD::GA:	223.237
y:DD::GA:	+84.421

3.2. Non-positional Flags

The remaining data flags give tremendous flexibility in the search capabilities. For example, by flagging which column contains the name of the associated image, data archives can be built to retrieve lists of images available in a given catalog. Likewise, by flagging which column is a magnitude column, plot routines can be automated to size the plot symbols based on magnitude. Other data flags can be added by the nodes as desired.

4. Summary

The ADS Mosaic Catalog Service offers data sites the opportunity to make their data available from a centralized location. We provide the software to search catalogs individually, and once the data dictionaries are in place, we will implement positional searches as well as multiple catalog searches. Output from the Catalog Service may be returned in a variety of formats (including ASCII, FITS, and ADS tables) giving users the necessary flexibility for manipulation of their results.

Putting a catalog on-line requires only four steps: (1) running an httpd, (2) installing SQL software[2] (we will assist), (3) writing catalog documentation, and (4) creating data dictionary (we will assist). The ADS provides WWW access to catalogs while still allowing the nodes to maintain ownership of all associated files for updating and maintenance. We offer a homogeneous Mosaic-based query mechanism which is already in place and ready to be expanded. For more information, please contact *ads@cfa.harvard.edu*.

Acknowledgments. This project is funded by the NASA Astrophysics program under grant NCCW-0024.

---

[2] http://adsdoc.harvard.edu/install.html

ADS Abstract Service Enhancements

A. Accomazzi, C. S. Grant, G. Eichhorn, M. J. Kurtz, and S. S. Murray

Smithsonian Astrophysical Observatory, 60 Garden Street, Cambridge, MA 02138

Abstract. The Astrophysics Data System is enhancing functionality and access to the Abstract Service in several areas. A new WAIS server running the freeWAIS search engine has been added to the existing HTTP and ANSA-based servers. Abstract coverage will be expanded to include more of the NASA "RECON" categories as well as abstracts obtained directly from astronomical journals. In addition, we are enhancing the service to include full journal articles in a suitable electronic format.

1. Introduction

The Astrophysics Science Information and Abstract Service (ASIAS) of the Astrophysics Data System (ADS), formerly known as "Abstract Service," has been very successful in providing the researcher and librarian the capability to search the astronomical literature. It currently provides access to over 160,000 astronomical abstracts with a sophisticated search engine.

Use of the service increased dramatically after it was made available on the World Wide Web (WWW), and now averages 30,000 queries and 500,000 retrieved abstracts per month. Its ease of use, flexibility, and data coverage have made it a very well known resource in the astronomical community. In this paper we discuss the current capabilities of the system and its planned enhancements.

2. The Database

The ASIAS provides access to astronomical abstracts since 1975 from the following five categories defined in NASA's Scientific and Technical Information (STI) Database: Astronomy, Astrophysics, Lunar and Planetary Exploration, Solar Physics, and Space Radiation.

The NASA database contains abstracts from over 200 journals, publications, colloquia, symposia, proceedings, and internal NASA reports. They include the majority of astronomical journals as well as several sources from journals loosely related to astronomy. Due to the incomplete coverage of some of these journals by NASA RECON and delays in providing us the abstracts, the database is at the time of this writing about 95% complete and one year behind current publications. As soon as NASA resumes the abstracting, which was interrupted for almost a year, we hope to provide on-line papers which are only a few months behind publication date.

3. The Server

The ASIAS is part of a larger set of programs that can read, format and index the abstracts compiled by NASA RECON and then search them according to queries generated by users on the Internet. The server itself is composed of a module that accepts network requests and a second one that implements the search algorithm. This design has allowed us to keep our efforts focused by having to maintain only one core software system (along with the set of auxiliary files needed to perform a search), instead of having to maintain one for each protocol supported. We will refer to this system as the "ADS search engine" in the rest of the paper.

The abstract server was designed to allow queries on separate "fields" in the documents. For instance, by defining separate fields for abstract text and paper title, a query can specify a term to be searched for only in the title but not in the abstract text. Complex queries can be composed by searching for different terms in some of the fields and then combining the results according to their relevance with respect to the original query. Query results are ranked by a score determining how they match the input query.

3.1. Access Methods and Protocols

As part of the ongoing effort to migrate the functionality and services of the Astrophysics Data System to protocols and software systems in the public domain, we now provide WWW and WAIS access to the ASIAS.

The server itself was originally written as part of a service in an earlier EOS/ADS system based on the ANSA (Advanced Network System Architecture) protocol. WWW access (based on the HTTP protocol) has been available since February 1994. As new HTML features have been implemented in WWW clients, we have been able to provide additional capabilities to our server, which now provides greater functionality and better performance than the ANSA-based one. Users can reach the Abstract Service WWW query form[1] using any browser that supports HTML forms.

The WWW-based server has been implemented by writing a Common Gateway Interface between the HTTP daemon and the server search engine which carries out the query and returns the results. Similarly, the ANSA based server starts a session manager which then runs the same search engine.

Recently we have brought on-line a WAIS server[2] running the freeWAIS-sf[3] search engine. Because of the limitations in the freeWAIS software and the Z39.50-1988 protocol it uses, only a portion of the functionality and flexibility offered by the WWW server are provided to WAIS clients, so whenever possible the WWW server should be used instead. The ADS WAIS server is currently used by the NASA Technical Report Server (NTRS)[4].

---

[1] http://adsabs.harvard.edu/abstract_service.html

[2] http://adswww.harvard.edu/abs_doc/wais/

[3] http://ls6-www.informatik.uni-dortmund.de/freeWAIS-sf/README-sf

[4] http://techreports.larc.nasa.gov/cgi-bin/NTRS

3.2. The ADS Search Engine

The ADS search engine currently provides great flexibility in searching and scoring documents by allowing users to change, among others, the following settings: the relative field weights to be used when computing the score of a document; what boolean logic is to be used when combining the results for each field; and whether synonym replacement is to take place on selected fields.

A query typically returns a list of documents ranked by relevance. The user can then choose whether to retrieve one or more of the selected abstracts. When a single abstract is retrieved, different portions of the abstract itself can then be selected to perform a "relevance feedback" query by requesting papers similar to the current one.

An additional feature that we recently introduced is providing links to data tables published with the paper, when available. The links are returned as part of the selected documents. Currently the only source of such tables is the Strasbourg Astronomical Data Center (CDS).

3.3. ADS Search Engine vs. freeWAIS

As part of our effort to stay abreast of the technology in the field of distributed database systems and search methodologies, we decided to invest some time using a WAIS-based system to index and search our dataset. We selected freeWAIS-sf among the WAIS packages currently available in the public domain since we found it to be the most advanced; in particular, it is superior to the CNIDR freeWAIS[5] version in that it introduces the concepts of search fields and document structure.

Some time was spent in enhancing the freeWAIS code to run faster by storing some of the frequently accessed data in shared memory, to allow better control of what words would be ignored when indexing, and to support extended headlines (the document identifier strings returned by the server upon completion of a query). These changes have since been incorporated in the freeWAIS code.

As a result, we were able to compare the performance and features of our search engine vs. the freeWAIS-sf one on the same set of data. Our tests have shown that our search engine is from 5 to 20 times faster than the freeWAIS-sf one, while the indexing process needed to create the inverted indexes used by our search engine is about 15 times faster than the freeWAIS-sf version.

This dramatic difference in performance is largely due to the fact that the ADS search engine was designed to work well on the particular dataset at hand, rather than to be a general purpose software package. In particular, the use of small inverted indexes kept in shared memory and the inclusion of publication years in the indexes allows searches to be carried out very quickly.

In addition, many of the search features that are available when using the WWW version of our server could not be implemented because of limitations in the freeWAIS-sf package. For instance, synonym replacement in freeWAIS applies indiscriminately to all of the document's fields, and cannot be selectively turned off at query time since it is built into the indexes. Similar problems exist

---

[5] http://cnidr.org/cnidr_projects/freewais.html

for the list of "stop words" (words to be ignored when indexing the document), and the minimum length of words to be indexed.

Despite these problems and reservations, we believe that freeWAIS-sf remains the best public-domain general purpose full-text indexing and search engine available today.

4. Planned Enhancements

The emphasis of the ADS in the coming years will be to utilize modern information systems technologies like WWW to provide increased access to its services for a wide variety of users through public domain client software. The protocols that are currently used for search and retrieval are Z39.50 and HTTP. While we do not expect that to change in the next few years, we intend to keep providing a reliable abstract server capable of understanding whatever protocols are in use in the astronomical community on the Internet.

We plan on expanding the abstract database to cover more topics, e.g., Instrumentation and Space Physics. We are also examining adding to the functionality of the Abstract Service by including a citation index that will allow users to browse through the abstracts of references associated with the current abstract.

Development work will include cooperation with the publishers of astronomical literature to provide access to the original author abstracts. We will also work on providing access to the full articles as image bitmaps. As a first step in this direction we plan to bring on-line several years of the Astrophysics Journal Letters as a test case. User response to having full journal articles available and linked with the abstracts will be evaluated. If it proves to be a valuable service, we will work with publishers to digitize more of the old literature and to see whether we can provide access to electronic forms of new articles.

Recently it has become possible to "publish" electronically data tables from a journal article. We have started work on linking these data tables to the abstracts of the articles. Currently we are making use of the on-line data available through the CDS. Our objective is to provide additional links to any data source which is relevant to the documents we have on-line. Obviously this is a long-term goal and will require a lot of work until standards for the access, classification and retrieval of astronomical data available on the network are agreed upon and implemented.

We believe that the efforts described above will extend the scope of the Astrophysics Science Information and Abstract Service and expand it into a wide ranging system with enhanced utility for the astronomical community.

Acknowledgments. We are grateful to Ulrich Pfeifer for his work in enhancing the freeWAIS software and for being receptive to suggestions and proposed changes to the package. This work is funded the NASA Astrophysics program under grant NCCW-0024.

WWW as a Support for the Long Term LBT Archive

L. Fini

Osservatorio Astrofisico di Arcetri, Largo Enrico Fermi 5, I-50125 Firenze, Italy

Abstract. Any large and complex project has the need to build and maintain an archive of all of its related information throughout its life. The archive content is typically multimedia in nature, consisting of drawings, pictures, papers and documents with different formats, spreadsheet data files, plots, and so on.

When a project is the common effort of a number of institutions, as in the case of the Large Binocular Telescope Project, where several institutions in different countries are cooperating, the most natural structure of the archive is a distributed one, where each party has responsibility over a part of the database, but shares the data with the others. A completely distributed system, i.e., a system where a given data item is stored in a single site, would be impractical for both security and access efficiency reasons. A mixed approach, where stable data items are duplicated at all the participating sites, while the work-in-progress ones are stored only where they are generated, seems to be a safer one. WWW services can then provide a consistent access to the whole archive for all the involved parties.

1. The LBT Project

The Large Binocular Telescope (LBT) is a telescope for infrared and optical wavelengths, equipped with two 8.4 m mirrors on a single mounting, to be located in the northern hemisphere at Mt. Graham (Arizona, USA). The technical details of the project are outside the scope of this presentation and can be found elsewhere (Salinari & Hill 1994; Hill & Salinari 1994).

The project is jointly developed by a number of institutions in different places: the University of Arizona, the Arcetri Astrophysical Observatory in Florence, and the Research Corporation in Tucson, together with some contractors both in USA and in Italy. All the involved parties will need to interact, cooperate, and exchange information throughout the project life. The LBT project is thus "distributed" in nature; pieces of information such as papers, reports, drawings, etc., are produced at far apart places and must be efficiently shared among all the involved parties.

2. The LBT Archive

The LBT archive is an information system which stores a huge number of items (or data files) from various sources and of different nature, gathered during the development phase and the operating life of the telescope. These include: (1) scientific papers, such as scientific justifications of the project, results of the preliminary studies, and so on, (2) technical reports, (3) drawings: mechanical, optical, electrical, etc, (4) datasheets of the commercially available parts and subsystems, (5) software documentation, including the full source code, (6) manuals, with different purposes: operating, troubleshooting, maintenance, observing, etc., probably in hypertext format, (7) statistical data, to maintain an historical log of weather conditions, seeing, instrument related data, etc, (8) observing Programs, to maintain an historical log of telescope usage, and (9) operation log: complete data on telescope operation to support troubleshooting, to improve usage efficiency, etc. All the above data items must be accessible "on-line" throughout the entire telescope life. A considerable fraction of them will continuously grow in size.

The LBT archive will be of central importance for a number of activities: coordinating telescope design and development, supporting instrumentation design and construction, properly operating the telescope, troubleshooting and maintenance, supporting astronomers who want to submit observation programs and the board which must review submitted programs, and refurbishing and modifying the telescope during its operating life.

3. Data Types

Due to the number of different sources of the data files to be stored in the archive, it will not be possible, or even desirable, to define a fixed list of supported formats. even though a set of standards can be agreed upon in order to avoid an uncontrollable explosion of formats. In the future, new formats may be adopted, perhaps requiring the conversion of existing data files; new tools could be developed; sounder standards could emerge; and so on. Although a part of the archive could be stored in the form of "finalized" documents (e.g., using PostScript, GIF, etc.) many data files will need to be stored in some "source" form (e.g., TeX/LaTeX, HTML, and various native formats for word processors, spreadsheets, CAD systems, databases, etc.) because they will be subject to modifications and updates.

All the supported formats will require the availability of adequate "viewers" or browsers to be easily used; this will likely set the ultimate constraint to the number of supported formats, especially when the long life span of the archive is considered.

4. Archive Requirements

The LBT archive system must fulfill a number of conflicting requirements:

- **Generality**. It must support and manage a number of different standards and file formats.
- **Durability**. It must follow the entire project life, which is a pretty long time, especially when the development speed of the supporting technologies is considered. It must, therefore, survive a number of technological updates and changes.

- **Security.** The archive must survive various kind of failures, both those of the supporting hardware and those due to human errors.
- **Flexibility.** It must allow complete reconfigurations of the supporting hardware/software structures. It must allow easy introduction of new formats.
- **Expandability.** It must allow the growth in size of its content, possibly by adopting new storage technologies as soon as they are needed and available.

5. A Distributed Mirrored Archive

The LBT archive will be hosted on many nodes of a LAN. Some are the nodes used for the development of the data items themselves, and so store the dynamical part of the archive. One or more other nodes may have only archiving functions, storing the static data files.

The "active" workstations (i.e., those where the development is actually performed) are integral parts of the archive. In this way a single, unified approach can be used to access both the "dynamic" and the "static" pieces of information.

6. Data Duplication

Most of the archive contents will be duplicated at the participating sites in order to obtain faster access to data, to lower the demand for network bandwidth and, at the same time, to increase the global system security; storing many copies of sensible data in far apart places is traditionally the best way to guarantee data integrity when faced with hardware and human failures and environmental threats.

7. Data Dissemination

The implementation and maintenance of the LBT archive will require the development of procedures to ensure the proper synchronization of the copies of the archive, i.e., to provide for the distribution across the network of new data items from the sites where they are generated to the other sites.

A number of measures can be adopted in order to increase the efficiency of the process and/or minimize the usage of the network bandwidth. Files to be downloaded could be compressed before transmission and decompressed before storing (although in many cases the compressed form could be stored instead). Sites could agree upon best time for downloading based on locally defined tables of "transmission costs." Huge downloads could be split across several days or, perhaps, delayed to next holiday.

Proper crosschecks must be performed after each downloading session to ensure that each updated file is received and that its content is not corrupted. For this purpose, the update lists will contain checksum information about the files so that file integrity can be verified.

8. Data Access and Browsing

The most important function of the archive, at least from the point of view of operations, is the ability for users to navigate through lists of data items, perform searches, access and view single data items, etc. Hypertext documents will impose a logical structure on the actual physical layout of the archive. This scheme allows many independent logical structures to be defined, each suited to the particular needs of a different kind of user. Other functions are needed in order to allow efficient retrieval of data items, such as searching through lists with various selection functions, textual searches in documents, and so on.

9. Tools

The main procedure for updating the archive is structured as a client-server system; each node will periodically run the client procedure, requesting update lists from the other nodes and downloading the required files. Each node, when requested, will start the server procedure to fulfill the request. The client-server mechanism is supported by the FORM capability of the *http* protocol, and is implemented with *Perl* scripts. Further work will include development of a user interface based on the *Mosaic* browser, and the implementation of searching functions, for which the *GlIMPSe* system is being explored.

References

Salinari, P., & Hill, J. M. 1995, Proceedings of the SPIE Conference on Large telescopes (Hawaii, SPIE), in press

Hill, J. M., & Salinari, P. 1995, Proceedings of the SPIE Conference on Large telescopes (Hawaii, SPIE), in press

The World Wide Web: Cornerstone of the EUVE Science Archive

K. McDonald, B. Stroozas, B. Antia, B. Roberts, K. Chen, N. Craig, and C. Christian

Center for EUV Astrophysics, University of California 2150 Kittredge Street, Berkeley, CA 94720-5030

Abstract. The Science Archive for the Extreme Ultraviolet Explorer (*EUVE*) satellite is using the World Wide Web (WWW) client/server software model as the cornerstone for the dissemination of *EUVE* archival material. A number of on-line services are available for accessing the large amounts of *EUVE* data that have been released publicly in the past year; additional services provide access to *EUVE*-related software and information. This paper outlines the current Archive WWW services and discusses plans for the future.

1. EUVE and the Science Archive

NASA's *Extreme Ultraviolet Explorer* (*EUVE*) satellite was launched on 1992 June 7. In the more than two years since, *EUVE* has performed exceedingly well, obtaining hundreds of gigabytes of scientifically invaluable data, which researchers will continue to analyze for years to come. Among the exciting early scientific results are the first complete EUV all-sky survey, the first EUV images of extended objects (e.g., the Moon, the Vela and Cygnus supernova remnants, and the Jupiter-Io plasma torus), and the first detection of helium on Mars.

The *EUVE* Science Archive at the Center for EUV Astrophysics (CEA) at the University of California, Berkeley, has been established to efficiently archive and disseminate to the public the large amounts of *EUVE* data and its associated software and documentation. The Archive has been actively working to implement innovative ideas and technologies, building the infrastructure to store and provide easy access to *EUVE* archival material. One such innovative idea is the use of the World Wide Web (WWW).

2. EUVE Science Archive WWW Services

The WWW provides a simple, fast, efficient, and user- and developer-friendly environment for the *global* dissemination of *EUVE* archival material. The CEA WWW site[1] offers a variety of Archive services for *EUVE* data, software, documentation, and general information, as detailed in the following sections.

---

[1] http://www.cea.berkeley.edu/HomePage.html

2.1. Data Services

One of the major functions of the Archive is to provide researchers with access to *EUVE* data. Proprietary data rights for *EUVE* observations began to expire in early 1994. In 1994 August, the data from the survey phase of the mission was released on the WWW via the following services:

- catalogs — In cooperation with NASA's Astrophysics Data System (ADS) project, the *EUVE* source catalogs (Malina et al. 1994; Bowyer et al. 1994; McDonald et al. 1994) have been installed in the new ADS WWW catalog service. This service provides users with immediate access to published scientific results on over 400 confirmed *EUVE* sources.

- source count rates — The *EUVE* Count Rate Service is a form-based service that analyzes the survey skymaps to search for any significant source detections near an input location. The count rate information for such detections are returned via e-mail to the user within hours of the request.

- skymaps — The *EUVE* Skymap Request Service is a form-based service that allows users to request small sections of the *EUVE* survey skymaps. After some initial processing based on the full skymaps at CEA, the requester is notified via e-mail within hours of the request that the FITS format skymap sections are available for pick-up in the CEA anonymous `ftp` site.

- "pigeonholes" — An *EUVE* pigeonhole is a file that contains the time-tagged information for photons within a small radius of a given position on the sky. The *EUVE* Pigeonhole Request Service is a form-based service that allows users to request *EUVE* pigeonholes. As with the skymaps above, the requester is notified via e-mail within a day or two of the request when the FITS Bintable format pigeonholes are available for pick-up in the CEA `ftp` site. This service is innovative in that *the requester is actually remotely accessing the raw EUVE telemetry and invoking the required processing to create the pigeonhole data products.*

- calibration data — In order to properly analyze *EUVE* pigeonholes, researchers require some supporting calibration data: instrument effective areas, vignetting maps, and point-spread functions, all of which are available on-line as FITS format files.

The proprietary data rights for the individual guest observer (GO) observations began to expire in 1994 April; additional data sets continue to be released on a monthly basis. Since the GO data sets typically require hundreds of MB of disk space to process and analyze, the Archive has developed the *EUVE* Spectral Data Browser service that allows users to browse, preview, and retrieve the public one-dimensional spectra. The spectral browser provides search capabilities (e.g., search by position and/or source classification) that enhance the usability of the browser by helping researchers quickly and easily locate those sources that match their particular interests. For those researchers requiring additional data for these observations (e.g., images, QPOE files, and telemetry tables), the full GO data sets may be ordered via WWW forms and are delivered on magnetic tape via postal mail.

Work is in progress to provide a variety of additional services including a complete standard set of skymap images, a standard set of pigeonholes for cataloged sources, more complete on-line access to the GO data sets (e.g., images and telemetry tables), and access to long-exposure imaging observations from the *EUVE* Right Angle Program (RAP—those observations taken concurrently during GO observations using the "scanning" telescopes, which are mounted at right angles to the GO spectrometers). Analysis services (e.g., light curves from pigeonhole data) are also under development.

2.2. Software Services

In addition to providing *EUVE* data, the Archive is committed to implementing various software services to complement the available data. The long-term goal is to serve as a "clearing-house" for software contributed by external users to support EUV-related research. Toward that end, the following tools are currently available:

- ISM transmission — Based on the model of Rumph, Bowyer, & Vennes (1994), this tool calculates the transmission of the interstellar medium (ISM) at EUV wavelengths.

- neutral hydrogen column density — Based on the published data of Fruscione et al. (1994) and Diplas & Savage (1994), this tool returns a table of neutral hydrogen (H I) column density information for the ten sources nearest to a given position.

- optically thin plasma spectra — This tool applies the Landini-Fossi X-ray/EUV spectral code (Landini & Fossi 1990) to create a spectrum for a specified emission measure, stellar distance, and interstellar H I, He I, and He II column density.

As this pool of unique utilities grows with contributions from the astronomical community, it will serve as a shared resource of useful software tools for researchers, fostering cooperation and enhancing the scientific return from the *EUVE* mission.

2.3. Information Services

A wide variety of documentation and information is also available to round out the public *EUVE* archival material. This material includes (1) general information on *EUVE*, CEA, and the Archive, (2) CEA publications, including the *EUVE* bibliography, journal and conference abstracts and papers, the special *EUVE* edition of the *Journal of the British Interplanetary Society* (JBIS, 1993), and past editions of the *EUVE* electronic newsletter, and (3) form-based services for such activities as ordering *EUVE* CD-ROMs and archival data sets or for contacting Archive personnel. Examples of some of the information services under development include a "meta-index" for the Archive and a Guest Investigator (GI) Program. The Archive meta-index will fully describe and link together all the public *EUVE* and associated data (e.g., finding charts and spectra from the optical identification program), providing an efficient and easy-to-use database containing all the available information for *EUVE* sources. The GI program is being implemented to provide data analysis services to the research community by offering a standard package of support (e.g., computer access and scientific/technical personnel) to assist researchers in exploring the scientific potential of the *EUVE* data sets.

3. Summary

The *EUVE* Science Archive is using the WWW as the cornerstone for disseminating *EUVE* archival material. A wide variety of on-line electronic services provide access to large amounts of public *EUVE* data as well as to related software and information. As the project continues to mature, additional WWW services will be made available in order to assist the research community with the efficient and maximal use of *EUVE* data. For additional information, contact the Archive at the following address:

> EUVE Science Archive
> Center for EUV Astrophysics, University of California
> 2150 Kittredge Street, Berkeley, CA 94720-5030
> 510-642-3032 (voice) or 510-643-5660 (fax)
>
> *archive@cea.berkeley.edu*
> *http://www.cea.berkeley.edu*

Acknowledgments. The authors would like to thank Prof. Stuart Bowyer, Dr. Roger F. Malina, and the *EUVE* science team for their general support. Special thanks go to Bill Boyd and Steve Chan for their assistance in the Archive WWW efforts. This work has been supported by NASA contract NAS5-29298.

References

Bowyer, S., Lieu, R., Lampton, M., Lewis, J., Wu, X., Drake, J. J., & Malina, R. F. 1994, ApJS, 93, 569
Diplas, A., & Savage, B. D. 1994, ApJS, 93, 211
Fruscione, A., Hawkins, I., Jelinsky, P., & Wiercigroch, A. 1994, ApJS, 94, 127
JBIS 1993, Journal of the British Interplanetary Society, 46
Landini, M., & Monsignori Fossi, B. C. 1990, A&AS, 82, 229
Malina, R. F., Marshall, H. L., Antia, B., Christian, C. A., & Dobson, C. A. 1994, AJ, 107, 751
McDonald, K., Craig, N., Sirk, M. M., Drake, J. J., Fruscione, A., Vallerga, J. V., & Malina, R. F. 1994, AJ, 108, 1843
Rumph, T., Bowyer, S., & Vennes, S. 1994, AJ, 107, 2108

The EINSTEIN On-Line Service

D. E. Harris, C. S. Grant

Center for Astrophysics, 60 Garden St., Cambridge, MA 02138

H. Andernach[1]

Observatoire de Lyon, F–69561 Saint-Genis-Laval Cedex, France

Abstract. The *Einstein* On-Line Service (EOLS) is a simple menu-driven system which provides an intuitive method of querying over one hundred database catalogs. In addition, the EOLS contains over 30 CD-ROMs of images from the *Einstein* X-ray Observatory which are available for downloading. The EOLS provides all of our databases to the NASA Astrophysics Data System (ADS) and our documents which describe each table are written in the ADS format. In conjunction with the IAU working group on Radioastronomical Databases, the EOLS serves as an experimental platform for on-line access to radio source catalogs. The number of entries in these catalogs exceeds half a million.

1. Introduction

In 1989 January, SAO established an on-line service to help astronomers prepare *ROSAT* proposals by providing access to the preliminary source list from the "*Einstein* Observatory Catalog of IPC X-ray Sources." In the intervening years, we have updated the source list, added to the documentation, included many more *Einstein* databases as well as a number of tables from other wavebands, provided access to images for downloading from all of the *Einstein* CD-ROMs, and installed new software for more sophisticated filtering and retrieval. Although we have improved the functionality and made significant additions to the databases, we still maintain a simple menu interface accessible from any type of terminal. An instruction manual does not exist: the on-line help facility provides the necessary information.

2. Database Retrieval Options

The EOLS provides three options for data retrieval. The "database query" is a menu-driven system which allows users to build their own query and select which output columns should be retrieved. The "quick query" provides an output format which requires only one specification of selection criteria, and returns up to 79 characters per row (pre-selected by us). The "multiple quick query" is a

---

[1]Current address: IUE Observatory/Villafranca, Apartado 50727, E–28080 Madrid, Spain

positional search on several tables. We have endeavored to make the selection process simple and the easiest option is to select one or more affinity groups. However, the user can specify particular tables from different groups by choosing the "Custom" option or choose to select all tables.

Here are six of the 30 rows retrieved from 105 catalogs with a single query:

```
------------------------------ eos_source ------------------------------
| SEQ |FLD|CAT |    RA     |    DEC   |+/-|COR C/S|         |      |          |     |
|  #  | # | #  | h  m   s  | d  '  "  |asc|cts/sec|   +/-   | S/N  |SIZCOR|RECO|ID   |
|-----|---|----|-----------|----------|---|-------|---------|------|------|----|---  |
| 3042|  1|4594|22 17 41.3|+08 44 55| 36|2.0e-02|4.4e-03|  4.4 |  1.1 |   0|Q    |
```

Total: 1 rows retrieved.

```
------------------------------ qso_veron85 ------------------------------
|            |    RA     |    DEC   |    |        | radio |  fdr6   | fdr11  |
|        name| h  m   s  | d  '  "  |  z |  vmag  |source?|  (cgs)  | (cgs)  |
|------------|-----------|----------|----|--------|-------|---------|--------|
|            |22 17 39.4|+08 44 56|0.228|17.60|      yes|7.50e-25|0.00e+00|
|            |22 17 42.5|+08 45 24|0.623|18.60|      yes|7.00e-25|0.00e+00|
```

Total: 2 rows retrieved.

```
------------------------------ rad_6cm_bwe91 ------------------------------
|            |    RA     |    DEC   |S(4.85GHz)|Extension|Spectral|Separation|
|   Name     | h  m   s  | d  '  "  |   (mJy)  |   Flag  |  Index |   Flag   |
|------------|-----------|----------|----------|---------|--------|----------|
|2217+0844   |22 17 42.0|+08 44 44|   121    |         |  -0.6  |          |
```

Total: 1 rows retrieved.

```
------------------------------ rad_gb87 ------------------------------
|    RA    |RA+/-|  DEC    |DEC+/-|S(4.85)| S +/- |EXT|WARN|CONF|FWHM|FWHM| PA|
| h  m  s  | sec |d  '  "  | arcs | (mJy) | (mJy) |FLG|FLAG|FLAG|MAJ |MIN |   |
|----------|-----|---------|------|-------|-------|---|----|----|----|----|---|
|22 17 41.7| 0.8|+08 44 45|  13  |  125  |  18   |   |    |    |1.20|1.01| -71|
```

Total: 1 rows retrieved.

```
------------------------------ rad_pks90 ------------------------------
| +/- |   |    |       |fdr1410 |fdr2700 |fdr5000 |    RA     |    DEC    |       |
| arcs| id| mag|    z  |  (Jy)  |  (Jy)  |  (Jy)  | h  m   s  | d  '  "  | alias |
|-----|---|----|-------|--------|--------|--------|-----------|----------|-------|
|  20 | Xs|    |       |        | 0.230  |        |22 17 39.7|+08 44 58|4C08.66|
```

Total: 1 rows retrieved.

3. Downloading X-Ray Images

All of our *Einstein* CD-ROMs are permanently available for downloading images. They may be obtained with ftp or with the VMS "copy" command. As of 1994 June 30, we have:

- the 2E Catalog of IPC X-ray Sources: FITS smoothed arrays

- the Database of HRI X-ray Images: FITS photon arrays
- the IPC Slew Survey: 1991 Jan 1
- the HRI Images in Event List Format: FITS binary tables
- the IPC Images in Event List Format: FITS binary tables
- the IPC Unscreened Data Archive: FITS binary tables

4. Other Features

There are several important features of the EOLS service, including a low maintenance operation, and the ADS connection. The EOLS provides all of our databases to the ADS, and our documents that describe each table are written in the ADS format. Documentation is available for each database table and provides extensive detail in some areas (e.g., Vol. 1 of the IPC catalog), archival material such as the "*Einstein* Revised User's Manual," and "help" files which provide information on system operations. A communication facility is also available, and features a bulletin board (including *ROSAT* news) and an e-mail facility for sending yourself documents and the results of database searches. Finally, a set of tools for precessing coordinates and column densities of galactic H I (north of Dec = $-40°$) are available.

There are four methods to log onto the EOLS:

1. Internet — telnet to *einline.harvard.edu*
2. decnet — set host 6714
3. modem — dial (617) 495–7047 or (617) 495–7048
4. mosaic — http://hea-www.harvard.edu/einline/einline.html

Once successful, the login name is "einline"; no password is required. Note: since UNIX is case sensitive, make sure you use lower case when logging in!

5. The Radio Source Catalog Initiative

For many years we have noticed the difficulty of finding radio source data in public archives or on-line databases. In 1991, the IAU Working Group on "Radioastronomical Databases" was formed with one of us (H.A.) as the chair. By now the Working Group has secured some 130 catalogs of radio sources with a total of $\sim 560,000$ entries. Although the working group recognized that a long term solution to this problem should be provided by an on-going commitment from the National Radio Astronomy Observatory (NRAO) or the Strasbourg Astronomical Data Center (CDS), the EOLS volunteered to serve as an experimental platform for access to the catalog collection. This "provisional" solution is now by far the most complete on-line facility for radio source data, offering access to 63 radio catalogs with 520,000 searchable entries. Contributions of new data tables are welcome, but can only be incorporated into the system if authors provide proper documentation and formatted ASCII tables.

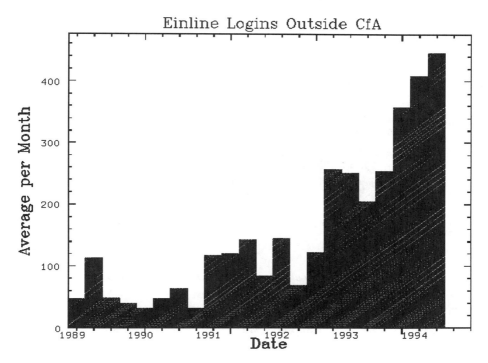

Figure 1. Logins per month from outside CfA: 1989–1994

6. Statistics of Usage

The histogram in Figure 1 shows the average number of logins per month from outside the Center for Astrophysics from 1989 to mid-1994.

AstroWeb – Internet Resources for Astronomers

R. E. Jackson
Computer Sciences Corporation/Space Telescope Science Institute, 3700 San Martin Dr., Baltimore, MD 21218

H.-M. Adorf
Space Telescope - European Coordinating Facility, Karl-Schwarzschild-Str. 2, Garching b. Muenchen, D-85748 Germany

D. Egret
Centre de Données de Strasbourg, 11, rue de l'Universite, Strasbourg, F-67000 France

A. Heck
Strasbourg Astronomical Observatory, 11, rue de l'Universite, Strasbourg, F-67000 France

A. Koekemoer
Mount Stromlo and Siding Spring Observatories, Private Bag, Weston Creek, A.C.T., 2606 Australia

F. Murtagh
Space Telescope - European Coordinating Facility, Karl-Schwarzschild-Str. 2, Garching, D-85748 Germany

D. C. Wells
National Radio Astronomy Observatory, 520 Edgemont Road, Charlottesville, VA 22903-2475

Abstract. AstroWeb is a World Wide Web (WWW) interface to a collection of Internet accessible resources aimed at the astronomical community. The collection currently contains more than 1000 WWW, Gopher, Wide Area Information System (WAIS), Telnet, and Anonymous FTP resources, and it is still growing. AstroWeb provides the additional value-added services: categorization of each resource; descriptive paragraphs for some resources; searchable index of all resource information; 3 times daily search for "dead" or "unreliable" resources.

1. What Is AstroWeb?

The goal of AstroWeb is to be the most complete, up to date, and useful listing of Internet accessible astronomical resources available anywhere. AstroWeb currently contains 1123 WWW, Gopher, WAIS, Telnet, FTP, and Usenet News resources, and 1235 Universal Resource Listings (URL's). As a value-added service, AstroWeb also provides: (1) one or more standardized Categories for each resource, (2) descriptive text for many resources, (3) searchable index of all resource information, (4) thrice daily search for "dead" or "unreliable" resources, and (5) the most complete listing of Astronomical resources. These value-added services set AstroWeb apart from other, more static lists of Internet resources.

2. Where Is AstroWeb?

AstroWeb is available at: ST ScI[1], NRAO[2], CDS[3], ST-ECF[4], and MSSSO[5]. Each version is based on the same resource database, but each site has a different structure or format. For example, ST-ECF only displays the title and acronym for each resource, while ST ScI displays the descriptive text as well. Each site updates and formats their version daily from the central listing at ST ScI. Some sites, in addition, will have "mirror" copies of the other sites' versions.

3. When Was AstroWeb?

On 1994 January 24, Jackson sent e-mail to Wells and Adorf suggesting that they collaborate rather than maintain separate resource listings. A standard interface file format was defined and some software tools were written. Additional members joined the AstroWeb Consortium[6] and on 1994 April 6, AstroWeb was announced to the public via NCSA Mosaic "What's New[7]" and several Usenet newsgroups.

4. Why Was AstroWeb Created?

Jackson, Wells, and Adorf noticed that they and others had created largely overlapping lists of Astronomical resources. New resources were being discovered every day and it seemed more productive to coordinate efforts and use a common master resource listing. From the early size and growth rate of the listing, it was also clear that standardized categories and search tools would be needed to allow the user to quickly find the desired resource. The volatile nature of URL's required that resources be frequently checked for "aliveness" in order to prevent the listing from accumulating "pointers to nowhere". Most importantly, it was a fun and useful thing to do.

---

[1] http://www.stsci.edu/net-resources.html

[2] http://fits.cv.nrao.edu/www/astronomy.html

[3] http://cdsweb.u-strasbg.fr/astroweb.html

[4] http://ecf.hq.eso.org/astro-resources.html

[5] http://meteor.anu.edu.au/astronomy.html

[6] http://fits.cv.nrao.edu/www/astroweb.html

[7] http://www.ncsa.uiuc.edu/SDG/Software/Mosaic/Docs/whats-new.html

5. How To Add Or Change AstroWeb?

To preserve the homogeneity and integrity of AstroWeb, the central listing at ST ScI can be edited only by AstroWeb Consortium members. However, we welcome comments, suggestions, new resources, corrections to existing resources, etc. from everyone. Consortium members can be reached via e-mail at: *astroweb@nrao.edu*[8] There are also HTML forms available to: (1) add a Resource,[9] (2) add a Personal Resource,[10] (3) and submit a Correction.[11]

References

Jackson, R., Wells, D., Adorf, H.-M., Egret, D., Heck, A., Koekemoer, A., & Murtagh, F. 1994, Bull. CDS, 45, 21

Jackson, R., Wells, D., Adorf, H.-M., Egret, D., Heck, A., Koekemoer, A., & Murtagh, F. 1995, A&AS, 108, 235

Adorf, H.-M., Egret, D., Heck, A., Jackson, R., Koekemoer, A., Murtagh, F., & Wells, D. 1994, ST-ECF Newsletter, 22

---

[8] mailto://astroweb@nrao.edu

[9] http://fits.cv.nrao.edu/www/astroweb/aref.html

[10] http://fits.cv.nrao.edu/www/astroweb/paref.html

[11] http://fits.cv.nrao.edu/www/astroweb/adcrf.html

Indexing and Searching Distributed Astronomical Data Archives

R. E. Jackson

Computer Sciences Corporation/Space Telescope Science Institute, 3700 San Martin Dr., Baltimore, MD 21218

Abstract. The technology needed to implement a Distributed Astronomical Data Archive (DADA) is available today (e.g., Fullton 1993). Query interface standards are needed, however, before the DADA information will be discoverable.

Fortunately, a small number of parameters can describe a large variety of astronomical datasets. One possible set of parameters is (RA, DEC, Wavelength, Time, Intensity) × (Minimum Value, Maximum Value, Resolution, Coverage). These twenty parameters can describe aperture photometry, images, time resolved spectroscopy, etc.

These parameters would be used to index each dataset in each catalog. Each catalog would in turn be indexed by the extremum values of the parameters into a catalog of catalogs. Replicating this catalog of catalogs would create a system with no centralized resource to be saturated by multiple users.

1. Available But Not Discoverable

The widespread availability of Internet access has made astronomical catalogs—or even astronomical data—available interactively via FTP, Telnet, Gopher, Wide Area Information System (WAIS), or World Wide Web. These tools have solved the access and navigation problem. However, there is no available system which can answer a query like: "All data for NGC1073 taken between 1989 Jan 1 and 1989 Aug 1 in the wavelength range 0.4–0.6 μm with spatial resolution less than 2 arcseconds." The ADS and ESIS attempt to provide this ability, but they only support cross catalog queries on RA, DEC, and wavelength region. They do not provide the ability to constrain the search to data with a specified spatial resolution, spatial extent, spectral resolution, etc. It is not easy to add new data to ADS or ESIS, and they both use a centralized resource which limits their scalability.

If the wealth of astronomical catalogs or data is to be really useful, the information must be easily discoverable.

2. The WAIS Solution

Fortunately, a similar problem has already been solved by the WAIS. There is a central (although replicated) directory-of-servers, which contains a manually generated description of each WAIS index. The user queries the directory-of-servers to find which indices should be searched. The user then queries a user-specified set of indices for the desired information. The key elements of the WAIS solution are: a standard query protocol, distributed WAIS index servers, a directory-of-servers, and a client which can query multiple servers. WAIS has solved the problems of scalability and ease of adding new information. However querying the entire system is still a two step process, and the information in the directory-of-servers is not always accurate or current.

3. Parameterizing Observations

The problem of accurately describing different resources is actually relatively easy for astronomical data. Astronomical observations can be described by the following parameters: (1) right ascension, (2) declination, (3) wavelength/frequency, (4) date, and (5) flux. Each parameter has a (1) maximum value, (2) minimum value, (3) resolution/sampling, and (4) coverage/filling factor. These twenty parameters can describe observations ranging from aperture photometry to time-resolved spectral imaging. A few additional parameters may be needed to describe information like the position angle of a rectangular region or the shape of a non-rectangular bandpass.

4. Individual Archives

Each archive site would index their observations by the twenty parameters. Observations with similar extremum values would be combined into a "catalog" described by the extremum values of (1) X width, and X resolution; (2) Y width, and Y resolution; (3) minimum wavelength, maximum wavelength, and wavelength resolution; (4) time duration, time, and resolution; and (5) minimum flux, maximum flux, and resolution. The catalog could be simply a list of observations or it could contain HTML links to the actual data. By having each archive site do its own indexing, the conversion to the standard representation is done by the people with the most knowledge of the data and its limitations.

5. Catalog of Catalogs

Individual archive sites would "register" their catalogs with the "Catalog of Catalogs" central repository site. This would provide a single point from which to announce new catalogs and obtain a list of existing catalogs. Each archive site would have a local copy of the "Catalog of Catalogs," updated daily from the central repository. This local copy would be used for user queries—not the one at the central repository. From the perspective of user queries, the system is completely distributed and there is no centralized resource to saturate. The central repository would query each catalog daily to verify its availability and mark "dead" catalogs in the "Catalog of Catalogs". It would also obtain the current values of the individual catalog extrema during the daily query.

6. Query Interface

The Query Interface would be a HTML form with the following fields: (1) what catalogs to query and what catalogs *not* to query; (2) RA, DEC, TARGNAME, and Search Radius; (3) X-Width and Y-Width; (4) X-Resolution, Y-Resolution, and Coverage; (5) Wavelength Center and Wavelength Width; (6) Wavelength Resolution and Coverage; (7) Time Center and Time Width; (8) Time Resolution and Coverage; (9) Minimum Flux and Maximum Flux; and (10) Flux Resolution.

The underlying server script would sanity check user input, determine which catalogs to query, query each catalog, and combine the results. The local copy of the "Catalog of Catalogs" would be used to determine which catalogs to query, and to allow a query at one site to query all the sites.

7. Query Fan-Out and Combination

The same query interface server script could be used to query a local index search engine, query a remote index search engine, and query another query interface. The additional level of indirection provided by the third case would allow each archive site to relocate or subdivide the catalogs to meet the changing user load, hardware availability, or catalog structure. Since the "Catalog of Catalogs" has virtually the same fields as an actual catalog, the same indexing and search software could be used for both purposes.

8. Index Search Engine

Each catalog would be served by an Index Search Engine which could be a relational database, freeWAIS-sf, a custom tool, or whatever software was suited to that archive site. The standard set of parameters combined with a standard query protocol does not force the archive site to store their information in a particular system or database.

Hopefully, a set of public domain software could be assembled to provide an Index Search Engine for those sites not wishing to buy a relational database.

9. Conclusions

The technology is available today to perform cross-catalog queries, to fetch the data at the click of a mouse, quickly to add new observation catalogs, and to distribute the load across multiple machines. The challenge is indexing observations by the standard parameters.

References

Fullton, J. 1994, in Astronomical Data Analysis Software and Systems III, ASP Conf. Ser., Vol. 61, eds. D. R. Crabtree, R. J. Hanisch, & J. Barnes (San Francisco, ASP), p. 3

An Information System for Proposal Submission and Handling

A. M. Chavan and M. A. Albrecht
European Southern Observatory, Karl-Schwarzschild-Str. 2, D-85748 Garching, Germany

Abstract. The Proposal Handling and Reporting System (PHRS) is a software system aimed at supporting ESO's Observing Programme Committee (OPC) during the entire review process of the Observing Time Proposals. Proposals are written in a mark-up language based on LaTeX and are submitted via e-mail. PHRS maintains a database of validated proposals, and operators are enabled to browse it via user-friendly GUI tools, developed using Tcl/Tk. Referees receive PHRS generated printed reports, and submit proposal ratings via e-mail. Panel and OPC meetings are supported by interactive data entry tools and printed documents; in order to obtain high-quality output, all printed reports are processed via LaTeX. The final telescope schedule is published both on-line (World Wide Web) and in printed form.

PHRS was already successfully employed twice (for Periods 54 and 55), handling over 250 (peak) proposal submissions per day.

1. Introduction

Astronomers who wish to use ESO's facilities in La Silla (Chile) must submit Observing Time Proposals (henceforth simply *proposals*), indicating which facilities they want to use, when and for how long, and the science they want to perform. All submitted proposals are peer-reviewed and ranked by the Observing Program Committee (OPC) of ESO; the best proposals are finally assigned observing time at one of the telescopes in La Silla. This process takes several weeks, twice each year, and involves tens of people both within and outside ESO; more than one thousand investigators submit a total of over five hundred proposals each observing period.

The Proposal Handling and Reporting System (PHRS) at ESO is a software system aimed at tackling this problem and supporting the OPC. An entirely new version was developed in 1994, drawing on the experience gained during some years' experience with a previous, less comprehensive system. PHRS handles proposals throughout the entire review process: submission, storage, referee evaluation, panel discussion, OPC recommendation, and time assignment (scheduling).

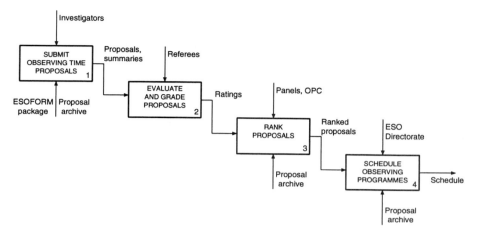

Figure 1. The proposal submission and review process.

2. Proposal Review Process

Figure 1 shows how the review process proceeds from proposal submission to publication of the final schedule, which is described below.

2.1. Proposal Submission and Archiving

Proposals need to be both computer-readable and nicely formatted on paper. We achieved both goals by developing a mark-up language based on LATEX macros: the new commands give the proposal its appropriate look and are easily parsed to extract relevant information. The *ESOFORM* package, which can be downloaded to each investigator's site, contains all necessary style files, template Observing Time applications, user manuals and period-related technical information. When printed at ESO, a submitted proposal looks identical to the proposal on the investigator's desk—thus eliminating the need for paper copy submission. Investigators write their proposals, print them at their institute for verification, and then send them via e-mail to ESO.

Here, a "receiver" program verifies that all mandatory information was provided, then stores valid proposals in a database—the "proposal archive." No manual intervention is necessary: in fact, PHRS can operate unattended around the clock, and it proved able to cope with over 250 e-mail messages per day (usually on the last day before the deadline). Investigators normally get an acknowledgment message back within a minute or so of their submission; errors found in proposals are reported in detail. The authors went to great lengths to avoid data loss, even in the case of hardware failure.

The proposal archive is based on relational data base management technology, the same used in the STARCAT system (Pirenne et al. 1993). Operators interface with the archive with user-friendly, GUI-based tools, developed using Tcl/Tk (Ousterhout 1994). Since most of the data in the archive is classified, at least until schedule publication, we had to insure that only authorized operators could access it.

The system is network oriented, but it is flexible enough to allow for regular (post) mail submission of proposals as well. In some special cases, investigators submit printed proposals, and operators need to type the proposal's main data into the database.

2.2. Peer Review

Valid proposals are peer-reviewed. Each referee reviews several proposals of the same category (for instance, category C groups proposals dealing with "Interstellar and intergalactic mediums"): he/she is required to rate them, by giving a grade—expressing the proposal's scientific merit—and a recommended number of nights (and, often, explanatory remarks). The same proposal is refereed by two or three different people.

Referees receive a printed copy of the proposals they must review and a set of summary reports generated by PHRS; these are printed via LaTeX, with the goal of providing appropriate documents, where scientific symbols and non-English names are correctly printed. Referees are also provided with a pre-initialized, e-mailed form (called a "report card") which they must fill in with ratings and comments. In order to eliminate possible errors, completed report cards are then returned via e-mail, and later processed by PHRS to extract and store ratings in the proposal archive.

2.3. Discussion and Ranking

The final step in the review process is the ranking of proposals, and it is a two-phase activity. Initially, all referees of the same category meet in a panel; they discuss each proposal belonging to the category, and agree on a final grade and recommended number of nights. The goal of panel discussion is to rank all proposals in one category according to their scientific value: better proposals are more likely to be assigned observing time, and telescopes are often oversubscribed by a factor of four or more.

PHRS supports the panels with more summary reports and interactive tools: since panels can directly update the proposal archive, there is no need to re-type information, thus eliminating possible errors; and panels can explore alternative rankings with respect to the number of available observing nights. Some data summaries are generated in spreadsheet format, for further processing and chart generation.

When the panels have completed their task, it is the OPC's responsibility to harmonize rankings across different categories and telescopes. A "cutoff line" separates the best proposals from those that will receive telescope time only if there is any time left, and the OPC ensures that proposals of different categories are evenly distributed around the line. PHRS provides per-telescope and per-category cutoff line reports to support OPC discussion.

2.4. Scheduling

Once the OPC has finalized its decisions, it is the ESO's directorate responsibility to distribute observing proposals (which are now called "observing programmes") over the range of available nights. This is a very complex and delicate process, and it is currently performed by hand; we plan to integrate telescope scheduling within PHRS as our next project. The final schedule is both stored in the proposal archive and published. ESO distributes the schedule document to all interested investigators, and we also developed a World Wide Web interface to the schedule archive[1] for on-line browsing.

3. Conclusions and Future Directions

Some recent advances in technology have made PHRS possible: (1) widespread Internet access enables most investigators to use FTP and e-mail for their submission, (2) "client-server" software techniques increase the system's modularity, reliability and efficiency, (3) easy to develop, "point-and-click" Graphical User Interfaces (GUI's) minimize operator training and reduce errors, and (4) centralized information storage, coupled with appropriate software tools, enable support staff to meet deadlines, providing timely and accurate information to the OPC.

A number of problems have arisen with the use of this new system, the main one being the inability to check the correctness of some user supplied information: for instance, investigators writing a (first) name where a surname (family name) is needed, and vice-versa.

Future projects include support for Observation Preparation (Phase II proposals), and both long- and short-term scheduling (telescope scheduling and observation scheduling). Finally, in order to reduce bulky paper shipments, we are investigating the possibility of having referees download proposals and reports from an FTP account.

Acknowledgments. The authors are grateful to J. Breysacher, C. Euler, E. Hoppe and G. Meylan, whose feedback was invaluable, and who guided our understanding of the issues involved in the design of PHRS.

References

Pirenne, B., Albrecht, M., Durand, D., & Gaudet, S. 1993, in Astronomical Data Analysis Software and Systems II, ASP Conf. Ser., Vol. 52, eds. R. J. Hanisch, R. J. V. Brissenden, & J. Barnes (San Francisco, ASP), p. 95

Ousterhout, J. 1994, Tcl and the Tk Toolkit (Reading, Addison-Wesley)

---

[1] http://arch-http.hq.eso.org/cgi-bin/wdb/eso/sched_rep/form

Design of a Remote Proposal Submission System

A. Richmond, M. Duesterhaus, S. Yom
Hughes STX, GSFC/NASA, Greenbelt, MD 20771

E. Schlegel, A. Smale
NASA, Research Scientist, USRA, Greenbelt, MD 20771

N. E. White
NASA/Goddard Space Flight Center, Laboratory for High Energy Astrophysics, Greenbelt MD 20771

Abstract. Many astrophysics and astronomy tools have been developed utilizing NCSA Mosaic and the World Wide Web (WWW). Several tools have been developed at the High Energy Astrophysics Science Archive Research Center (HEASARC) at NASA's Goddard Space Flight Center, including a Remote Proposal Submission[1] (RPS) system for NASA satellite missions. The RPS provides facilities for submitting Observation Request forms. It is an automated service, designed to support several missions. Currently only the *XTE* mission is supported. Other projects that will be supported in the future include *ASCA*, *GRO*, and *ROSAT*.

1. The High Energy Astrophysics Science Archive Research Center

HEASARC was created by NASA in 1990 as a site for X-ray and Gamma-ray archive research. The total data volume will be of the order of 1 Terabyte by 1995, available on-line for immediate access. HEASARC provides a multi-mission archive for data from the *Röntgen Satellite*, the *Compton Gamma-Ray Observatory*, the *Astro 1 Broad Band X-ray Telescope*, *Advanced Satellite for Cosmology & Astrophysics*, and *XTE* missions. This co-exists with the archival data from past missions such as *HEAO 3*, *OSO 8*, *SAS 2* and *3*, *Uhuru*, *Vela5B*, the second *High Energy Astrophysics Observatory*, and the first *High Energy Astrophysics* Observatory. Data from non-US missions (e.g., the *European X-ray Astronomy Satellite* and *Ginga*), is also provided as international agreements allow, along with data from *Ariel-V*, *COS-B*, *EUVE*, *HST*, *IRAS*, *SMM*, and *TD1*. The HEASARC is located at the Goddard Space Flight Center and is a collaboration between Goddard's Laboratory for High Energy Astrophysics (LHEA, responsible for the science content of the archive), and the National Space Science Data Center (NSSDC, responsible for the data archive management).

---

[1] http://heasarc.gsfc.nasa.gov/cgi-bin/RPS.pl

2. What is XTE?

XTE is the *X-ray Timing Explorer*, categorized by (1) 1 microsec time resolution, (2) 2–200 keV energy band covered, (3) 1.06 microJy sensitivity at 5.2 keV, (4) 1° FWHM field-of-view, (5) all-sky monitor capability, (6) having more than 93% of sky visible, and (7) near real-time operation.

The WWW was chosen as the user interface to the *XTE* RPS to promote

Familiar interfaces. The learning curve is very small. NCSA Mosaic is free and many users already have it installed. The WWW interface has point-and-click functionality making it very user-friendly.

Support All Platforms. The WWW was chosen to allow us to support all platforms with minimal effort. This minimizes the amount of software that needs to be maintained. And since the server is centralized, it does not have to be ported to every type of machine and every operating system version used in the astronomy community.

3. The Major Implementation Problems: User/Browser Side

Loss of Control. When we were writing GUI's we had complete control over our interface and could simply create user-side code as we saw fit, e.g., to pop up some dialog. We have given up that flexibility, for the time being, because of the significant advantages we gained. But we hope that ultimately there will be some client-side equivalent to server scripts, so that we can, for example, interface data analysis tools to browsers.

Layout. One of the major implementation problems was getting satisfactory layout. We experimented with this and finally settled on parsing a LaTeX template to get position information and making the screen resemble the actual paper form by using preformatted layout. And if you do not—well, the widget and its label may get separated by a line break.

Lack of Browser Uniformity. Another problem is variations in presentation via different browsers and platforms. We really would like to see more uniformity between browsers. Often one may spend considerable time getting the layout to look good on one (with all defaults selected), only to find that it is a horrible mess on another.

4. The Major Implementation Problems: Server Side

The major limiting characteristic of HTTP is statelessness. For our StarTrax interface we want to engage in a dialog with the user which progresses from nothing, and builds up to a state where we have identified a target satisfying given constraints, and can offer associated data products. The WWW offers the opportunity to "mix and match" reusable components in a rapid prototyping approach; we were astonished in our early days (only 1 year ago!) at how quickly we could deliver useful functionality. Later we began to realize that some of our early optimism was perhaps too naïve: the need for good software engineering practices is probably even greater than before. There are many more components being interconnected and interfaced, and exponentially more failure points. Perl helps you to hack up something real fast without worrying too much about the engineering aspects until much too late.

5. General Observations

In spite of the sometimes severe disadvantages, we believe the WWW is a very effective platform for rapid information systems development in astronomy and astrophysics. The WWW provides a substantial pre-existing infrastructure supporting a client-server architecture on the Internet, so one can deliver application functionality at a much faster rate—say, an order of magnitude faster. In the "classical" mode—in which we started this project—the developer(s) spend a great deal of time wallowing in relatively low level code. We have not yet achieved the goal of "reusability," in spite of the promises of modern software engineering methodologies. Using WWW and HTML (Hypertext Markup Language) (and ideally, Perl), since you build on that infrastructure, you can deliver functionality very quickly. This not only delivers the promises of rapid prototyping (throw one away), but also of rapid development. The iteration cycle time was often of the order of minutes, rather than hours or days, as in the classical approach. This was partly due to the pre-existing WWW functionality, but to our choice of Perl rather than C. Perl is ideally suited to this kind of work, because it, too, provides a great deal of ready-made high-level functionality. For example, we converted some 30 lines of C code, for decoding URLs, into 3 lines of Perl.

Electronic Submission of HST Phase I Observing Proposals

H. E. Payne and D. J. Asson

Space Telescope Science Institute, 3700 San Martin Dr., Baltimore, MD 21218

Abstract. The process of proposing for *Hubble Space Telescope* observing time has been revised, to make it simpler for proposers. We describe the simple, flexible, and robust prototype system created to handle proposal submission. Using tools available in the public domain, we created our prototype in a remarkably short period of time.

1. Introduction

Proposing for *Hubble Space Telescope* (*HST*) observing time proceeds in two phases: the scientific merits of the proposal are considered in Phase I; only accepted proposals enter Phase II, where the observations are described in complete detail. The procedure for submitting Phase I proposals has been revised, to simplify it. Our contribution has been to implement a system for electronic proposal submission. Our goals were (1) to provide a single form for producing both the formatted, paper output required by the reviewers and a "database" (flat files, really) used by ST ScI staff to track proposals, (2) to minimize the amount of typing required by the proposer, (3) to provide a robust, secure system for archiving and analyzing proposals, and (4) to provide report generators for tracking the proposals.

The single form turned out to be a LaTeX template that proposers can obtain from a mail server, edit, and e-mail back. An accompanying style file allows proposers to format and print their proposals, since we generally require paper copies, as well (although we did run a successful pilot program for paperless, fully electronic submission). On our end, the e-mailed proposals are carefully archived and sent to a "stripper," which parses them, checks for a variety of errors and omissions, sends an acknowledgment back to the proposer, and sends a report to someone on staff. The `procmail` program allows most of these activities to be automated. Many aspects of our system were inspired by the Kitt Peak proposal handling system.

2. From the Proposer's Perspective

We asked proposers to fill out a heavily commented LaTeX template. All input destined for our electronic database is entered as tagged items, as shown in Figure 1. Proposers can process the same document to obtain formatted hardcopy. The layout was designed to give time allocation committees the information they require, with a minimum of distracting clutter.

```
%       4. PRINCIPAL INVESTIGATOR
%
%       Identify the PRINCIPAL INVESTIGATOR (PI). If you wish to
%       include your title (Dr., Prof.,...), include it in \PIfirstname,
%       as in
%                 \PIfirstname{Dr. Bob}.
%
%       Use \\ to break lines in your address, as in
%
%       \address{Science Hall\\4321 University Avenue\\Anytown, MD 21218}
%
%       but limit yourself to four lines (three \\'s).
\PIfirstname      {Michael}
\PIlastname       {Shara}
\institution      {Space Telescope Science Institute}
\address          {3700 San Martin Drive\\Baltimore, MD USA 21218}
\telephone        {410-516-4543}
\email            {mshara@stsci.edu}
\country          {USA}   % Country of above named institution.
\USstate          {MD}    % 2-letter code required for US proposers only
%\ESAmember       {yes}   % Uncomment this line only if you are an ESA member.
```

Figure 1. A fragment from the observing proposal template.

We (HEP) used the programming capabilities of LaTeX to format the printed output according to the proposer's input. For example, the observing proposal format is different from the archival research (AR) proposal format—AR proposals have a budget and do not have observing time requests. By eliminating the table of observing time requests from the formatted output we spare the reviewers from a possible distraction. Typesetting tagged items as tables was an interesting challenge; column widths and formats are calculated on-the-fly to improve readability, and the Jurriens and Braams **supertabular** style is used to break long tables. Default strings highlight omissions, and the proposer's arithmetic is checked. Although these checks are extremely simple, they did reduce errors.

3. The Processing

Proposal files received at the ST ScI are "stripped" by Perl scripts, extracting the values of the tagged items, and writing them to a file. A thorough check for missing or illegal inputs is performed, and the complete report is sent to a staff member. An abbreviated report is sent back to the submitter, as an acknowledgment. A variety of report generators make use of the information stripped from all of the proposals. In addition, we generate partially completed Phase II proposals, to save proposers from entering information they already gave us for Phase I.

We (DJA) chose an interpreted language to speed development, debugging, and testing, and at the time, Perl was the obvious choice. The entire processing system was assembled in a period of weeks. Its regular expression handling and report generation tools made parsing the LaTeX file quite easy. The report writing tools (**format** and **write**) simplified the task of generating mail messages, analysis reports, time allocation committee reports, and partially completed Phase II templates. Again, the reports vary according to the user's input. Perl's associative arrays and interpreted nature allow actual data values to invoke the routines needed to process them in the parsing or archiving steps, making it a truly data-driven system! Perl regular expressions were used to replace the user specified home institution with a canonical form.

The code is quite generic, and will run on the many platforms that support Perl. The code was adapted to process the abstracts for this meeting, and the core processing software is being used in RPS2, the new Phase II proposal preparation and submission system. In hindsight, the tools might better have been written in Tcl/Tk, to provide an easier path to expanding the system to a graphical interface for generating a Phase I proposal and submitting it.

4. The Glue

The pieces of this system are all held together by the public domain UNIX procmail program, a simple utility to trigger various actions via e-mail, depending on the subject line or contents of the message. Users can request proposal templates and they are automatically sent. Proposal submissions are automatically recognized, carefully archived, and sent off for processing. Bounced mail messages and "vacation" messages were automatically set aside. All activity was logged.

5. Status

From the beginning of operations until the Phase I proposal deadline, the system handled a total of 1360 e-mail messages, which included 350 requests for proposal templates, 863 proposals (not counting proposals that were corrected and re-submitted), and 48 PostScript proposals (in lieu of paper copies) in the all-electronic pilot program. Only 2 all-paper (i.e., non-electronic) proposals were received. Proposers and staff agree that Phase I went smoothly. A survey of 125 proposers found most users were favorably impressed with the proposal templates. There are still a few reservations about moving to paperless submission, however, most having to do with PostScript figures and the need for signatures.

Acknowledgments. Our work took place in the context of a committee chaired by Mark Johnston. Many improvements were made at the suggestion of other members of the committee, and it is our pleasure to thank them.

A Generalized Mosaic-to-SQL Interface with Extensions to Distributed Archives

F. Pasian and R. Smareglia

Osservatorio Astronomico di Trieste, Via G B Tiepolo 11, Trieste, I 34131 Italy

Abstract. A graphical user interface based on Mosaic, and allowing form-based access to SQL-driven relational databases, is described. The system is general, due to some simple naming conventions for database tables. This allows a description of the database structure (contents of tables, joins, and views) to be stored in the database itself, and the Mosaic pages to be prepared on-the-fly, with no need for external definition files.

Furthermore, the interface allows X-Y plotting of any two numerical quantities retrieved from the database, and access to data archives distributed over Internet, following the WWW concept. Quick-look data can be displayed, and observational data can be retrieved in their original format (FITS) inside compressed `tar` files. All of these operations are performed on the archive hosts by proper server processes, and no additional software, besides Mosaic, is required at the user's site.

The functionality of the system has been tested on a small archive of objective prism data (images and extracted spectra) and will be implemented for a distributed archive of ISM data.

1. Introduction

The World-Wide-Web (WWW) and NCSA Mosaic can be considered a *de facto* standard for information retrieval, and their wide acceptance is leading to a different and more homogeneous design of interfaces to archives and databases in astronomy. The advantages of using Mosaic as an interface are numerous: distribution and development of the software are taken care of at NCSA, the Mosaic client handles all the intricacies of the man-machine interface, the development rules are easy to learn, and data centers do not need to bother about distributing and maintaining user interface or commercial software at a number of external sites.

The information providers can therefore concentrate on building servers able to guarantee efficient access to their data, without the need of building *ad-hoc* interfaces to their systems. It is essential that data access tools are flexibly built and can easily communicate with this *standard* user interface. These are the basic concepts we followed while implementing the system described in this work.

2. System Architecture

The user interface allows access to the underlying structure of the archive: a catalogue managed by a commercial SQL-based database management system, and a data bank holding the original data and compressed quick-look data.

The interface itself, which is based on NCSA Mosaic, allows a number of operations on the archive: selection of specific database tables or views to be accessed; browsing of the catalogue (database tables or views) by specifying query constraints inside forms to be filled by the user; possible editing of the resulting SQL query; display of the results in a HTML page; graphical visualization of the numerical results derived from a query; display of quick-look 2-D images and finding charts; display of quick-look 1-D graphs (e.g., extracted spectra); and retrieval of observational data from the data bank.

The structure of the system is shown in Figure 1. Users access the archive through a Mosaic interface; at the host node, the http messages coming from the interface are interpreted by a **Protocol Handler**, and different specialized servers can be started:

- The **SQL Server** builds a SQL query from information received from the interface, and passes back to the interface the result of the query or quick-look data (if any).

  - The **DBMS Server** extracts from the database information on the table or view to be accessed and builds the Mosaic interface page; in other cases passes the SQL query on to the DBMS and gets results.

- If the data are stored locally, the **Data Server** performs a username/password check through the DBMS Server, accesses on-line databank files and sends them to the user, or sends retrieval requests for off-line data to the archive operator.

- The **Graphic Server** plots the numerical results of a database query (or of archive data, in the case that quick-look data do not exist), and passes the graphic files back to the interface.

In keeping with the concept of the WWW, this system allows access to archives of data distributed over the Internet. From a logical point of view, it makes no difference if the archive is local or remote: the pointers to the data are stored in the database and, once accessed, are fed back to the interface for standard WWW access. If necessary, the Protocol Handler sends a message asking the remote archive to perform some specific action which may be required (e.g., storing files to be transferred in a specific world-readable directory, running a dedicated process to perform format conversion on archive data, or packing data in a compressed `tar` file).

3. Database Structure and Naming Conventions

The SQL Server uses *standard* SQL calls to access the database in retrieve mode, therefore guaranteeing independence from the SQL dialect used by the specific DBMS. It is, however, another feature that allows the system to be fully general: all of the interface information is dynamically stored in the database and is retrieved when needed to build on-the-fly Mosaic interface pages.

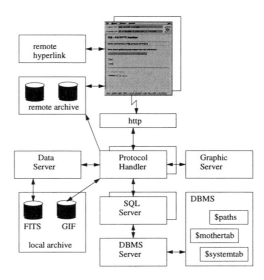

Figure 1. Structure of the system, based on a number of different servers, and allowing access to remote archives and hypertext media via the HTML protocol. The possibility of sending communications to remote archives through the Protocol Handler is shown. In the database, the three special tables allowing system portability are evidenced.

This mechanism works as follows: when it is necessary (i.e., when a user has issued a request to access data stored in the database), the DBMS server first retrieves from the database information on the table (or view joining a number of different tables) that needs to be accessed. Such information is used to build on-the-fly the form the user will to use to browse the contents of the database. The form, in HTML format, is then fed to the Mosaic interface.

The mechanism can be considered similar to the one described by Rasmussen & Pirenne (1994) for Sybase, but in our case form definition files (FDFs) are never built, nor stored in the database. This design feature allows modification of the database structure without changing the archive software, keeping description files up-to-date, or maintaining additional software to be run on the user's computer.

To allow independence of the structure from the specific DBMS used, some conventions are enforced. First, no DBMS system table is directly used; a table called $SYSTEMTAB is used instead. This table contains the names of the various tables stored in the database, together with their fields, descriptions, and formats; information on joins and views are also stored. $SYSTEMTAB may be built from a DBMS system table by means of a specific utility.

Two additional tables must be available in the system. $MOTHERTAB contains the list of all tables and views which are explicitly available to the user when accessing the archive. $PATHS contains information on the location of both quick-look and original FITS data files.

Some additional naming conventions are enforced on fields. If a field in a table is called $tablename, it contains names of other tables stored in the database; this convention allows the creation of a hierarchy of database tables, if needed. The $filename field contains names of files stored in the archive, either on disk, on a juke box, or off-line. The $hostname field contains the name of a (remote) archive server. The $path_id, $gif_path, and $fits_path fields contain the location of quick-look and FITS data files in terms of directory paths; $path_id is given as a numerical value corresponding to a directory path, and is used to save space in the database.

4. Applications

The scheme developed allows the interface, and the overall system, to be completely independent from operating system, DBMS, archive structure, and database structure (except for the conventions described in the previous section).

These concepts have been tested on a prototype archive containing a few digitized objective prism images, related tables containing the position of detected spectra, and the extracted spectra themselves. Both original and quick look data are contained in the archive. Details on the implementation of the archive are given in Pasian & Smareglia (1994b). The related URL is SQL-ARCHIVE interface[1].

A distributed archive containing interstellar medium spectra is currently being designed (Porceddu et al. 1994), and an initial version should be available before the end of 1994. This archive will use the interface described here; its metadata will be contained in a database in Trieste, while the data will be stored in Cagliari, Pisa (Italy) and Torun (Poland). Finally, this architecture is planned to be used (Pasian 1994) as the interface to the technical archive of TNG, the Italian National 3.5 m telescope to be located at La Palma, in the Canary Islands.

Acknowledgments. M. Albrecht, A. Balestra, P. Marcucci, B. Pirenne, G. Russo, and C. Vuerli are gratefully acknowledged for many interesting discussions on the topics of archives and information retrieval.

References

Pasian, F. 1994, Archives at the TNG Telescope – Architecture Design Document, TNG Project Technical Report (draft 0.9)

Pasian, F., & Smareglia, R. 1994a, in New tools for network information retrieval, eds I. Porceddu, & S. Corda (Cagliari, Oss. Astr. Cagliari)

Pasian, F., & Smareglia, R. 1994b, Int. Jour. of Mod. Phys. C - Physics and Computers, in press

Porceddu, I., Corda, S., Pasian, F., & Smareglia, R., 1994, Proc. Conf. Boulder, Colorado, in press

Rasmussen, B. F. 1995, this volume, p. 72

---

[1] http://atlantis.oat.ts.astro.it

WDB—A Web Interface to Sybase

B. F. Rasmussen
*Space Telescope-European Coordinating Facility,
Karl-Schwarzschild-Strasse 2, D-85748 Garching bei München,
Germany*

Abstract. WDB is a software toolset that tremendously simplifies the integration of Sybase databases into the World Wide Web. WDB lets you provide WWW access to the contents of Sybase databases without writing a single line of code, yet provides the flexibility to easily add new capabilities as needed. This article describes the technical implications of implementing such an Interface. The tool itself, and full documentation is available on the WWW at theWDB Home Page[1].

1. Introduction

To run WDB, all that is needed is the WDB script (written in **sybperl**) and a set of high-level *form definition files*, each describing a different view on the database. WDB automatically creates HTML forms, *on-the-fly*, to allow users to query the database, and given these query constraints, it will query the database and present the result to the user. WDB even comes with a utility to automatically extract information about a table from the database and create a working template *form definition file*.

A number of conversions are possible on the data coming from the database before they are shown to users: formatting of coordinates into hours, minutes, and seconds; formatting dates; etc. The most noticeable feature is the possibility to convert data from the database into hypertext links—and as it is possible through WDB to access any database element directly via a Web URL—the entire database can then be turned into a huge hypertext system. These hypertext links can be links to other elements in the database, thereby providing a simple way of jumping between related information. There could also be links to other documents on the Web, providing easily integration between data in the database and related documents on the Web. One could also have links to other databases with a WDB or similar interface—providing a simple mechanism for cross-database links.

2. Architecture

WDB is implemented as a CGI script which can generate forms and send them to the user, retrieve the contents of the form, process it and return the results

---

[1] http://arch-http.hq.eso.org/bfrasmus/wdb/wdb.html

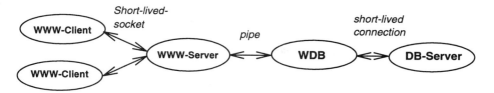

Figure 1.   A WDB Connection.

to the user. The forms and description of how the individual fields in the form links to fields in the database are defined using *Form Definition Files (fdf's)* located on the server.

Due to the nature of a WWW connection, the WDB system has to be designed differently from a normal database application (see Figure 1). As there are no long-lived connections, the server has no notion of what has been going on earlier in a session. The WDB process is exited immediately after the results are send back to the user. This means that all information needed to perform a given action has to be sent from the client each time. Of course, the client will have this information from the last result it received from the server.

The way this information is sent back and forth between the WWW client and the server is by the URL. It is a standard for specifying a piece of information on the Internet, and is composed of a protocol name (ftp, http, gopher), the name of the server, and a path on that server pointing to the piece of information needed. This can be a file or, on an http server, a CGI script such as WDB. The trick is that if there is any extra information in the URL after the path, then it is passed to the script as what is called PATHINFO.

(ex. http://arch-http.hq.eso.org/cgi-bin/wdb/*PATHINFO*)

2.1.   Flow of Control in WDB.

The actual flow of control is better illustrated in Figure 2. What happens is the following :

1. One "clicks" on a URL leading to the WDB script. The URL contains extra information in the path (PATHINFO), specifying what information in the database the user would like to query. This is typically obtained from a normal hypertext page with a menu of available query forms.

2. The server receives the URL request, starts the WDB script and passes the extra information in PATHINFO, actually the name of the *Form Definition File (fdf)* to use, and the word "form".

3. WDB reads the *fdf* and generates an HTML form according to the specifications in the *fdf*. The form is returned to the user and WDB exits.

4. The user fills out the form with the query constraints and presses the "Submit" button. This sends the WDB URL (with PATHINFO) plus the contents of the form (encoded) to the WWW Server.

5. The server receives the URL, starts the WDB script and passes the extra information in PATHINFO, plus the form content.

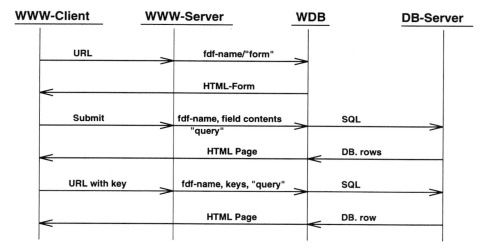

Figure 2.   Flow of Control

6. WDB parses the PATHINFO which in this case is the name of the *fdf* plus the word "query". This tells WDB to construct a SQL query from the information in the *fdf* file and the user's constraints.

7. The SQL query is send to the database server.

8. The database server processes the query and returns a number of rows.

9. WDB retrieves the rows from the database server and after having converted the data according to the specifications in the *fdf*, constructs an HTML page with the result. If more than one row was returned, a tabular output format is chosen, allowing the user to select a row, via a hypertext link, to get all the details.

10. The HTML page is returned to the user, and WDB exits.

11. The user clicks on a hypertext link for a row. This link contains the WDB URL with the *fdf* name, the word "query" and the key(s) for the current row in the PATHINFO.

12. The WWW server receives the URL, starts the WDB script and passes it the extra information in PATHINFO.

13. WDB constructs a new database query, this time using the key(s) instead of the form contents.

14. The database processes the query and returns the result.

15. WDB retrieves the row from the database server, converts the data according to the specifications in the *fdf*, and this time, as only one row is returned, constructs a HTML page with all the details. The HTML page is returned to the user and WDB exits.

3. Conclusions

WDB has proven very successful—mainly because of its simplicity and the ease with which it can be extended. One of the goals of WDB was to provide an easy, yet flexible, way to make database information available. In the simple case, no coding should be necessary. However, if special behavior is needed, it should be possible to add it *without changing the core system*. Also, if some special behavior has been defined for one field in a form, it should be easy to reuse this in other forms.

This has been achieved by allowing extensions to be added in the individual *Form Definition Files* using the powerful Perl language. This makes local customization a lot easier than with most existing WWW/Database gateways.

WDB already has a large user base. At the time of writing, more than 30 sites world-wide have installed it. An e-mail mailing list is maintained for announcements of new releases. To be added to this list, send e-mail to the author (*bfrasmus@eso.org*).

References

McCool, Rob 1993, The Common Gateway Interface[2], National Center for Supercomputing Applications (Champaign, University of Illinois)

McCool, Rob 1994, NCSA httpd Overview[3], National Center for Supercomputing Applications (Champaign, University of Illinois)

Brenner, Steven E. 1994, CGI Form Handling in Perl[4] (Cambridge, University of Cambridge)

---

[2] http://hoohoo.ncsa.uiuc.edu/cgi/overview.html

[3] http://hoohoo.ncsa.uiuc.edu/docs/Overview.html

[4] http://www.bio.cam.ac.uk/web/form.html

Two Small Astronomical Catalogs Available In Hypertext

K. M. Strom
Five College Astronomy Department, University of Massachusetts, Amherst, MA 01003

Abstract. The *Third Catalog of Emission Line Stars* (Herbig & Bell 1988) and the *General Catalog of Herbig-Haro Objects* (Reipurth 1994) have both been translated into hypertext and linked to other on-line material. As a byproduct of this work, a method for displaying subscripts and superscripts using Mosaic has been found.

1. Introduction

In recent years astronomical catalogs have disappeared from the archival literature. A few years ago, these could be found among the Observatory publications (see the *Third Catalog of Emission Line Stars,* Herbig & Bell 1988). Today catalogs are issued in electronic form only (see A General Catalogue of Herbig-Haro Objects[1], Reipurth 1994), usually via anonymous FTP.

For the past two years astronomers working in star formation have been kept abreast of developments in their field by the *Star Formation Newsletter*, edited by Bo Reipurth of ESO. This newsletter is distributed monthly by e-mail to over 800 astronomers worldwide. A LaTeX form is used, into which astronomers enter abstracts of their recently accepted papers. These are e-mailed to a central collection point and assembled into a monthly newsletter. The recipients strip off the e-mail header and process the LaTeX file to read the newsletter. The newsletter eliminates the previous hit-and-miss circulation of preprints, especially for younger workers.

2. Catalogs

This spring I received a request from Bo Reipurth of ESO to review his preliminary *General Catalog of Herbig-Haro Objects* (Reipurth 1994) before it was made available via anonymous FTP from the ESO FTP server. The imminent appearance of a new catalog of objects associated with the star formation process, solely in electronic form, motivated me to think of making another database, the *Third Catalog of Emission Line Stars* (Herbig & Bell 1988), available in electronic form as well. I elected to create hypertext documents, linking them not only internally, but also to the on-line database of abstracts of papers in astronomy journals and to each other.

---

[1] ftp://ftp.hq.eso.org/pub/Catalogs/Herbig-Haro/

2.1. The Herbig-Bell Catalog of Emission Line Stars

The *Third Catalog of Emission Line Stars* (Herbig & Bell 1988, HBC) was distributed as a Lick Observatory Bulletin. It is a catalog of pre-main-sequence stars which have had slit spectra taken to confirm their nature. Included in this catalog of approximately 750 objects is the complete coordinate information, magnitude, color, and variability in bands from the X-ray through the radio (less the *IRAS* data), spectral type and emission line data, both radial and rotational velocity information, references for all of this information, and the identification of the molecular cloud in which the star is found. A few copies of the catalog were distributed as 80 column card images on 9 track tapes, but the paper version of the catalog has been the basic format used.

The HBC has always been somewhat awkward to use because the data for an individual star is spread over two pages bound one above the other. There are large gaps across any line, except for the most frequently observed stars, making it difficult to follow the correct line across the page. In making the electronic version[2], I have linked the catalog number of the object to the corresponding entries on the second page so that the top of the window will act as a guide line across the columns. An asterisk in the notes column is linked to the note for that object. The references were linked to the abstracts contained within the database of journal article abstracts used by the ADS Abstract Server[3].

Because of the difficulty in reading the table, and the unlikelihood of the need for *all* of the information spread across the two pages, I also provided a forms-based access[4] to the data. This form requests the catalog number of the object desired and then returns all of the most commonly desired data for the object, as well as any other data requested, in a nicely formatted page.

2.2. The Reipurth Catalog of Herbig-Haro Objects

Upon release of the public version of the *General Catalog of Herbig-Haro Objects*, I began the process of its conversion to a hypertext document.[5] The Catalog was formatted in LaTeX and consisted of a table of the basic data and extensive notes on each object. The catalog contains approximately 250 objects, and for each object lists the catalog number, any previous designations, the best available positions, the source of the outflow (if known), the name of the star formation region in which it is found, and the best distance to the object. Each catalog number was linked to the notes for that object. When the suspected source of the outflow was listed in the HBC, the source was linked to its entry there.

The notes for each catalogued object are heavily referenced. The majority of the references are available in the ADS Abstract Server, and the notes have been linked to those abstracts. Because this field is extremely active, there were a large number of papers (25–30% of the references) not yet available in the abstract database. Many of these abstracts were available in the *Star Formation*

---

[2] http://www-astro.phast.umass.edu/latex/HBC/tab1a0-5.html

[3] http://adswww.harvard.edu/abs_doc/abstract_service.html

[4] http://www-astro.phast.umass.edu/latex/HBC/HBCform.html

[5] http://www-astro.phast.umass.edu/latex/HHcat/HHcat.html

Newsletters [6]. Therefore we placed the entire archive of the *Newsletter* on line, using LaTeX2HTML[7] and Perl scripts described later. I then linked the catalog references to these abstracts. This procedure allowed us to provide abstracts for almost 90% of the references, leaving only those papers which are still in preparation and the papers published before 1975. Since the papers which were still in preparation appear in the *Star Formation Newsletters*, the catalog references are linked to them.

2.3. The Star Formation Newsletters

The hypertext versions of *Star Formation Newsletters* were placed on line in order to have immediate, searchable access to abstracts of the most recent papers. We found it necessary to integrate into the text a large number of symbols for the proper display of the content of the abstracts. We constructed the table of contents for these newsletters in two forms:

1. a single document containing all of the papers found in the newsletters; each title is linked to the corresponding abstract (this document is large but allows a search through the entire database).

2. individual table of contents files for each issue of the newsletter. These documents are much smaller but require knowledge of the publication date to be used effectively. The titles are also linked to the abstracts.

When a new issue arrives, it is added to the available database. When papers are published, the final references are added to the newsletters.

3. Symbol Library

In response to the practical and perceptual problems of displaying mathematical symbols using the Mosaic browser[8], I developed a library of transparent GIF images of the most commonly used subscripts and superscripts in astronomical manuscripts. This library was designed with several considerations in mind:

- The symbols should blend into the text as inconspicuously as possible.

- The total size of the images to be transferred should be as small as possible, to minimize load time and network traffic.

- The reading of an entire paper should be accomplished without the need to discard and reload either symbol or figure images, if at all possible.

When an entire expression is captured as an image, the expression may look out of place in the rest of the text displayed by Mosaic, no matter which font is selected. However, if only the sub- and superscripts are generated as images, this problem is greatly alleviated. Since sub- and superscripts are always generated

---

[6] http://www-astro.phast.umass.edu/latex/sfnews/no1/toc2.html

[7] http://cbl.leeds.ac.uk/nikos/tek2html/doc/manual/manual.html

[8] http://www.ncsa.uiuc.edu/SDG/Software/Mosaic/Docs/helpabout.html

in a different font and are much smaller, few differences in the structure of the character are possible or noticeable.

The second and third items take into consideration the default image cache size of Mosaic—2048 kB[9]—and the fact that the least recently accessed image will be the first discarded. The typical size of a 2–5 color image used as a figure in a preprint is 4–8 kB. The typical size of an image to be used as a sub- or superscript is 0.06 kB.

These three considerations work in concert to provide a clean-looking, easily readable paper, taking minimal time to load each page, thus allowing browsing as well as reading in depth.

This library includes the Greek letter set, numerals positive and negative, the entire alphabet (both capital and lower case), and some special symbols and letter combinations of common use in astronomy. To implement this set of transparent GIFs, Perl scripts were written to preprocess the manuscripts, inserting the in-line images as required. LaTeX2HTML was used to convert the manuscript to HTML format, and a post-processing Perl script was used to clean up the few things that may have been affected by LaTeX2HTML. The manuscript is then ready for the insertion of the images, tables, and reference links.

Acknowledgments. I wish to express my gratitude to my summer student and aide, Jessica Norman, without whose help much of the work on the *Star Formation Newsletters* and the *Catalog of Herbig-Haro Objects* would remain incomplete.

References

Drakos, N. 1994, LaTeX2HTML manual[10]

Herbig, G. H. & Bell, K. R. 1988, Third Catalog of Emission-Line Stars of the Orion Population, Lick Observatory Bulletin No. 1111

Reipurth, B. 1994, A General Catalogue of Herbig-Haro Objects[11]

The *Star Formation Newsletter*, ed. Bo Reipurth, is available as PostScript files[12], or as hypertext[13]

---

[9] http://www.ncsa.uiuc.edu/SDG/Software/Mosaic/Docs/d2config.html

[10] http://cbl.leeds.ac.uk/nikos/tex2html/doc/manual/manual.html

[11] ftp://ftp.hq.eso.org/pub/Catalogs/Herbig-Haro

[12] http://http.hq.eso.org/star-form-newsl/star-form-list.html

[13] http://www-astro.phast.umass.edu/latex/sfnews/no1/toc2.html

User Interfaces to HST Data: StarView and the World Wide Web

J. J. Travisano

Computer Sciences Corporation, Space Telescope Science Institute, 3700 San Martin Dr., Baltimore, MD 21218

Abstract. This paper describes two approaches for user interfaces to the *Hubble Space Telescope* (*HST*) archive. StarView is a custom interface with many interactive features for the serious archive user. An experimental World Wide Web (WWW) interface is also described, which supports some of the basic features of StarView available via standard WWW clients such as NCSA Mosaic.

1. Basic Features

There is a set of core features we provide to *HST* archive users in the user interfaces. These are: Forms-based catalog queries, for ease of use and to cover different areas of interest effectively; coordinate lookup of target names via SIMBAD[1] and NED[2], to insure completeness in searching for *HST* exposures by target; preview of public datasets (images and spectra), to be able to make some judgment of data quality before actually retrieving the data; and submission of retrieval requests to the archive (registered users only). Beyond these, additional features are supported in StarView.

2. StarView

StarView is the primary user interface to the *HST* archive (Williams 1993; Long 1994). It runs as a client software package on SunOS and VMS currently, supporting both CRT and X/Motif displays. Over 35 forms have been developed by archive scientists for catalog queries in a number of areas, such as: completed and planned science exposures; proposal information and abstracts; *HST* instrument and calibration parameters; and astrometry and engineering parameters. In addition, it is possible to build custom queries simply by choosing catalog field names from a list. Filtering and sorting operations are used to navigate this list of over 3000 fields to find those of interest. An automatic query generation facility then turns this chosen set of database fields into a reasonable SQL query (Silberberg 1994).

Using the cross correlation function, a list of targets—RA, Dec, and a description—is provided to StarView. The catalog is then searched for matching

---

[1] http://cdsweb.u-strasbg.fr/Simbad.html

[2] http://www.ipac.caltech.edu/ipac/projects/ned.html

records of any of the targets in the list using a fuzzy join by RA and Dec. For any set of query results, from pre-defined forms, custom queries, or cross correlated queries, the values can be exported to an ASCII file or FITS ASCII table for use outside of StarView. Try it out on: *stdatu.stsci.edu* (UNIX)[3] or *stdata.stsci.edu* (VMS).[4] Use account: `guest`, password: `archive`. Enter `starview` for the CRT version and `xstarview` for X/Motif. We do not recommend running `xstarview` across the Internet—get the distributed client version instead. Please send mail to *archive@stsci.edu* for more information.

3. World Wide Web

An experimental WWW interface was developed in September 1994 to explore the many aspects of making archive information available on the Internet using standard protocols and methods. It was shown at the conference.

We started with the NCSA httpd server and GSQL, which is a simple gateway to SQL databases. GSQL was greatly modified, primarily to support the processing of search constraints on input and the formatting of astronomical values on output. Pieces of StarView code, StarView utilities, and additional programs and scripts were integrated via the Common Gateway Interface (CGI)[5].

Two primary search screens were implemented: Completed *HST* Exposures and Planned *HST* Exposures. HTML Fill-Out Forms[6] are used whereby the user can enter a set of search constraints. The subsequent search results screen contains hypertext links to related proposal information and to the preview datasets. Individual datasets can be selected and submitted for retrieval from the archive.

By integrating a diverse set of programs, tools, and scripts, the basic features listed in Section 1 are supported. There are efficiency issues with the support of the FITS preview data. Another issue is security, in protecting the user account information that is needed for dataset retrieval.

This WWW interface is experimental, and is not publicly available.

4. Experiences

4.1. StarView

We have a large investment in our custom, full-featured interface. The user can be given immediate feedback on any function, there are pop-up dialogs for options and warnings, pull-down menus, etc. StarView can deal with complicated input and data interaction, for example, taking a long list of targets to cross-correlate with the *HST* catalog.

Many changes can be accomplished by editing ASCII configuration files. These include changes to the form definition files defining the screen layout,

---

[3] rlogin://guest@stdatu.stsci.edu/

[4] rlogin://guest@stdata.stsci.edu/

[5] http://hoohoo.ncsa.uiuc.edu/cgi/

[6] http://www.ncsa.uiuc.edu/SDG/Software/Mosaic/Docs/fill-out-forms/overview.html

the data definition files describing the details of the catalogs, help files, etc. Such changes are often done by the archive scientists and support staff without software code modifications. Still, major enhancements require C++/C code changes by the developers.

4.2. World Wide Web

Anyone who has surfed the *Web* can see that it is quite easy to make data available via the WWW. Pretty pictures, movies, sound files, even video snapshots of an office or a fishtank are just a hypertext link away. User interface clients are freely available for all the major platforms (UNIX, VMS, Windows, Macintosh, etc), making it easy to support your intended audience without distributing software.

However, the stateless nature of the WWW architecture, using HTTP and HTML, requires one to rethink how to present a complex catalog or archive system to the end user. Tricks can be used to maintain state, but essentially every mouse click is a new transaction to be processed. This translates directly to issues of database loading. For example, StarView displays records from the database as soon as the first record is found. The user can then get more records or cancel once sufficient additional records have been seen. But in the case of the WWW, the HTTP connection will go away after each transaction, so it is necessary to complete the query and return all of the results (up to some maximum record count).

The current version of HTML provides limited control of form presentation and interaction. Proposed enhancements, such as HTML 2.0[7] and the Common Client Interface (CCI)[8] for NCSA Mosaic, are attempting to deal with some of the issues of improved formatting and interaction.

5. Future Plans

StarView continues to be maintained and enhanced to support the *HST* archive. We plan to look at better ways of supporting catalog navigation and custom queries, possibly with more graphical techniques. Access to other catalogs and sky survey data, which will be useful to researchers, is also being investigated.

Further assessment of WWW access to the *HST* archive is planned. More experimentation and thought is needed to address issues of performance, loading of database servers, security for retrievals, and user support. We have already incorporated ideas from the WWW into StarView, such as using the HTTP protocol to serve files easily and quickly to the StarView client.

Acknowledgments. A number of engineers and astronomers have contributed to the successful development of StarView in the past few years. The core parts of the experimental WWW interface were implemented by Shian Lin.

---

[7] http://www.hal.com/users/connolly/html-spec/index.html

[8] http://yahoo.ncsa.uiuc.edu/mosaic/cci/cciTalk.html

References

Long, K. S. 1994, in Astronomical Data Analysis Software and Systems III, ASP Conf. Ser., Vol. 61, eds. D. R. Crabtree, R. J. Hanisch, & J. Barnes (San Francisco, ASP), p. 151

Silberberg, D. P., & Semmel, R. D. 1994, in Astronomical Data Analysis Software and Systems III, ASP Conf. Ser., Vol. 61, eds. D. R. Crabtree, R. J. Hanisch, & J. Barnes (San Francisco, ASP), p. 92

Williams, J. 1993, in Astronomical Data Analysis Software and Systems II, ASP Conf. Ser., Vol. 52, eds. R. J. Hanisch, R. J. V. Brissenden, & J. Barnes (San Francisco, ASP), p. 100

ISSA-PS, The Postage Stamp Server for IRAS Imaging Data

D. Van Buren, R. Ebert

Infrared Processing and Analysis Center, MS 100-22, Caltech, Pasadena, CA 91125

D. Egret

Centre de Données de Strasbourg, 11 rue de l'Universite, 67000 Strasbourg, France

Abstract. The ISSA Postage Stamp Server[1] is a Web-accessible service that delivers *IRAS* images to users with a minimum of effort. It is unique in that it will operate by object name as well as celestial position. Users can connect using custom clients to automatically make requests and fetch images for local use.

1. Concept

The basic idea of the ISSA Postage Stamp Server is to provide small subsets of the *IRAS* imaging data to users with minimal fuss and bother. Our philosophy is to deliver the data and not do any processing at all; in particular we do not perform any coordinate transformations, because then we will no longer be delivering the product itself but rather some untested derivative.

Targets are specified by name or by position using NCSA Mosaic, or an equivalent Web browser, on an introductory html page describing the service. The search is launched, and when completed the user is presented with four postage stamp representations of the extracted FITS data. A number of caveats are provided on the use of ISSA images, which can contain traps for the unwary. Users retrieve the FITS images by clicking on the postage stamps, provided their Web clients are properly configured (we give instructions on how to do this).

2. Implementation

The service is managed by the AstroVR server (see the paper in this volume by Van Buren, Curtis, Nichols & Brundage). Web clients dial into AstroVR, where a general purpose Web Server handles the request by dispatching it to the postage stamp generating object. There, the target string is taken and recast into the query grammar of the low-level postage stamp server running at IPAC, which requires a valid position. If an error results, a query to SIMBAD is made

---

[1] http://astrovr.ipac.caltech.edu:8888/ISSA-PS

using the target string as an object name. If successful, this query returns a position and the low-level server is tried again.

The low-level server determines which ISSA plate contains the position of interest, reads the data off CD-ROM and extracts a 2-degree square field from each of the four *IRAS* bands. We chose this size because it is the largest size that guarantees that the entire field will be on a single plate, and our philosophy is not to do any further processing of the data, such as creating mosaics, which might alter its reliability and usefulness. Users are, of course, free to do so at their end.

Once extracted and saved to a public ftp area, gif representations of the postage stamps are made, and pointers to all these files are returned to the managing server. There an html page is generated, containing the gifs as in-line images, along with text containing the target coordinates and a number of caveats as to the use of ISSA images. The html page is sent back to the user and formatted by their Web client.

3. Usage

The service gets from a few to 50 requests a day, averaging about 10. We log connecting site names, and find that there are many casual users as well as obviously astronomical users. A few sites are responsible for large numbers of requests, and we surmise that they are conducting large scale surveys. Large surveys were very difficult to conduct prior to the existence of this service, unless one was willing to write sophisticated software to manage and query the CD-ROM image dataset.

One can connect to the service using not only NCSA Mosaic or other Web browsers, but also with custom client software that sends queries and interprets the returned html to then transfer the FITS files back to the user's disk space. Operating in this mode, users can have an "*IRAS* in a box" for which they can define their own interface.

Acknowledgments. We acknowledge the support of US taxpayers through a contract to the Jet Propulsion Laboratory from the National Aeronautics and Space Administration.

Part 3. Graphical User Interfaces and Visualization

A Portable GUI Development System—The IRAF Widget Server

D. Tody[1]

National Optical Astronomy Observatories, P.O. Box 26732, Tucson, AZ 85726

Abstract. We describe a new GUI (Graphics User Interface) development environment which extends the X window system and the X Toolkit (Xt) with a high-level, object-oriented, interpreted programming language. This approach will allow GUIs written using standard Xt-based widget sets to be constructed without requiring any window system programming on the part of the programmer. The architecture of the resulting program completely separates the GUI from the applications code, allowing the GUI to be developed separately and replaced at will without modifying the application. Despite the separation of the GUI from the application the two are tightly integrated using an asynchronous, event-driven messaging system based on requests and client events, with the remote client application appearing as just another class of object within the GUI. This approach maximizes window system and toolkit independence and is well suited to distributed applications, allowing the GUI and client application to be run easily on separate processors or computers.

1. Introduction

Most window system programming today is done via one of two approaches. The first approach is to code directly in C or C++ at the window system toolkit level, usually integrating the user interface code and applications code within the same program. The second approach, which is really just a variation on the first, is to use one of the many commercial GUI builders available. This simplifies the programmer's task by allowing the user interface to be interactively designed, relying on the GUI builder to generate the window system code required to implement the user interface specified by the programmer and requiring the programmer to code only the application functionality. As with the direct coding approach, the applications code is often integrated with the GUI in the same program.

Recently a third approach has been used. This employs a high-level, interpreted language in which the programmer codes all or part of the application, with at least the GUI being implemented in the high-level interpreted language. Tcl/Tk is a recent example of this approach. Other examples of the high-level

---

[1] The National Optical Astronomy Observatories are operated by the Association of Universities for Research in Astronomy, Inc. (AURA) under cooperative agreement with the National Science Foundation.

interpreted approach from outside the Unix/X world include Apple's HyperCard/HyperTalk, and the Visual Basic facility of Microsoft Windows.

This paper presents a new, high-level, interpreted window system toolkit which like Tcl/Tk is also based on the Tcl interpreter. Unlike Tcl/Tk, which is by definition Tk specific, our approach tries to maximize window system and window system toolkit independence to ease future upgrades to new window systems or toolkits. The initial implementation, which is for the X Window System, is based on the X Toolkit (a part of the X Consortium X11 release) and uses standard Xt-based widgets. Support for an asynchronous, event-driven messaging system is an integral part of the design, allowing the GUI to be isolated from the functional part of an application and making the facility well suited to distributed applications, e.g., where a GUI executing locally talks to application code executing remotely, with the application downloading the GUI to be run at startup time. The core facility is implemented as a simple C-callable library which can be used to implement new GUI-based programs without having to write any window system-level code.

An important aspect of the facility described in this paper is that it represents a general user interface management system (UIMS) tailored for astronomical GUIs. The intent is that by providing a high-level facility tailored for our applications, we can simplify the task of developing GUIs for astronomical applications while providing a greater degree of consistency between applications since they will share the same GUI components. This is particularly important in the area of 2D graphics and imaging, including presentation and user interaction with such data, since the standard toolkits do not address this area.

2. The Widget Server

2.1. Overview

"Widgets" (window objects), are the basic building blocks of graphical user interfaces. A window system toolkit provides a selection of widgets that the programmer can use to construct a user interface. Typical widgets include things like push buttons, scrollbars, pop-down menus, or scrolling text regions. The *widget server* serves up widgets to a remote client process in much the same way that the X display server serves up windows and other low-level display resources to a remote client application.

When a client application starts up it connects to the widget server and downloads its GUI. The GUI is a simple block of text which is interpreted and executed by the widget server. The GUI text contains a description of the widgets comprising the user interface, a number of interpreted action procedures to be called to process user interface or client events during execution, and any code needed to initialize the GUI. During execution the client application waits for and executes requests from the GUI, and sends messages to the GUI to inform it of any "client events," or changes in the state of the client. The conventional compiled client application implements all the application-specific functionality, but does not communicate directly with the user.

To the client application the GUI is merely a block of text to be passed on to the widget server; the client code knows nothing about the GUI other than the name of the GUI file to be sent to the widget server. The client knows only about the applications functionality which it implements, and the messages and

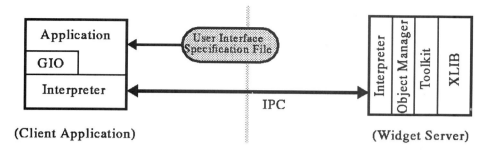

Figure 1. Widget Server Architecture

requests used to communicate with the GUI. The GUI is completely isolated from the client application. While to the user the GUI appears to be an integral part of the application, the actual compiled client application has an interpreted command-line interface and can be executed stand-alone without any GUI.

2.2. Widget Server Architecture

The architecture of an application which uses the widget server is summarized in Figure 1. The typical widget server-based application consists of two processes, the process containing the client application, and the widget server process itself which executes the client's GUI. All user interface functionality resides in the GUI and all application-specific functionality resides in the client. During execution these processes communicate via an asynchronous object-based messaging system.

2.3. Advantages of the Widget Server

The widget server architecture separates the user interface from the application-specific code and provides a high-level interpreted language for developing GUIs. This approach has significant advantages, including the following:

1. The high-level, interpreted nature of the widget server makes it much easier to develop GUIs than is the case with toolkit-level programming or the conventional, compile-link approach.

2. The use of an interpreted runtime language (Tcl) to compose the GUI is more powerful than the visual programming approach used in GUI builders, since the latter only address the appearance of the user interface. User interface builders can still be used with the widget server to interactively layout the user interface, although this is less important than it might be given the interpreted nature of the GUI (i.e., changes can be made and the GUI redisplayed very quickly).

3. The high-level, integrated nature of the widget server-based GUI development environment makes it straightforward to customize the environment to support a particular class of application. This is important for large systems where GUIs are developed by many people working independently, to reduce the overall effort and improve consistency.

4. The widget server isolates all window system and toolkit code into a single executable which can serve any number of applications. This greatly reduces disk space consumption in a large system that has many application GUIs.

5. Since the widget server is a single executable it is easy to have multiple versions, e.g., supporting different toolkits or incorporating proprietary software to optimize performance for a particular class of workstation.

6. Since all the window system code is isolated into a separate process the client application is completely window system independent, allowing the same client application to be used with a widget server GUI executing on any local or remote platform running any operating system.

7. Porting a whole system full of GUIs to a new platform can be done by a single individual since only the widget server itself need be ported.

8. No window system libraries, and indeed no compilation, is needed to develop GUIs. All that is needed is the widget server executable. This makes it much more feasible to use commercial or platform specific libraries should this be desired.

9. The GUI is completely isolated into a small text file separate from the application, allowing the GUI and the application to be developed separately, or several alternate GUIs to be used with a single application.

10. The widget server is well suited to distributed applications since the widget server can be run on the local workstation while the client application executes remotely. This allows the entire GUI to execute interactively on the local workstation, a much more efficient approach than, for example, running an X application over the network.

Possible disadvantages of the widget server approach are its relative complexity and the possibility of inefficiency when the application is distributed over two or more processes. For a small system where only a few GUI-based applications are needed many of our big-system concerns are unimportant and it might be simpler to program directly at the toolkit level, especially if a user interface builder tool is available. Efficiency can be a problem if the client code is required to respond in real time to user interface events, however this is rarely a problem in well designed applications since the more interactive portions of the program can be moved into the GUI, implemented as interpreted GUI procedures or as calls to the compiled functions in the widget server itself. The asynchronous nature of the messaging system ensures that the user interface will always be responsive even when the client is busy computing.

2.4. Platform Independence

The widget server automatically provides a high degree of platform and window system independence since the GUI is isolated from the client application; in the worst case only the GUI file has to be changed to use a GUI-based application on a new platform. The current implementation provides full platform independence for platforms which run the X Window System since the current widget

```
reset-server

appInitialize hello Hello {
   Hello.objects:\
        toplevel        Form        helloForm\
        helloForm       Label       helloLabel\
        helloForm       Command     quitButton

    *helloForm*background:                      bisque
    *helloForm*helloLabel.label:                Hello, world!
    *helloForm*quitButton.fromHoriz:            helloLabel
    *helloForm*quitButton.label:                Quit
}

createObjects
proc quit args { send client gkey q; deactivate unmap }
send quitButton addCallback quit
activate
```

Figure 2. A Simple GUI: The "Hello, world" Application

server implementation is X-based. No changes to the GUI files are needed for these platforms.

The current widget server implementation does not, however, provide full window system toolkit independence for window systems other than X. Ideally the widget server should define a virtual set of widgets which can be implemented on a variety of window systems and window system toolkits; not only X but also Windows, Windows NT, Macintosh, etc. This problem has partially been solved in that the language used in widget server GUIs isolates the widget-dependent code into a portion of the GUI which describes the widget hierarchy, assigning widget classes to named GUI objects. The runtime part of the GUI, i.e., the interpreted action or callback procedures called at runtime as the GUI executes, is already almost completely widget- and toolkit-independent. Defining a fully toolkit-independent virtual widget set is a future problem which cannot be attempted until we have a better idea what widgets are needed for our applications. Several commercial window system toolkits or GUI development environments exist which have already attempted to address this problem, at least for the standard toolkits.

The portion of the widget server which interfaces to the underlying window system and window system toolkit (widget set) is the *Object Manager*. This is discussed in the next section.

3. The Object Manager Library

3.1. Overview

The widget server is actually just a shell around the Object Manager library (OBM). The widget server extends the Object Manager by providing a way for external clients to connect to the Object Manager to download and execute a GUI. All of the real work of creating and executing the GUI is done by the Object Manager library. The widget server adds a client-server communications method.

The Object Manager provides services for creating, deactivating, reactivating, or destroying a GUI, creating or destroying objects, and delivering messages and events to objects within the GUI. The OBM provides the framework within which GUIs execute, including the interpreter, automatic memory allocation, and a library of runtime services. The Object Manager defines four main classes of objects: *Server*, *Client*, *Parameter* (for client events), and *Widget*, for the graphical elements of the interface. Within the Widget class are many subclasses, one for each type of widget. All Object Manager execution is event driven and asynchronous and is based on messages (requests), callbacks, and events. For example, defining a new GUI is done by sending a message to the server object.

The set of widgets implemented by the Object Manager is not fixed, i.e., new widgets can be added or existing widgets removed to meet the requirements of the applications which will be using the widget server. The base Widget class provides a generic set of methods usable with all widget subclasses. Complex widgets subclass the base Widget class to add their own methods. The current Object Manager provides a mixture of Xt-based widgets which provide a Motif-like appearance but which are publically available and redistributable. Source for these widgets and for all code used in the widget server is included in the distribution. The current widget set includes the base Xt widgets, the 3D Athena widgets, selected FWF (Free Widget Foundation) widgets, plus a few others such as the Layout widget from MIT, the HTML hypertext markup language widget from NCSA *Mosaic*, and the gterm-image widget from the IRAF project. Additional Xt-based widgets (including Motif, OLIT, and commercial widgets) can easily be added.

3.2. The OBM Library

The main entry points of the OBM library are shown in Figure 3. The library is very simple since everything complicated is done by the interpreted GUI code. The main runtime function of the OBM library, from the point of view of the application which uses the library, is messaging. A window system application (such as the widget server) calls *ObmDeliverMsg* to deliver a message from the client application to a GUI object. A callback function is registered with the Object Manager to intercept client requests and pass them on to the client.

3.3. Using the OBM Library to Build Standalone Applications

Our discussion has thus far concentrated on the widget server and distributed applications, because the widget server provides the best architecture for adding GUIs to tasks in an existing, large data processing system. Another important class of applications are window system applications, where the focus is on the window system functionality implemented by the application. Most conventional X window system applications fall into this class.

An important use of the Object Manager library is to implement such applications. A stand-alone, single-process window system application can be built using the OBM library. In this case the "client" is not a separate process, but an application-specific interpreter within the same process as the OBM library.

```
           obm = ObmOpen (app_context, argc, argv)
                ObmClose (obm)
           ObmInitialize (obm)
             ObmActivate (obm)
           ObmDeactivate (obm, unmap)
        status = ObmDeliverMsg (obm, object, message)

          id = ObmAddCallback (obm, callback_type, fcn, client_data)
            ObmRemoveCallback (obm, id)
```

Figure 3. Principal routines of the Object Manager library, *libobm.a*

The program could be written as a conventional window system program making direct calls to the underlying window system toolkit, but by using the OBM library virtually all window system specific code is eliminated and the GUI is isolated to a high-level, interpreted GUI file. The only compiled code required is that which implements the functionality of the application itself. The resulting task is almost completely window system independent.

A good example of an existing stand-alone window system task built around the OBM library is *ximtool*, the IRAF image display server program. This is a stand-alone X window system application used for image display and image interaction. *Ximtool* contains only about one page of C code which has anything to do with X. All of the remaining C code handles window system-independent raster image processing and the fifo or socket-based binary protocol used to communicate with remote clients for image display. The *ximtool* GUI is an interpreted GUI text file, identical to what one might use with a widget server-based task. The widget server itself, of course, is another example of a stand-alone X application based on the OBM library.

3.4. The Object Manager Shell

Another stand-alone host application built around the Object Manager library is *obmsh*, the Object Manager shell. This is a Unix shell which executes OBM windowing scripts. It can be used to execute GUI files from the Unix command line, or be used in OBM-based scripts to write stand-alone Unix shell scripts that can be called as commands from the Unix environment. For example, the "hello, world" GUI shown in Figure 2 could be converted to a Unix command *hello* by changing the file name to "hello" and adding something like "#!/usr/local/bin/obmsh" to the file header.

4. Messaging

4.1. Messaging Fundamentals

The key to isolating the GUI from the client code of an application, while providing a tightly integrated, efficient application, lies with messaging. The messaging scheme used determines how objects within the application interact with each other during execution. This includes the interaction of the client code (client object) with the GUI. Our discussion here will concentrate on how messaging is used to link the client to the GUI, but it should be noted that the same messaging scheme is used for all object-to-object communications within the GUI as well.

Messaging as defined by the OBM consists of two parts: requests and client events. Requests are commands send to the client (or any other object). The recipient is free to modify or ignore the request as it wishes. Client events are messages sent to the GUI when the client state changes in any way. The same mechanism is also used to deliver other forms of client information to the GUI, e.g., in response to requests. The GUI is free to ignore client events. It is not unusual for a given GUI to be interested in only a portion of the client events generated by a client.

A client event is a message sent to an object of the OBM class *Parameter*. A parameter object is very simple, consisting of a name and a string value. The GUI registers callbacks with the user interface (UI) parameters, i.e., client events, that it wishes to know about. The string value of a UI parameter can be anything, for example a number, a structure, a list, or a large block of text. It is common for multiple callbacks to be registered on a single UI parameter by independent elements of the GUI.

Messaging is fully asynchronous. Both requests and client events are queued and buffered, and periodically flushed to the process on the other end. Synchronization occurs automatically when the client waits for input from the server (GUI). The GUI never waits for a request to complete, nor does it check to see that a request has been honored. Rather, when the client processes the request it sends client events back to the GUI to inform the GUI of any actions performed by the client. The same thing happens when the client performs actions for any other reason, hence the GUI always reflects the true state of the client.

Client events are an important abstraction mechanism. Client events and messages allow the client to provide the GUI with all the information it needs to function, without the client code having any knowledge of the nature of the user interface. Yet, since requests and client events are decoupled, the client will function even if the client events and messages it sends are discarded, as when running the client code without a GUI or with radically different GUIs.

4.2. Simple Messaging Example

Messaging is one of those things which is fundamentally simple, yet surprisingly hard to explain. As a simple example, consider what happens when the user selects a frame to be displayed in *ximtool*:

1. The user pushes the next frame button.

2. The callback procedure (in the GUI) for the nextFrame button sends the command "nextFrame" to the client.

3. The client receives and processes the request, changing the frame, sending the new frame number to the UI parameter "frame".

4. A GUI callback procedure registered on the "frame" parameter is called, updating the GUI to indicate the new display frame number.

This example simplifies things considerably but is accurate so far as it goes. In the real program there are a number of different ways the frame can be changed, e.g., by the next frame or previous frame buttons, by menu selection, keyboard accelerators, the blink timer, IRAF running in another process, and so on. All of these end up sending a request to the *ximtool* client which directly or indirectly results in a frame change. When the display frame is changed a number of client events are generated to inform the GUI not only of the new frame number, but also the new frame title, zoom, pan, and frame flip values, type of enhancement used for the frame, and so on. Each of these items represents a separate client event. Although the action of the program may be arbitrarily complex in real world examples, the basic messaging mechanism on which this is all based remains very simple.

5. Software Products

5.1. The X11IRAF Package

All of the software described in this paper is packaged in a single distribution called *x11iraf*. This includes *xgterm*, *ximtool*, the Object Manager library, sources for all the third party widgets used in OBM, and assorted demo applications. Everything needed to build the package is included, including compatible versions of some publically available libraries, e.g., Tcl and Xpm. Despite the name the software is not tied to IRAF in any way, other than that it is a product of the IRAF project and is used for IRAF GUIs.

5.2. Xgterm

Xgterm is an upwards compatible version of the popular *xterm* with the *xterm* graphics ripped out and replaced with an OBM-based GUI which uses the gterm-image widget for graphics. The graphics supports a number of extensions, including full color support, an integrated imaging capability, dialog interaction, intelligent unconstrained resize, and a full crosshair cursor. Although *xgterm* is often used as a simple terminal emulator it is also a general widget server since it contains the full OBM library, and it can be used to execute arbitrarily complex OBM-based GUIs and manage the communications with the remote client. The current version of *xgterm* is based on the X11R6 version of *xterm*.

5.3. Ximtool

Ximtool is an image display server, used by remote client applications such as IRAF to display and interact with images. Several frames can be loaded and independently displayed in a full-frame or tiled configuration. Display frames can be any size. *Ximtool* is a good example of a conventional single process windowing application which uses the OBM library for the GUI and the window system interface.

5.4. The Object Manager Library

The Object Manager library (OBM) is a high-level, interpreted window system toolkit that is used to implement arbitrary graphics user interfaces. The OBM library uses Tcl as the interpreter. The current version of the OBM library is based on the X toolkit and can be used with any Xt-based widget or widget set.

5.5. The Gterm-Image Widget

The gterm-image widget is an X Toolkit-based widget for general 2D graphics and image display. This is a complex widget and a full description of its capabilities is beyond the scope of this paper. The Gterm widget provides a general GKS-like vector graphics and text display capability. An integrated image display capability allows any number and size of image rasters to be created within the widget or in the X server. *Mappings* can be defined to map one raster to another, permitting general graphics pipelines involving scaling and other geometric transforms to be set up. Colormap support is included for grayscale and pseudocolor rendering of raster data. An interactive *graphics marker* facility is provided for interaction with the displayed graphic. The Gterm widget is used for all graphics and imaging in *xgterm* and *ximtool*.

6. Adding GUIs to IRAF Applications

A major application of the widget server and the other software described in this paper is in adding GUIs to IRAF applications. In this case the IRAF task is the client: when the IRAF task starts up it downloads its GUI to the widget server, and during execution the IRAF task and the GUI communicate via the messaging facility described earlier. The changes required to add a GUI to an IRAF task are minor, ranging from changing a single line of code to cause the GUI file to be downloaded, to defining a set of client events and adding *gmsg* calls to allow more complete integration of the GUI. Adding a GUI to a task increases the system size by only the 10 Kb or so required for the GUI file.

7. Availability and Further Information

Further information on the software described in this paper, including more detailed documentation, full sources, and executables for a variety of platforms can be found in the X11IRAF Web page[2]. Further information on IRAF itself and other IRAF products can be found in the IRAF archives[3]. Documentation on Tcl, Xt/XLIB and the other standard software products used in X11IRAF is available from many sources.

---

[2] http://iraf.noao.edu/x11iraf

[3] http://iraf.noao.edu

The AstroVR Collaboratory, An On-line Multi-User Environment for Research in Astrophysics

D. Van Buren

Infrared Processing and Analysis Center, MS 100-22, Caltech, Pasadena, CA 91125

P. Curtis, D. A. Nichols

Xerox Palo Alto Research Center, 3333 Coyote Hill Rd., Palo Alto, CA 94304

M. Brundage

Dept. of Mathematics, GN-50, Univ. Washington, Seattle, WA 98195

Abstract. We describe our experiment with an on-line collaborative environment where users share the execution of programs and communicate via audio, video, and typed text. Collaborative environments represent the next step in computer-mediated conferencing, combining powerful compute engines, data persistence, shared applications, and teleconferencing tools. As proof of concept, we have implemented a shared image analysis tool, allowing geographically distinct users to analyze FITS images together. We anticipate that AstroVR[1] and similar systems will become an important part of collaborative work in the next decade, with applications in remote observing, spacecraft operations, on-line meetings, and day-to-day research activities. The technology is generic and promises to find uses in business, medicine, government, and education.

1. Introduction

Collaborative research in astronomy is the norm. Most papers published in the Astrophysical Journal have multiple authors, and most multi-author papers involve collaborators at different physical locations. The reasons for this are many: the breadth of astronomy, the limited resources available to any one institution, and the social nature of the enterprise. In any case there is a qualitative difference in how a project proceeds when the collaborators are geographically distinct compared to when they are at the same place. Many of the differences are barriers: of communication, of access to data, of access to software, and of access to expertise. Researchers now cope with these barriers in a number of ways: telephone, e-mail, ftp, giving each other local computer accounts for access over the network, and more recently the creation of Web-based information services.

---

[1] http://astrovr.ipac.caltech.edu:8888

We were motivated several years ago by advances in network connectivity and multi-user database technologies to experiment with a new mode of remote collaboration. Our idea was to build a multi-user, network-accessible environment which we could populate with tools useful for astrophysics research. We thought such a system would be useful not only for distributed groups undertaking research projects, but also for teleconferencing, small seminars, browsing the astronomical portion of cyberspace, and providing a social space for participants. Eventually this technology could be hooked up to telescopes and other facilities; it could evolve in the direction of a "collaboratory," wherein geographic location becomes unimportant, and access to a wealth of services and tools is immediate.

During this interval a number of commercial vendors and NCSA have developed distributed whiteboard and teleconferencing applications, but these only superficially meet the needs of scientific workers. These rudimentary shared environments often require particular hardware, the purchase of proprietary software, have a limited toolset (NCSA Collage supports shared HDF data browsing and is probably the most useful) and are not extensible. Nor do they have persistence (i.e., maintenance of state between invocations), nor continuity of availability. We were interested in developing a more science-fiction style cyberspace where participants could interact with each other and the environment at any time, where changes to the environment could be persistent, and where new tools and facilities could easily be added, often synergistically extending the environment's potential.

At IPAC, Van Buren was monitoring the development of multi-user networked games called MUDs (Multi-User Dungeons). These client-server games were usually written by graduate and undergraduate students, and plain telnet was often the client. Generally they lacked the stability, support, extensibility, and/or persistence of data needed for a true collaborative environment, but they clearly were headed in the right direction as the idea of multi-user game servers became more and more sophisticated. One particular server, called MOO (for MUD Object-Oriented), written by Stephen White of the University of Waterloo, was chosen by Curtis to form the basis of his social virtual reality study at Xerox PARC, and he took over its further development. This server suddenly became professionally supported and maintained. In all other respects it was superior to the other MUD servers as well, so it became the choice for the underlying software for AstroVR. At the same time, the Xerox workers were thinking of extensions to the MOO technology that would support real-world applications. They initiated the Jupiter project, which is further described in Section 2.2. At this point we began a formal collaboration where AstroVR incorporates and tests the Jupiter technology, as well as prototyping new Jupiter extensions. AstroVR further extends the MOO and Jupiter technologies to include astronomically useful tools.

2. Architecture (with Examples)

2.1. The MOO Server

AstroVR's MOO server manages the many network connections and communications streams making up the environment. Contained in the server's database are objects, i.e., data structures containing code and information implementing the various capabilities. For example, one object generates "post-it" notes that users can pop onto each other's screens from a distance. Another object implements a shared Mosaic Web browser, and yet another allows a group of astronomers to join in the analysis of a single image, even though they may be sitting at workstations thousands of miles apart.

The basic architecture is client-server. In fact, the environment is heavily distributed, in that it makes use not only of the MOO server, but also auxiliary servers, remote information services, and a potentially distributed multi-process client. AstroVR also includes the entire public portion of the World-Wide-Web in its data space. The MOO server provides the persistent database and manipulative functions that manage the environment. It comprises a general purpose, multi-user, object-oriented database and an embedded C-like language: the MOO language. A "core" database, the LambdaCore, served as our starting point. This database included a general-purpose object-class library and sufficient code to further extend the environment. The MOO server and LambdaCore, as well as a programmer's manual[2] for the MOO language, are available in their most recent form via ftp[3].

The embedded MOO language makes AstroVR an extensible, evolving system. The behavior of all objects in the AstroVR database is defined by verbs (methods) and properties (data). Objects, verbs, and properties are created and altered from within the environment, making it easy to add and test new functionality without recompiling the server. An object attains its behavior in several ways. First, it inherits the behavior of its parent object, and the parent's parent, etc. Secondly, new verbs (possibly overriding inherited verbs) can be attached to objects. Thirdly, there is a large class of utility objects holding large libraries of general-use verbs. Finally, objects have data attached to them which represent their state, which of course affects their subsequent behavior.

The act of extending the functionality of existing objects and creating new objects in AstroVR is called "building." All users are potentially builders. Because the environment also facilitates communication between users, building becomes a collaborative effort itself. Many of the tools that exist are in fact building tools. In this sense, AstroVR is a meta-tool: it is used to create itself!

Building often takes the form of creating an instance of a previously existing object and then defining new behaviors by attaching verbs and properties to the new object. As a hypothetical example (the MOO server does not yet support floating-point math), consider how we might create an astrophysically interesting calculator, the H II region calculator. We want this specialized calculator to know how to figure Strömgren radii. Suppose that there is already a calculator object, `$calculator`, embodying the behavior of a general-purpose calculator. We create a child of `$calculator` to be our starting point because all of behavior of `$calculator` will be inherited. The new object is placed in our "namespace" and we operate on it, referring to it by a unique identifier. In particular, we next add a verb "r_stromgren" to the calculator which will do the work in calculating Stromgren radii, but which is initially empty. To create the verb code we invoke the AstroVR verb editor GUI and edit the verb code in a textedit widget. The textedit widget supports many **emacs** commands and is used in many places

---

[2] ftp://parcftp.parc.xerox.com/pub/MOO/ProgrammersManual.texinfo_toc.html

[3] ftp://parcftp.parc.xerox.com/pub/MOO

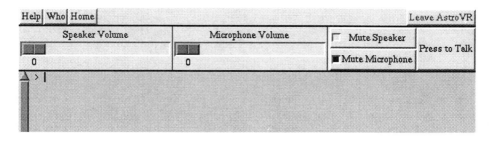

Figure 1. A particular client window. Buttons along the top are: *Help* to invoke the built in help browser, *Who* to pop up a constantly updating window that shows who else is on-line, *Home* to move to your default location, and *Leave AstroVR* to shut down your user object, client and network connection. Below are the audio gain and muting controls. The bottom panel, which is truncated for display, is the main input/output typescript where users issue AstroVR commands and receive messages.

inside AstroVR for data entry. The MOO server side of the verb editor is implemented entirely in MOO code.

There is a sophisticated permissions system inside AstroVR, because in a multi-user environment not everyone should be able to do anything at any time. It turns out that writing secure verbs is not that difficult, but it is necessary to guarantee that objects behave properly. For example, suppose users Galileo and Copernicus were happily working out Strömgren radii with the H II region calculator. Then user Herschel, who is not participating in the collaboration, should not be able to punch numbers into the calculator, turn it off, or otherwise alter its state.

2.2. The Jupiter Client

The Jupiter user client is the same client that Xerox is using for its Jupiter project, where a virtual environment is being overlaid on the physical environment at PARC, using a MOO-based cyberspace. For most game MUDs, telnet is an (almost) adequate client, providing a single typescript where users issue commands and receive output from the server. But in a collaborative environment, where a number of tools may be in simultaneous use, a more sophisticated client is needed. The Jupiter client meets this need by supporting window interfaces to AstroVR objects; other capabilities include multicast audio and video support, transparent file transfer, and a generic interface for local applications to be shared with other AstroVR users.

The Jupiter window support is similar in architecture to the GUI layer for IRAF (Tody 1995), but with a slightly different widget set and window definition language. In a nutshell, the client builds and manages GUIs after receiving text strings from the MOO server which encode their character and behavior. Subsequently, only significant events (such as mouse clicks, but not mouse motions) initiate messages to be sent back to the MOO server. This approach represents a drastic reduction in network traffic compared to running X windows across the network.

Since the same network connection passes in one direction both plain text meant to be read by the user and requests for the client to do something, and in the other direction both plain text representing user commands to the server and client commands to the server to do something, we use an out-of-band (OOB) protocol. OOB messages across the network connection are #$#-escaped, newline-terminated strings. When sent to the server they are intercepted by a special object for further processing and dispatch. For example, the user may have changed the microphone gain using the volume slider on the main client window. In this case, the following OOB command is sent to the AstroVR server:

```
#$#win-event id: ''315'' widget: ''mgain'' value: ''27''
```

The OOB protocol in the client-to-server direction is defined by #$# followed immediately by a request type string, then a set of key and quoted value pairs. An extension allows an arbitrarily long list of newline-terminated strings to be sent. In the other direction, from server to client, there is an additional datum required to provide some level of security. An authentication key known only to the trusted portion of the AstroVR server and client is included, and must be correct for the request to be carried out:

```
#$#audio-set-microphone-gain ''10565-258830'' value: ''27''
```

In this case the client is being requested to change the microphone gain. Note that the request to change the gain was made in the main client GUI, but the actual change was made in response to the server's notification to the client. In Figure 1 we show a screen dump of a particular AstroVR client GUI. It is only a particular one because users are free to redefine their client GUI's (and many other GUIs) according to taste. The user's ability to customize the interface is a key feature of our environment.

2.3. Auxiliary Servers and Remote Information Services

The MOO server is capable of opening arbitrary network connections. In principle this allows the integration of network based astronomical data services in AstroVR. We created simple interfaces to NED, SIMBAD and the STELAR abstract services to demonstrate the concept. Consequently, the data contained in these remote services is available for further computing by AstroVR. If the remote service's protocol consists only of newline terminated strings of printable characters, like the WAIS protocol for example, AstroVR can make the connection directly. Otherwise, as in the case of SIMBAD, a simple intermediate server is created as a stand-alone process that does the appropriate protocol translations.

One interesting auxiliary server is used to overcome the MOO server's current lack of floating-point support. We have the Unix calculator program **bc** running as a server that can be connected to directly using MOO code. AstroVR makes use of this by loading to the **bc** server a list of strings in the **bc** language representing the calculation to be performed, and then reading the results as a list of strings. Although the floating-point numbers are represented in AstroVR as strings, they are computed as reals. This service suffers some overhead per use, but a calculation proceeds quickly once the connection is made.

2.4. Multicast Support

Audio, video, and dynamic screen broadcast make use of the multicast network, often called the "mbone." This is a broadcast technology (one copy out) which sends packets to destinations in a way that eliminates redundancies and is very network friendly. The AstroVR server does not handle any of this data. Instead it computes and sends switching information to the Jupiter client, telling it to listen to a particular (or set of) mbone channel(s). For example, all AstroVR users in the same virtual conference room are all tuned to the same mbone channel. The mbone audio gives telephone-quality sound.

Microphones and speakers/earphones are inexpensive and readily available for practically all workstations, and allow users to make use of this one feature that goes the farthest in creating the impression of a virtual space. We also have the capability for reduced frame rate video conferencing, but the expense of a video board and camera currently restricts its use. Users without this hardware can still receive video. Users with video boards can also define an area or areas on their screens to broadcast to collaborators, allowing some sharing of data for which a multi-user interface is not otherwise available, or to show the output of a program that for security reasons a shared interface is unwise—for example a telescope control system.

2.5. Client-Side Shared Applications

A large class of stand-alone applications may be shared through AstroVR and its client. One can imagine two modes for sharing an application: in one mode a single copy of the application is executing, with fan-in of user commands and fan-out of outputs. This method requires a rather restrictive set of circumstances because of the large variety of platforms and operating systems. In the other mode each user sharing an application has a local copy which is synchronized via AstroVR messages with those of the other users. Both modes are supported, and we have implemented instances of each. There is a shared MONGO interface which requires one collaborator to have MONGO locally. The other members of the work group can issue commands, for example to overlay their own data points on a plot being collaboratively constructed. The shared application interface knows how to properly distribute the Tek4010 commands to display a plot on an X windowing device (though more generic X windows are not supported). What the user sees is an xterm for entering mongo commands and a Tek4010 window where the plot is constructed. The user commands are prepended by the user-name of the person issuing the command to maintain accountability.

The second mode of sharing, where multiple local copies are synchronized via AstroVR client requests issued by the server, is used for sharing the image browsing and analysis tool "SkyView." This tool was created originally for analysis of *IRAS* images and is quite powerful, yet has a fairly simple command set and grammar. Inside AstroVR we have built a shared interface that operates with the shared application client software to present a SkyView GUI. Even in single user mode, the GUI provides much added value to SkyView users: it allows the user to define macros and assign buttons to them, to manage separate image frames easily, and generally to program SkyView in arbitrary ways using MOO code attached to GUI callbacks.

A particularly useful shared application is NCSA Mosaic. AstroVR can start a local Mosaic for document viewing in single-user as well as synchronized/shared mode. At least in terms of data display, this gives AstroVR users a vast cyberspace, but with the ability to do computations on URLs, and so implement intelligent Web navigators. (In fact, we had implemented many of Mosaic's transparent fetch and display functions inside AstroVR and its client, but the advent of the NCSA software means we can switch over and inherit all their updates and be assured of a well maintained package.)

An issue that arises when sharing applications is exactly how are they to be shared. One possibility is that commands are executed in the order they are typed, no matter which participant does so. This can be thought of as a first-come, first-served model, and is adequate for small (1–4 person) groups working collaboratively as equals. When a large group is sharing an application, this can break down and the users might want a different sharing model to be implemented. An extreme model is leader-follower, where only a single user is allowed to execute commands, and any other synchronized processes duly follow suit. In this mode, commands issued by the other users are ignored. This model is useful for doing software demos or when giving a seminar or on-line lecture. One of the strengths of the AstroVR system is that both models are readily implemented, and anything in between the two. We will discuss these models a little more in the section on teleconferencing.

Finally, we show in Figure 2 the overall architecture of the AstroVR system: server, clients, auxiliary and remote services, client-side processes, and the multicast network.

3. A Sought-for Richness

The goal of building is to create a large enough set of objects to be useful for conducting collaborative research. The technologies upon which AstroVR is based allow and encourage users to contribute to defining and building this toolset in a collaborative fashion. Our hope is that AstroVR users themselves will generate much of the environment's richness and functionality. But even a small toolset allows interesting combinations and we fully expect that a much larger set will allow for extremely powerful synergisms. Our thinking is based on the experience leading to the ISSA Postage Stamp Server[4], described elsewhere in this volume by Van Buren, Ebert, & Egret (1995).

Almost two years ago a simple text interface was built in AstroVR for SIMBAD; at the same time we were experimenting with transparent AstroVR-mediated file transfer, and separately providing on-line access at IPAC to all the ISSA[5] data. These ideas came together when we constructed a "virtual *IRAS* satellite" inside AstroVR that would deliver to users small pieces of the infrared sky identified by the name of a target or celestial coordinate. In the meantime the idea of the World Wide Web exploded, leading us to build a general-purpose Web server inside AstroVR to provide Web access to portions of the environment. The virtual *IRAS* satellite was rewritten to make use of this interface, and was recast as the ISSA Postage Stamp Server. Inside AstroVR users now have new tools to further manipulate this service. One drawback of using NCSA "Mosaic"

---

[4] http://astrovr.ipac.caltech.edu:8888/ISSA-PS

[5] http://www.ipac.caltech.edu/ipac/iras/issa.html

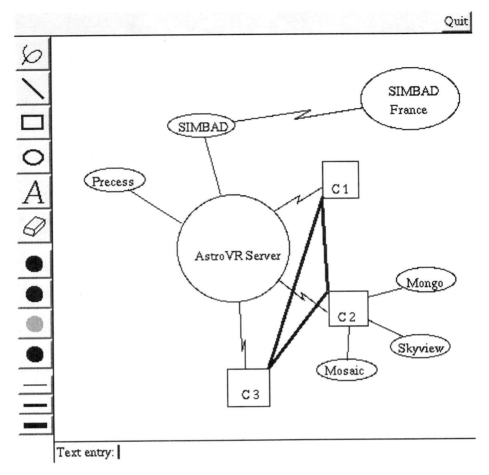

Figure 2. The AstroVR architecture, composed and displayed using the AstroVR shared whiteboard application. One connects to the main server via user clients (C1, etc.), which in turn run local shared applications and tune in to the multicast network (*heavy lines*). The AstroVR server also makes connections to other local and remote services, perhaps through intermediate protocol-translating servers (e.g., SIMBAD). The whiteboard itself can be shared by an arbitrary number of users. Commands are serviced on a first-come, first-served basis. The buttons along the left affect the drawing mode: *curlique* for freehand drawing with spline smoothing; *line* to draw a line between two points given by mouse clicks; *box* to draw a rectangle; *oval* to draw an ellipse; the letter "*A*" to write some text; *parallelopiped* to erase; four *dots* to specify drawing color as black, red, green or blue; and three line-width selectors. An arbitrarily large number of whiteboards may be active at any one time. Whiteboard data is also easily saved and restored, giving an archive capability.

to access the postage stamps is that it requires repetitive human input to get data for a number of objects (unless one wants to write Mosaic drivers). The ISSA Survey Engine was therefore created. It takes a list of target names or positions edited with the built-in AstroVR text editor and delivers all the FITS files to the user's local system for further study. The lesson learned from this exercise was that tools can be combined in new ways that are very powerful. In the past, large resources at IPAC were spent delivering equivalent data to users that is now automatically provided at miniscule expense.

4. Teleconferencing

The technology of teleconferencing is still in relative infancy. Most people have experience with telephone conference calls. For small groups this can be an effective mode of communication since we are all fairly well socialized with correct phone behavior. But with larger groups, the ambiguity of meaning, confusion of speaking order, missing facial expressions, body language and other signals that help facilitate a large conversation reduce the effectiveness of conference calls drastically.

4.1. Floor Control

Floor control can be improved using in-server software to manage audio and other data streams. Each user has as part of the client an audio "receive" channel and possibly a different audio "send" channel. For example, each distinct location (or "room" in the VR metaphor) inside AstroVR normally has a unique send/receive broadcast channel. As a user changes location, her audio channel changes transparently so she can participate in whatever audio activities are taking place in the new location. Other audio behaviors are possible because the "send" and "receive" channels do not have to be the same. In lecture mode, all the users' "receive" channels are set to the lecturer's "send" channel, but everyone else's "send" channels are unconnected, so their audio data are not broadcast to the group. In "talking stone" mode, an object is passed from user to user, essentially giving them temporary lecturer status. Or participants could register an interest to speak with a meeting-chair object, and then talk when their turn comes up in a first-come, first-served fashion. Another mode is where a meeting facilitator services speaking requests with the assistance of a GUI that keeps track of outstanding requests. Arbitrary floor control methodologies are possible, so depending on circumstances the best one available can be chosen for any given purpose.

One situation where floor control becomes very important is sequence planning for flight projects. The science team typically has a phone teleconference with a dozen or so participants. Each member rightfully advocates a particular course of action and all must be reconciled. Without proper floor control such meetings are inefficient, unsatisfying and sometimes the results are incorrect due to errors arising from attention lapses, confusion or uncorrected misunderstandings. Facilitator mode conferencing, especially if augmented with minute-taking software (possibly working with audio records), shows great promise in increasing the effectiveness of such meetings.

4.2. Human vs. Virtual Presence

For the forseeable future computer-mediated conferencing and on-line collaborative environments will not take the place of actually being with someone. There is no substitute for seeing a person's body language, feeling a casual touch or sharing a meal. These kinds of social interaction are crucial to the quality of many professional relationships and cannot be recreated in a virtual environment. On the other hand, much effective work can be undertaken on-line between visits, and if the environment is easy enough to use, the work can proceed as any other work, unconstrained by geography.

5. A Rosy Future for Collaborative Environments?

This past summer the National Institute for Standards and Technology issued a call for proposals to develop ideas for distributed multi-user software technologies in the field of manufacturing, explicitly targeting MUD and derivative technologies. AT&T television advertisements feature multi-user, on-line environments as what we can expect in the future. As network bandwidth increases and the computing power of desktop machines does likewise, the technical ability to create an immersive on-line environment will lead to the creation of multi-user virtual spaces in many disciplines. Some obvious applications include medicine, where consultations with distant physicians will be possible; education, where classes and seminars can be held in fields that are too small to support a local effort; government, where many meetings will no longer require travel; and business, where far-flung operations can keep in touch and conduct business in a virtual environment. The technology is a humanizing technology, enabling people to come together for work or play from all over the world, to develop new connections and to discover new ideas.

Acknowledgments. We acknowledge the support of US taxpayers through a contract to the Jet Propulsion Laboratory, California Institute of Technology from the National Aeronautics and Space Administration. The original author of the MOO server is Stephen White. The LambdaCore software was written as a collaborative effort by a large group of pioneering MOOers.

References

Van Buren, D., Ebert, R., & Egret, D. 1995, this volume, p. 84
Tody, D. 1995, this volume, p. 89

A Graphical User Interface for a Development Environment (GUIDE)

A. Bhatnagar

The Aerospace Corporation, PO Box 92957-M2/259, Los Angeles, CA 90009

Abstract. "GUIDE" (Graphical User Interface for a Data Environment) is an interactive IDL widget-based package that enables users to retrieve selected sets of satellite instrument data and perform various types of graphical analyses on them, without writing a specialized program. GUIDE simplifies the task of laying out multiple plots on a page by (1) using a page-editor to define and laying out multiple "panels" which contain the "plot objects", and (2) integrating data retrieval functions.

1. Introduction and Design Concept

GUIDE was developed in support of the POLAR CEPPAD (Comprehensive Energetic Particle Pitch Angle Distribution) experiment. Its fundamental software design was created with the following goals and requirements in mind:

- The system should require little or no knowledge of the underlying application software (IDL).

- The software routines should be as modularized as possible.

- The software should be machine independent to allow for portability and a consistent user interface. It should also be device independent to permit various types of output, such as publication-quality hardcopies or slides.

- The analysis and presentation routines should be time-span independent.

- The system should be capable of accessing archived event data given a time interval reference (i.e., start/stop time, orbit, date, etc.) as well as energy levels, channel number, etc.

- The system should offer the capability to retrieve, analyze and compare data from ancillary sources including ephemeris, geomagnetic indices, and data from other experiments.

2. Architecture and Function

A "page editor" was devised to create and modify three components: the Page Description, the Panel Description, and the Plot Object. These components characterize the layout and content of all the plots grouped within the given

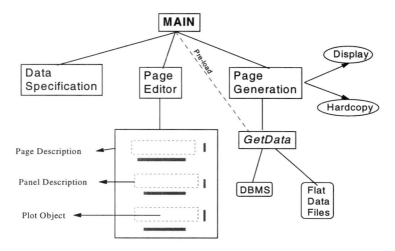

Figure 1. System design.

page. Figure 1 illustrates this subsumptive architecture with respect to other functional aspects of GUIDE, described below.

The Page Description File contains information pertaining to the overall number, size, and relative location of panels for the given page, but not the actual contents of those panels—that information is in the Panel Description Files. This natural separation facilitates a more flexible system by allowing panels to be reused.

The Panel Description File contains all the necessary content descriptors for that particular panel including the name of the routine generating the plot for that particular panel and the input parameters associated with it. The input parameters required by the plot object are divided into two categories—fixed and run-time. The distinction is based upon the stage at which their values are specified or altered. Fixed parameters include plot labels and titles, the number of series (lines) in a plot, or minimum and maximum flux values in a spectrogram. Run-time parameters, on the other hand, might include items whose values are typically altered for each run such as the start and stop times for a given event analysis.

Finally, the Plot Object is a procedure, or collection of procedures, containing the actual (IDL) code responsible for generating the specified plot, image, or any other pre-defined graphical object. At run-time, this module is called by the main GUIDE processing routine after it has set up system environment variables that define, for instance, window or viewport boundaries.

2.1. Page Generation

Using point-and-click mouse functions, GUIDE steps the user through the process of creating and laying out a Page Description File embedded with pointers describing the contents of each of its panels (see Figure 2). Once this is done and the user has specified values for the run-time parameters (see Figure 3), the user selects the menu option "Go!". The main GUIDE routine interprets

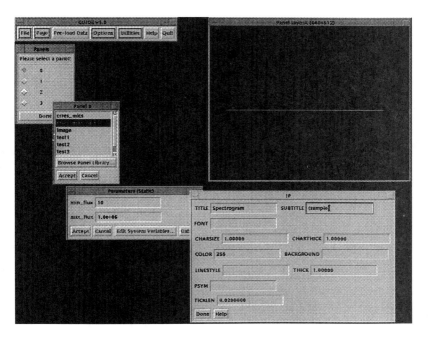

Figure 2. Creation of a new Page File.

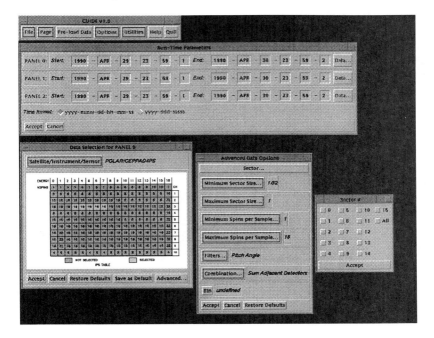

Figure 3. Run-time parameter specification with optimal filters.

the contents of the Page File, and for each panel (1) gets the location, and size of each panel from the Page Description File, (2) extracts the parameter list, parameter values, and plot object name associated with the current panel from the Panel Description File, (3) sets the plot viewport region based on the panel sizes, (4) invokes the appropriate plot object with the corresponding parameters to retrieve and process the pertinent data and display the plot, and (5) returns, restoring the default system environment.

2.2. GetData

In step (4) above, the plot object routine will typically make calls to a GetData procedure. These external procedures act as an application program interface (API) to the satellite/instrument specific dataset or database. For every distinct data type, there would be a separate GetData procedure which handles all the retrieval and processing specific to that data type. GetData is responsible for returning the data to common blocks accessed by the display routines, after reduction and filtering steps.have been performed. If no data match the selected criteria, or in the event of some other error condition, GetData returns a message to the calling program, the user is notified, and the generation of that panel ceases. Decoupling GetData from the GUIDE framework adds to the generality of the overall system by allowing GUIDE to transparently support multiple data types. GetData may be developed off-line and independent of GUIDE. There are, however, two basic input/output constraints imposed on GetData: (1) GetData must accept its inputs in a format pre-defined by GUIDE and its plot object routines, so that parameter lists match, and (2) the structure and location of the common blocks filled by GetData must be consistent with those utilized by GUIDE's plot object routines.

The user may also use the GUIDE menu option of "Pre-loading". This is essentially a direct top-level call to GetData for the purpose of verifying the existence of data within a given period and having it placed in common blocks so that it is readily available when needed.

3. Implementation Issues

GUIDE was developed using IDL 3.5.1 and uses the EXECUTE command to interpret the contents of the Panel Files at run-time and to call the Plot Objects. This facilitates accommodating new Plot Objects: (1) regardless of whether the code is all written in IDL or contains embedded CALL_EXTERNAL calls to C or FORTRAN images, all graphical output is done by IDL commands, (2) all Plot Object Files in the current directory are automatically added to the list used in building the "Browse Panel Library" widget, (3) when a user selects the new Plot Object in the panel definition stage, GUIDE automatically parses the Plot Object File and constructs the custom input widget, eliminating the need for hard-coding input widgets, and (4) GUIDE takes the parameter values provided by the user, stores them in the Panel Definition File, and passes them to the given Plot Object at run-time.

Acknowledgments. This work was supported under NASA Contract No. NAS5-30368 and through the co-development efforts of members of the Space Sciences Department of the Aerospace Corporation.

Tcl/Tk with DRAMA - A Natural for Building User Interfaces to Instrumentation Systems?

T. J. Farrell, J. A. Bailey, and K. Shortridge

Anglo-Australian Observatory, P.O. Box 296, Epping N.S.W. 2121 Australia

Abstract. DRAMA allows you to build distributed real-time systems consisting of a set of event-driven tasks. The Tk windowing system is also event-driven—like most windowing systems. The similarities between the two have allowed them to be merged extremely effectively, providing an almost seamless interface between the two. The control system for the AAO's 2dF project has instrument control tasks written using DRAMA and running on VMS, UNIX, and VxWorks systems. These are tested and controlled by higher-level tasks whose user interfaces are provided by Tcl/Tk and which communicate with the rest of the system using purpose-built Tcl commands that invoke the DRAMA system routines. This gives the critical instrument control tasks the fixed nature and reliability that comes from a compiled, linked system, while allowing a very flexible user interface and test system that takes advantage of the ease of modification and the flexibility of an interactive language like Tcl/Tk.

1. DRAMA - A Quick Introduction

The AAO's Two degree Field (2dF) project, involving two robotic positioners manipulating two sets of 400 optical fibers and multiple detector systems, will be the AAO's most complex user instrument. The DRAMA system that has been developed for its control is a portable environment designed for writing instrumentation software. It has similarities to the Starlink ADAM environment (Kelly 1992), but is written entirely in C and has been ported to VMS, various flavors of UNIX and the VxWorks real-time kernel.

Its basic unit is the *Task*. A DRAMA Task is normally implemented as a separate process within a multi-process operating system. A Task may send or receive messages of various types. The most fundamental message type is the "OBEY" message. An associated name specifies the name of an "Action" that the task will perform. The simplest DRAMA task will set up relationships between "Action" names and C routines. It then enters a loop where messages are received and dispatched to the routine that is the current "handler" for that action.

A DRAMA task may have parameters associated with it which may be read or written by other tasks. So we have "GET" and "SET" messages which get and set the value of a named parameter. We also have "MONITOR" messages which allow a task to be notified if the value of a parameter in another task changes. Task parameters may be hierarchical structures of considerable complexity (although they are often simple scalar values) and a DRAMA message may contain such a structure, so this provides a very flexible way of transferring information around a DRAMA system.

DRAMA is well suited to both the lower level control of individual instruments and to the coordination of the overall system. What was lacking a year ago was a clear direction for the user interface. Although we had made it possible for a DRAMA task to include an X-based interface, and had some tasks written with Motif interfaces, it was clear that this was not an easy system with which to write flexible or experimental user interfaces.

2. DRAMA and Tcl/Tk

Tcl (Ousterhout 1994) is an easily extensible scripting language that is becoming very popular. Tk is an extension to Tcl that supports building graphical user interfaces under X-Windows. Initially we considered using Tcl to provide a scripting language for automatic test procedures for DRAMA. It turned out that because both Tk and DRAMA are event-oriented systems, the basic DRAMA messaging operations fit very well into the Tcl/Tk approach. For example, to allow Tcl/Tk applications to send the message types mentioned above, we implemented new Tcl commands named obey, pget, pset and monitor. (get and set are already defined by Tcl.) These are all similar. Let us consider the obey command.

The basic form of the obey command is

obey task action [args] [options]

Where task is the name of the DRAMA task to send the message to and action is the action name. Args holds the optional arguments to the action. Various options are available including

- -success *command*. This specifies a Tcl command to be executed when and if the obey completes successfully.

- -error *command*. This specifies a Tcl command to be executed if the obey completes with an error.

As a very simple example let us consider a one button Tcl/Tk application that sends a single DRAMA obey message when the button is pressed. The Tcl code is

```
button .a -text "my button" -command {obey CAMERA EXPOSE \
        -success { puts "Expose action completed ok";# } \
        -error   { puts "Expose action completed with error";# }}
pack .a
```

The Tk command button creates a button widget and pack makes it appear. The label of the button is specified with -text. A Tcl command to be executed when the button is pressed is specified with -command. In this case, it invokes the action EXPOSE in a task named CAMERA. Note that this all happens asynchronously. The process running the Tcl script continues running, and the user interface remains responsive, once the obey command is sent. Eventually, the message from the camera task indicating either success or failure will arrive and the appropriate action will be taken by the Tcl script.

It is possible to replace the argument to the button's -command option and obey's -success and -error options with the names of Tcl procedures to be invoked, allowing operations of any required complexity to be performed. In this style of application Tk events such as button presses are seen as events that initiate DRAMA messages. The responses to the DRAMA messages can then trigger any required changes in the user interface. A user interface may add C code to implement additional Tcl commands. An example would be code to perform mean to apparent place conversions, which can be done in Tcl, but for which C routines already exist.

3. Monitor Messages

Monitor messages have proven particularly useful. The monitor mechanism allows one task to "express an interest" in a parameter of another task. When this parameter changes, the interested task receives a message identifying the parameter and containing its new value. This solves a common problem quite neatly. Often, a low level task, controlling some part of an instrument such as a spectrograph, will have information such as grating angles, filter positions, etc. that is of interest to other parts of the system such as the user interface. However, the modularity of the system is broken, awkwardly, if such a task has to make assumptions about the higher levels of the system. You would not want to write a spectrograph control task with explicit code that always sent a new filter position to a task called "USER_INTERFACE", since you may be running with a completely different user interface task.

It would be possible for the spectrograph task to have an mechanism coded into it whereby it could be sent the name of a task which it was to notify whenever a filter position changed. This solves the modularity problem, but requires a lot of explicit code in the low level task. Alternatively, the user interface could poll the filter value, but this is inelegant.

Under DRAMA the combination of the monitor facility with the very flexible task parameters normally supported allows a neat solution to the problem that is event-driven and which requires no specific code in the low-level task. The user-interface code that "expresses an interest" in the low-level task's parameter is just a few lines of Tcl. The user interface can even direct that an image parameter maintained by a camera task be monitored by an image display task, with no specific code required in either the camera task or the image display task.

This could also be used to help implement remote observing. We envisage a scheme where two copies of a user interface are run, one local and one remote. The lower level DRAMA control software would be run locally and commands sent from either the local or remote user interface using normal DRAMA messages. Monitor messages would allow both user interfaces to be kept up to date with what the other was doing to the system using a minimum of network traffic.

4. 2dF Commissioning

The combination of DRAMA and Tcl/Tk let us put together simple user interfaces for testing individual components of the 2dF system. We were also able to produce high quality overall system interfaces, controlling multiple low-level tasks.

During the initial commissioning run for 2dF, we found ourselves struggling with a VME system that would not reboot cleanly. This discouraged us from making any changes which might require a restart of the system. The ability to dynamically change a Tcl/Tk program and execute simple DRAMA commands proved invaluable. When we devised fixes or improvements to the user interface, it was easy to load and test them without having to risk rebooting the faulty VME system.

5. Conclusion

DRAMA and Tcl/Tk seem to work together extremely well. The original DRAMA design provided a good environment for writing distributed real-time systems running on a disparate set of machines. With Tcl/Tk it is now easy to produce powerful and flexible user interfaces that can take advantage of the underlying DRAMA system.

References

Kelly, B. D. 1992, ADAM—Guide to Writing Instrumentation Tasks, Starlink User Note 134

Ousterhout, J. K. 1994, Tcl and the Tk Toolkit (Reading, Addison-Wesley)

A Method for Visualizing Time Variability in X-Ray Images

P. Giommi

ESIS, Information Systems Division, ESA/ESRIN, Frascati, Italy

N. E. White, L. Angelini

HEASARC, NASA/GSFC, Greenbelt, MD 20771

Abstract. We present a new method for visualizing flux variations in X-ray images. Starting from event list files, images are created where the pixel intensity is proportional to the probability that a strong variation in flux (approximately a factor 2 or more) occurred during the observation. The method is based on the Kolmogorov-Smirnov test and is not sensitive to data interruptions due to temporary loss of telemetry, earth occultations etc. Since only strong variations can be detected this approach is also insensitive to most spurious events induced by satellite pointing instability. This method has been implemented within the XIMAGE package and was used during a systematic analysis of all publicly available *ROSAT* PSPC images. The results of this analysis were used to construct the WGA catalog of X-ray sources. In this paper we describe the method and give some specific examples. The results presented here make use of the *ROSAT* PSPC data obtained from the *ROSAT* public archive accessible via the ESIS or the HEASARC systems.

1. Introduction

Systematic studies of luminosity variability in data obtained from low orbit satellites often require binning and are complicated by the frequent data interruptions due to Earth occultation and other causes. We present here a method to detect X-ray variations that does not require binning, is simple, and numerically stable, and is not sensitive to data interruptions. Variability on time scales ranging from a few seconds to the actual observation duration can be detected. We have used this technique to systematically analyze of all PSPC images available in the *ROSAT* public archive. This project led to the construction of a catalog including more than 40,000 X-ray sources (the WGA catalog, White, Giommi, and Angelini 1994). Several hundred of these sources have been found to be rapidly variable.

2. The Method

The method consists in comparing the time arrival distribution of the photons collected in each pixel with the corresponding distribution of the entire image using a Kolmogorov-Smirnov (KS) test. The result of the KS test is a χ^2 value (with 2 degrees of freedom) that is used to assign an intensity value to each pixel. In this way, pixels where the distribution of photon arrival times is not consistent with that of the entire image are given high intensity (χ^2) values. The *time variability image* so constructed visually shows area where strong time variation occurred. To calculate the χ^2 values the event list must be *time sorted* and must be read twice. The first time the event list is read a normal intensity image is built and is stored in memory as an array of integers. This array is used during the second pass to provide the normalization values (i.e., the number of detected photons during the full exposure) in each pixel and in the entire image. The second time the event list is read, the cumulative distributions for each pixel and for the entire image are calculated. When a photon i is considered the cumulative distribution at time t_i in pixel (x_i, y_i) is obtained as the ratio of the number of photons detected up to time t_i divided by the total number of photons detected at position (x_i, y_i): $D(t_i, x_i, y_i) = N(t_i, x_i, y_i)/N_{tot}(x_i, y_i)$. The cumulative distribution of the whole image is calculated as the ratio of all the photons arrived up to time t_i, and the total number of photons in the image accumulated during the entire exposure is $D_{image}(t_i) = N(t_i)_{image}/N_{tot}$. The distance $\Delta = |D_{image}(t_i) - D(t_i, x_i, y_i)|$ between the two distributions is written into a new array at location (x_i, y_i) only if it is greater than the value previously stored in that pixel. These steps are repeated for every photon in the event list. At the end of this process the array containing the (maximum) distances between the arrival time distribution of all pixels and the entire image is converted into an array of χ^2 values using the usual formula of the KS test.

This method has been incorporated within the XIMAGE package (Giommi et al. 1991). In this particular implementation the KS test is applied only to pixels where more than 10 photons were detected during the entire observation. The resulting χ^2 values are multiplied by a factor 10, to increase the dynamic range, and are limited to a maximum value of 200 to avoid that a bright variable source with a very high χ^2 dominating the entire image. By this method, all pixels with values between $\approx$ 100 and 200 represent variable sources.

3. The Limits

The method is based on the comparison between the source light curve and that of the rest of the image. To properly detect source variations it is necessary that the image background be not strongly variable. This condition is generally satisfied in most *ROSAT* images. For the case of observations with highly variable background it is necessary to remove all the time intervals where the background was variable. This obviously reduces the sensitivity of the test. A second limitation of this method is that it is not sensitive to small amplitude variability (less than about a factor 2) and to weak periodic oscillations.

Given the limitations described, this method cannot be used to perform *uniform* variability surveys. It is, however, very useful for detecting large flares (on time scales of hours or days) or strong periodic sources. In all cases a careful check must be carried out to remove spurious events. This is an ongoing activity on the candidate variable sources included in the WGA catalog.

4. Some Examples

In this section we give some examples of rapidly variable sources found in *ROSAT* public data using the ESIS system (Giommi et al. 1994) which makes use of XIMAGE and several other Xanadu routines. Figure 1. shows the *intensity* (left) and time variability (right) image of the *ROSAT* field centered on the X-ray source MKW 3. A strongly variable source is clearly visible in the bottom-right part of the right image; its light curve is shown in Figure 2. Figure 3 shows a PSPC exposure of the Orion region where several sources have been detected (left). The corresponding time variability image is shown to the right. At least three sources varied during the observation. The light curves of two of these variable sources are shown in Figure 4. These examples clearly show the power of the method described in evidencing variable sources in crowded X-ray images.

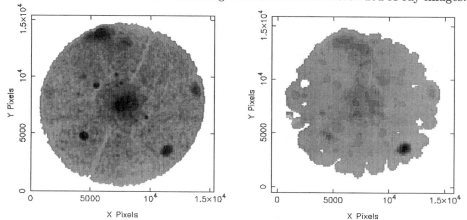

Figure 1. Intensity and time variability image of the *ROSAT* field centered on MKW 3.

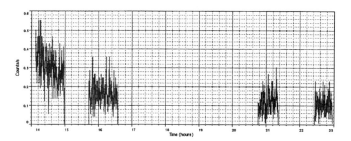

Figure 2. The light curve of the variable source shown in Figure 1.

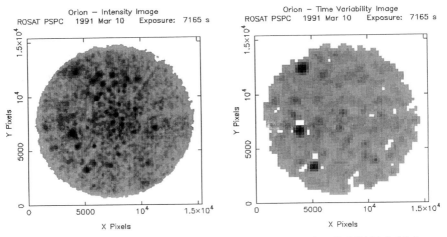

Figure 3. Intensity and time variability image of a *ROSAT* field in the Orion region.

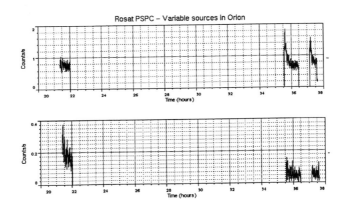

Figure 4. The light curves of the top left (upper panel) and middle left (lower panel) variable sources of Figure 3.

References

Giommi, P., Ansari, S. G., Donzelli, P., & Micol, A. 1994, Experimental Astronomy, in press

Giommi, P., Angelini, L., Jacobs P., & Tagliaferri, G. 1991, in Astronomical Data Analysis Software and Systems I, ASP Conf. Ser., Vol. 25, eds. D. M. Worrall, C. Biemesderfer, & J. Barnes (San Francisco, ASP), p. 100

White, N. E., Giommi, P., & Angelini, L. 1994, Proceedings of the AAS-HED meeting, The Multi-Mission Perspective (Napa Valley, AAS-HED)

Astronomical Data Analysis Software and Systems IV
ASP Conference Series, Vol. 77, 1995
R. A. Shaw, H. E. Payne, and J. J. E. Hayes, eds.

ASpect: A Multi-Wavelength Spectrum Analysis Package for IRAF

S. J. Hulbert, J. D. Eisenhamer, Z. G. Levay, and R. A. Shaw

Space Telescope Science Institute, 3700 San Martin Dr., Baltimore, MD 21218

Abstract. ASpect is spectrum analysis package being developed at ST ScI. We discuss the implementation of a GUI using the IRAF Object Manager, the ASpect task design, and the mechanics of performing a fit to a spectral feature with ASpect.

1. Introduction

ASpect is a spectrum and line analysis package being developed at ST ScI. ASpect is designed as an add-on package for IRAF and incorporates a variety of analysis techniques for astronomical spectra. ASpect operates on spectra from a wide variety of ground-based and space-based instruments, allowing simultaneous handling of spectra from different wavelength regimes. It accommodates non-linear dispersion relations. ASpect provides a variety of functions, individually or in combination, with which to fit spectral and continuum features. It allows for the masking of known bad data. Most importantly, this tool provides a powerful, intuitive graphical user interface (GUI) implemented using the IRAF Object Manager and customized to handle the burden of data input/output (I/O), on-line help, selection of relevant features for analysis, plotting and graphical interaction, and database management. ASpect is scheduled for release in late 1995.

2. ASpect GUI

The functional need for a full set of graphical I/O capabilities originally led us to implement the ASpect GUI with the public-domain software, Tcl/Tk. While Tcl/Tk provided the GUI functionality required for the task, this was accomplished at the expense of working outside of the IRAF environment. With the availability of the IRAF Object Manager, however, we now have access to the necessary GUI elements within IRAF. As originally envisioned, ASpect is an IRAF application–it is written is SPP and uses the IRAF Object Manager to provide the GUI. In this scenario the ASpect task is the client; the Object Manager is the server. At startup, the client application downloads a text file that defines the GUI. The Object Manager interprets this GUI file using the Tcl interpreter and then creates the GUI objects (each widget is an object). For the duration of the task execution, the Object Manager manages the messages between the graphical objects and the client application.

3. ASpect Task Design

The ASpect task has been organized as a group of objects: each object is implemented by a separate code library. Figure 1 shows a graphical representation of the relationships between the ASpect objects and code libraries. In this figure, the connections show which objects are available for "use" by any given object, i.e., the information "flows" towards the small circles at the ends of the connections. For example, the fit library "knows" about the component and spectrum objects but a given spectrum or component object knows nothing of fitting. This approach enforces modularity and by requiring a set of well-defined interfaces makes reuse of code more practical.

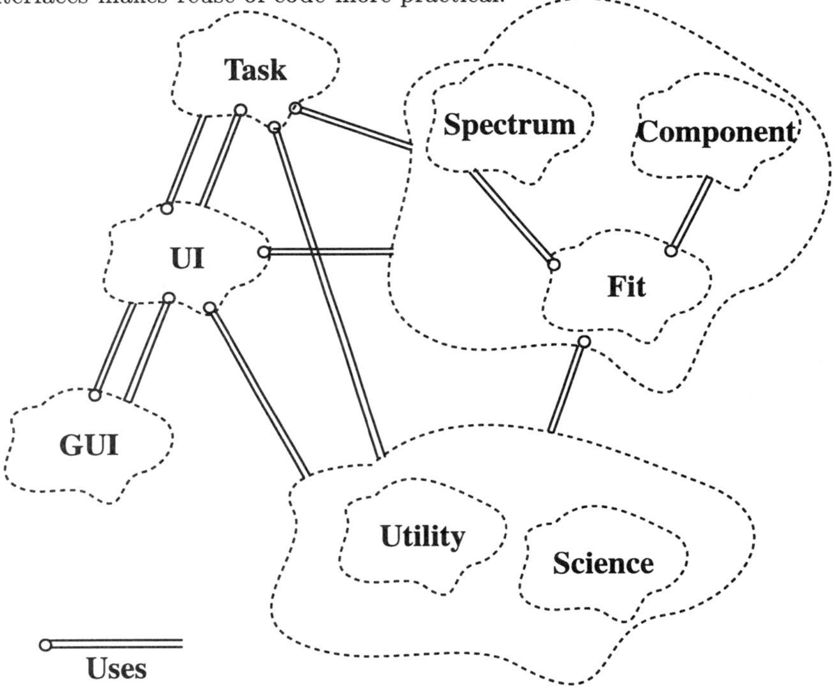

Figure 1.   Relationships between ASpect Objects and Code Libraries.

The highest level of objects are designed to bridge the gap between the user and the actual task. The GUI object, which is created by the IRAF Object Manager, manages all information passed between the UI object and the GUI. The UI object consists primarily of the "event loop", which handles events from the GUI. The UI object is solely responsible for setting UI parameters and all GIO calls are isolated here.

The basic task functionality is implemented in the task object, which is in turn dependent on a host of other objects and libraries. The task object is the main procedure and all global variables are maintained here. The task object manages a series of spectrum and component objects in the course of

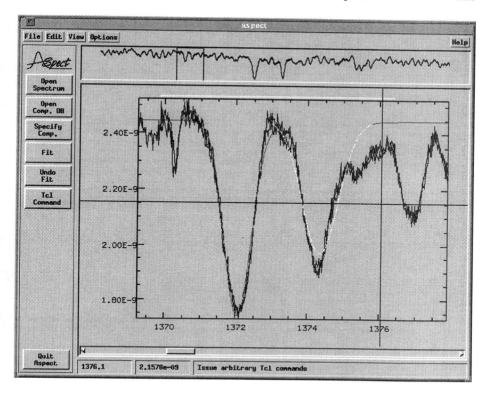

Figure 2. The ASpect workspace consists of a main window showing a zoomed portion of the spectrum bordered by a menu bar and spectrum overview window across the top, a button bar to the left, and a scroll bar, cursor readback display, and context-sensitive help box along the bottom.

the analysis. The spectrum object merely defines a spectrum: a flux array, a wavelength array, an error array, and a data mask array. A component object is a collection of models for various features in a spectrum which are to be fit. The actual fitting process is controlled by the data contained in the fit object.

The lowest level of objects provide support for all other parts of the ASpect task. The science libraries consist of a set of routines that implement the basic scientific functionality. In the case of fitting these consist of the actual fitting functions, e.g., Gaussian absorption line or black-body continuum. The utility library is a set of generic routines used by more than one part of ASpect.

4. Fitting Procedure

The set of actions that results in a fit to features in a spectrum consists of: (1) selecting the region of the spectrum to be analyzed, (2) selecting the type of component, (3) marking the region of the spectrum where the component is

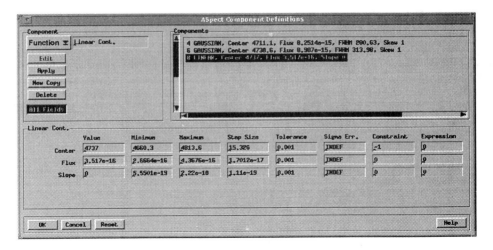

Figure 3. The component dialog box is used to display and edit components and the parameters that define them.

relevant, (4) marking the region of the spectrum over which to do the fit, and (5) initiating the fit. The initial spectral region selection is accomplished with the pan and zoom capabilities of ASpect. The marker in the overview window (see Figure 2) can be moved or resized. Alternately, a marker can be drawn in the main window to zoom in on a particular part of the spectrum. The type of component is selected in the component dialog box which is shown in Figure 3. A pull down menu labeled "Function" is used to reveal a set of line and continuum functional models. This function is mapped to a particular feature of the spectrum by drawing out a box around the feature. The size and placement of this marker is used to generate an initial guess for the parameter values of the component. Once defined, the component is added to the component list (seen in the top right-hand corner of the component dialog box). Additionally, the guess is displayed graphically in the main window as a fit (of sorts). At this point the user may edit the parameter values of an individual component as well as certain fit-related constraints and limits. Any number of components can be defined in this manner. When all components have been defined, one or more fit regions are marked. These fit regions are the parts of the spectrum that will be used in evaluating the fit. Finally, the fit is initiated using the fit dialog box. Generic fit parameters, such as number of iterations and type of fit algorithm, are also controlled from the fit dialog box. Once the fit is complete, the fit is overplotted in the main window. As the fitting process is iterative, the user may decide to repeat the fit after modifying any of the components or fit regions.

Acknowledgments. This ASpect project is funded under contract with the NASA Astrophysical Data Program.

Astronomical Data Analysis Software and Systems IV
ASP Conference Series, Vol. 77, 1995
R. A. Shaw, H. E. Payne, and J. J. E. Hayes, eds.

Applying Public Access Programming Techniques To SAOimage

E. Mandel

Smithsonian Astrophysical Observatory, 60 Garden St., Cambridge, MA 02138

D. Tody

National Optical Astronomy Observatories, P. O. Box 26732, Tucson, AZ 85719

Abstract. This paper describes our application of the X Public Access (*XPA*) interface to the new version of *SAOimage*. *XPA* allows an Xt program to define named public access points through which data and commands can be exchanged with external programs. It makes possible the external control of the program's main functions, including image display, image zoom and pan, color map manipulation, cursor/region definition, and frame selection. It also supports "externalization" of internal algorithms such as file access and scaling. Finally, we describe how *XPA* is used to support user-configurable "quick-look" analysis of image data and bi-directional communication with other processes.

1. Introduction

Astronomical software development needs to adopt more "open systems" concepts and designs. Users increasingly want to extend existing software and to combine tasks from different systems in order to create their own heterogeneous research environments. Our software designs must reflect this increasing need for flexibility and extensibility without sacrificing functionality and without violating budget constraints.

The *SAOimage* display program has been very popular with astronomers over the past decade. In the age of "open systems", it needs updating to support extensible features. These features include providing an easy way to display arbitrary data file formats, control image display from an external process, and integrate analysis routines into the image display.

In order to update *SAOimage* efficiently and cost-effectively, SAO and NOAO will collaborate to develop a high-quality image display program, basing our work on the *ximtool* program developed by D. Tody for IRAF. *Ximtool* uses the NOAO widget server and "gterm" image widget to support customized graphical user interfaces (GUIs), image and line graphics, multiple frame buffers, user-defined color maps, region markers, etc. It provides a high-level image display programming interface as well as low-level access to gterm-image widget functions. Our aim is to layer open-ended functionality on top of *ximtool* in order to develop *SAOtng* (SAOimage: The Next Generation). In this way, SAO and NOAO can share development of image display, while satisfying our individual needs.

2. The X Public Access Mechanism

To add open systems features to *SAOtng*, we will utilize the X Public Access mechanism (*XPA*), described in Mandel & Swick (1994). *XPA* allows an Xt program to define named points of public access, through which data and commands can be communicated to/from external programs:

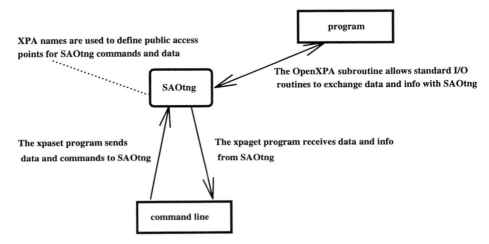

Figure 1. We have used *XPA* to extend the *ximtool* image server into an open-ended *SAOtng* display service that can cooperate with other processes and programs.

3. Controlling SAOtng

The standard *SAOtng* graphical user interface is a simple menu-based GUI that parallels the original *SAOimage* menu/command paradigm. It provides menus for image and frame selection, scaling, color maps, zoom/pan, regions, etc. All of these menu functions also are defined to be *XPA* public access points. Thus, *SAOtng* can be controlled directly through its menu-based GUI or externally through the *XPA* mechanism (see Figure 2).

For example, the *xpaset* program can be used to change the *SAOtng* scaling algorithm, and the *xpaget* program can be used to retrieve the current scale:

```
csh> echo "scale log" | xpaset SAOtng
csh> xpaget SAOtng scale
log
```

In fact, the *SAOtng* GUI uses *XPA* to send commands to itself. This implementation ensures that support for external control of *SAOtng* does not lag behind the internally-supported user interface: the GUI initiates commands as if it were just another external process!

4. External Support For File Access

A FITS file can be sent directly to *SAOtng* for display using *XPA*:

```
cat foo.fits | xpaset SAOtng
```

Other file formats are supported by writing two file access programs: a header access program to generate a FITS header of the full image, and a data access program to generate a FITS image of a specified data section (given the center, dimensions, and block factor). These access programs are defined at start-up time in a user-configurable ASCII file. To load an image, the programs are run externally (using the *system()* subroutine) to create FITS-format data for a specified data section, which are passed back to *SAOtng* for display using *XPA*. When a new file name is sent to *SAOtng*, the appropriate header access program is run to gather the overall image dimensions. Next, the data access program is run to extract the desired image section, which then is scaled to 8 bits and displayed. Finally, the data access program again extracts the data section and stores it as raw data for later use (e.g., re-scaling the image). Note that adding new formats does not require re-compilation of *SAOtng*.

5. Integration of Analysis Routines

Each file type can have user-defined analysis commands associated with it. These analysis commands are defined at start-up time by means of ASCII descriptions. The analysis commands associated with the currently displayed image are available for execution via an analysis menu. When selected from this menu, an analysis command is macro-expanded and then executed externally. Results can be displayed in a separate window or even can be sent back to *SAOtng*, i.e., a command can create an image and then send it to *SAOtng* for display:

```
$data | smooth ... | xpaset SAOtng "frame new"
```

In the example above, the "$data" macro is expanded into an *XPA* command that retrieves the FITS representation of the displayed image. The FITS image generated by the smooth program is piped into a new frame buffer of *SAOtng*.

6. Cooperation With Other Processes

XPA allows the re-use of GUI programs so that new systems can be built from existing high-level components. For example, the *SAOtng* "Load Image" menu option pops up the *XDir* program, a directory and file browser that also offers *XPA* services (see Figure 2). The *XDir* program then can be used to browse through directory trees using template filters. Double clicking on an image file causes *XDir* to send an *XPA* command to *SAOtng* to load the new image. Thus, *SAOtng* can make use of sophisticated *XDir* capabilities without linking them explicitly: GUI programs can be re-used in the same manner as subroutines!

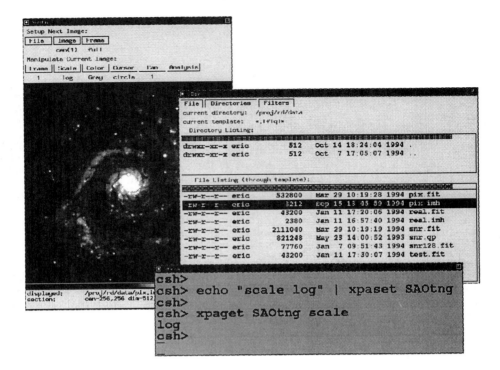

Figure 2. *SAOtng* has a simple menu-based GUI. It can be commanded externally (*lower right*) and can communicate with programs such as *XDir* (*upper right*).

7. Conclusion

SAOtng exemplifies a new type of astronomical analysis program that is extensible and that cooperates with other processes. It takes us closer to an era in which our heterogeneous analysis systems can work as an integrated whole.

Acknowledgments. This work was supported under NASA contracts to the *IRAF* Technical Working Group (NAGW-1921), the *AXAF* High Resolution Camera (NAS8-38248), and the *AXAF* Science Center (NAS8-39073).

References

Mandel, E., & Swick, R. 1994, PASP, 96, 198

WIP – An Interactive Graphics Software Package

J. A. Morgan

Astronomy Program, University of Maryland, College Park, MD 20742

Abstract. WIP[1] is an interactive package with a simple to use interface designed to produce high quality graphical output. WIP was developed as part of the Berkeley-Illinois-Maryland Association[2] (BIMA) project and is available via anonymous ftp. Details are presented about the WIP package along with a few examples.

1. Introduction

Astronomers need to generate high quality graphics. There are, in general, two reasons why existing graphics software is limited. First, the software often used to reduce the data can also generate graphics but not of publication quality. Either it generates graphics at a low to moderate quality or it produces figures that contain ancillary information that is not desirable in the final draft. Second, astronomers collaborating with astronomers at other institutions find it helpful to have a common plotting environment at each institution. However, commercial graphics products can not be freely passed from one institution to another making it impossible for non-licensed sites to be able to contribute to the graphics directly.

WIP was written to address these concerns. WIP was developed to be used along with BIMA's Miriad[3] data reduction package (Sault et al. 1995) and is based, at the lowest level, on Tim Pearson's PGPLOT subroutine library.

Because WIP is public domain software and can easily be installed on various machines (and architectures), it provides a common interface to high quality graphics that can be readily available to every researcher. The WIP distribution is available on anonymous ftp[4]. Included in the WIP distribution is a complete manual[5] which contains several sample figures with associated commands, a frequently asked questions section, and a descriptive listing of all available commands.

---

[1] http://bima.astro.umd.edu/bima/wip/wip.html

[2] http://bima.astro.umd.edu/bima/home.html

[3] http://bima.astro.umd.edu/bima/miriad/miriad.html

[4] ftp://ftp.astro.umd.edu/progs/morgan/

[5] http://bima.astro.umd.edu/bima/wip/manual/wip.html

130   Morgan

2. The WIP Package

WIP has a very simple user interface which easily provides high quality graphics output. However, in addition to the simplicity of the command line interface, there are four characteristics that make WIP quite powerful. The first of these is probably most important to astronomical researchers: WIP has the ability to read two dimensional images and can sense the image type (Miriad, FITS, IRAF, etc.) automatically. In addition, most image types contain headers that WIP can use when displaying the image as a halftone or contour plot. This greatly simplifies one of the most difficult plotting tasks astronomers face: overlaying of images with different resolutions and spatial extents.

Next, WIP has variables that may be assigned to arbitrarily difficult expressions. These variables allow the user to write command files in a generic way so that they evaluate expressions internally rather than work with hard coded numerical values. Also, the default variable list can be expanded by dynamically creating or freeing additional variables as needed.

WIP also supports user definable macros. Macros can be defined and edited at run time or defined in, and read from, external files. These files can also be automatically loaded each time WIP is started providing a way to expand easily the number of available commands. Additionally, macros (along with the ability to pass arguments) provide a way to write, in a compact way, a collection of repetitive commands and also provide another way to simplify the process of writing, debugging, and enhancing plot files.

Finally, plotting scripts may be controlled internally by a conditional command and a looping construct. The *if* command permits conditional execution of commands. The *loop* command simplifies the commands needed to execute a macro or command multiple times. Both of these constructs (along with the user variables) permit complex graphics to be written quite easily.

2.1. Examples

There are many other features available within the WIP package. Rather than describe or list each feature, some will be illustrated in the examples that follow. Space limitations make it impossible to explain each example in depth; the interested reader will find more details in the manual distributed with the package.

Figure 1 illustrates three of the four major characteristics described above: user variables, macros, and looping. The use of these traits minimizes the number of commands that would otherwise be necessary to generate the figure.

Commands needed to generate Figure 1

```
define flippanel         # $1=Panel #; $2/3=Nx/Ny.
set \0 ($1 % $2) + 1     # Index in the x-direction.
set \1 ($1 \ $2) + 1     # Index in the y-direction.
set \0 \0 + ($2 * ($3 - \1)) # Flip the y index direction.
panel -$2 -$3 \0         # Set the panel.
end

define dosymbol          # $1=Counter; $2=Nx; $3=Ny.
set $1 $1 + 1            # Increment the Counter.
flippanel $1 $2 $3       # Set up the panel.
```

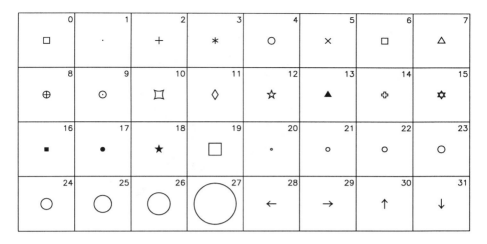

Figure 1. A plot of the standard symbol markers used by WIP.

```
box bc bc              # Draw the frame for this panel.
expand 0.5             # Set a small character size.
mtext T -1.1 0.97 1.0 \[$1]
expand 1.0             # Reset the character size.
symbol $1              # Chose this symbol.
move 0 0               # Move to the center of the panel.
dot                    # Draw the symbol.
end

limits -1 1 -1 1       # Set the limits.
set nsig 0             # Use integer format.
set \10 -1             # Initialize the loop counter.
loop 32 dosymbol \10 8 4   # Draw each symbol.
```

Figure 2 illustrates WIP's ability to handle images. One of two data sets used to generate this figure is an image from the VLA (FITS format) and the other from the BIMA array (Miriad format). The point of this figure (de Geus 1994) is to illustrate the ease at which two images with different resolutions and spatial extents can be overlaid.

Commands needed to generate Figure 2

```
levels -1 1 2 3 4 5 6 7 8 9 10 11 12 13 # Set the contour array.
slev A 25.0            # Contours now at -25, 25, 50, etc.
ticksize 10 10 60 6    # Change the default tick intervals.
image s127.6cm         # Read in the VLA FITS image.
winadj 0 nx 0 ny       # Set the window's aspect ratio.
header rd rd           # Read the image header and limits.
halftone 0.003 0.1     # Draw the halftone figure.
image s127.co 1 0.0    # Read the BIMA Miriad image.
header rd rd           # Read the image header and limits.
contour                # Draw the contour plot.
box bcnsthz bcnstvdzy  # Draw a box with RA/Dec labels.
```

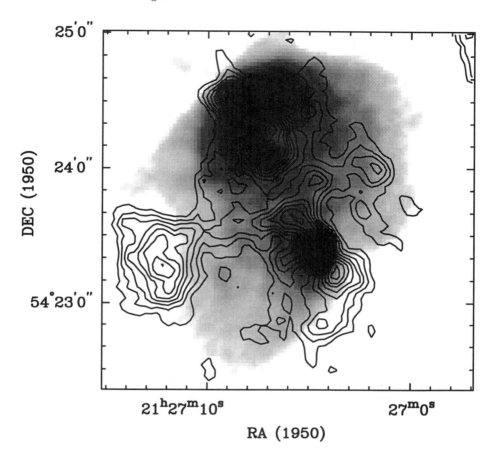

Figure 2. A 6 cm VLA image of the HII region S 127 overlaid with contours of the integrated CO emission observed with the BIMA array.

```
xlabel RA (1950)
ylabel DEC (1950)
```

References

de Geus, E. J. 1994, in 3rd Annual October Astrophysics Conference in Maryland, Back to the Galaxy, AIP Conference Proceedings, Vol. 278, eds. S. Holt and F. Verter (New York, American Institute of Physics), p. 481

Sault, R. J., Teuben, P. J., & Wright, M. C. H. 1995, this volume, p. 433

Astronomical Data Analysis Software and Systems IV
ASP Conference Series, Vol. 77, 1995
R. A. Shaw, H. E. Payne, and J. J. E. Hayes, eds.

Providing a Common GUI to Image Processing Tasks under PCIPS

O. M. Smirnov

*Institute of Astronomy of the Russian Academy of Sciences
48 Pyatnitskaya Str., Moscow 109017 Russia*

N. E. Piskunov

*Joint Institute for Laboratory Astrophysics
University of Colorado, Boulder, CO 80309-0440*

1. Introduction

The PCIPS image processing system (Smirnov & Piskunov 1994) has been in development since 1991; the current version is 2.01. PCIPS is a PC-based platform for all kinds of astronomical image processing and data-analysis. Key features of the system include: (1) an intuitive graphical user interface (GUI), (2) all functionality is provided by external application modules, which are easily added to the system as they are developed, (3) there is an application program interface (API) for development of custom applications, and (4) the code runs on practically any Intel-based PC.

As of now, several application packages have been developed under PCIPS. These include: (1) basic image processing operations (mathematics, statistics, filtering, fitting, geometric transforms, export and import of standard data formats, etc.), (2) PCDAOPHOT II, a GUI-equipped version of the DAOPHOT II (Stetson 1991) stellar photometry package (finding objects, aperture photometry, PSF construction, simultaneous PSF fits), (3) CCD/Echelle processing package (flat fielding, cosmic ray hit detection, finding and extracting spectral orders), (4) a spectral analysis package (dispersion curve fitting, continuum normalization, multiple line profile fits), (5) a Fourier analysis package (FFTs, power spectra, filtering in the Fourier domain), and (6) DASHA, a globular cluster/field photometry package (Smirnov & Ipatov 1995).

All of these very different packages have a consistent and powerful GUI that is common to all PCIPS applications. This paper reviews some issues that arose when developing the GUI.

2. Primary Design Goals

Several design goals were formulated at the initial stage of development. First and foremost was *expandability:* the system would not be limited to a set of built-in image processing operations. All image processing functionality was to be provided by external modules (called *applications*), loadable on demand. The upshot of this decision is that the user interface for the applications became the interface of the whole system, since there is not much the user would do in PCIPS apart from running applications. Thus the second design goal emerged— making the user interface consistent across all applications, both existing ones, and those yet to be written. A third goal was making the interface as simple and easy to use as possible (substituting development time and CPU load for astronomer, or end user, time), while still providing enough features to support a great variety of applications.

A simple solution such as a single (usually huge) form with all the possible parameters, to be filled out by the user when starting an application, was unacceptable. A lot of applications (or, to be exact, their users) would benefit from a truly interactive operation. Finally, a fourth goal was to provide future compatibility with a visual programming metaphor (Smirnov & Piskunov 1993) to be implemented in an advanced version of PCIPS, which requires elements of the user interface to be modular and connectable by pipes.

3. GUI Implementation

Who Provides the GUI? A fundamental design decision was to off-load the task of providing a user interface from individual applications onto the system itself. The decision killed two birds with one stone: interface consistency was assured, since only one common program was handling the GUI, and application development time was cut dramatically by freeing the programmer from the mundane task of baby-sitting the user.

A PCIPS application interacts with the system using a set of AIS (Application Interface Services, the application program interface, or API, of PCIPS) calls. It never communicates with the user directly. Instead, PCIPS translates AIS calls into user interface objects, maintains them, and reports the results back to the application. At the core, AIS is object-oriented. However, we did not want to implement a true object-oriented (OO) application interface in C++, for the sake of compatibility with non-OO languages such as C and FORTRAN. The application programmer works with the more conventional concept of handles, while inside PCIPS the handles are converted into true objects.

Example 1: Viewports. To display images on the screen, an application first makes one AIS call to create a viewport object. Next, it calls another function to display an image in the viewport. PCIPS displays the image, and automatically outfits the viewport with GUI elements that provide visualization tools (zooming, re-coloring, etc.) The application need not know anything of the GUI that goes with a viewport; all it needs are two handles returned to it by AIS, one to a viewport object, and the other to an image object. The functionality of a viewport is completely hidden inside the viewport object.

Example 2: Dialog Boxes. Another example is user input. When the application needs a set of parameters, it calls AIS to create a dialog box object, then requests the necessary parameters, then tells AIS to realize the dialog box. PCIPS decides where and how to display the dialog box, outfits it with standard buttons and tools, allows the user to enter the parameters (providing all sorts of useful gizmos to make the task easier), then reports their values back to the application. All the application has to specify are the names and types of the parameters (and where to store their values), and the name of the dialog box, if necessary. All it needs to know from AIS is the handle of the dialog box object, which it uses in the call to realize the dialog box. It does not need to know what a dialog box looks like, and what the user can do with it.

A big advantage of isolating the user from the application in this manner becomes apparent when we consider the visual programming metaphor, in which individual applications are connected by pipes that move data between them. With AIS, it does not matter to an application whether a parameters was actually requested from the user, or arrived via a pipe from another application. Therefore, all existing packages need little or no modification to adapt to the new metaphor.

4. Future Plans

A completely revised version of PCIPS is currently in development, targeted at UNIX/X11, Windows NT and DOS/Windows95. Some features of the new system are: (1) an unconventional object-oriented database, supporting all sorts of data objects and relations between them. It has a fully extendible data interface, and new data classes will be easily derivable from existing basic classes (i.e., images, arrays, tables, lists, etc.); (2) more visualization capabilities, implemented as properties of each data class (and thus easily extendible as well); (3) an overhauled API, with a separate, truly object-oriented C++ version, with support for more user interface objects and features; and (4) use of the visual programming metaphor.

4.1. Feedback

PCIPS is a commercially distributed product. For details, contact Oleg Smirnov (e-mail: *oms@inasan.rssi.ru*). We welcome any comments and questions both regarding the current version of PCIPS, and the one in development. We are maintaining an e-mail distribution list for the PCIPS Electronic Bulletin (Peb), ask Oleg Smirnov if you want to sign up. Also, visit our anonymous ftp host, *pcips.inasan.rssi.ru* (193.232.30.12). The directory */pcips* contains everything related to the system. A trial version can be picked up from */pcips/trial*.

Acknowledgments. We wish to thank ST ScI for providing the financial support that made this presentation possible. DAOPHOT II was ported with the kind permission of the original author, Dr. P. Stetson.

References

Smirnov, O. M., Piskunov, N. E. 1993, in Astronomical Data Analysis Software and Systems II, ASP Conf. Ser., Vol. 52, eds. R. J. Hanisch, R. J. V. Brissenden, & J. Barnes (San Francisco, ASP), p. 208

Smirnov, O. M., Piskunov, N. E. 1994, in Astronomical Data Analysis Software and Systems III, ASP Conf. Ser., Vol. 61, eds. D. R. Crabtree, R. J. Hanisch, & J. Barnes (San Francisco, ASP), p. 245

Stetson, P. B. 1992, in Astronomical Data Analysis Software and Systems I, ASP Conf. Ser., Vol. 25, eds. D. M. Worrall, C. Biemesderfer, & J. Barnes (San Francisco, ASP), p. 297

A Graphical Planning and Scheduling Toolkit for Astronomical Spacecraft

S. C. Kleiner

Smithsonian Astrophysical Observatory, 60 Garden Street, Cambridge MA 02138

Abstract. A small yet powerful planning and scheduling toolkit has been built for the *Submillimeter Wave Astronomy Satellite (SWAS)* Small Explorer spacecraft. It makes extensive use of graphics to illuminate the planning and scheduling process. The simple design, minimal resource requirements and easy extensibility of the *SWAS* planning and scheduling toolkit should make it useful for other space astronomy missions. A release of the toolkit for general use is planned shortly.

1. The SWAS Mission

SWAS is a NASA Small Explorer spacecraft to be launched in low Earth orbit in 1995. It will investigate the chemistry and energetics of star forming molecular clouds via the simultaneous observation of the O_2, C I, H_2O and ^{13}CO spectral lines in the 487–557 μm (538–615 GHz) range. The mission was proposed by the Smithsonian Astrophysical Observatory in Cambridge MA, which has the responsibility for the scientific component of the mission. The mission is managed by the Goddard Space Flight Center (GSFC).

The science instrument consists of a 0.65 m dual receiver radio telescope with an acousto-optical spectrometer backend. The spectrometer is read out every two seconds for the life of the mission, producing 100 MB of raw data every day. *SWAS* will observe 50–100 targets per day. The minimum planned mission duration is two years.

SWAS is the first astronomical Small Explorer, a series of missions to be developed under a "smaller, cheaper, faster" imperative. The turnaround time for *SWAS*, for example, should be about five years from acceptance of proposal to launch. The SAO Science Operations Center responsible for the development and operation of the science ground system consists of six scientists, including Principal Investigator Gary Melnick and Project Scientist John Stauffer. The planning toolkit described below was designed and written in two years by the *SWAS* Planning Scientist.

2. The SWAS Planning and Scheduling Toolkit

This stand-alone toolkit provides all the planning and scheduling functions for the *SWAS* spacecraft, including processing of the NASA predictive ephemerides, target visibility calculations, long range planning and short term (orbit-to-orbit) scheduling, slew constraint checking, nominal roll calculations, guide star selec-

Planning/Scheduling Toolkit for Astronomical Spacecraft

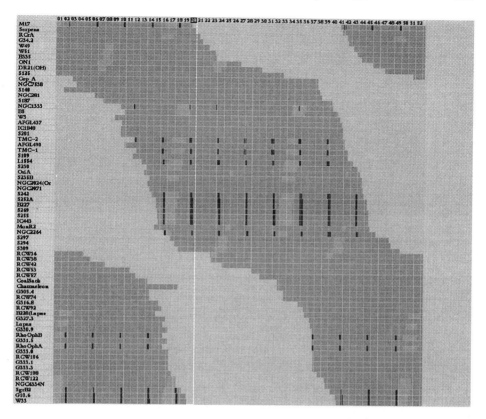

Figure 1. One-year planning display for week 20. Fifty-two weeks run across the top of the display. The shading indicates days when a target is visible or when Earth, Moon or Sun constraints are violated.

tion, and generation of detailed spacecraft timelines for conversion into command uploads. The toolkit displays its calculations graphically and makes extensive use of coordinate transformations in order to avoid any brute force calculations, a concept recognized by David Koch of NASA/Ames Research Center in his development of a prototype scheduler for *SWAS*. A new guide star catalog for CCD star trackers has also been developed (Stauffer 1993). The toolkit is currently generating timelines to support pre-launch testing of the flight operations facilities at GSFC.

The toolkit has a minimalist design, consisting of independent tools or 'filters' which operate on a single stream of scheduling events. Events include orbital ascending node crossings, the rising or setting of a target above the Earth horizon and the entry and exit of a target into pointing avoidance regions around the Sun and the Moon. The toolkit is extended by defining new events and adding the appropriate filters. Since *SWAS* uses the planets for calibration, the toolkit can also schedule planetary pointings. Calculations are done in orbit relative time rather than absolute time to minimize the effect of uncertainties in

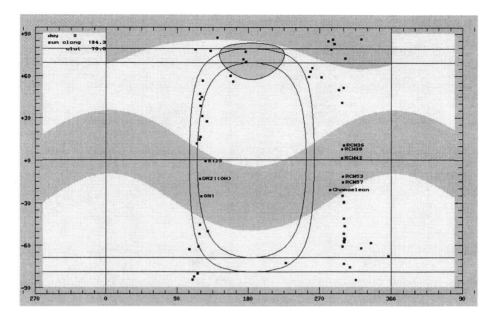

Figure 2. Pointing Constraint Display. The horizontal scale is orbital longitude, the vertical scale orbital latitude. The labeled targets in the central swath satisfy the Sun, Earth and Moon pointing constraints.

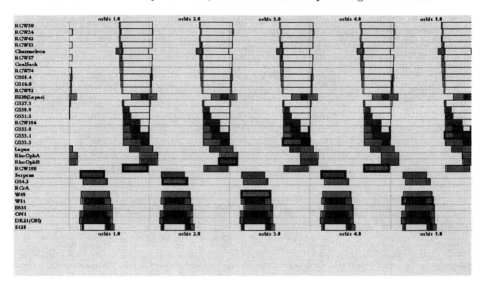

Figure 3. Scheduler Display for Five Orbits. The rectangles are target rise-and-set events shaded according to their scientific value and scheduling efficiency. The heavily outlined targets have been selected for scheduling.

predictive ephemerides. The toolkit is fast enough that the same tools are used both for long range planning and short term scheduling. The planning toolkit consists of about a dozen tools, less than ten thousand lines of ANSI C code in total. It makes only plain Xlib calls for the graphics and does not make any Unix system calls.

3. SWAS Planning and Scheduling Graphics

Figure 1 is a planning display showing target visibilities over the course of the year. The lighter bands running from top left to bottom right are days in which the target is too close to the Sun. (*SWAS* must point within 75°–105° of the Sun, and more than 40° from the Earth and more than 15° from the Moon.) We have found by experience that long term planning is driven primarily by the position of the Sun with respect to the target. Therefore, for any given week targets going into the Sun soonest are given the highest scheduling priority.

Figure 2 shows pointing constraints as seen from the orbital plane. The Sun is near the north orbital pole and the lighter swaths are regions of the sky which violate the *SWAS* Sun pointing constraint. The small black squares are potential targets, but only those which satisfy the pointing constraints at some time in the orbit are labeled by the software.

Figure 3 is the interactive scheduling display showing five orbits. Each rectangle represents the rise and set of a target during an orbit. The rectangles are filled and shaded according to the target's scientific interest and scheduling efficiency. The scheduling scientist uses a mouse to point and click to select or de-select a target for inclusion into the timeline sent to the spacecraft, as indicated by their heavy black outline around selected targets.

4. Distribution of the SWAS Planning and Scheduling Toolkit

The relatively severe time and manpower constraints for the development of the *SWAS* Planning and Scheduling toolkit have forced it to remain small and simple. Nonetheless, it had to be powerful enough to support production scheduling for the *SWAS* spacecraft and flexible enough to do year-long mission planning.

During the development of the toolkit, we realized that these attributes, together with its ease of modification and extension, should make it useful for other space missions and for astronomers developing future missions. We have received a small NASA Astrophysics Data Program grant to package the toolkit for general distribution. We anticipate a release of the toolkit shortly. Please contact the author for sample products and more information.

References

Stauffer, J. 1993, "Creation of a Guide Star Catalog for the BASG CT-601 Startracker," SWAS Technical Memorandum

A Graphical Front End for the WIYN Telescope Engineering Data System

J. W. Percival

Space Astronomy Laboratory, 1150 University Avenue, Madison, WI 53706 USA

Abstract. The WIYN 3.5 m Telescope Control System features a platform independent, network transparent protocol for distributing telescope control data to both local and remote sites. This Engineering Data Subsystem (EDS) can be run in both a burst mode, providing a short (10 s) sampling of 5-axis servo data at the full 200 Hz bandwidth of the system, and a slower (1 Hz) archive mode, which is used to monitor and record control system data during a whole operating session.

The EDS provides over 200 telescope monitors of use to astronomers, engineers, and system designers including pointing performance data, servo and mechanical data, and software state data. The general popularity of this data resource called for a easy way for casual users and experts alike to select data files and particular telescope monitors, graph time-series data, and do simple data manipulations such as pan, zoom, differentiation, and Fourier Transforms. In addition, the inherently distributed nature of the EDS called for a solution that was portable across platforms and operating systems that were in common use by any of the WIYN consortium members.

We have developed a simple but powerful Graphical User Interface to satisfy these needs. It uses the Tk graphical toolkit language and common UNIX tools such as shell scripts, grep, and awk to achieve a lightweight but effective solution to the problem of accessing the EDS. The data selection portion of the interface is separate from the graphical manipulation part, which allows both standalone use of the graphing widget in other applications as well as using it with different data selection front ends designed for other telescope systems and scientific instruments.

1. Summary of the WIYN Engineering Data Subsystem

The WIYN Engineering Data Subsystem (EDS) is a spacecraft-style telemetry system that sends telescope telemetry (encoders, currents, voltages, pointing data, etc.) in a machine-independent format to clients connected across a network. It has these features:

- It is content-programmable. The data are arranged into Engineering Data Records (EDRs) according to data type (e.g., azimuth servo data share a record, secondary mirror data share a record, etc). Any combination of EDRs can be turned off.

- It is rate-programmable. Each EDR can be programmed to appear at a certain rate, (e.g., pointing data once per second, temperatures once per minute, etc).

- A 10-second, 20 Hz "burst mode" can capture all pointing data in the outer pointing loop.

- A 10-second, 200 Hz "burst mode" can capture all servo data in the inner pointing loop.

- All-night telemetry archives contain every command and error message that passed through the system, and once-per-second sampling of over 200 telescope telemetry monitors.

The various data file types (2 burst modes, plus archive mode) and large number of telemetry monitors, not all of which are in each file type, required a tool to lead engineers and scientists through the dazzling array of choices, and zero in on what they wanted to see. The design requirements were: that the software be lightweight, without a large budget; it must be portable, and run on a number of platforms (i.e. SunOS, DECstations, and Vaxen); and that it must be flexible, so that quick changes and rapid prototyping is possible.

We chose Tk by John Ousterhout at UCB for the overall GUI capability and **blt_graph** by George Howlett at AT&T for the graphing widget. For maximum flexibility, we broke the problem up into two parts: The first is the *chooser*, which helps the user select a file type, file name, data type, and data name; and the second is the *grapher*, with which the user plots, pans, zooms, and prints data. The grapher is a separate program, which allows it to be used apart from the chooser, i.e., as a stand-alone department utility.

2. Off-the-Shelf Construction

Tk and Tcl are ideal for this sort of low-effort, high-impact tool. We already had differentiators and FFT programs laying about that accepted **graph(1)**-like ASCII input, and a graphical scripting language living close to the UNIX shell allowed a quick, painless integration of these existing tools into readily available free-ware. The *chooser* uses simple filename globbing to fill the listbox, and uses **grep(1)** and **awk(1)** to cut and paste the ASCII data file. The *grapher* uses standard Tcl string operations to test for two-column (suffix .xy) data or one-column (suffix .y) data, with an optional .Z denoting compression.

3. tkeds: File Name and Telemetry Chooser

tkeds helps the user choose a telemetry file and a specific telemetry monitor to examine. The left panel presents the three file types: 20 Hz burst file, 200 Hz burst file, and all-night archive. The left-hand listbox shows all choices for the selected file type. The listbox changes when the file type changes. The right panel presents the major data classifications: telescope axes, mount control electronics, Optical Support Structure, Instrument Adapter Subsystem, and miscellaneous data. The right-hand listbox shows the specific telemetry monitors available for the chosen data type. The listbox changes when the data type changes. The "plot" button produces a simple ASCII data file, one coordinate pair per line, and then launches the graphing application.

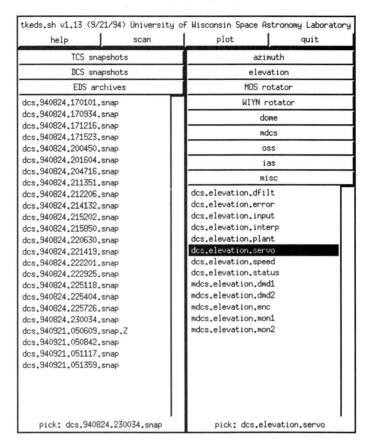

Figure 1. The chooser. This application lets the user select a data file and telemetry monitor, and then launches a graphing application for data display and processing.

4. tkgraph: Data Graphing Application

tkgraph is the graphing application. Its primary input data format is like the UNIX **graph(1)** program: simple ASCII files, one coordinate pair per line. It also can read ASCII files with one value per line, assuming a simple sequential abscissa. Finally, compressed versions of each of these file types are allowed. Plot styles include unconnected points, connected points, and a histogram-style plot, emphasizing the sampled nature of the data. Data processing includes pan and zoom, differentiation, and Fast Fourier Transform. Output options include PostScript disk files and line printer spooling.

WIYN Telescope Engineering Data System 143

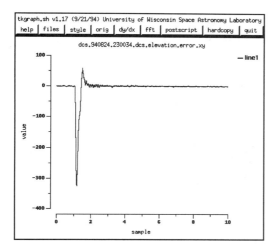

Figure 2. The grapher. This application supports panning and zooming, as well as differentiation and Fourier transformation. The data set shown here represents the elevation servo error (85 counts = 1 arcsecond) during a disturbance test in which a 20-lb weight was suspended then cut from the telescope during tracking.

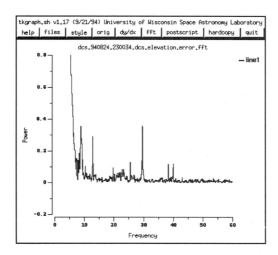

Figure 3. This figure shows a zoomed detail of the Fourier Transform of the data in Figure 2. Structure resonances are clearly displayed for the servo and mechanical engineers.

Space and the Spaceball

R. Gooch

Australia Telescope National Facility, CSIRO, P.O. Box 76, Epping, N.S.W., 2121, Australia, and Macquarie University

Abstract. The vast quantities of data produced by modern radio telescopes have outstripped conventional visualization techniques available to astronomers. ATNF staff have developed new visualization techniques to give astronomers a greater intuitive insight into their data. While visualization techniques in other areas find some application in astronomy, problems peculiar to the field require new techniques, such as methods for identifying three-dimensional regions. This paper presents an overview of some of the problems of visualization for astronomy and describes experiments with the Spaceball, a three-dimensional pointing device.

1. Volume Rendering

Visualization of three-dimensional data sets has already been researched and implemented in fields such medicine, for visualizing three-dimensional CAT scans. Great progress has been made through the use of volume rendering tools. These tools allow a three-dimensional data set to be displayed on a two-dimensional display (the computer monitor), with controls which the user can rotate and so obtain different views.

These techniques may also be applied to astronomy. While medical visualization deals with data which truly represent three spatial dimensions, radio astronomy spectral-line data sets represent two spatial dimensions and one frequency dimension. This does not prevent the data from being displayed as if they were a three-dimensional object, but the astronomer needs to be aware that the display is merely a representation of the data.

At the Australia Telescope, we have had considerable success using volume rendering. While we have experimented with both surface rendering as well as volumetric rendering tools, we have found the latter to be more successful. The focus of this technique is to make best use of the astronomers' spatial recognition functions. By presenting data three dimensionally, the astronomer may identify structure in the data which is not obvious when using conventional tools. Once this structure is identified, the astronomer may then proceed to further analysis.

1.1. Limitations of Standard Volume Renderers

Many existing volume-rendering tools are designed for visualization. The objects of interest are solids and fluids which are purely absorptive. In contrast, objects observed in radio astronomy contain regions of emission and absorption. Volume-rendering tools for radio astronomy need to take account of this fact. Furthermore, most astronomical data contain noise, which further differentiates them from medical data. However, even simple shaders are far more effective in revealing structure than the techniques traditionally used by astronomers. While these techniques are suitable for revealing two-dimensional structure, the astronomer missed much of the three-dimensional structure. We find that a radiative transfer ("hot gas") shader is particularly helpful for volume-rendering of astronomical data.

Another limitation of standard volume renderers is the lack of analytical software. While in many cases medical specialists are content with a qualitative assessment and rely solely on visual inspection of images, the astronomer depends far more on quantitative analysis. Once a feature is identified in a multi-channel data set, quantitative measurements are required to determine the physical processes at work. Some of these measurements are simple (such as identifying the frequency extent of an emission), while others require complex processing to obtain a meaningful result.

1.2. Identifying Structure

Once the astronomer has visually identified a feature of interest in the data set, some means of defining that feature is required. Therefore, a way of identifying points in three-dimensional space is needed. Merely using a two-dimensional pointing device (e.g., a mouse) in conjunction with a two-dimensional projection (a "view") is insufficient. Some means of moving and displaying a three-dimensional cursor is required. Once this problem is solved, a visually identified feature may be related to the analysis software.

2. The Spaceball

The Spaceball is a three-dimensional pointing device, available from Spatial Systems, Inc. It is a force-sensing device which provides six parameters: three orthogonal forces and three orthogonal torques. I have coupled the Spaceball to a volume-rendering tool, allowing the user to rotate the volume in an intuitive manner. In addition, users can move a three-dimensional cursor through the volume, allowing them to identify three-dimensional regions of interest. While the positioning tools available in immersive virtual reality environments (such as a high-quality data glove) are far more advanced, they are also far more expensive. The Spaceball and similar devices provide a cost-effective means to position cursors in three-dimensional space.

When the user pushes the Spaceball in a particular direction, a three-dimensional cursor is seen to move inside the volume. From experiments, we have found that depth placement of the cursor is rather difficult. A simple solution is for the user to set the placement in the X and Y directions, then rotate the cube by 90° and set the placement in the remaining direction. Clearly, this is still a somewhat cumbersome interface. An improvement may be obtained by rendering the cursor as part of the data, rather than overlaying the cursor on top of the rendered volume. This method is particularly effective when moving the cursor behind thin opaque regions, as the placement is directly tied to the data, which is the ultimate goal. For placement relative to larger regions, other methods must be used to give depth cues to the user.

To assist the user, a wire frame is displayed, with color-coded lines projecting from the cursor (a simple three-dimensional crosshair) to the three orthogonal corner planes. By upgrading to a stereo display, we expect to provide the necessary depth cues to enable the cursor to be placed in three-dimensional space.

3. Applications of a Three-Dimensional Cursor

3.1. Extracting Quantitative Information

To address the problem of extracting quantitative information, the user must be able to define and extract sub-regions of the data set and process these with a wide variety of algorithms that provide measures of the physical processes in the observed astronomical object. Work is in progress to allow subsets of data to be passed seamlessly to the analysis tools (such as AIPS++).

3.2. Viewing Small-Scale Structure

To expose small-scale structure the astronomer needs to isolate a region of interest. One method is to integrate a "slicer" tool which allows the astronomer to view the three orthogonal slices which intersect at a specified position. While slicer tools have been available in some astronomical analysis packages for a few years, specifying the slices was done using three separate linear controls (knobs or sliders). To my knowledge, this is the first time an integrated input device such as the Spaceball has been used to control a slicer tool for analyzing astronomical data sets. Coupling this slicer tool with a volume-rendering tool allows the astronomer to specify a point in three-dimensional space relative to the overall structure while at the same time displaying small-scale structure.

3.3. Understanding Data

Most radio astronomy is not solely a matter of collecting, viewing and interpreting data. Many theoretical models exist which attempt to explain observed phenomena. Once an astronomer has identified structure in the data and proceeded to perform quantitative analysis, the data need to be compared with existing models. Various astronomical data analysis packages provide tools to do this. However, they are limited to one- or two-dimensional data sets.

A greater challenge will be to provide model-fitting tools for three-dimensional data sets using algorithms tuned to such data. New model-fitting algorithms need to be developed to take full advantage of emerging visualization technologies.

4. Results

The feedback we have obtained from astronomers indicates that they can extract more science from their data using the visualization tools we have developed than was previously possible. By taking advantage of the spatial recognition and integration powers of the brain (powers which are tuned for three-dimensional moving objects), features that would not appear using conventional techniques become readily apparent using volume-rendering tools. Objects that would appear as a number of faint, disjoint fuzzy patches in individual images appear as a clearly defined object in three dimensions. In a number of instances astronomers have found previously unknown features in their data when they used the new visualization tools.

Using the Spaceball to identify regions of interest is proving an effective means to extract quantitative information and focus on small-scale structure, especially when coupled with a slicer tool.

5. Future Work

Further into the future is the possibility of experimenting with fully immersive virtual reality environments. Virtual reality offers the potential to present far more information to the user's brain than current video display technology. With the capability to present more information will come the challenge to structure that information in a cohesive, meaningful way. For example, the current tools we are developing allow the user to view the data from the outside, using the controls to enhance and suppress regions of the data. Using virtual reality techniques, astronomers could walk through the data to regions of interest; this would give them more selectivity in viewing data.

6. Summary

Visualization of three-dimensional data sets for radio astronomy will continue to develop over the years. The field is currently in its infancy. While more-general visualization of three-dimensional data sets is at a slightly more advanced stage (but by no means mature), the problems unique to radio astronomy are challenging. Experiments with stereo displays, Spaceballs, analysis and modelling software may open up new vistas for the astronomer, and provide interesting and perhaps unexpected challenges for those developing visualization systems.

Acknowledgments. I thank Ray Norris and Tom Oosterloo for ideas and contributions to the visualization project.

Visualization of GBT Geometry

D. C. Wells
National Radio Astronomy Observatory, 520 Edgemont Road, Charlottesville, VA 22903-2475

Abstract. The Advanced Visual Systems (AVS) has been used to explore lines of sight through the complex geometry of the Green Bank Telescope (GBT). Perspective images computed by AVS from geometric descriptions of the components of the telescope have been used when selecting locations for the installation of laser rangefinders and when checking geometrical relationships along the optical path. The success of this application demonstrates the flexibility of the toolkit-plus-visual-programming paradigm for scientific visualization.

1. Introduction

NRAO has developed a laser ranging system[1] which will be used to measure the active surface of the GBT (Green Bank Telescope)[2] and to measure the orientation of the structure in order to improve the pointing. Some of the ray paths between the laser rangefinders and the retroreflectors are sometimes blocked by the structural elements of the GBT. The author conjectured that visualization software could be an aid in planning placement and operational strategy for the rangefinders and retroreflectors.

The output from the MSC/NASTRAN structural model of the GBT was loaded into the AVS (Advanced Visual Systems)[3] installation in the Charlottesville image processing laboratory, using "UCD Builder", an interactive Motif module for importing FEA and CFD data into AVS, which NRAO purchased from SciViz, Inc. (Concord, MA). The AVS software was configured, using its graphical programming tool, to render images of the structure. The azimuth and elevation of the models could be controlled. Objects could be introduced to represent laser rangefinders on the ground around the GBT. The AVS "camera" could be commanded to look from any location in the scene toward any another location.

The images produced by the AVS "geometry viewer" are in full color, at high resolution. In this paper, we can show only selected images, and only in monochrome.

---

[1] http://sadira.gb.nrao.edu:80/~laser/

[2] http://info.gb.nrao.edu/GBT/GBT.html

[3] http://avs.ncsc.org/HTML/ITD/IAC/IAC.html

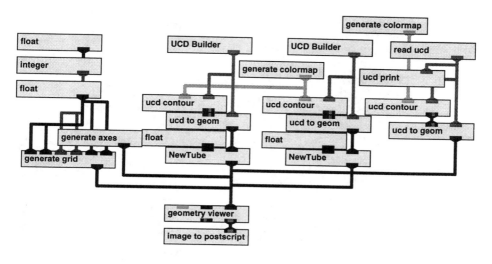

Figure 1. The AVS Network

2. The AVS Network

AVS is controlled by a graphical interface. Named boxes represent modules (programs), and they are connected by pipes, with the color code indicating the types of the data being passed from module to module. In the network shown in Figure 1, two different structural models (the alidade structure and the tipping structure) are being read by the two executions of the "UCD Builder" module, which converts NASTRAN output files to AVS "unstructured-cell-data" (UCD). Module "read ucd" reads a data file which describes a set of cubical objects which represent laser rangefinders. Several generator modules add axes and a grid. The outputs of these pipelines are combined as input to the "geometry viewer", which renders the imagery. AVS has a graphical editor tool with which the network can be modified interactively; this is the toolkit-plus-visual-programming paradigm which is used in several other scientific visualization systems (e.g. Khoros and IBM's Data Explorer) in addition to AVS.

3. Viewing a Rangefinder from an Elevation Bearing

In Figure 2, the GBT model is pointing northeast, resting on the grid of 25 m squares. The spot marked "laser" is a 1 m cube 1.5 m above the ground, 74 m from the pintle bearing, and about 90 m from the "camera". The camera is looking from the neighborhood of one of the alidade elevation bearings toward a single rangefinder on the ground. The simulated "rangefinder" is visible between the beams of the box structure which carries the backup structure and the feed arm of the GBT. The beams are being approximated by cylinders (using a version of the AVS "tube" filter), and the closeup view of the end of one of the principal beams of the alidade demonstrates that the cylinders are hexagonal. The sizes of the hexagonal cylinders of the alidade and tipping structures are

Figure 2. Viewing a Rangefinder from an Elevation Bearing

independently controllable, and are about 0.9 m and 0.4 m in this image. It appears that this potential ray path would have a duty cycle of at least 50% while the GBT is tracking.

4. Sample GBT Images Available on the Web

Color GIFs of GBT visualization images[4] are available. The first is an overview of the telescope model[5] as seen from several hundred meters distance and a height of about 100 m. There are 18 cubical objects in the scene, with sides

---

[4] http://info.gb.nrao.edu/GBT/Visualization.html

[5] http://info.gb.nrao.edu/GBT/visual_3_lg.gif

20 inches (0.5 m) long. The telescope is pointing at the zenith, and is resting on a grid with a spacing of 1000 inches (about 25 m). The second scene is an oblique view of the feedroom area from below[6]. The prime focus boom and the subreflector actuator model are above the room, and there are four 0.5 m cubes on the edges of the roof of the feedroom, representing suggested locations for rangefinders. The third scene shows a side view of the feedroom[7], as seen from a point at about the same height as the room. In this scene the two cubes at the rear corners of the feedroom represent proposed locations for rangefinders. The AVS "camera" was moved to these cube locations and pointed toward the big primary mirror in order to examine the field-of-view (which is partially blocked by the feedroom) as seen from these locations. It was possible to confirm that the two rangefinders, operating together, will be able to view the entire mirror from the locations shown in this scene.

5. Checking Feedroom-to-Primary Clearance

An important specification of the GBT is that the edge of the beam formed by the primary mirror should clear the feedroom by at least 20 cm, in order to eliminate sidelobes produced by scattering. In order to check this feedroom-to-primary clearance condition[8] and to check that two proposed rangefinders would also be clear of the beam, the camera was commanded to look almost straight down from the tip of the prime focus boom (i.e., the prime focal point of the primary), toward the roof of the feedroom and the inner edge of the big dish. The fact that the beam was *not* clear in the visualization scene led to an engineering review of the detailed construction drawings to assure that the as-built telescope will indeed have a clear beam. (The geometry of the feedroom as used in the structural model, and therefore in the AVS rendering, was a simplified abstraction.) In the author's opinion, this important case justified the money and effort expended to make the AVS visualization software work with the GBT structural model.

---

[6] http://info.gb.nrao.edu/GBT/visual_2_lg.gif

[7] http://info.gb.nrao.edu/GBT/visual_4_lg.gif

[8] http://info.gb.nrao.edu/GBT/visual_1_lg.gif

Part 4. Archives and Databases

What Happened to the Results?

S. G. Ansari

ESIS, Information Systems Division of ESA, ESRIN, Via G. Gallilei, C.P. 64, Frascati, Rome 00044 Italy

A. Micol

ISO Project Science Team, Astrophysics Division of ESA, ESTEC, P.O. Box 299/SA, Noordwijk, 2200 AG The Netherlands

Abstract. The archiving process stops at the point where a Principle Investigator program or a survey that produces a catalogue has been carried out. No further contact to the actual observer(s) is kept after they have received their data. In recent years only one single archive has made an effort to scan the publications and collect references for each dataset to build a publications archive. The *International Ultraviolet Explorer* (*IUE*) database provides these links between the actual observation and the publication reference. In this paper we propose the establishment of a link between the actual observation made and the science that has been carried out with it. The merit of having such a link is obvious. The current trend has only gone into providing quality flags in mission logs giving some indication as to whether an acquisition is useful or not. If, however, a publication is linked to the actual observation, an archive user may immediately refer to a particular paper that may provide more details as to what science may have been achieved with it. This way, archives can be used more efficiently and a judgment on using data in an archive can be made on the basis of previous work carried out.

1. Introduction

Today's astronomical mission archives tend to provide data pertaining to the actual acquired observation. Object names, instrumentation used, and observation dates are typical entries found in almost all mission archives. Even future mission archives tend to concentrate more on what can be extracted out of a proposal or a Guaranteed Time Observation log than the actual extracted science out of the data itself.

The archives of today stop after the Principle Investigator has acquired the data. A year later the data is then publicly available. An archive researcher has then to go through a mass of data to find what may be useful without any indication as to what the actual observer's results were while using the data except the proposal, which in most cases is very general.

2. The Example of IUE

The *IUE* mission has one of the oldest digital archives in astronomy (Barylak 1988). The effort has not only gone into providing the *IUE* mission log, but also to scan all the publications that referred to any *IUE* data product. As a practical example, we searched the *IUE* database for all the spectra of the Andromeda galaxy using the on-line database at the VILSPA observing station and at ESIS (Giommi et al. 1994). We list in Table 1 each spectrum found and in which article it has been referenced. Scanning the on-line abstracts in the ESIS bibliographic service, we find several interesting papers that shed some light on what parameters were derived and what science was achieved with the data.

3. A Proposed Policy

The only way to achieve this connection between a satellite mission and the science it produces is to make it a mission's policy that every data product be *explicitly* referred to in a publication. It would, however, simplify matters for archivists if Principle Investigators can provide *at least one* publication for each usable observation made.

Present mission archives should invest more time in accumulating published results. A common nomenclature such as that adopted by SIMBAD (Egret et al. 1992) could be used to refer to papers. These publications can then be easily used by systems like ESIS and ADS (Eichhorn 1994) to access the abstracts and from there on the printed publication.

4. Conclusions

From the *IUE* example it is obvious what the merits are, not only archiving the actual observation, but also archiving the publications related to it. Any future investigator will have a better understanding of the data and how reliable they may be. Another side effect is the justification of usage of an ongoing mission. It is not sufficient only to provide access statistics, since casual or curious users will normally be in the majority. The true usage of an archive can only be measured by the amount of scientific output achieved.

References

Barylak, M. 1988, The VILSPA Database Users' Guide, ESA *IUE* Newsletter, No. 37

Egret, D. et al. 1992, SIMBAD User's Guide & Reference Manual

Eichhorn, G. 1994, in Astronomical Data Analysis Software and Systems III, ASP Conf. Ser., Vol. 61, eds. D. R. Crabtree, R. J. Hanisch, & J. Barnes (San Francisco, ASP), p. 18

Giommi, P., Ansari, S. G., Donzelli, P., & Micol, A. 1994, Experimental Astronomy, in press

R.A.	Decl.	Object	Disp.	Observation	Spectrum	Publication
00 40 30.01	+40 50 59.8	M31	L	02-SEP-80	LWR8699L	A&A Vol. 0145, pg. 0296, 1985
00 40 30.01	+40 50 59.8	M31158	L	04-OCT-80	SWP10280L	APJ Vol. 0261, pg. 0077, 1982
00 40 03.01	+40 58 59.8	NGC224	L	23-JUN-79	SWP5610L	APJ Vol. 0259, pg. 0077, 1982
00 40 03.01	+40 58 59.8	NGC224	L	23-JUN-79	LWR4853L	A&A Vol. 0106, pg. 0016, 1982
00 40 00.01	+40 59 00.2	M31	L	03-DEC-78	SWP3520L	APJ Vol. 0230, pg. L137, 1979
00 40 00.51	+40 59 04.0	NGC224	L	01-JUN-79	LWR4665L	A&A Vol. 0145, pg. 0296, 1985
00 40 00.51	+40 59 04.0	NGC224	L	10-DEC-79	LWR6343L	APJ Vol. 0259, pg. 0077, 1982
00 40 00.51	+40 59 04.0	NGC224	L	10-DEC-79	LWR6343S	APJ Vol. 0328, pg. 0440, 1988
00 40 00.51	+40 59 04.0	NGC224	L	11-DEC-79	SWP7376L	APJ Vol. 0259, pg. 0077, 1982
00 40 00.51	+40 59 04.0	NGC224	L	12-DEC-79	LWR6368L	APJ Vol. 0259, pg. 0077, 1982
00 40 00.51	+40 59 04.0	NGC224	L	12-DEC-79	LWR6368S	APJ Vol. 0328, pg. 0440, 1988
00 40 00.51	+40 59 04.0	NGC224	L	12-DEC-79	SWP7376S	APJ Vol. 0259, pg. 0077, 1982
00 40 00.51	+40 59 04.0	NGC224	L	11-JUL-80	SWP9494L	APJ Vol. 0259, pg. 0077, 1982
00 40 00.51	+40 59 04.0	NGC224	L	12-JUL-80	SWP9502L	APJ Vol. 0259, pg. 0077, 1982
00 40 00.51	+40 59 04.0	NGC224	L	13-JUL-80	LWR8236L	APJ Vol. 0259, pg. 0077, 1982
00 40 00.51	+40 59 04.0	NGC224	L	14-JUL-80	SWP9519L	APJ Vol. 0259, pg. 0077, 1982
00 40 00.51	+40 59 04.0	NGC224	L	14-JUL-80	LWR8244L	A&A Vol. 0145, pg. 0296, 1985
00 40 02.01	+40 59 59.8	M31	L	03-SEP-79	LWR5502L	A&A Vol. 0106, pg. 0016, 1982
00 40 02.01	+40 59 59.8	M31	L	03-SEP-79	SWP6378L	APJ Vol. 0259, pg. 0077, 1982

Table 1. A list of *IUE* spectra of the Andromeda galaxy derived from the VILSPA archives and the ESIS system. For each entry, the Right Ascension, declination, name, dispersion, observation date, spectrum number, and literature reference are given.

The Hubble Space Telescope Data Archive

K. D. Borne, S. A. Baum, A. Fruchter, and K. S. Long

Space Telescope Science Institute, 3700 San Martin Drive, Baltimore, MD 21218

Abstract. We describe recent progress at the Space Telescope Science Institute[1] in the development of the *HST* Data Archive[2] and of StarView[3], the user interface to the *HST* archive system. Access to this system is available to users through one of our *HST* archive host machines[4] or through the local use of the distributed version of StarView[5].

1. Introduction

The *Hubble Space Telescope* (*HST*) Data Archive has been open for archival research since early in 1993. It contains all of the observational data, calibration files, and related catalog information produced by the *HST* since its deployment. The archive currently comprises 1.2 TB of data, of which 60% are science data, more than 80% of that being publicly available. A new archive engine, ST-DADS (Space Telescope Data Archive and Distribution Service), is now in use. ST-DADS is designed to maintain 3 TB, or about 8 years' worth, of *HST* data on-line in four optical disk jukeboxes. ST-DADS stores all important science data files internally in FITS-compatible formats and will eventually be able to deliver data directly to remote workstations. The archive also includes an on-line catalog, which can be browsed by any user via a connection to one of two archive host machines (VMS and Unix). StarView is the user interface to the *HST* catalog at ST ScI, and it operates on Sun Unix and DEC VMS machines. There are two versions of the interface available: one is CRT-based (VT100-compatible) and the other is X-based. The latter includes a data-previewing capability and all of the point-and-click features typical of Motif-based graphical user interfaces. A distributed version of StarView allows users to run StarView locally on their Sun machines. It has the same functionality as the version of StarView running on the ST ScI archive host Unix machine and reduces network loading by creating X-windows locally and by accessing ST ScI machines only to query the database.

---

[1] http://marvel.stsci.edu/top.html

[2] http://www.stsci.edu/archive-html/archive.html

[3] http://www.stsci.edu/archive-html/starview.html

[4] http://www.stsci.edu/archive-html/host_mach.html

[5] http://www.stsci.edu/archive-html/distributed_sv.html

2. What was DMF? What is DADS?

The Data Management Facility (DMF) was the prototype archive for *HST* data from launch until 1994 September 21. The system was developed by ST ScI, ST-ECF, and CADC, and it had a 170 GB on-line data capacity (Long et al. 1993).

The Space Telescope Data Archive and Distribution Service (ST-DADS) is the permanent archive for *HST* data. It has been operational since 1993 December (the time of the First Servicing Mission to the *HST*). The system was developed by Loral Aerosys and ST ScI for NASA, and it has a 3.4 TB on-line data capacity, which corresponds roughly to 8 years' worth of *HST* data.

3. DMF-to-DADS Transition

The transition to full archive operations using ST-DADS has involved a number of steps. First, a new user interface (StarView) was developed at the ST ScI to replace the previous interface (Starcat). Next, all *HST* data obtained prior to 1993 December were transferred from DMF into ST-DADS (900 GB total). While the DMF-to-DADS data transfer was taking place, DMF and ST-DADS were operated in parallel for an extended period (from 1993 December through 1994 September). As part of the data transfer project, the completeness and integrity of the ST-DADS data and catalog were verified. We determined that the first pass through the data (in the DMF-to-DADS data transfer) was 99.5% complete and 99.99% accurate (i.e., roughly one file in 10,000 had some kind of error in ST-DADS). The cleanup of these data got underway in 1994 August and is continuing.

As part of the transition, a retrieval mechanism for the data in ST-DADS was provided to users through StarView, and a bulk copying mechanism was also developed for the generation of ST-DADS optical disk platters for use at other *HST* data archive sites: currently, the ST-ECF and CADC.

4. Contents of the HST Archive

The *HST* Data Archive consists of observational data (the *HST* Archive) and derived data (the *HST* Catalog). We plan in the near future to add user-derived *HST* data and supplemental non-*HST* data which will enhance the scientific use of the *HST* data (e.g., the Guide Star plate scans, *HUT* data, VLA FIRST survey data, and other astronomical catalogs). All of the data stored in ST-DADS are in either FITS or binary format (e.g., GEIS and ASCII files are converted to FITS before archiving).

The *HST* Catalog consists of ~50 tables, including a science table, instrument tables, observation tables, target tables, engineering tables, and our internal archive bookkeeping tables. All of the scientifically interesting fields in these tables are accessible through the StarView user interface, either through fixed predefined forms or through the custom query mechanism in StarView (see below). As part of its extensive on-line help, StarView provides a description of the fields that it uses from the various database tables.

The current *HST* Archive data volume is 1.2 TB, comprising approximately 1.7 million individual files. The monthly ingest rate (for new data) into ST-DADS is ~30–50 GB, corresponding to a yearly ingest rate of ~300–500 GB.

The science data volume in ST-DADS is nearly 0.7 TB (60% of the total), comprising 65,000 science observation datasets. Approximately 83% of these science datasets are public. Since its launch, the *HST* has obtained data on ~5000 distinct astronomical targets. The number of science datasets in the *HST* Archive, by instrument, is: FOC: 4900; GHRS: 14,700; WFPC: 15,800; FOS: 14,000; HSP: 5200; and WFPC2: 10,500.

5. StarView—The HST Archive User Interface

StarView is the user interface to the *HST* Archive. It supports VT-100 compatible CRTs and X-windows, for both VMS and SunOS. A distributed version is also available for SunOS—this version uses network bandwidth only for database calls, not for any of the intensive X-window interactions.

User interactions with the StarView interface are through a variety of data screens. A "Quick Search" screen is provided (with direct access from the StarView "Welcome" screen), which can be used to initiate the most common type of informational searches of the *HST* Catalog—it is the most basic archival search screen, including only a few simple fields by which one can constrain a catalog search. After initiating a quick search of the catalog, the results are returned on a separate screen that contains numerous catalog entries related to each observation. In addition to the "Quick Search" screen, there are separate StarView screens for exposure parameters, target properties, proposal information, planned exposures, instrument parameters, calibration files, and engineering files. Each screen contains a wide selection of user-qualifiable fields connected to the archive database. For ease of maintenance and development of the user data screens, the screen definitions, formats, and contents are all kept "outside" of StarView software. StarView screens are used to search for specific datasets, to view the results of a query, to preview the data, to mark datasets for retrieval, and to submit the retrieval request. The results of a query may be viewed one record at a time (in portrait form) or many records at a time (in tabular form).

The principal science observation catalog information is collected into a single "science" table in the *HST* Catalog (to minimize database joins). StarView uses that table for most of its basic exposure search screens. Target searches are simplified in StarView by allowing the user to get target coordinates from either the NED or SIMBAD databases using a user-specified target name. An additional feature of StarView allows a user-provided target search list of RA and Dec pairs to be cross-correlated with the contents of the *HST* Catalog. This is particularly helpful when planning *HST* observing proposals or *HST* archive research proposals.

One of the more powerful features of StarView is a "custom query" interface, which allows the user to define (i.e., customize) their own query of the archive database, dynamically selecting any database fields of their choice to be included in the query. A corollary aspect of this feature is an on-line editor that allows the user to compose a personalized SQL query and to submit that query directly to the database server.

Of particular interest to researchers is a public data-previewing capability available through XStarView. Compressed images and spectra of all public *HST* data are available at the push of a button, allowing the user to see the stored image or spectrum prior to submitting a data retrieval request. (The compressed data are provided courtesy of the CADC.)

StarView provides users with a variety of on-line help, including "Strategy" help, which is available at the push of a button, for each StarView screen.

6. Access to the HST Archive

Two archive host machines are made available for external user logins (for VMS: *stdata.stsci.edu*, and for Unix: *stdatu.stsci.edu*). The username is **guest** and the password is **archive**. Both CRT-StarView and X-StarView are available on the host machines.

The number of logins to the ST ScI Archive hosts is currently 2500–3000 per month. There are ∼800 individual archive user accounts (∼600 of which are for non-ST ScI users), from which 15–20 GB of *HST* data are retrieved each month (∼30% by non-ST ScI users). Note: one must be a registered user of the ST ScI archive in order to retrieve data, but not to search the catalog or to preview public data.

Science data are retrieved for users in FITS format. Currently, retrievals are only supported on the two archive host machines: *stdata.stsci.edu* and *stdatu.stsci.edu*. In the future, ST-DADS will be set up to deliver data (via ftp) to a user-specified destination anywhere on the Internet.

To support users of the *HST* Archive, several forms of documentation are available: the *HST Archive Primer*, the *HST Archive Manual*, and the *HST Data Handbook*. In addition, archive hotseat support is available, either by sending e-mail to *archive@stsci.edu* or by phoning (410) 338-4547.

7. Current Status

As of this writing, StarView has been in successful operation for over a year. ST-DADS has archived all *HST* science, calibration, astrometry, and engineering data since 1993 December (i.e., since the First *HST* Servicing Mission). All of the science and calibration data stored in the DMF since launch have been transferred into ST-DADS. The transfer of all astrometry and engineering data from DMF into ST-DADS will be completed by 1994 December. Finally, archiving of new *HST* data to DMF was turned off in 1994 September. As a consequence of these developments, all access to the *HST* Data Archive has been through ST-DADS since 1994 October. The *HST* Data Archive is thereby operational and ready to support the *HST* data needs of the astronomical research community.

References

Long, K. S., Baum, S. A., Borne, K. D., & Swade, D. 1993, in Astronomical Data Analysis Software and Systems III, ASP Conf. Ser., Vol. 61, eds. D. R. Crabtree, R. J. Hanisch, & J. Barnes (San Francisco, ASP), p. 151

The Hubble Data Archive: Opening the Treasures of the HST to the Community

J. A. Pollizzi, III

Space Telescope Science Institute, 3700 San Martin Dr., Baltimore, MD 21218

Abstract. This paper briefly summarizes the history and present state of the archive systems for the *Hubble Space Telescope*.

1. DMF as the Hubble Data Archive - Launch through Present

The data to be collected by the *Hubble Space Telescope* (*HST*) promised to be a significant new resource to the astronomy community, even before *HST*'s launch. In plans made to capture data from *HST*, the capacity to accommodate the expected data volume and the astronomers' accessibility to the data were both considered important.

Prior to launch, the Space Telescope Science Institute (ST ScI) initiated a prototype program to explore technologies that could be used in archiving *HST* data. The prototype, called the Interim Data Management Facility (hereafter, DMF), was developed, collaboratively with the Space Telescope European Coordinating Facility (ST-ECF). DMF experimented with optical disk media, jukeboxes, database indexing of the data, and user interfaces. The ST-ECF took the lead in developing the user interface, and the well-known STARCAT program was the result of their efforts. It was expected that DMF itself would only be short-lived, with plans for a major software development contract to deliver a permanent archive system underway. However, when the contract on the permanent archive was delayed, DMF was pressed into operational use.

The DMF system, with a subsequent and substantive augmentation, operated as the primary archive from *HST* launch until early fall 1994. DMF was configured as a VAX-8650 with a Cygnet jukebox and LMSI optical drives. Each LMSI platter stored about 1 GB/platter; the jukebox held 76 platters, for a total on-line capacity of 76 GB. With the data rate from the *HST* at 1.5 GB day^{-1}, the DMF system could hold approximately two months of data on-line. As part of its ingest process, DMF would create a second "safe-store" platter that was shipped to the ST-ECF. An auxiliary process, developed by the Canadian Astronomy Data Center (CADC), copied selected data from DMF for CADC platters.

Within DMF, platters were in one of two states: they were either writable, or they were closed for writing and were only available for retrievals. Since platters generally filled-up with data within a day, there was only a slight (roughly 24 hour) delay in being able to access new data through DMF. Had a larger capacity disk been available for DMF, this delay might not have been as acceptable.

DMF was restrictive in that it only took data in the native ground-system format (GEIS). This would typically have to be converted to some standard format (i.e., FITS) for use by astronomers. DMF could only deliver data to internal Institute systems, which required remote users to log-in to an ST ScI system (called a "host" system) to retrieve their data and then manually copy the data back to their home site. Nonetheless, DMF well supported users from around world in accessing *HST* data.

2. STDADS, the New Hubble Data Archive

The Space Telescope Data Archive and Distribution Service (STDADS) was developed as the permanent archive by Loral Aerosys, under contract to NASA. Its initial version was released to the ST ScI in the spring of 1993. Since then, a joint ST ScI–Loral team, under the supervision of ST ScI, has been evolving and deploying the system. STDADS has been routinely archiving *HST* data since the first *HST* servicing mission in 1993 December. From 1994 February to 1994 November, all of the DMF data were installed into STDADS. STDADS was enabled as the operational archive in 1994 October.

STDADS is based on a cluster of Digital Equipment Corporation VAXes and Alphas. It utilizes four Cygnet 1803 Jukeboxes and nine SONY 930-series large-format optical disk drives. The SONY platters can each store about 6 GB of data; the new jukeboxes each hold 130 platters. This gives STDADS 3.1 TB of on-line capacity or about 5 years' worth of *HST* data at the current data rates.

Unlike DMF, STDADS converts all the data (as practical) to FITS format, prior to writing it to optical disk. This eliminates the need for astronomers to convert the data when they retrieve it. As with DMF, STDADS writes to two platters (a primary and a secondary) as it ingests data. Unlike DMF, STDADS can retrieve data from a platter while it is still open for writing. Once closed, the secondary platter is kept at the Institute at an on-site safe-store. An auxillary subsystem with STDADS, called "bulk distribution," prepares copies of the data for the ST-ECF and CADC in an off-line fashion from the secondary platters. This subsystem can be configured to recreate platters for any of the three sites (ST ScI, ST-ECF or CADC). It can also be modified to support future additional distribution sites, should that be required.

STDADS will also have the ability to deliver data, using standard ftp, to any system that is accessible via the Internet. This will eliminate the need for remote users to have to manually copy data from ST ScI "host" systems, although these systems will remain to support those users that require them. We have adopted the use of e-mail as the mechanism for requesting data from STDADS. A properly formatted mail message is sent to a defined account name and read by the STDADS software. Using this approach, STDADS is open to many different forms of interfaces as well as to our own; a paper by Richon (1995) more completely describes this interface.

3. StarView - the HDA User Interface

As mentioned above, ST-ECF developed STARCAT as a user interface to DMF. However, STARCAT was limited to principally being a CRT interface, and was layered on top of a large body of public code that was deemed unmaintainable. In the early 1990's, the ST ScI negotiated with NASA to develop a new user interface to the *HST* data archive (HDA). This interface, named StarView, was designed from its beginning to be a highly portable tool: specifically able to support both CRT and X-Window devices, and both VMS and UNIX systems.

StarView was developed using object-oriented techniques and coded in the C++ language. The use of these advanced tools has given StarView the flexibility it needs to meet the current and future environments of *HST* researchers. StarView has been supporting access to the HDA for about two years.

StarView employs many notable features, but perhaps the two most significant are the use of ASCII text definitions for all database and display definitions, and the use of a universal-relation SQL generator. By keeping all the definitions of databases and displays in text files, StarView is adaptable to any SQL database that accepts a callable interface. Currently, StarView can interface to any Sybase database. Changes to displays or to the database (or even to new databases) can be made without modifying the code. Moreover, the definitions can be made and maintained by non-programmers.

The universal relation SQL generator within StarView uses information in the database definition files to automatically determine the appropriate joins to be used in making a query. This eliminates the need for either the general user, or the form designer, to have knowledge of the database relations (or even of SQL) to construct working queries. Using these features, the displays, forms, menus, and even help have been carefully constructed by HDA Astronomers to best aid the *HST* GO or researcher in locating and using data within the archive.

4. HDA and the World Wide Web

The use of the Internet, and specifically the World Wide Web, has grown explosively over the past year. It is now expected that most of the information resources available to the public are reachable through some of the more popular WWW tools (e.g., Mosaic, Netscape, etc.). The ST ScI has long been a proponent of using the Internet for connectivity to the astronomy community (in particular) and those interested in *HST* in general. For some time, the Institute has supported WWW, Gopher, WAIS and basic ftp access to its resources. The HDA is also accessible through these resources. For example, pointers on the Institute's home page[1] provide links to the *HST* Data Archives. From there, the user is lead to information allowing them to start-up StarView on one of the host systems. They can initiate either a CRT or an X session.

Beyond using WWW as a reference pointer, we have experimented in providing basic archive services directly through a WWW supported interface. We developed a prototype WWW interface (Travisano 1995) that could initiate queries of the Science or Proposal tables (queries of the Science table can be based on an area about a sky position), and display data in tabular format for the science table (in portrait format for proposal data). Where the science data is public, it can be previewed if the appropriate supporting browser is installed on the user's system (e.g., SAOimage for images, and "xmgr" for spectra). Users

---

[1] http://marvel.stsci.edu/top.html

with archive accounts can mark selected public science data for retrieval from the archive.

This prototype was developed using the new forms features of HTML, with a custom "common gateway interface" program which was developed to accept the form's parameters. While the prototype was successful, with the easy availability of StarView, we are currently not pursuing an extension of the prototype into actual use.

5. HDA Futures

Certainly STDADS will continue to evolve in meeting the needs of *HST* researchers. In the near term, the ability to deliver data directly to users' home sites, and to respond to requests for data via magnetic tape are the next features to become available. Beyond that, we also have plans to provide access to the Guide Star catalog and other *HST*-related datasets as they become available to ST ScI.

StarView's features will also evolve. With the generic nature of StarView's interface to databases and forms, the Institute is seeking to connect StarView to other catalogs (and possibly other archives) that further enhance a researcher's ability to work with *HST* data. New means of specifying and viewing queries, especially with an eye to using the advanced graphical features of X, are planned. While the current focus will be on evolving StarView, continual attention will also be paid to the evolving nature of the WWW and its browsers, and its potential use with the HDA.

The collection of *HST* observations is a treasure trove for astronomers of today and a generation to come. The Hubble Data Archive is vested with the responsibility for gathering and disseminating this collection to the community. As an integrated set of systems, the HDA will grow and change to meet this responsibility and to participate in its own way, in mapping the future of astronomy.

References

Richon, J. 1995, this volume, p. 166
Travisano, J. 1995, this volume, p. 80

Designing an Open E-Mail Interface to a Data Archive

J. Richon

Space Telescope Science Institute, 3700 San Martin Drive, Baltimore, MD 21218

Abstract. An active and useful data archive often has to support two conflicting goals: providing an easy way to retrieve data, and preventing people from getting proprietary data. The retrieval method should be insulated from the user interface so that changes to one system do not force changes to the other. This paper discusses the issues associated with data retrieval and system isolation, and how the Space Telescope Science Institute developed a flexible and open means to retrieve data from the Hubble Data Archive that can support almost any user front-end system.

1. Introduction

In 1993 the Hubble Data Archive group reviewed the original design of the Space Telescope Data Archive and Distribution System (STDADS) for retrieving data. The design called for any user interested in retrieving data to have an account on the operational archive system, and to use a command line syntax to retrieve data. This made it impossible to use a query tool to easily retrieve data. After reviewing the current system, we decided to create an open interface to the retrieval system for the archive.

2. Design Goals

Our first goal was to make the new system simple so that it would be easy to get public data out of the archive. Next, it had to be open so that almost anybody could get public data. Yet it had to be secure so that proprietary data would never be released to the wrong people. Finally, we wanted a unified system that worked with StarView, our user interface for querying the database of observations. While not a design goal, we also wanted to find a solution that would meet our goals quickly with a minimum impact on the current system.

3. Assumptions

In creating our design we recognized two assumptions we had made. First, that the user, or the user's system, knew what datasets they wanted. In other words, the retrieval system would not support queries, only requests for specific datasets. The second assumption was that the user had access to the Internet. Since our initial design, access to the Internet has become even easier.

4. The Design

To achieve our design goals we defined a fairly simple syntax to be used to request datasets. Basically, the user must specify an archive username, where the data is to be delivered, and a list of datasets to be retrieved. Additionally there are a few global settings that can be used to simplify the request of datasets.

Once the request text is complete, it is sent as an e-mail message to an account at ST ScI where a mail daemon reads, parses, validates, and converts the message into the STDADS commands to retrieve the data. The daemon will send a message to the archive user's registered e-mail address with the request identification number if all is successful. After the datasets have been retrieved from the optical disks, they are then transfered, using FTP, to either the user's host computer, or to one of the staging systems (*stdatu.stsci.edu* or *stdata.stsci.edu*) at ST ScI. When the last dataset has been transfered, a message detailing the status of the transfer is sent to the user's registered e-mail address.

To retrieve proprietary data, a password verifying the archive username must be supplied. No password is required for retrieving non-proprietary data. To ensure the security of the password, the message must be encrypted, and currently StarView is the only tool which can send an encrypted message to STDADS.

5. The Reasoning Behind the Design

We recognize that e-mail is not a secure way to send passwords in the clear. For access to public data we do not need to validate the archive username since any archive user can fetch public data, and so a password is not required. A password is required to verify the archive user to retrieve proprietary data. Additionally, the request must include a private host computer, account, and password as part of the specification of the destination for the data. It is the user's responsibility to properly secure the destination directory.

As mentioned above, we send all confirmation mail to a registered e-mail address. This is done so that if an unscrupulous person sends a request using another archive user's name, the rightful owner of the account will be informed of the activity. Users who know they are going to use StarView only, and do not want to worry about unauthorized use of their accounts, can request that a password be required for all requests made for their archive user name.

Once we have a valid archive user name, the mail daemon will perform several checks on the user account. It will determine the level of access the user has to the data. If the user is privileged, they can retrieve any data from the archive. This privileged access can be revoked for a request if the request attempts to transfer the data to a public system, or if the archive password is missing. In this case any non-proprietary data requested will be retrieved.

All user requests are then checked to see if they have exceeded certain limits. There is an expiration date for the account, and a lifetime limit on the amount of data one can request. There is a limit on the total number of requests a user can currently have in the system, and a limit on the number of bytes that can be transfered in a day. These checks are done to protect STDADS from being overwhelmed by a single user.

6. How We Met Our Goals

Our design goals are to have a system that is simple, open, secure, and unified. The syntax we created is simple and straightforward, and can easily be generated by a user, StarView, or another query tool. By using e-mail and FTP for requesting and transferring data, the Hubble Data Archive is open to anyone on the Internet with an archive account at ST ScI. Security of proprietary data is preserved by various checks in the system, and the requirement for encrypted requests. Since StarView can generate and send the request for a user based upon the selections, we have achieved a unified query and retrieval system.

7. What We Learned

In adding a new layer or interface, we found that by making a clean break between the two systems we insulated the internal retrieval mechanisms from the external interface. This simplifies modifying the STDADS system without having to notify users of the changes, and to enhance the request interface without having to update STDADS. One of the advantages to our approach of writing a separate interface was that we did not have to make extensive changes to the STDADS system to support the mail interface. In doing this we were able to add a great deal of new capability without the risks associated with modifying large amounts of existing code.

In designing the message syntax, we asked the people who would be making requests of the archive to help us refine the syntax and define what the system should deliver. By getting users involved in the design, the system should meet the needs and expectations of most of our archive users. We intentionally kept the syntax of the request simple and obvious so that it would be easy for users to write their own requests, and so that other query tools could easily generate a request message. Having a simple syntax also makes it easier on our part to test and verify the correctness of the system.

As we moved from design to implementation we discovered that by taking advantage of other tools that we could easily add new functionality. For example, if the STDADS system is busy, we want to queue user requests to be processed later. Rather than writing our own message queuing routines, we take advantage of the mail system's natural queuing capabilities and move requests to various mail folders depending upon the action that needs to be performed.

Probably the most surprising thing we learned is that once you have created a new tool, other people will find new, and unexpected ways, to use that tool. Other groups within the STDADS project have started to incorporate the e-mail interface into their designs as an easy way to retrieve data. It is hoped that people outside ST ScI will also find interesting ways to use the e-mail interface as an extension of their catalog query tools.

As of this writing the the e-mail interface has not been made available to the public. The Space Telescope Science Institute Archive[1] home page will be updated to point to the specifications once they are made public.

---

[1] http://www.stsci.edu/archive-html/archive.html

An Object-Oriented Approach to Astronomical Databases

R. J. Brunner, K. Ramaiyer[1], A. Szalay, A. J. Connolly

Department of Physics & Astronomy, The Johns Hopkins University, Baltimore, MD 21218

R. H. Lupton

Department of Astronomy, Princeton University, Princeton, NJ 08544

Abstract. The rate at which we can acquire astronomical data outstrips our ability to manipulate and interpret it using standard techniques. Future digital astronomical surveys will enter the terabyte regime. In addition, the actual data themselves will be complex, comprising images and spectra, as well as the traditional measured attributes. We describe here an implementation of a prototype object-oriented database, distributed in a client-server framework that is capable of both handling these volumes of data and facilitating geometric queries in an efficient manner.

1. Introduction

With the size of astronomical databases increasing into the terabytes, new approaches must be taken to store, manage, and access these datasets. As an example, consider the Sloan Digital Sky Survey (Gunn & Knapp 1992), a multicolor digital survey of the north Galactic cap which will produce images in 5 bands of more than one hundred million objects, as well as a spectroscopic survey that will obtain more than a million spectra. A single archive site for the entire set of data produced by this survey will require tens of terabytes in permanent storage media. Under the current approach to astronomical databases, statistical studies will be hampered by both long query times and the enormous amounts of data that must be examined.

Our proposed solution is to store the data in an object-oriented database using a multidimensional indexing scheme: the $k - d$ tree. The data will be distributed in a client–server architecture to provide the maximum flexibility as well as retain portability. The user will formulate queries which are then transformed into data requests and routed to the appropriate databases where the requested data are extracted and returned. Section 2 discusses the general design principles behind our system. Section 3 describes the data organizational model we employ, and in particular the $k - d$ tree. Finally, section 4 outlines the data flow model and our prototype implementations.

---

[1] Department of Computer Science, The Johns Hopkins University, Baltimore, MD 21218

2. Design Principles

Our data archiving scheme is based on two fundamental principles: object-oriented programming (OOP), and distributed processing via a client–server architecture. Before describing the technical aspects of our system, a brief digression into the reasoning behind the choice of these principles should prove useful.

2.1. The Object-Oriented Philosophy

Historically, astronomical databases have been stored in a relational format. The previously 'flat' data was easily adapted to the tabular format of relational records. These databases would then be indexed by a limited number of key fields, providing rapid responses to queries involving indexed fields. The main advantages of this approach are the stability of the commercial products and the standard query language (SQL) available for interacting with the database.

The object-oriented philosophy is a fundamental shift away from the relational paradigm. Instead of following the traditional path of formulating an algorithm, and then forcing the data to work within the program, OOP utilizes the fundamental relationships within the data to naturally implement the appropriate algorithm. Object-oriented databases extend this philosophy, modeling the database on a schema (parameter lists). Thus, the database becomes a distribution of objects with associated properties instead of a table of values. The objects within the database are then compared, stored, and operated on as individual units instead of their fundamental parameters.

All object-oriented database management systems that follow the Object Database Standard (ODMG–93) must permit a C++ interface, providing efficient access to the database. As a result, these databases are extremely fast in extracting large amounts of information often approaching the inherent I/O bandwidth limitations.

Another ability of OOP is the ability to persistently model complex data structures. As an example, the schema for an object from an astronomical survey might consist of Galactic coordinates, multicolor fluxes, deblended multicolor images, associated spectra, related housekeeping information, and even references to objects in external databases. This object would then be manipulated as a fundamental unit in the same fashion as the base datatypes: char, integer, and float. The attributes of an object are extracted by methods, allowing for a dynamic update of the data without rebuilding the entire database. Therefore, the implementation of recalibrations of the astrometry or photometry is accomplished by merely modifying the appropriate access functions.

2.2. Client–Server Architecture

A client–server architecture serves to distribute the processing load over multiple computers in a redundant fashion. This results in a more flexible system which can be optimized to avoid congested archive sites in processing queries. Additionally, each process is isolated, allowing commercial products to be integrated into the system with minimal impact and providing cross platform connectivity. Such a system is easily expandable to more client/server processes or databases without disturbing the rest of the archival system.

Using multiple archive servers can extend the data retrieval into a parallel I/O operation, increasing the access efficiency. With the future development of massively parallel I/O machines, an inherently client–server architecture will be quickly adapted. Essentially, this approach hides the details of the processing from the user, providing only the results.

3. Data Organizational Model

The $k-d$ tree (Friedman, Bentley, & Finkel 1977; Samet 1990) is a multidimensional indexing scheme in which d dimensional data is indexed in k dimensions. Each tree node represents a subvolume of the parameter space, with the root node containing the entire k dimensional volume spanned by the data. A balanced binary tree is constructed by splitting each node into two sections along the median of the k dimension which will maximize the clustering of data in the two child nodes. The k dimensions can be actual parameters or the principal components of the parameter data.

The number of levels in the tree is determined by maximizing the number of objects in a leaf node (container) with the constraint that the $k-d$ tree must be small enough to reside in the memory of the user's computer. All objects in a leaf node are then stored together on the physical media. As a result, friends are inherently stored together, producing a built-in index.

Since each node knows its actual boundaries in the k dimensions, the $k-d$ tree can also serve as a coarse grained density map of the actual data parameters. Queries are first performed on the $k-d$ tree, limiting the query volume to those leaf nodes that fall within the query, prior to the entire database, resulting in predictions for the search time, and estimated number of objects satisfying the search. This provides a feedback feature for the user, which can minimize costly queries.

Database queries can be modeled as cuts within the parameter space of the database. These parameter cuts can be combined using boolean operators to create complicated query volumes. Geometric queries can be constructed through linear combinations of the parameters, proximity cuts, and k-nearest neighbor searches. These geometric queries utilize the nodal boundaries to determine the containers that intersect the query volume and must be searched in more detail. Using multiple linear combinatorial cuts, one can create convex hulls (polyhedra) that carve out a subspace of the parameter volume. This allows associative queries to be performed, by first constructing a convex hull from a given sample of objects, and then finding all other objects in the database that lie within the given convex hull.

4. Data Flow Model

The data flow model for our system involves four main components: the object-oriented database management system (OODBMS), the object request broker (ORB), the client process, and the graphical user interface (GUI).

Our prototype system currently employs Objectivity, a commercial object-oriented database management system, for maintaining the persistence of our objects. The $k-d$ tree and the actual data objects are retrieved from the OODBMS using an object request broker (ORB) which is the only process in our system that is syntactically tied to the OODBMS. The ORB becomes a server, awaiting TCP/IP socket connections from client processes. Together,

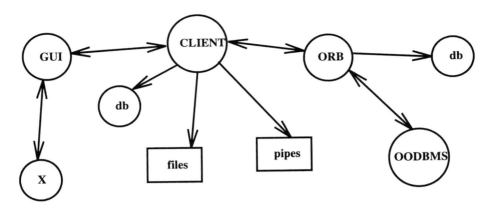

Figure 1. The data flow model.

the OODBMS and the ORB are the only process that physically reside, along with the actual data, at the archive site.

The client process is responsible for interpreting the queries using the $k-d$ tree to determine containers that satisfy the search, from which an estimate for the returned number of objects and search time are determined. The client also must establish the connection to the ORB, initially retrieving first the $k-d$ tree and then the desired objects. The extracted data can be piped directly into an analysis tool, saved into a file, or made locally persistent. The client interacts with the user via the GUI, which facilitates query construction and data retrieval.

We have adopted this model in two implementations. The first prototype involved an all sky simulation containing two million objects with eight parameters: right ascension (RA), declination (DEC), five colors (U, G, R, I, Z), and redshift. The $k-d$ tree was constructed with ten levels involving splits in the order: RA, RA, RA, DEC, DEC, DEC, G, G, R, and R. Complex queries involving multiple parameters are one hundred times faster than simple linear searches. Currently, only boolean combinations of parameter cuts and linear combinations of parameters are supported. A second prototype using the $IRAS$ point source catalog is under construction.

References

Friedman, J. H., Bentley, J. L., & Finkel, R. A. 1977, ACM Trans. Math. Software, 3, 209

Gunn, J. E., & Knapp, G. R. 1992, PASP, 43, 267

Samet, H. 1990, The Design and Analysis of Spatial Data Structures (Reading, Addison-Wesley)

The Object Database Standard: ODMG-93, ed. R. Cattel (New York, Morgan Kaufmann)

Data Archive System for Kiso Observatory and Okayama Astrophysical Observatory

S. Ichikawa

National Astronomical Observatory of Japan, Mitaka, Tokyo 181, Japan

S. Yoshida

Kiso Observatory, University of Tokyo, Mitake, Kiso, Nagano 397-01, Japan

M. Yoshida

Okayama Astrophysical Observatory, NAOJ, Kamogata, Asakuchi, Okayama 719-02, Japan

T. Horaguchi

National Science Museum, Shinjuku, Tokyo 169, Japan

M. Hamabe

Institute of Astronomy, University of Tokyo, Mitaka, Tokyo 181, Japan

Abstract. A data archive system for optical CCD data produced at Kiso Observatory and Okayama Astrophysical Observatory is discussed. This system, named MOKA, allows users to search data with some constraints, preview data, and view FITS headers. This paper outlines the MOKA system in some detail.

1. Introduction

Japanese optical astronomers have not made much effort to archive the data produced with telescopes in Japan because of their low quality. The construction of Japanese 8m telescope, SUBARU, is now in progress. SUBARU will be in operation by the end of this century and is expected to produce a large amount of high quality data which deserves to be archived. This motivated us to develop a data archive system for SUBARU, although we have little experience in developing or maintaining such a system. To overcome this, we decided to develop an archive system for the CCD data available now at the Kiso Observatory (Kiso) and the Okayama Astrophysical Observatory (OAO). This system, named MOKA (Mitaka Okayama Kiso data Archive system), is designed to be a prototype of the SUBARU data archive system, as well as a practical facility for effective archiving of the data produced at Kiso and OAO.

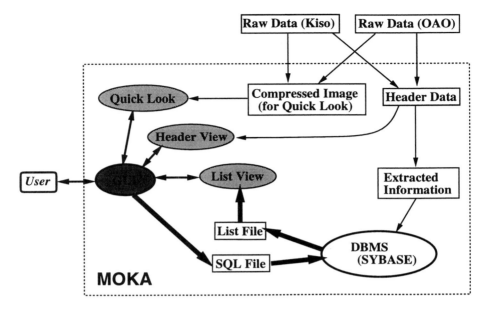

Figure 1. Structure of MOKA and data flow.

2. Outline of MOKA

At present, MOKA deals with raw observational data, not calibrated data. The data are produced from Kiso 105 cm Schmidt telescope prime focus camera CCD and OAO 188 cm telescope Cassegrain focus spectrograph CCD. MOKA offers data search, data preview (quick look), and FITS header view. MOKA does not handle the raw data directly, and the raw data itself is stored in each observatory on tapes or CD-ROMs. MOKA handles header information, and compressed images for previews extracted from the raw data. The structure of MOKA and the data flow are shown in Figure 1.

Unfortunately, we have a severe bandwidth restriction: 9.6 kbps to Kiso and 19.2 kbps to OAO. Users of MOKA must access NAOJ Mitaka, or visit Kiso or OAO. We anticipate most users accessing MOKA through Mitaka, although MOKA is installed in Mitaka, Kiso, and OAO.

FITS header data and compressed images are created from the raw data at each observatory, and sent to Mitaka, Kiso, and OAO. Information for a search is then extracted from the header data and input to a database management system (DBMS; in this case Sybase). A GUI has been written for MOKA that creates a SQL file according to search constraints set by the user. At present, possible constraints are instruments (observatories), a range of acquisition dates, types of frames, and a position (ranges of R.A. and Dec.). The SQL file is submitted, and the results of a search are stored in a file. The user can see the list of the frames found in the search, preview these images, and view the header data of the selected frame. MOKA sends requests for raw data to each observatory. The raw data are then sent by tape because of the network restriction.

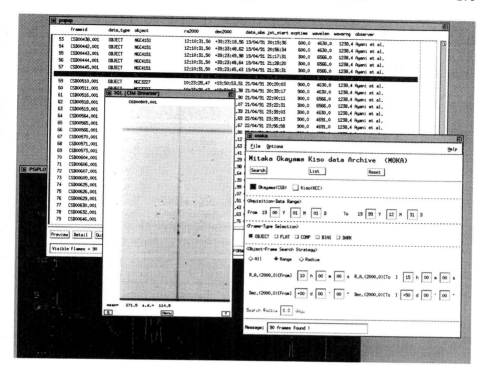

Figure 2. Sample screen of MOKA.

We use highly compressed images for the previewing, to save both disk space and a CPU overhead. The compression methods used are binning or discrete sampling, 8 bit-scaling, and gzip. A 500 × 500 OAO frame (about 500 kB) is compressed to about 10 kB, and a 1000 × 1000 Kiso frame (about 2000 kB) is compressed to about 6 kB. MOKA displays the compressed images for preview, and shows either a horizontal or vertical profile at the requested position.

3. Next Step of MOKA

We will soon be able to display many images at the same time in the preview mode, and to set more search constraints (wavelength, etc.). As MOKA is a prototype of the SUBARU archive system, we must address several tasks. For example; the handling of raw and calibrated data directly, dealing with data from other instruments or observatories, connecting with other astronomical databases, and so on. We expect our efforts will lead the better design of the SUBARU archive system.

Acknowledgments. This work was supported in part under the Scientific Research Fund of the Ministry of Education, Science and Culture (06554001) and in part by NAOJ. The authors thank to Prof S. Nishimura and Prof K. Sadakane for continuous encouragement.

Exploring Interactive Archive Data Presentation at the COSSC

J. M. Jordan, M. Cresitello-Dittmar, and J. S. Allen

Hughes STX/COSSC NASA/GSFC, Greenbelt, MD 20771

Abstract. As the *Compton* Observatory archive grows in size and complexity, it is increasingly difficult to present the archived data in a way which allows users to get a good grasp of what the archive holds. We are now using hypertext, presented through the World Wide Web, in an attempt to alleviate this problem. We are working on giving our users methods for searching the holdings of the archive.

Hypertext form pages have been created for two instruments: BATSE and EGRET. Each page lets users define the parameters of a search. The results of the user defined search are returned both graphically and as text hyperlinked to the relevant data in the archive.

As our understanding of the requirements grows, we hope to present views of the data in the archive which are better tailored to the ways in which our users look at those data.

1. Introduction

The exploration of hypertextual data presentation at the COSSC[1] was facilitated by the creation of tables of data, in electronic form, which describe the BATSE and EGRET datasets available in the *Compton* Observatory archive.

Bursts recorded by the BATSE instrument are characterized by a set of parameters stored in a catalog. We allow users to specify parameter values in a hypertext form. The values are then checked against the catalog, and qualifying bursts are selected. The selected bursts are plotted on an Aitoff projection. Hypertext links are provided, both to the data in the archive and to pages which show graphs of the bursts and burst specific information.

Data collected by the EGRET instrument can be selected from an Aitoff projection of the EGRET sources. In hypertext browsers supporting imagemap selection, the user can click on the skymap and links to pertinent data for that region of the sky are returned.

2. BATSE

The BATSE hypertext search page[2] is based on the BATSE 2nd Burst Catalog, created by the BATSE instrument team. The information in the set of seven

---

[1] http://enemy.gsfc.nasa.gov/

[2] http://enemy.gsfc.nasa.gov/cossc/batse/burstcatalog/batseform.html

tables which make up the catalog provides a wide variety of characteristics parameterizing the detected gamma ray bursts. The tables include, among other things, data on:

duration The duration on the burst at 50% and 90% burst fluence levels.

Cmax/Cmin The peak count rate in units of the threshold count rate on 64 ms, 256 ms and 1024 ms time scales.

fluence The fluence of bursts in channels one through 4 in units of $\mathrm{ergs\,cm^{-2}}$.

location The galactic longitude and latitude of the bursts.

The BATSE page allows a user to select which parameters will be used for comparisons. The parameters which are of interest can then be characterized with $>$ and $<$ combinations and values. Once active parameters and their values are selected, the hypertext form is submitted, and processed by a Perl script.

The Perl script searches the various tables for matching bursts, graphs those bursts on an Aitoff projection of the sky, and lists the data for which the user was searching as hyperlinks to the data files in the anonymous ftp area of the *Compton Observatory* archive. A C program plots the bursts on an Aitoff projection in a GIF file for inclusion on the HTML page. Coordinate transformations are performed by FORTRAN functions.

3. EGRET

Two resources gave us the ability to put together a hypertext page for searching EGRET data[3]: a skymap of the objects that EGRET has observed, and a set of tables containing some of the interesting parameters which characterize the EGRET observations. The EGRET search opens with an Aitoff projection of the sky. The skymap is an imagemap which is part of an HTML form; clicking on the skymap returns (x, y) coordinates. Below the map on the page are buttons allowing the selection of objects of interest (pulsars, quasars, or unidentified sources). The width of the sky search is also selectable with a fill-in-the-blank area of the form.

After clicking on the imagemap, the (x, y) coordinates, objects of interest, and the width of the search are sent to a Perl script which converts the coordinates to galactic longitude and latitude, and then checks through the EGRET tables for the specified object types, within the specified search radius. The results are returned on an Aitoff projection of the sky, displaying the selected objects. Along with the graph, the information selected from the EGRET data tables is displayed as text, with hyperlinks to the data in the *Compton* Observatory archive.

An alternative way of using the EGRET form is to click on the Aitoff projection without selecting any objects of interest. When this is done, the object nearest the selected location is returned, with the data about the selected object presented in the same manner as when multiple objects are selected.

---

[3] http://enemy.gsfc.nasa.gov/cossc/egret/egretcatalog/egretform.html

4. Some Words on the Software

The software driving the BATSE and EGRET hypertext search forms is built in several languages, with the intent of reusing as many of the software modules as possible. We also sought to perform each of the required functions in the language most appropriate to the task.

The main program for each form is written the scripting language Perl. We have found that Perl is excellent for interfacing with hypertext forms and for rapidly creating complex software. C was used to create the GIF images which are inlined in the hypertext pages. FORTRAN is used for routines doing coordinate transformations. Each of the C and FORTRAN routines is completely modular and works as a command line routine. Each of the routines was used in the development of both the EGRET and BATSE forms.

5. Conclusions

The hypertext forms for searching the EGRET and BATSE databases provide added value for users of the archival data of the *Compton* Observatory by allowing users a better method for directing a search through the data. As our store of data continues to grow, we will continue to explore new and novel ways in which hypertext allows the data in the COSSC archive to be presented and selected.

Accessing the Digitized Sky Survey

J. E. Morrison

Space Telescope Science Institute, 3700 San Martin Drive, Baltimore MD 21218

Abstract. The Digitized Sky Survey is now available on CD-ROMs and will be on the ST ScI archive. This is a whole sky survey constructed from the Palomar E and SERC J Schmidt surveys. We present an overview of software produced by the ST ScI STSDAS group to access and manipulate these images. We have developed portable code to use in the IRAF environment, hopefully in the future this we be available as a stand alone package.

1. Introduction

ST ScI has digitized Schmidt plates of the entire Sky. Due to the massive volume of data involved (a total of about 600 Gbytes) the scans have been compressed by a factor of 10 and are made available in the form of 100 CD-ROMs. Software to give the astronomical community access to these scans is being developed by the STSDAS and Guide Stars groups at ST ScI.

2. Background Information: Plates and Scans

The northern-hemisphere data are from scans of the E plates of the National Geographic Society – Palomar Observatory Sky Survey (NGS-POSS) obtained using the Oschin Telescope on Palomar Mountain. These plates are deep 103a-E exposures obtained through a red plexiglass filter. The southern-hemisphere plates were obtained using the UK Schmidt Telescope while it was operated by the Royal Observatory Edinburgh (ROE). These are deep IIIa-J exposures obtained through a GG 395 filter. The scans were made at the Space Telescope Science Institute using the scanning microdensitometers, which are described in Lasker et al. (1990). A pixel size of 25 μm with a 50 μm apodized aperture was used throughout.

3. Software

The original plate images were composed of 14,000 by 13,999 pixels (a pixel is 1.70 arcseconds square). To allow fast access to any portion of the plate, the image was divided into blocks of 500 by 500 pixels that were separately compressed by a factor of ten. Experiments to study the degradation of astrometry on the compressed images compared to the original image were performed by White et al. (1992), who concluded that the astrometry is hardly affected by the compression for modest compress factors (up to about a factor of 20).

The software to read the compressed images is provided with the set of CD ROMs and is also a new task in the GASP/STSDAS package called **getimage**. In the **getimage** software the user specifies the region of interest. The software then accesses the data, decompresses the scans and returns the data in a form of a photographic intensity as a function of x and y plate pixel coordinates.

New tasks are being added to the GASP package that will allow the user to manipulate this data. At this time these tasks use the original plate parameters produced by a single global model reduction of the plates. In the foreseeable future the astrometry will use a more sophisticated method.

4. New Tasks in or to be Added to GASP/STSDAS

getimage accesses the Digitized Sky Survey (DSS) CD ROMs and decompressed the data returning a photographic intensity as a function of x and y pixel coordinates. This software package is available on the CD ROM set and was written by Jesse Doggett and placed into GASP/STSDAS by Allen Farris.

dss_targets extracts Guide Star Catalog (GSC) stars from the CD-ROM Guide Star Catalog in the field specified by the user and overlays the these stars on an DSS image produced by **getimage**.

dss_pxcoord given a table of GSC stars (or other table of equatorial coordinates) finds the pixel coordinates corresponding to the DSS image.

dss_xyeq procedure to calculate the equatorial coordinates from a set of (x,y) pixels using the original plate parameters.

dss_eqxy procedure to calculate the pixel coordinates from a set of equatorial coordinates using the original plate parameters.

Acknowledgments. I would like to acknowledge Jeffrey Hayes, Brian McLean, and Jesse Doggett for their scientific assistance.

References

Lasker, B., Sturch, C. R., McLean, B. J., Russell, J. L., & Jenkner, H. 1990, AJ, 99, 2019

White, R. L., Postman, M., & Lattanzi, M. G. 1992, Digitized Optical Sky Surveys, 167

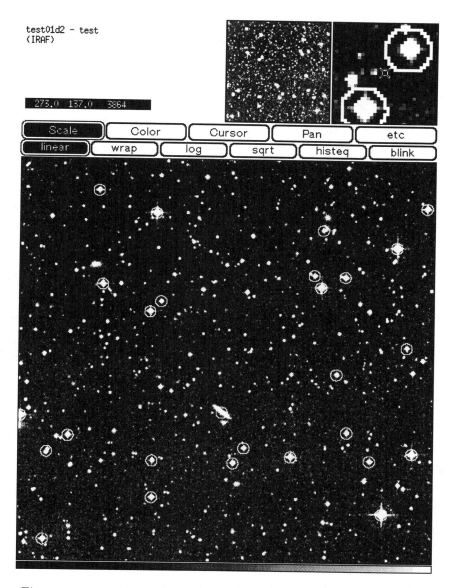

Figure 1. An image from the Digitized Sky Survey: center ($0^h.0$, $-89°.0$), and size: $0°.2$. Produced by **getimage** and **dss_targets** tasks. **Getimage** was used to extract the data, while **dss_targets** was used to display the image, search the Guide Star Catalog for stars in the region, and then plot white circles around the GSC stars found in the image.

GUIDARES: Reading the Guide Star Catalog in Very Many Ways

O. Yu. Malkov, O. M. Smirnov

Institute of Astronomy of the Russian Academy of Sciences, 48 Pyatnitskaya Str., Moscow 109017 Russia

Abstract. We present some current and future variations of our Guide Star Catalog Data Retrieval Software (GUIDARES). Topics include the latest DOS version, a Unix/DOS C library, an IDL widgets version, and some future developments, such as an intelligent reclassifier, a Pocket GSC and a Pocket GUIDARES, and a WWW version which will supply finder charts through Mosaic.

1. GUIDARES, the Software of Many Faces

The GUIDARES software was born in 1991 (Malkov et al. 1992) as a PC-based tool for easy access to the Guide Star Catalog. Since then, the program has survived one major overhaul and a lot of minor tweaking (Malkov & Smirnov 1993, 1994a), and has even inspired a new project for automatic improvement of the original GSC material (Malkov & Smirnov 1995).

In its most current DOS form (version 2.∞), GUIDARES (also known as G2) is an interactive, GUI-style program that can retrieve GSC data for a specified area of the sky, and optionally produce a sky map in a celestial or Aitoff projection. GUIDARES supports four coordinate systems (equatorial, ecliptic, galactic, supergalactic) at any equinox, and automatically manages conversion between them and J2000.0 (the native GSC system); the user may specify rectangular or circular areas, set a magnitude range, and decide what to do with multiple-entry objects. The resulting data is written to an ASCII table; an optional SkyMap module produces plots, which is very useful for finder charts. This paper describes some recent evolution of the software in other directions.

2. GUIDARES as a C Library

In the past two years, it has become apparent that GUIDARES was needed not only as a stand-alone tool. Several applications arose (e.g., Malkov & Smirnov 1994b) where it was very useful for a program to be able to read the Guide Star Catalog easily, without the burden of sorting through its (very complicated) internal structure. For this reason, GUIDARES was recently recast into a library of several C functions (GUIDARES/Lib), versions of which now exist both under MS-DOS and Unix.

With GUIDARES/Lib, a C or FORTRAN program can sample the GSC without any burden on the programmer's part. One function call sets up the CD-ROM drive, another selects the coordinate system, and a third retrieves data for any area. If the GSC disc needs to be changed, GUIDARES/Lib can report this event back to the program for it to handle.

3. GUIDARES for IDL

GUIDARES/IDL is an example application of GUIDARES/Lib. A short C program wraps around GUIDARES/Lib to handle communication (using the SPAWN procedure of IDL) between it and a few IDL routines. GUIDARES/IDL provides a way for an IDL program to read easily a sampling of any GSC area (as with the original DOS GUIDARES, the area may be specified in any coordinate system, etc.) into an array of structures. At ADASS IV, we demonstrated a GUIDARES interface based on IDL widgets.

4. Intelligent GUIDARES, Pocket GSC, and the WWW

The Guide Star Catalog is the biggest source of all-sky data to date. However, most of its data was produced by automatic classifiers, so it is prone to various errors and defects. The GSC Object Classifier (Malkov & Smirnov 1995) aims to extend GUIDARES with some sort of intelligence that would allow it to filter and improve GSC data automatically.

It it possible to compress the GSC significantly to a more reasonable size (for practical disk-space considerations) without any loss of data (see, e.g., Preite-Martinez & Ochsenbein 1993). If the GSC Object Classifier lives up to expectations, we will produce a compressed and re-classified GSC of our own. Finally, we will make a GUIDARES/WWW, which will provide access to this Pocket GSC through the World Wide Web. Using a WWW browser that supports query forms (such as NCSA Mosaic), it will be possible to submit a query to a version of GUIDARES running on our machine (using all of the features described above), and receive a sampling of the data as a FITS or plain ASCII table, or even a complete finder chart in PostScript, GIF, or JPEG format. This is but one of the possible applications of a GUIDARES for the WWW.

5. Conclusion

GUIDARES is an ongoing project and new ideas are constantly being born. Any comments, questions, or suggestions will be appreciated: please contact the the authors by e-mail at *omalkov@inasan.rssi.ru* or *oms@inasan.rssi.ru*. Some flavors of GUIDARES can probably be retrieved by anonymous ftp from *ftp.inasan.rssi.ru*, in the directory */pub/guidares*. As of 1994 October, GUIDARES for DOS, being the only product mature enough, was the sole application in */pub/guidares*. However, some additional platforms may be supported by the time this paper is published; interested parties are invited to browse our ftp area for updates.

Acknowledgments. This presentation was made possible by financial support from ST ScI. Many thanks to M. Pucillo of the Osservatorio di Trieste for inspiring a large portion of GUIDARES/Lib and GUIDARES/IDL.

References

Malkov, O. Yu., Kulkova, L. I., & Smirnov, O. M. 1992, in Astronomical Data Analysis Software and Systems I, ASP Conf. Ser., Vol. 25, eds. D. M. Worrall, C. Biemesderfer, & J. Barnes (San Francisco, ASP), p. 79

Malkov, O. Yu., & Smirnov, O. M. 1993, in Astronomical Data Analysis Software and Systems II, ASP Conf. Ser., Vol. 52, eds. R. J. Hanisch, R. J. V. Brissenden, & J. Barnes (San Francisco, ASP), p. 504

Malkov, O. Yu., & Smirnov, O. M. 1994a, in Astronomical Data Analysis Software and Systems III, ASP Conf. Ser., Vol. 61, eds. D. R. Crabtree, R. J. Hanisch, & J. Barnes (San Francisco, ASP), p. 183

Malkov, O. Yu., & Smirnov, O. M. 1994b, in Astronomical Data Analysis Software and Systems III, ASP Conf. Ser., Vol. 61, eds. D. R. Crabtree, R. J. Hanisch, & J. Barnes (San Francisco, ASP), p. 187

Malkov, O. Yu., & Smirnov, O. M. 1995, this volume, p. 257

Preite-Martinez, A., Ochsenbein, F. 1993, in Handling and Archiving Data from Ground-Based Telescopes, ESO Conference Proceedings, eds. H. Albrecht & F. Pasian (Garching, ESO), p. 199

Storing and Distributing GONG Data

M. Trueblood, W. Erdwurm, and J. A. Pintar
National Solar Observatory, National Optical Astronomy Observatories[1], P.O. Box 26732, Tucson, Arizona 85726-6732, USA

Abstract. The Global Oscillation Network Group (GONG) helioseismology observing network will consist of six instruments deployed worldwide to provide nearly continuous observations of the Sun beginning in 1995.

Data reduction is performed on a network of high-performance UNIX workstations to process the data and store them on Exabyte 8-mm cartridges. The single observed object (the Sun) and other constraints imposed by the nature of the project permitted developing a more robust and less expensive DSDS than is possible for open-ended general institutional support systems performing similar functions. For example, the data product file catalog was compressed by a factor of over 160 to a series of bitmaps that permit the DSDS to provide good query response to several simultaneous users on a workstation. UNIX interprocess communication and networking were used to develop a mirrored database between two DSDS workstations, providing a high level of DSDS availability to support data reduction pipeline operations.

The Global Oscillation Network Group (GONG) will record Dopplergrams of the Sun once per minute over three years. The data will be reduced at the Data Management and Analysis Center (DMAC) in Tucson, Arizona through a pipeline consisting of a network of workstations. The approximately 3 TB of data will be stored on approximately 10,000 8-mm Exabyte cartridges and will be managed and distributed by the Data Storage and Distribution System (DSDS). Details of the expected science return and data products to be produced are given in Kennedy & Pintar (1988).

1. DSDS Design

Unlike most other astronomical observatories, GONG observes only one object with a single name, and the celestial coordinates of that object at the time of observation are not important to the user community. This permits us to limit query keys to data product type and time of data acquisition. Some users want the ability to query the catalog for data products taken at the same time as

---

[1] The Global Oscillation Network Group (GONG) is an international community-based project funded principally by the National Science Foundation and administered by the National Solar Observatory. NSO is a division of the National Optical Astronomy Observatories, operated by the Association of Universities for Research in Astronomy, Inc. under a cooperative agreement with the National Science Foundation.

certain solar events. Consequently, we provide the means for users to add their own software to the query system to support such "correlated queries".

Wide area networks are used to distribute up to 100 MB of data per user per day. Requests for more data are distributed using removable media. Since the GONG image file size is relatively small (130 kB) and to keep data distribution straightforward, the DSDS designers decided to refrain from making spatial subsets of data, and to make the process of assembling temporal subsets easy by placing each separate data product instance (e.g., a one-minute image) in a separate disk file. This reduces the problem of distributing data to merely copying files from one cartridge to another using operating system utilities. The DSDS software developers needed no knowledge of data file internal formats, since they had no need to develop custom software to read individual data files.

Operators of other data centers suggested that we keep file names short and meaningful to minimize operator errors. GONG data file names are formed from the data product type and time of data acquisition using only lower case letters and numbers. The first file on each library (archive) cartridge contains a "table of contents" listing the names of all data files on the cartridge.

The approach of placing each data product instance in a separate file means that the file catalog will contain almost 75 million data file name entries after a three-year project. Conventional commercial DBMS products would require an unacceptably long time to execute a simple query on this many table entries, even if the file catalog were divided into several smaller tables. Furthermore, there is sufficient user interest in a catalog that can be queried on users' home institution computers to justify a query system that does not require each user to purchase a commercial database product.

To solve these problems, the DSDS designers "compressed" the file catalog by defining a file for each data product containing a bit for each possible time slot over the three-year GONG project. A bit is set in the file if the library contains that time slot's data product file. That is, in the case of images, since there are 1440 time slots (minutes) in a day, 1440 bits are used to represent a single day of a single image data product. A three-year period of a single image data product can be represented by 1440 bits × 365 days × 3 years = 1,576,800 bits or 0.2 MB. With approximately 200 data product types defined to date, many of which are produced less frequently than each minute, the entire data archive can be represented by a collection of files, one per data product type, with a total storage requirement of about 20 MB. Users without network access to the DMAC Users' Machine who have ported the query software to their own computers need to have on their home systems only those bitmap files corresponding to data products of interest, so a single investigator's collection of bitmaps might consume only 5 MB of disk space, which can fit on a few floppy diskettes. Performing a query consists of specifying a data product and a time, opening the bitmap file for that data product, and checking the single bit corresponding to the specified time.

The menu query system consists of three parts: a program that generates the menus, receives user keyboard input, and generates a "query file" as its output; a program that takes a query file as input and generates a "hits list" and an optional "misses list"; and a UNIX script that calls these other two parts. The query file defines the scope of a query. Each record or row of the file is a file name that specifies the data product type and a single time slot. Although the menu system generates the query file, a user could generate the query file using custom software or even a simple text editor. The hits list is a file of similar format containing all the file names that satisfy a query, and the misses file lists the files in the space defined by the query not in the library. When the hits list reflects the data subset the user wants, the user runs a program that reads the hits list and generates a data request. The DSDS operator fills the data request by extracting the appropriate cartridges from the library, and copying the requested files to the distribution medium that was requested.

In addition to the menu system, the DSDS provides a query tool written in C that can be used interactively to see if a particular data product for a single time is in the database. It also generates hits and misses list files. The same program can be called from a UNIX script (command file) written by the user to form the query. The query tool uses a typical UNIX command form in which the data product name and the date and time are specified using input parameters in the UNIX command line that invokes the program.

The file catalog for tracking the location of each data file was designed around the method of storing data files on Exabyte cartridges. Pipeline operators store groups of disk files on tape using the UNIX *tar* program. Each *tar* file contains files of only one data product type for a 24-hour period, up to 1,440 image files. The file catalog, instead of listing each individual data file, lists only each *tar* file on a tape and the beginning and ending dates/times of the data. When a new data tape is checked into the DSDS, a row representing each *tar* file on the tape is inserted into the file catalog database table and the bits in the bitmap files corresponding to each new data file are set. The bitmap files are then copied over to the Users' Machine for immediate use in queries. The bitmap "fills in" the individual time slots between the beginning and ending dates/times in the file catalog database table, listing the availability of a data file for each possible time slot. This design compressed the file catalog by a factor of 160 over using RDBMS tables to store all file names. On-line data are handled in a similar way, with the pipeline collecting on-line files into pseudo-*tar* groups.

Another feature of the file catalog design is the ability to store all versions of a data file. The bitmaps reflect only whether a file exists in the archive, not its version, so queries on version numbers are not permitted and only the latest version of data are distributed in routine operations. But if a situation arises in which the science requires access to a previous version, this can be handled as a special request that must be approved by project management.

2. Database Mirroring for High DSDS Availability Levels

The DSDS is the central hub of the DMAC through which all Exabyte data cartridges must pass from one pipeline stage to another. If one stage of the pipeline goes down, other stages can continue processing data until no more input tapes are available. If the DSDS goes down, the entire DMAC stalls. To provide a DSDS with a high level of availability to the DMAC, the database is mirrored on two workstations, and DSDS applications are designed to run on either workstation. During normal DMAC operations, an operator on either DSDS workstation can perform any DSDS function and the results appear on both workstations in near real-time (within a few seconds). This is achieved using UNIX message queues and sockets (communications links). When an application reads from the database, the read is performed only on the local workstation. But if an application writes to the database, it first writes to the local database. If the local database is updated without error, then the application places a Structured Query Language (SQL) command or a bitmap update command in a buffer and gives the buffer to a routine that places it on a local message queue. A database mirror daemon removes the buffer from the queue and sends it over a socket to the other (remote) DSDS Operations Machine (OM), where a daemon receives the buffer from the socket and places it on another message queue on the remote OM.

On the remote OM, one of two daemons removes the buffer from the message queue. If the buffer is a bitmap update command, the bitmap daemon dequeues the buffer and updates the bitmap. This daemon processes bitmap update commands from both the local and remote OM's, and sends a new bitmap to the Users' Machine when it receives a buffer coded to indicate that updates to the current bitmap are complete. If the command is an SQL command to the database, another daemon removes the command buffer from the queue and performs an SQL EXECUTE IMMEDIATE on the buffer. This process is repeated in the opposite direction, enabling either OM to perform any DSDS function and keep the other OM's database current.

When a socket goes down (such as when an OM fails), or if an OM daemon tries to send a buffer to the remote OM, it receives an error from the socket. The daemon then places the buffer in one of two files on the local OM, depending on whether the message is an SQL message or a bitmap update message. When the remote OM comes back up, the remote database is brought up, and all sockets and daemons are reestablished. The DSDS operator on the remote OM then copies over these files and runs a recovery program that reads the files and places the buffers back on the message queue for processing. This permits one OM to keep processing by itself while the other OM is down, and permits the DSDS to weather most single points of failure (excluding power failures, which bring down the entire DMAC, but which are rare enough and of short enough duration to be no more than a temporary nuisance).

References

Kennedy, J. R., & Pintar, J. A. 1988, in Astronomy From Large Databases, eds. F. Murtagh & A. Heck (Garching, ESO), p. 367

Part 5. Data Models and Formats

What is an Astronomical Data Model?

A. Farris and R. J. Allen
Space Telescope Science Institute, 3700 San Martin Dr., Baltimore, MD 21218

Abstract. Using techniques from semantic data modeling and object-oriented data modeling, an astronomical data model is identified as a kind of conceptual analysis. The general characteristics of such a model are presented, as well as its relationship to computer-based data structures and the FITS standard for transporting data. The effective use of data models in relation to instrument specific data, in conjunction with data archives, and in fundamental areas of the data analysis is discussed.

1. Origins of the Concept of a Data Model

The basic concept of a data model originated within the database community as the logical model of an integrated database management system (DBMS). The early DBMSs employed hierarchical, network, and relational data models that were directly implemented by the DBMS software. The entity-relationship model was introduced as a generalization of these previous models and is a more intuitive approach to modeling data relationships. It is widely used as a conceptual tool for relational database design (Batini 1992). Semantic data modeling is an outgrowth of the entity-relationship model in which additional structures are employed to portray more complex data relationships. Object-oriented modeling techniques (Booch 1994) are closely related to semantic data modeling, the major difference being that the object-oriented techniques add behavior to characterize entities rather than being restricted to data attributes.

A semantic data model or an object-oriented data model portrays named objects and relationships between objects. Each object has a set of specific attributes characterized by data items of a specific type. The relationships between objects also have names and quantitative aspects, one to one, one to many, many to many, etc. Relationships may have data attributes as well. For example, if Book and Student are objects in a library data model, then Borrows represents a relationship between Book and Student and the data item, due-date, belongs to the Borrows relationship. Most relationships tend to be binary, but they can be n-ary as well. For example, Meets may be a relationship between three objects, a Course, a Room, and a Time ("a course meets in a room at a time"). There are two other important ways in which objects may be related. One object may be a part of another object (an Engine is part of a Car) and one object may be a specialization of another object (a Rectangle is a kind of Polygon).

There are a number of graphical techniques for representing both semantic data models and object-oriented models. These techniques include representations for all of the basic features used in modeling objects and their relationships, including whole-part and generalization-specialization relationships. They have the advantage of being easily understood and, yet, can model data complexity to arbitrary levels of detail. They are also independent of any specific computer language, enabling them to be implemented in many software environments.

2. Characteristics of an Astronomical Data Model

A data model, as depicted above, is a representation of some state of affairs that exists within the domain of the problem under consideration. It is the result of a conceptual analysis of that problem domain. An astronomical data model is the result of a conceptual analysis of the characteristics of astronomical data and the relationships that obtain between those kinds of data. This analysis is mapped onto a set of graphical or linguistic conventions, able to faithfully represent the characteristics and complexity of the data. Each component of the mapping must have a physical interpretation. The entire model designates a state of affairs that exists, has existed, or might possibly exist in reality.

There has been a tendency within astronomy to confuse computer-based data structures with a data model. Data structures are used to implement a data model. In themselves, apart from any physical interpretation, they are merely abstract data structures. The physical interpretation is an essential part of the concept of an astronomical data model. It is in virtue of this aspect that a data model can be said to be true or false. In other words, data models have meaning; they make assertions about the nature of reality.

It should be clear, at this point, why FITS (NOST 1993) is not an astronomical data model. In its current form as a standard transport mechanism, FITS does not require a physical interpretation. All astronomical keywords and even units are optional. FITS is merely a convention for exchanging bits in a manner that is independent of hardware. A FITS image might have nothing to do with astronomy; it might be a bit mapped image of Greek text. However, the FITS standardization process can become a vehicle for defining standard models of the basic astronomical data concepts. To be a data model of an astronomical image, a FITS image must *require* sufficient astronomical keywords to provide a meaningful interpretation, including units and a world coordinate system.

3. Uses of Data Models within Astronomy

In considering astronomical data handling there are three specific areas where data models can make an effective contribution: in characterizing instrument specific data, in providing a higher conceptual view for searching data archives, and as the basis for implementing a more open approach to the data analysis process.

One of the most vexing areas of modern astronomy, for both users and software developers, is in dealing with modern astronomical instrumentation. Whether it is optical, radio, or X-ray astronomy, instrumentation is complex, with many different modes of operation, each having its own capabilities, calibration procedures, and data formats. Moreover, newer instrumentation will become more complex as basic hardware shrinks in size and increases in functionality. The use of object-oriented modeling techniques can greatly aid in coping with this increased complexity. Graphical representations of data relationships, corresponding to modes of operation of astronomical instrumentation, can be used as an effective means of communication between instrument scientists developing the hardware, end-users attempting to understand how to use the instrument, and software developers responsible for developing the calibration and data analysis procedures. Since such representations are independent of specific computer languages and software development environments, they can serve to capture fundamental design features in an implementation independent manner and in a manner that makes data relationships intelligible to a broader spectrum of people.

Large archives of astronomical data already exist and will become increasingly significant for research in astronomy. The current archives are difficult to use, with no standardization in interface software. While most archives return data in the form of FITS files, there is no standardized way of presenting the data within the FITS file. In short, there is no standard data model. This situation is exacerbated by the fact that these archives, in many cases, store data internally in small, independent FITS files with little or no way of dealing with collections of files or forming relationships between those files. This latter point is particularly acute in storing calibration data, which is usually tied to specific instruments. As archives grow in size and complexity, better methods must be found to store and access data within these archives.

One approach to dealing with this data complexity is illustrated by the Space Telescope Data Archive and Distribution Service (ST-DADS) (Schreier 1991). Scientific and engineering data from the *Hubble Space Telescope* are archived on WORM optical disks in the form of self contained FITS data sets. The data in the FITS keywords is automatically captured and used to populate a relational database that forms a catalog describing the contents of the data archive. This relational database is used to find and request that data be placed on-line for access. The complexity of the *Hubble Space Telescope* data is reflected in the complexity of the relational database that is the catalog portion of ST-DADS. This database has over 1500 attributes distributed among more than 40 relational tables. Using SQL, the standard relational database interface, for a database of this complexity is a challenging experience. Writing a correct SQL query to satisfy even a simple request, i.e., simple from an astronomer's point of view, is a feat that few can master.

To cope with this complex catalog, an object-oriented user interface to ST-DADS, called StarView (Williams 1993), was developed. At its heart is an automated query generator that is based on a higher-level view of the database using the extended entity-relationship model. This query generator uses a system called QUICK (Semmel 1993) which creates contexts for generating SQL statements based on the conceptual model. Typical queries generated vary in size from a few lines to as much as seven pages of SQL statements. A example of a query from an astronomer might be: "find dataset name, archive class, and date of archives containing information about pulsars in a specified region of the sky." In this particular case, because of the large number of tables in the database, the generated SQL statement is over one page of text. Without the automated query generator, it required a designer who is intimately familiar with the database more than twenty minutes to construct a rough draft of the same query and this did not include the time to debug it. This approach is a good illustration of using high-level models as vehicles to cope with underlying complexity.

As instrumentation becomes more complex, the calibration and data analysis process becomes correspondingly more complex. It is not an easy matter to discover what users want in an "ideal" data analysis system. (It is much easier to find out what they do not like.) However, users appear to want the following general features: (1) They want the easy tasks done in a straightforward and intuitive manner. (2) They want considerable flexibility in doing the difficult tasks. (3) They want the entire data analysis scheme wrapped in an intuitive graphical user interface. (4) They want to change what they do not like. (5) They want to be able to add their own custom developed software tasks to the analysis scheme without being told what computer language to program in or what packages they can or cannot use.

One approach to providing such a scheme is to view the system as consisting of a loosely coupled set of independent tasks. Such a system could be implemented provided there is a common data model recognized by all the tasks. If such a data model were defined within a given analysis environment, it would serve as the mechanism that unifies the set of independent tasks. New tasks could be added, provided they implemented the same data model. Such an approach would result in a much more open analysis environment than exists at present. A first step in such a process would consist of defining standard models of basic astronomical concepts, such as "image" and "spectrum," that are independent of particular computer languages and particular data analysis systems.

References

Batini, C., Ceri, S., & Navathe, S. 1992, Conceptual Database Design, An Entity-Relationship Approach (New York, Benjamin/Cummings)

Booch, G. 1994, Object-Oriented Analysis and Design with Applications, second edition (New York, Benjamin/Cummings)

NASA Office of Standards and Technology 1993, Definition of the Flexible Image Transport System (FITS) (Greenbelt, NASA/OSSA)

Schreier, E., Benvenuti, P., & Pasian, F. 1991, in Databases & On-Line Data in Astronomy, eds. M. Albrecht & D. Egret, (Dordrecht, Kluwer), p. 47

Semmel, R., & Silberberg, D. 1993, Telematics and Informatics, 10, 301

Williams, J. 1993, in Astronomical Data Analysis Software and Systems II, ASP Conf. Ser., Vol. 52, eds. R. J. Hanisch, R. J. V. Brissenden, & J. Barnes (San Francisco, ASP), p. 100

Propagating Uncertainties and Units in Data Structures

J. McDowell and M. Elvis

AXAF Science Center/Smithsonian Astrophysical Observatory, 60 Garden Street, Cambridge, MA 02138

Abstract. We describe a possible data structure designed to improve propagation of physics information in an analysis system, and an associated subroutine library for combining physical units.

1. Rationale

The Advanced X-ray Astrophysics Facility (AXAF) is a sophisticated X-ray observatory scheduled for launch in 1998. As part of the combined AXAF pipeline processing and data analysis system being developed at the ASC (AXAF Science Center), we are investigating possible data structures for the ASC data system. Our goal is to provide a system which treats scientific data as rigorously as possible, maintaining the integrity of the auxiliary information supplied with the data. Some existing systems propagate some kinds of auxiliary information (e.g., coordinate systems), with varying degrees of success; the multiwaveband quasar database analysis system developed by one of us (JCM) demonstrated the need for flexibility in handling and converting heterogeneous data in various units and coordinate systems when datasets taken in different wavebands are to be combined. Making the software do the bookkeeping reduces workload on the astronomer and improves the archival quality of the data products. These considerations led us to concentrate on some object-oriented concepts.

We now introduce a definition of a Physical Data Object. This prototype data structure is optimized for the requirements of our X-ray data analysis system, but has been designed with multiwaveband support in mind. It should be emphasized that this prototype is one of a number of possible approaches and does not represent our final design. We have proven some of the concepts discussed here via rapid prototyping and will implement full versions over the next year.

2. The Physical Data Object

The accompanying entity-relationship diagram shows the full structure of the Physical Data Object (PDO) prototype concept. (The diagram shows entities as rectangles, attributes as circles, two-way relationships as diamonds, and 'is-a' (A is an example of B) relationships as rounded rectangles.) The PDO is considered to be a 1-dimensional vector of length N each element of which is an M-dimensional array, together with associated attributes. The most common cases are M=0, i.e., a 1-D vector of scalars, and (N=1,M=2), i.e., an image.

A "Physical Data" Table is a collection of PDOs of the same length which share the same data subspace (see below); it is the analog of the FITS binary table. Any attribute of the object may have the value 'Not Present' and software will deal with that case. Support for Null (NaN) values and upper limit flags for numerical quantities will be provided at a low level. Arrays of data could be stored either explicitly or implicitly as coefficients of a functional expansion, as done in the UK's Asterix/NDF system. We would support three different kinds of uncertainty (statistical, and systematic scale and offset) to allow a greater degree of control over error propagation. Each data value is considered to be stored in a 'local' coordinate system, and the WCS (World Coordinate System) allows it to be registered on a 'global' system. We do not allow WCS to provide world coordinates mixing different physical objects; we instead support multidimensional objects so that, for example, X and Y positions are stored as a single 2-D object.

We also introduce a Data Subspace infrastructure which is a little different from WCS; it records from where the data was extracted (including coordinates which are no longer in the data). For X-ray data this will typically include an instrument ID, a set of good time-intervals, a set of detector coordinate regions versus time, and a pulse-height range. The Data Subspace allows one to combine different datasets. Note that in practice the detector coordinate regions will usually be specified with a single region fixed in world coordinates coupled with a WCS that varies with time (the 'aspect history'). The Data Subspace library will also support a Bad Value Mask, an integer vector which can be used to temporarily mask elements of the data vector, allowing rapid re-screening.

The PDO might also support an Exposure Map Array: this array is usually defined implicitly rather than carried around with the data. It has the size and shape of the physical data array, and its pixel values represent the exposure time for that element of the array. It is generated by projecting the data subspace on the relevant axes.

3. Software Library Infrastructure

To use the proposed data objects, we would require a set of libraries to provide services. A PDO I/O library would provide routines to get and put PDOs and each of their sub-objects. A Unit handling library parses physical unit definitions, and provides simple routines to combine them. An Uncertainty handling library handles simple cases of combining uncertainties, and contains output routines to combine available uncertainties into a single quoted error estimate. Applications would still need to provide their own formulation of the correct propagation of uncertainties and units, but the libraries provide a simplified notation for doing so (e.g., a routine unc_rms(A,B) might combine the uncertainties in a stack A of objects and assign the root mean square of the result as the uncertainties for the result object B). An Implicit Array library supplies an interface to make functionally defined data items look like a simple tabulated array. A WCS library will handle propagation of WCS data. Finally, a Data Subspace library would handle adding data subspaces, projecting them and extracting subsets of them.

4. Storing Physical Objects in FITS

We want to be able to archive PDOs in FITS format. One approach would be to use the ability of FITS BINTABLES to store multidimensional objects and keep the uncertainties together with the data values in a single 'Item' column. We feel this would make it harder for simple programs to access the data. We therefore intend to take the approach of storing each axis of the data (e.g., X, Y values) and each uncertainty in separate columns (or keywords for scalars), and encoding the connection between them in special header keywords. This means that other software systems reading the file will lose the PDO superstructure, but will see the actual data in a familiar form. Our goal is for the resulting data files to be valid HEASARC type FITS files where appropriate, with extra header keywords and columns that the current HEASARC software would ignore. Our design would enable backward compatibility with any system that accepts such HEASARC FITS files.

5. The Prototype ASC Units Library

Our prototype units handling library is now being tested. The prototype is written in Fortran and has been tested on a SPARC machine running SunOS Unix. The operations we support are: (1) multiplication and division of units, (2) raising of units to a real power, and (3) reducing compound units to the base units which define them. We use the following syntax to manipulate units:

$$m \times 10^{e_0} u_1^{e_1} u_2^{e_2} ... u_n^{e_n}$$

For example, a simple valid unit is cm and a more complicated one is

6.67x10^{-11} m^3 s^{-2} kg^{-1}:

The braces are optional, but allow the output string to be directly typeset by TeX or compatible output programs (e.g., the SM plotting package which supports TeX syntax in its axis labels).

Simple use of the program requires no knowledge of the units, treating the individual units as simple tokens. The user may supply a Unit Definition File, simply an ASCII file with a list of known units, some of which are defined in terms of others. This allows the program to parse SI prefixes (without this it doesn't know whether 'pc' is a parsec or a pico-c) and, on request, to resolve compound units.

Since the choice of which units are compound and which are fundamental is specified at run time with a Unit Definition File, the library is very flexible. This is illustrated by one example supplied with the library, in which the unit definition file specifies the base units as 'm' and 'yr', and defines 'pc' and 's' in terms of them. This allows the software to divide the units 1 and 50 km s^-1 Mpc^-1 to obtain the result 1.96x10^{10} yr, the Hubble time.

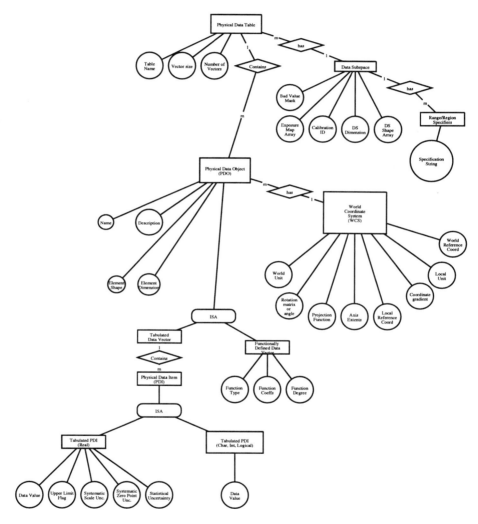

Figure 1. Structure of a Physical Data Object

An Abstract Data Interface

D. J. Allan

Department of Physics and Space Research, University of Birmingham, UK

Abstract. The Abstract Data Interface (ADI) is a system within which both abstract data models and their mappings on to file formats can be defined. The data model system is object-oriented and closely follows the Common Lisp Object System (CLOS) object model. Programming interfaces in both C and FORTRAN are supplied, and are designed to be simple enough for use by users with limited software skills.

The prototype system supports access to those FITS formats most commonly used in the X-ray community, as well as the Starlink NDF data format. New interfaces can be rapidly added to the system—these may communicate directly with the file system, other ADI objects or elsewhere (e.g., a network connection).

1. Introduction

The Asterix X-ray analysis package has hitherto dealt solely with data belonging to the Starlink Hierarchical Data System (Warren-Smith & Lawden 1993). In an attempt to broaden the scope of the package it was decided to add FITS (NASA Office of Standards and Technology 1993) compatibility at a low level, rather than simply providing format conversion software. As this addition involved a rewrite of the existing interface library to binned data, a general redesign of the way applications access their data was undertaken.

Two important goals of the data interface redesign were to remove all file-format specific code from general purpose applications, and to make adding new functionality to the data interface less time consuming. In the old system, FORTRAN common blocks would usually be changed, necessitating a rebuild of the entire data interface. It was also important to retain efficiency of access to the various data formats which have different natural modes of access. For example, HDS is poor at representing tabular data and software using HDS tends to be strongly column oriented, whereas FITS more efficiently supports row access. We also needed to improve fault tolerance. When a user has a file whose components do not match what the data interface expects, the system should be able to cope, preferably by the user editing an ASCII text file, rather than altering the dataset. We also wanted to simplify the programming interface to encourage novices to write analysis tasks, and to provide FORTRAN and C interfaces. (The demand for a C interface has been growing steadily.) Finally, we wanted to ensure that the redesign would not have to be repeated too soon!

2. Data Models

ADI is a system for defining "abstract data models." These models generally provide a "view" of some underlying object, such as a data file whose structure we want to conceal from application software. Different views of the same underlying object are possible, but the most useful feature as the ability to support the same view of different objects.

A data model consists of a number of "slots" which have a name and a value. Figure 1 shows a schematic view of the Array abstract model. The

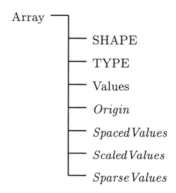

Figure 1.   Structure of the Array abstract model

names in upper case denote slots which are required to define a new instance of a particular model. The Array definition allows for the possibility of unset values, so the Values slot does not fall into this category. Names in normal type but capitalized are those slots which a user should expect to be defined for any instance of an Array. Names in italics mark slots which need not have defined values for any instances of an Array. The slot values may be any of the usual scalar types (integers, floats, strings and logicals), arrays of these scalars, or any other ADI object. ADI can chain together data models to provide further layers of abstraction, e.g., an "image" view of an event table is possible if there exist both spatial coordinate lists for the events and some default spatial bin size.

In Asterix a further layer of abstraction, this time a procedural one, is used to insulate the application from details of a particular abstract model. A data model object which is not linked to anything else exists purely as a construction in memory. This can be very useful as an aid to the application developer, as dynamic objects of arbitrary complexity can be created, obviating the need for the "make the arrays as big as we'll ever need" style of programming. An image processing application, for example, could work by maintaining a stack of X-Y image objects, allowing operations on the data to be undone if required.

3. Implementation

ADI was designed as an object-oriented system where data models are defined as classes. The object model chosen is that of the Common Lisp Object System (CLOS) (Bobrow 1988). The building blocks of the ADI system are primitives, classes, instances, generic functions and methods.

Primitives ADI supports most of the basic types used in FORTRAN and C, as well as arrays of these types with up to 7 dimensions.

Type Name	Description	Short-code
_UBYTE	Unsigned byte	UB
_BYTE	Signed byte	B
_UWORD	Unsigned word	UW
_WORD	Signed word	W
_INTEGER	Signed integer	I
_REAL	Single precision F.P.	R
_DOUBLE	Double precision F.P.	D
_CHAR	Character string	C
_LOGICAL	Logical	L

Table 1. Supported primitive data types.

ADI will perform type conversion when the access type for a primitive does not match its storage type. ADI also allows new primitive types to be defined.

Classes A "class" is a form of data structure which is a collection of "slots." Each slot has a name and a value (which may be null). A particular object of a class is called an "instance." ADI has several built-in classes which it uses to manage its own internal data storage. A class can inherit the slots and behavior of existing classes in the usual CLOS fashion.

ADI data models are implemented as ADI classes which are derived from the class `ADIdataModel` either directly by inclusion in their superclass list, or indirectly by inheriting a class which is thus derived. Inheriting this class enables a new class to act as the viewing object in a data model chain (classes do not have to be so derived to be the rightmost element in the chain).

Generic Functions The "generic" function in ADI is a function definition with no implementation. The term generic is used because the arguments to the function can be of any type. The implementation of a generic function is distributed over a set of "methods."

Methods ADI lets the user define methods to perform all the built in ADI generic functions (which are used to access data models and their slots), as well as defining new generic functions. A method can be specialized to work on particular classes, classes with particular objects to the right in the ADI chain, or even a specific ADI object.

The CLOS method forms "primary," "around," "before," and "after" are all supported. Methods are executed following the CLOS method combination rules (Bobrow et al. 1988; Keene 1989).

4. Status

A general purpose library for defining and using data models has been constructed. ADI is completely configurable at a low-level, but through its support for complex data models can provide a very high-level data interface. It encourages the application programmer to concentrate on the data objects being manipulated and the relationships between them. The result of the changes when applied to Asterix will be package which processes data according to a given set of data models, rather than a package which processes HDS files.

It is anticipated that a stand-alone release of ADI will be available by the end of 1994; releases of Asterix after that date will incorporate ADI in an increasing fraction of the applications.

References

Bobrow, D. G., DeMichiel L. G., Gabriel, R. P., Keene, S. E., Kiczales, G., & Moon, D. A. 1988, Common Lisp Object System Specification, X3J13 Document 88-002R

Keene, S. E. 1989, Object-Oriented Programming in COMMON LISP (Reading, Addison-Wesley)

NASA Office of Standards and Technology 1993, Definition of the Flexible Image Transport System (FITS) (Greenbelt, NASA/OSSA)

Warren-Smith, R. F., & Lawden, M. D. 1993, HDS—Hierarchical Data System, Starlink Document SUN 92

Astronomical Data Analysis Software and Systems IV
ASP Conference Series, Vol. 77, 1995
R. A. Shaw, H. E. Payne, and J. J. E. Hayes, eds.

Proposed FITS Keywords and Column Headers for ALEXIS Mission Data Files

J. Bloch and J. Theiler

Astrophysics and Radiation Measurement Group, Nonproliferation and International Security Division, Los Alamos National Laboratory, Los Alamos, NM 87545

Abstract. We propose a set of standard character strings to be used in FITS files generated by data from the *ALEXIS* (Array of Low Energy X-ray Imaging Sensors) satellite mission.

1. Introduction

The Array of Low Energy X-ray Imaging Sensors (*ALEXIS*) satellite contains two experiments: a VHF radio-frequency ionospheric experiment called "BLACKBEARD," and the *ALEXIS* experiment itself, which consists of six low-energy X-ray telescopes in three pairs: named 1A, 1B, 2A, 2B, 3A, and 3B. These are multilayer mirror telescopes with micro-channel plate detectors, and each is tuned to a relatively narrow ($\Delta E/E \approx 5\%$) energy bandpass (see Table 1).

Telescope:	1A	1B	2A	2B	3A	3B
Energy	93 eV	71 eV	93 eV	66 eV	71 eV	66 eV
Wavelength	130 Å	172 Å	130 Å	186 Å	172 Å	186 Å

Table 1. The bandpass energy and wavelength for each of the six narrow-band telescopes on the *ALEXIS* satellite.

The scientific objective of the *ALEXIS* experiment is to map out the sky in these narrow energy bands and search for transient EUV sources as well as map out the diffuse EUV background (Priedhorsky et al. 1990; Bloch et al. 1994). The project is a collaborative effort between Los Alamos National Laboratory, Sandia National Laboratory, and the University of California–Berkeley Space Sciences Laboratory. The satellite is controlled entirely from a small ground station located at Los Alamos.

The six telescopes are arranged in three pairs. The satellite is always in scanning mode, and during each (≈ 50 s) rotation period they scan most of the anti-solar hemisphere of the sky. No pointed observations are possible. Each f/1 telescope consists of a spherical, multilayer-coated mirror and a photon counting detector containing a curved microchannel plate located at the prime focus. In front of each detector is a thin metal or plastic UV rejection filter. The spacing in the periodic layers of the multilayer coatings determine the bandpasses of the telescopes. The field of view of each telescope is 30° with a spatial resolution of 0°.25, limited by spherical aberration. Peak effective on-axis collecting areas range from 0.05 to $0.25\,\text{cm}^2$, while the peak area-solid angle products of each telescope range from 0.01 to $0.05\,\text{cm}^2\,\text{sr}$.

Each X-ray photon or background event from each telescope is tagged with its time of arrival and location in the field of view. Ground processing must combine these event lists with the satellite aspect solution to place each photon back onto the proper place on the sky.

The mission was launched on 1993 April 25 by a Pegasus booster dropped from a B-52 into a 400 × 450 nautical mile, 70° inclination orbit. Unfortunately the satellite suffered damage to one of its solar arrays on launch, and the single spacecraft magnetometer, which was located on the damaged panel is dead. The missing magnetometer and the modified (and time varying) spacecraft mass properties have forced the project team to invent new methods for both controlling the orientation of the satellite and to determine spacecraft attitude. As of 1994 September, *ALEXIS* has generated over 55 CD-ROMs (650 MB each) of telemetry data. About 50% of that data are from the six telescopes. In the year since launch, the team has developed an attitude algorithm which is providing attitude solutions close to the originally desired 0.25° precision.

The current data processing system for *ALEXIS* relies on two internally generated data formats that date back to the beginning of the project in 1988, Generalized Data Format (GDF) files and IDL Data Format (IDF) files (Bloch et al. 1992, 1993). In order to use *ALEXIS* data with various software packages that are available in the high energy astrophysics communities, we need to define the correspondence between *ALEXIS* data in GDF or IDF files, and FITS files.

2. Proposed Keywords

The following is our current proposal for keywords to use for storing raw and processed *ALEXIS* mission data in FITS format.

TELESCOP= 'ALEXIS ' is the name of the mission.

INSTRUME= 'TPi ' corresponds to the three telescope pairs (i=1,2,3) which are the instruments in the *ALEXIS* mission. Data from a given region of the sky can be detected in either one or both elements of the binocular pair that is scanning that part of the sky. (The coverage of the different telescope pairs is non-overlapping.) We will use:

DETNAM = 'A ' to identify data from telescope A only,

DETNAM = 'B ' for data from telescope B only, or

DETNAM = 'AB ' for combined data from both A and B.

We will use the filter keyword to specify the fixed mirror, filter, and detector complement of each telescope, since they all determine the telescope's bandpass response. These will be specified by concatenating strings identifying the mirror ID, the filter material, the filter ID, the photocathode material, and the detector ID. The mirror/filter/detector combinations that are on the telescopes now (and therefore are fixed) are:

```
1A: FILTER  = 'XRO4770_Lex/Ti/B-03639-2_MgF2-AF01'
1B: FILTER  = 'XRO5371_Al(Si)/C-03479-1_NaBr-AF06'
2A: FILTER  = 'XRO4763_Lex/Ti/B-03606-1_MgF2-AF03'
2B: FILTER  = 'XRO5339_Al(Si)/C-03483-1_NaBr-AF08'
3A: FILTER  = 'XRO5369_Al(Si)/C-03479-2_NaBr-AF05'
3B: FILTER  = 'XRO5306_Al(Si)/C-03483-2_NaBr-AF07'
```

Three observation modes have been identified as relevant to the *ALEXIS* project. We will use:

OBS_MODE= 'SCAN ' to indicate the normal spinning, drifting satellite condition,

OBS_MODE= 'SPIN_UP ' to indicate that Torque Coil activity is in progress to change satellite spin period, and

OBS_MODE= 'MANEUVER' to indicate that a Torque Coil maneuver is in progress to change satellite spin vector orientation. We note that the SLEW keyword is usually reserved for satellites that are capable of a stable pointing.

The raw data can be output in three different formats (Klarkowski 1992):

DATAMODE= 'RAWTEL ' is the raw data mode which puts detector Wedge Strip ZigZag anode values for each event into the data stream. (It turns out that almost all flight data is in this mode).

DATAMODE= 'REDTEL ' is a reduced data mode which puts X,Y and timestamp values for each event into the data stream.

DATAMODE= 'COMPTEL ' is a compressed data mode which puts X,Y, and the delta times between events into the data stream.

3. Proposed Column Headers for Photon Event Lists

Currently, photon lists are stored in an in-house format (either GDF or IDF). In migrating these existing files to FITS formats, we propose the following non-inclusive list of generic column headers (these will be of the form TTYPEn = 'header').

TIME is the double precision time in seconds since 1993 January 1.

DETX and DETY are the integer X and Y detector locations, each ranging from 0 to 127.

PHA is the long integer (32 bit) pulse height value for each event. For our telescope, pulse height contains little or no photon energy information, but instead tells about microchannel plate gain and background performance.

PI will be used for *ALEXIS* to denote spatial gain corrected pulse height values.

RA and DEC are currently stored in terms of XYZ direction cosines in event GDF records, and will be converted to right ascension and declination as part of the file transfer.

ENERGY and WAVELENGTH are essentially fixed by the narrow bandpass in *ALEXIS*; we will occasionally use this column when combining photons from several telescopes into one data file. The units would be in eV or Å.

STATUS of the data is currently kept in a set of instrument flags and a set of event tagging flags; these bits will be combined into a single word for this column.

The following are *ALEXIS*-specific column header names:

- TELID identifies with which telescope a given photon was observed. This knowledge will be important since many of our data sets combine data from several different telescopes.

- RAWW, RAWS, and RAWZ are integer values for the raw Wedge, Strip, and Zigzag anode values.

- HIVOLT is the single precision high voltage on the detector at the time the photon was observed.

Acknowledgments. This work was supported by the Department of Energy.

References

Priedhorsky, W. C., et al. 1990, in Extreme Ultraviolet Astronomy, eds. R. F. Malina & S. Bowyer (New York: Pergammon Press)

Klarkowski, J. R. M. 1992, ALEXIS DPU Software Requirements Specifications, Sandia National Laboratories Document AD038

Bloch, J. J., Smith, B. W., & Edwards, B. C. 1992, in Astronomical Data Analysis Software and Systems I, ASP Conf. Ser., Vol. 25, eds. D. M. Worrall, C. Biemesderfer, & J. Barnes (San Francisco, ASP), p. 502

Bloch, J. J., Smith, B. W., & Edwards, B. C. 1993, in Astronomical Data Analysis Software and Systems II, ASP Conf. Ser., Vol. 52, eds. R. J. Hanisch, R. J. V. Brissenden, & J. Barnes (San Francisco, ASP), p. 243

Bloch, J., et al. 1994, in EUV, X-Ray, and Gamma-Ray Instrumentation for Astronomy V, SPIE Vol. 2280, p. 297

ASC Data Structures and Model

M. Conroy, R. Simon, J. McDowell

SAO/ASC, 60 Garden St., Cambridge, Mass. 02138

K. Barry

TRW/ASC, 60 Garden St., Cambridge, Mass. 02138

Abstract. The *AXAF* Science Center (ASC) is using an "open architecture" approach to develop the data analysis system. The Dynamic Data File (DDF) abstraction is being developed to provide sophisticated, format-independent filtering capabilities and support for multiple file formats. The DDF interface library will use the Open-IRAF data structure libraries to provide environment independent access to the DDF data structures from any C or FORTRAN host-level application.

1. Introduction

The characteristics of the X-ray data model and the goals of the ASC data analysis system, combined with the concept of an open architecture, lead to the design of the Dynamic Data Format (DDF) abstraction and interface library. DDF isolates the user from the details of the underlying implementation, while allowing the same analysis programs to accept a variety of data input formats, such as FITS and QPOE.

2. X-Ray Data Model

The definition of an X-ray data model consists of abstract data structures, access functions, and an astronomical interpretation (Farris & Allen 1992). Observational data in X-ray astronomy can be divided into the following data components (each decomposable into data structures): The primary data contain the science information and traditionally appear as either photon event-lists or image arrays. The ancillary data consist of supporting engineering and housekeeping observational information, such as voltages and satellite aspect. The derived data record the results of the analysis of the primary data, such as extracted spectra and detected source lists. The associated data represent information needed for correct interpretation of any of the above quantities, such as units, uncertainties, and errors. Another item, volumes, can be thought of as a quantity to represent the size of the sample space, e.g., exposure time. The meta-data are needed to link all of these data together.

3. Primary Data Access Requirements

The primary data-access functions have received the most attention in current data analysis systems. Most of these access requirements are now well understood, and a base of software tools is accumulating. For example, PROS uses the IRAF QPOE data format to store primary data and provides for most of the access requirements. These requirements include the data access functions for an event-list, which must support a photon record with arbitrary attributes. The photons of the event-list must be selectable both by attributes in the file, such as energy and time, as well as engineering attributes from ancillary files, such as the master-veto rate and the viewing geometry. The selection must be possible without intermediate files and the access must be able to produce both image and event-list output.

The data must also be interpretable in multiple World Coordinate Systems (WCSs), as well as in stored-data coordinates. Items such as exposure time and energy band must be retrievable dynamically, to reflect the selection criteria described above. The event structure must also be extensible, with derived quantities. Finally, derived data must be extractable from the primary events. Data such as light-curves and spectra are derived from the primary data by binning on one or more attributes.

4. Ancillary and Derived Data-Access Requirements

The ancillary and derived data-access functions have received less attention in current data analysis systems. The access functions that do exist are, for the most part, independent of the primary data. The software is less well developed and the access functions to connect these data to the primary data are less mature. For optimal results, the ancillary data lists must be accessible with the same filters as the primary data. Also, the WCS conventions must be extended from the primary data to these data, and include non-celestial coordinate systems such as interpolated functions and sampled points. As well, the observation-specific calibration data must be generated by matching data selections from both primary and ancillary data. The resulting calibration files will then correctly correspond to the filtered observation of the primary data. Finally, the low counting statistics prevalent in X-ray observations dictate the need to calculate uncertainties of all data quantities, and to propagate these uncertainties during the analysis process.

5. Dynamic Data Format (DDF)

One of the key developments of the ASC data system is the DDF data abstraction. The principal features of this abstraction are that the software abstraction allows applications to be completely independent of the underlying physical format. Changes required by a change in physical file format are isolated to the DDF library implementation. In turn, the DDF library implementation can support more than one physical format. The current design plans include support for FITS, QPOE, image, table, array, and ASCII list. The DDF library also makes extensive use of the Open-IRAF data structure libraries. These libraries de-couple the data structures from the IRAF environment, making them exportable and available to stand-alone applications.

Filtering capabilities must be an integral part of the data access I/O routines rather than a feature layered on top of the I/O. Otherwise, applications are required to have knowledge of the file format or contents. The DDF library design accepts a filter specification string with every data retrieval function, thus allowing the user to tailor filter specifications to the content of each file, without application code modifications.

The DDF abstraction, as well as the underlying physical formats listed above, support an extensible, self-defining data structure. This allows application code to be completely independent of the items or data-types contained in the physical file. The DDF design, by pairing a data filter expression with every data retrieval function, allows the application to process a virtual file. That is, the application sees a data file that contains only the records specified by the filter string, thus eliminating the need to create a physical instantiation of the filtered file, unless specifically desired. Many summary quantities describing a data file are functions of the file selections. Access to a virtual filtered file requires the ability to define summary functions and to evaluate them dynamically.

6. Technical Issues to Resolve

Many of the basic functional elements for the DDF implementation already exist in various forms. However, there remain some technical issues to resolve before the design is complete. For example, PROS uses the STSDAS TABLE format to store ancillary and derived data, but this format provides a different set of capabilities from the QPOE structure used for the primary data.

Some utilities that allow the extension of an event structure exist, but more sophisticated tools that allow creation of a new event structure by selecting and merging items from multiple existing lists are needed. Also, the mechanism for linking the primary event-list to ancillary lists and calibration files is missing. The mechanism for associating and propagating uncertainties is a topic of prototyping activities.

Work also continues on FITS/WCS support for image format in the FITS community (Greisen & Calabretta 1995). Work to extend the FITS WCS conventions to TABLE and BINTABLE formats is being pursued at SAO/ASC and HEASARC for the High-Energy Astrophysics domain (Corcoran et al. 1995). Finally, mapping the DDF data abstraction to FITS needs further development. Current work on FITS/HDF interoperability (Jennings, Pence, & Folk 1995), and the work in the AIPS++ group to map their class libraries to FITS are both promising directions.

7. Status

There are several efforts currently underway that represent necessary steps towards a complete DDF design and implementation. The ETOOLS ADP grant is supporting the development of Event Tools as a joint project between CEA/Berkeley and SAO/Cambridge. The products of this work will include: a package of QPOE support tools, a QPOE browser (with a GUI), and a delivered C-callable library for the QPOE API. Projects are also underway to integrate IRAF TABLES and QPOE data formats. The STSDAS Group at ST ScI is extending the existing tables support to store a TABLES-compatible data structure within an IRAF QPOE file. Similarly, the IRAF Group at NOAO plans to develop a Common Data Format (CDF) that will allow the storage of additional data structures within an IRAF physical data file, such as QPOE.

Work to produce a complete data model of the X-ray data is still preliminary, but data modeling efforts by the AIPS++ project (Farris 1993) offer a promising direction.

Acknowledgments. This work was partially supported by NASA contract NAS8–39073. We also thank A. Farris, B. Glendenning, and G. van Diepen of the AIPS++ project for their generous assistance with astronomical data modeling.

References

Greisen, E., & Calabretta, M. 1995, this volume, p. 233

Corcoran, M., Angelini, L., George, I., Pence, B., McGlynn, T., Mukai, K., & Rots, A. 1995, this volume, p. 219

Jennings, D., Pence, W., & Folk, M. 1995, this volume, p. 241

Farris, A. 1993, in Astronomical Data Analysis Software and Systems II, ASP Conf. Ser., Vol. 52, eds. R. J. Hanisch, R. J. V. Brissenden, & J. Barnes (San Francisco, ASP), p. 145

Farris, A., & Allen, R. J. 1992, in Astronomical Data Analysis Software and Systems I, ASP Conf. Ser., Vol. 25, eds. D. M. Worrall, C. Biemesderfer, & J. Barnes (San Francisco, ASP), p. 157

Astronomical Data Analysis Software and Systems IV
ASP Conference Series, Vol. 77, 1995
R. A. Shaw, H. E. Payne, and J. J. E. Hayes, eds.

Reformatting the Ginga Database to FITS and the Creation of a Data Products Archive

R. H. D. Corbet[1], C. Larkin

Pennsylvania State University, 525 Davey Lab., University Park, PA 16802

J. A. Butcher, J. P. Osborne

Department of Physics and Astronomy, University of Leicester, Leicester, LE1 7RH, UK

J. A. Nousek

Pennsylvania State University, 525 Davey Lab., University Park, PA 16802

Abstract. We present status reports on projects to make both raw data and products from the *Ginga* X-ray astronomy satellite available in FITS formats to the general astronomical community.

1. Introduction

The highly successful *Ginga* X-ray mission was developed in a Japanese/British collaboration under the leadership of the Institute of Space and Astronautical Science (ISAS) in Japan. The main Large Area proportional Counter (LAC) detector had a collecting area of approximately 4000 cm^2 and operated for almost 5 years producing approximately 40 GB of data. The data are, however, stored in a highly non-standard format known as FRF (first reduction file) which closely parallels the telemetry stream from the satellite. This seriously hinders the accessibility of these data to the majority of astronomers. We are therefore undertaking a project to convert these data from the FRF format to FITS. The intention is to thereby enable the data to be analyzed using a variety of existing software such as Xanadu, IDL, and PROS.

The FITS converter is based on software developed at Leicester University known as "`sortac`". The details of the FITS formats to be used are still being debated but will be based on existing formats such as the ASCA GIS MPC mode in order to simply the task of analyzing data using existing software. The FITS conversion software is being developed in close collaboration with the HEASARC at the Goddard Space Flight Center which will be responsible for archiving the FITS files and making these data accessible to the astronomical community. In parallel with the data reformatting project, a database of software products (light curves and spectra) is being produced at Leicester University which will also be made widely available.

---

[1] Now NASA/GSFC, Code 666, Greenbelt, MD 20771

2. The Ginga Mission

Ginga was launched on 1987 February 5 and remained operational until reentry on 1991 November 1. The two X-ray detectors carried were a large area proportional counter (LAC) with approximately 4000 cm^2 collecting area (Makino 1987) and an all sky monitor (ASM). The LAC (Turner et al. 1989) was developed by ISAS in collaboration with the University of Leicester and the Rutherford Appleton Laboratory in the UK and the University of Tokyo and Nagoya University in Japan. The LAC consisted of eight identical proportional counters and covered the energy range of approximately 1.5 to 30 keV with a field of view of $1° \times 2°$ (FWHM). Background rejection was obtained from several guard counters and anti-coincidence among individual LAC modules.

During operation the LAC observed approximately 350 targets. To date, over 190 papers in refereed journals have been written based on these observations. Notable highlights include the detection of X-rays from SN1987A, the discovery of six new pulsars, the identification of three new black hole candidates, discovery of cyclotron lines in the spectra of several pulsators, and astrophysically important spectral and timing information on a variety of active galaxies.

The ASM (Tsunemi et al. 1989) collected data typically once per day when *Ginga* executed a 360° rotation. Long term light curves for a large number of bright X-ray sources were thus produced. However, our project does not include the archiving of these data.

The *Ginga* data set totals 40 GB. To date, the principal means of access by US astronomers has been the *Sirius* database system, available only at ISAS. A few investigators from outside Japan and the UK have made use of the *Ginga* data, but this has been limited by the general requirement that U.S. scientists, for example, must undertake the majority of their data reduction in Japan. The *Ginga* data sets provide an opportunity to cross-compare with data from other satellites, but due to the accessibility difficulties, the existing data have not yet been fully exploited.

3. Data Products Archive

The *Ginga* data archive at Leicester is currently held in FRF form, and guest investigators visiting Leicester are supported in their analysis of *Ginga* data on DEC Alpha workstations running OSF. The existing *Ginga* data analysis software at Leicester is being used to produce a products database of spectra and lightcurves. A spectrum and lightcurves in two colors will be available for each observation. The lightcurves will be at the lowest time resolution for each data mode (e.g., 16 s for MPC1 data).

The products will be cleaned and background subtracted. Stringent cleaning criteria and quality control will be applied and a quality assessment will be provided. Background subtraction of *Ginga* observations is not a trivial process. The LAC had no facility for simultaneous background measurement and backgrounds for source observations have to be reconstructed from contemporaneous blank sky observations or from a more general model. Fluctuations in the diffuse X-ray background can, however, cause significant problems. Wherever possible, the products database will provide products produced with the two different methods of background determination to provide the user with an estimate of the importance of these systematic effects.

To ensure maximum utility of the products database, we intend that users will be able to extract spectra from different time intervals. Thus, although access to the raw data will be required to answer detailed questions, the non-expert user, and users requiring answers to basic questions, will not need to go back to the raw data.

The data products will be written in OGIP FITS format and will be made available through the Leicester Data Archive Service (LEDAS), which incorporates the HEASARC BROWSE database system. In this way they will be easily accessible over the Internet. The raw FRF data are now held in a CD-ROM jukebox, and will very shortly also be available over the net to expert users (until the FITS-converted data become available). Up to date information on the status of the *Ginga* products archive project can be obtained via the LEDAS[2] World Wide Web home page.

4. Reformatting to FITS

To minimize the amount of code that needs to be written the FITS reformatter takes *sortac* as the starting point. In parallel with creating hypercube format data products, the output routines are being rewritten to write FITS files as well. These routines make heavy use of the FITSIO library written by Pence (1992). The status of the *sortac* modifications is that header information is now written to FITS files and we have created a parameter file interface (cf. IRAF parameter files and the ASCA reformatter) which facilitates running in batch mode.

Science information will be stored in a binary table extension similar to that used with ASCA (cf. Corbet et al. 1992). The precise FITS formats will be defined in close collaboration with a project to reformat the HEAO-1 data base to ensure maximum compatibility. It has not yet been decided whether to store house keeping information as separate files or as separate extensions within science files. It is expected that the FITS creation software will be run once on the entire *Ginga* data set. The resulting FITS files will then be made available to the community by the HEASARC[3] at the Goddard Space Flight Center.

The *Ginga* LAC could be operated in several modes. These made trade-offs between time resolution and detector information (which affects the accuracy with which spectra can be measured). The FITS formats for the various modes will be similar, with the principal difference being the number of columns in the binary tables.

MPC1 mode: mainly used for spectral studies of faint sources. Events are accumulated in sixteen separate spectra of 48 channels each. The sixteen spectra comprise the top and middle layers from each of the 8 detectors. The separation of top and middle layers improves the signal to noise ratio for weak sources and helps the background estimation.

MPC2 mode: provides compression of data by a factor of eight enabling better time resolution. Combines top and middle layers from four detectors into

---

[2] http://darc-www.star.le.ac.uk/public

[3] http://heasarc.gsfc.nasa.gov/

one and gives two separate spectra. The combination of layers decreases the signal to noise ratio and background estimation is less precise.

MPC3 mode: carries the process further by combining the 48 energy channels into twelve and grouping all eight detectors together, giving a further factor of eight compression.

PC mode: used for timing studies; it by-passes the ADC to avoid dead time effects. Signals are divided into two energy bands by three discriminators (lower, middle and upper) and no other energy information is retained. In this way the dead time is reduced to $16.5\mu s$/event and time resolution down to $976.6\mu s$ ($= 1/1024$) is obtained from two energy bands per detector group. The lower discriminator is the same as that used for the pulse height spectra, while the other two discriminators have two commandable levels.

5. Other Tools

In addition to the reformatting work tools will be required to make the *Ginga* data useful to astronomers. The main additional software that is required is: (1) Response matrix generator, and (2) Background subtraction. It is envisaged that these will be provided as "FTOOLS" (e.g., Pence et al. 1993). Response matrix generation has already been implemented by K. Ebisawa for use with XSPEC and it is expected that this software will be converted to an FTOOL. Background subtraction for *Ginga* is either done making use of source free regions of sky observed close to the target or by making use of relations between parameters such guard counter rates, time since SAA-passage and other factors which are correlated with the non-X-ray background (e.g., Hayashida et al. 1989).

Acknowledgments. This work is funded in part by NASA contract NAS5-32489.

References

Corbet, R. H. D., Larkin, C., & Nousek, J. A. 1992, in Astronomical Data Analysis Software and Systems I, ASP Conf. Ser., Vol. 25, eds. D. M. Worrall, C. Biemesderfer, & J. Barnes (San Francisco, ASP), p. 106

Hayashida, K., et al. 1989, PASJ, 41, 345

Makino, F., and the Astro-C Team 1987, Astrophys. Let. Commun., 25, 223

Pence, W. D. 1992, in Astronomical Data Analysis Software and Systems I, ASP Conf. Ser., Vol. 25, eds. D. M. Worrall, C. Biemesderfer, & J. Barnes (San Francisco, ASP), p. 22

Pence, W., Blackburn, J. K., & Greene, E. 1993, in Astronomical Data Analysis Software and Systems II, ASP Conf. Ser., Vol. 52, eds. R. J. Hanisch, R. J. V. Brissenden, & J. Barnes (San Francisco, ASP), p. 541

Tsunemi, K., et al. 1989, PASJ, 41, 373

Turner, M. J. L., et al. 1989, PASJ, 41, 739

ns IV
Source-searching in Photon-event Lists without Imaging

C. G. Page

X-ray Astronomy Group, University of Leicester, UK

Abstract. Many methods have been proposed for the detection of point sources in images. In re-processing data from the *ROSAT* extreme ultra-violet sky-survey we have made use of two different methods. One of these works directly on an event list, avoiding the losses which occur when events are binned into images.

1. Introduction

The first complete survey of the sky in the extreme ultra-violet (EUV) band was carried out by the Wide Field Camera (WFC) on *ROSAT*. The data, collected from 1990 August through 1991 January, were analyzed jointly by the five institutes in the UK *ROSAT* Consortium. The aim of our initial reduction (Page & Denby, 1992) was to obtain results without delay, so we did not expect to achieve the ultimate in sensitivity and completeness. The resulting *Bright Source Catalogue* (Pounds et. al. 1993) listed 383 EUV sources, a notable increase on the mere handful known previously.

The entire data-set of about 4×10^8 events has now been re-processed to get the best possible sensitivity, reliability, and completeness. Software improvements have been made in many areas, especially in determining the telescope aspect from the star-tracker data, in filtering out data with high background levels, and in searching for point sources.

2. Source Detection Methods

Textbooks on signal detection theory tell us that the matched filter is the optimum detector of a signal of known shape buried in stationary, Gaussian, additive noise. Even with our count-rates, clearly low enough to be in the Poissonian regime, matched filtering is still an attractive technique. Its simplest implementation involves binning the events into an image, convolving this with a unit replica of the point-spread function (PSF), and searching for peaks in the resulting array. This is relatively easy to compute, but there is one notable drawback: if the raw data have a Poissonian distribution there is in general no analytic function for the distribution of the convolution products. This makes it hard to set confidence limits for detections. A Gaussian approximation (e.g., Marshall 1994) is really only good enough to set a detection threshold for an initial screening pass.

This problem can be overcome by fitting the PSF to each peak in the raw image by least-squares and using a χ^2 test for significance. Unfortunately if, as often happens, some pixels have zero counts, determining the data variance presents a problem. Again there are ways around this (e.g., Kearns, Primini, & Alexander 1995), but the proper solution is to use maximum likelihood.

Maximum likelihood techniques are now widely used in astronomy: Cash (1979) pointed out the valuable properties of the C-statistic, which has a χ^2 distribution under almost all circumstances. This makes it easy to set a detection threshold, and to determine confidence limits in the fitted parameters. For an image where e_{ij} counts were expected and n_{ij} counts observed in pixel (i,j), the C-statistic to be minimized is:

$$C = -2 \sum_{ij} (e_{ij} - n_{ij} \ln e_{ij})$$

The expected count in each pixel depends, of course, on the size and position of the source, the shape of the PSF, and on the background level.

Although this method produces good results, it is much slower. In order to process a large data volume the usual practice has been to scan the sky using a matched filter, then refine the search around each apparent peak and evaluate its significance using maximum likelihood. The PSS program, part of the Starlink ASTERIX package, uses just such a hybrid technique. PSS is widely used in the UK to analyze data from both the WFC and the XRT telescopes on *ROSAT*, and the author of the program, David Allan, has supplied a version to CEA Berkeley for handling *EUVE data*.

3. Source Searching without Forming an Image

When a telescope is scanned across the sky, as in a sky-survey, and it has a resolution which varies across the detector surface, some information is inevitably lost in forming an image. In such cases PSS uses an averaged point-spread function: for the WFC the resolution (FWHM) ranges from around 1 arc-minute at the center to 3 arc-minutes at the edge of the field of view, so the loss from using the mean PSF is small but not negligible.

The resolution range of the *ROSAT* XRT is rather larger. The solution proposed by Cruddace et al. (1988) for XRT survey analysis was to consider the detector surface divided into many concentric annuli, and form separate images from each one, before applying the maximum likelihood technique to the ensemble. For the WFC, however, such a division of the detector surface would not work as well because its off-axis PSF is far from circular.

A much better solution would be to avoid forming images at all and apply maximum likelihood to the raw event list. The C-statistic is actually simpler:

$$C = 2 \left(E - \sum_k \ln e_k \right)$$

where e_k is the normalized probability density where each event was detected, and E is their sum, equal to the observed total count.

The main advantage of this is that each event can be processed using the PSF corresponding to its point of incidence on the detector. This should make better use of the higher spatial resolution at the center of the detector and avoid any loss of information from binning the events into an image.

A disadvantage is the number of computations required, given the need to cover the sky using a one-arcminute grid in two spectral bands. Close attention therefore had to be paid to efficiency, with a prime target being the function minimization, an inherently iterative technique. After examining the many methods listed by Press et al. (1992), it was found the downhill simplex method of Nelder & Mead (1965), said to be reliable but slow, was in fact among the fastest. This speed arises from the fact that it does not require the computation of derivatives of the function being minimized. An even greater performance gain resulted from using a source of strength S above zero but below the expected detection threshold in the first test at each grid point. Since the χ^2 surface of C against S is fairly smooth and roughly parabolic, the direction of the change in S after the first iteration shows whether there is any significant flux or not near that point. At the majority of points in the sky, therefore, only one iteration was required.

Even so it was still necessary to evaluate the kernel, and hence the PSF, over 10^{11} times. To make this efficient, several megabytes of memory were used to hold the PSF as a look-up table, sampled at suitable intervals in all four dimensions, and use quadrilinear interpolation. Another useful step was to evaluate the sum of the logarithms in the kernel as the logarithm of their product, with careful checks to avoid arithmetic overflow or underflow.

Rapid access to the event data was also essential. The raw data were stored on disk in regions about $2° \times 2°$ across with the events in time order. Each region was processed by reading all events into a table, with one column per attribute. When testing each grid point one needs to access all events in its vicinity: in our case those in a circle of radius 8 arc-minutes. The problem could have been solved by sorting the table into spatial order (as in an IRAFQPOE file) but sorting is inherently rather slow. Our solution was to turn the table into a set of linked-lists by adding two more arrays. The first array was 2-dimensional, like an image, but holding for each pixel just a pointer to the first event in the list which had a celestial position within that pixel. An extra column vector was then used to hold, for each event, a pointer to the next event located within the same pixel. The end of each pointer chain was set to zero. Thus all events in a small region could be accessed just by following, for each pixel, a short chain of pointers. Note that the event coordinates were left as floating-point numbers, they were rounded to integers only to determine membership of a particular pointer chain. The benefit of using this structure was that it could be created on a single pass through the table.

As a result of these changes, we were able to achieve a similar execution speed to that of the established PSS program. It is important to note, however, that in our data-set there are only a few events per pixel: your mileage may vary.

Determining whether this method was more or less sensitive that the standard PSS program turned out to be much more difficult. Since the differences were clearly small, the relative performance could be established only by carrying out extensive simulations. Simulating a suitably realistic sky with added sources takes long enough in the image domain, the generation of a sufficiently large set of photon-event lists with similar properties would have taken even longer.

We decided, therefore, to process the entire sky-survey data set using both the PSS program and an event-based maximum likelihood program in parallel, in the hope that the results from one would be clearly better than the other. The results were compared frequently during the processing, and this made it possible to identify and solve many problems which might otherwise have escaped notice. With these problems fixed, each program detected a small number of sources near the detection threshold which were missed by the other, but the results generally showed very good consistency, with very similar sensitivity. The two programs used somewhat different background estimation algorithms, and this may have masked any small difference in their inherent detection sensitivities.

The results, which have just been submitted for publication, are a substantial advance on our earlier work: there are now nearly 500 sources in total, with most of them detected in both filter bands.

4. Conclusions

Our main conclusion is not that one technique was clearly better than the other, but that we gained a lot from using the two of them in parallel. Some other projects, for example HIPPARCOS (Hog et al. 1992), have recognized the value of passing their data through two independent processing paths.

Acknowledgments. I acknowledge the contributions of many members of the UK *ROSAT* Consortium to the analysis of the WFC sky survey. The work was supported by the UK SERC/PPARC.

References

Cash, W., 1979 ApJ, 228, 939

Cruddace, R. G., Hasinger, G. R., & Schmitt, J. H. 1988, Astronomy from Large Databases, ESO Conference and Workshop Proceedings No. 28, p. 177

Hog, E., Kovalevsky, J., & Lindegren, L. 1992, ESA Bulletin, 43H

Kearns, K., Primini, F., & Alexander, D. 1995, this volume, p. 331

Marshall, H. 1994, in Astronomical Data Analysis Software and Systems III, ASP Conf. Ser., Vol. 61, eds. D. R. Crabtree, R. J. Hanisch, & J. Barnes (San Francisco, ASP), p. 403

Nelder, J. A., & Mead, R. 1965 Comp. J., 7, 308

Page, C. G., & Denby, M. 1992, in Astronomical Data Analysis Software and Systems I, ASP Conf. Ser., Vol. 25, eds. D. M. Worrall, C. Biemesderfer, & J. Barnes (San Francisco, ASP), p. 351

Pounds, K. A., et al. 1993, MNRAS, 260, 77P

Press, W. H., Teukolsky, S. A., Vetterling, W. T., & Flannery, B. P. 1992, Numerical Recipes 2nd edition (New York, Cambridge University Press)

The OGIP FITS Working Group

M. F. Corcoran, L. Angelini, I. George, T. McGlynn, K. Mukai, W. Pence, A. Rots

Office of Guest Investigator Programs, Goddard Space Flight Center, Greenbelt, MD, 20771

Abstract. We present an overview of the workings of the OGIP FITS Working Group (OFWG).

1. Introduction

Nearly all high-energy astrophysics projects provide data in FITS format (Wells, Greisen, & Harten 1981). However, arbitrary and/or inconsistent use of file formats and header keywords leads to confusion and requires that specialized software be used for correct data interpretation and analysis. The need to support specialized FITS formats and software systems is a significant (sometimes unmanageable) burden for data archives. However, adherence to simple, agreed-upon conventions can alleviate this burden at little cost.

The Office of Guest Investigator Programs (OGIP) at the Goddard Space Flight Center has organized a working group to identify appropriate conventions, and, with community support, to encourage adherence to these conventions.

2. The OFWG

The OFWG was set up to (1) ensure that new FITS definitions used within the OGIP do not violate any established FITS standards/conventions; (2) encourage standardization of keyword usage, data types, and FITS file formats used within the OGIP (i.e., the so-called Rationalized Data File or RDF format); and (3) disseminate OFWG recommendations to the broader High Energy Astrophysics and FITS communities to promote community-wide standardization.

3. Community Involvement

The OFWG recognizes that many of the issues it considers are of potential interest to the wider FITS community. The OFWG encourages community involvement by announcing proposed standards/conventions to the HEA and FITS communities to gauge community reaction, and by disseminating accepted recommendations for possible use in the community. The OFWG uses two e-mail exploders to distribute proposals or recommendations and gauge community response:

HEAFITS - which primarily serves the high energy astronomy community

FITSBITS - serving the entire FITS community

as well as contributing to discussions on the newsgroup *sci.astro.fits*.

4. For More Information

The minutes of all OFWG meetings, along with the text of all full recommendations and pending proposals, are publically available via the anonymous ftp account on *legacy.gsfc.nasa.gov* in two areas directly related to the OFWG:

fits_info/ofwg_minutes containing ASCII files of the minutes from all OFWG meetings

fits_info/ofwg_recomm containing ASCII (and occasionally also PostScript) files of OFWG recommendations, pending proposals, etc.

Information regarding the OFWG is also available via the WWW. The most useful starting point is probably the URL

   http://heasarc.gsfc.nasa.gov/0/docs/heasarc/ofwg/ofwg_intro.html

All of the documents available via anonymous ftp are also accessible via the WWW.

References

Wells, D. C., Greisen, E. W., & Harten, R. H. 1981, A&AS, 44, 363

Organizing Observational Data at the Telescope

M. Peron, D. Baade, M. A. Albrecht, and P. Grosbøl

European Southern Observatory, Karl-Schwarzschild-Straße 2, D-85748 Garching, Germany

Abstract. The MIDAS Data Organizer, a customizable utility to analyze and identify associations in a database of astronomical observations, is described. Its implementation in a data acquisition environment is discussed as a particular application.

1. The Problem

There is a number of situations where users of astronomical observations would benefit from advanced software tools which can analyze the composition of the available data pool:

Data Acquisition Control: Spectroscopists may wish to take after each scientific spectrum a new arc spectrum if, for instance, the spectrograph tilt angle has changed by more than a certain amount since the previous arc exposure. (Presumably, a new arc spectrum would not be required following acquisition images.) Given the numerous other on-line activities, a system knowledgeable enough to issue an automatic reminder to the observer, when and if necessary, would be very advantageous.

On-line Data Reduction: The main difficulty of automatic on-line reduction procedures is to identify the optimal calibration files that are applicable to a given science exposure. These data may be either acquired during the observing run or supplied by an observatory-maintained database. If calibration data are temporarily unavailable, the on-line reduction tasks may also need to be suspended and resumed when all the necessary data become available.

Off-line Data Reduction: This situation is very similar to the on-line case, except that the volume of data is larger and the user may no longer remember (or not know) the structure of the database. A system that can group science exposures and calibration data according to user-definable criteria would be a considerable asset, especially if it can also be interfaced to subsequent standard reduction procedures.

Archival Research: The lack of a suitable overview of the structure and contents of the database is the standard problem of archival researchers who need to specify the calibration files they want to extract. Observatory staff are facing a similar situation when they wish to perform trend analyses for instruments.

2. Derived Requirements

The above scenarios have in common that the user wants to define a context-specific structure on the observational database. From this, the following requirements on the desired tool can be extracted: It must be able to classify each file into user-definable categories, and it must be able to establish relations between files of different categories (e.g., associate to a given science frame a set of suitable calibration files). The Data Organizer (DO) (Peron et al. 1993) allows users to perform these tasks in an easily customizable way. The main characteristics of the on-line implementation of the DO at the ESO New Technology Telescope (NTT) are described below.

2.1. Observation Summary Table

The database for all operations is a MIDAS (ESO-IPG 1993) table, the so-called Observation Summary Table (OST) which contains all relevant parameters extracted from the FITS keywords. Because the Data Organizer is built on existing capabilities of the MIDAS Table File System (Peron et al. 1992), standard relational database operations (e.g., select, merge, copy, and project) may be used.

2.2. Classification of Exposures

The information contained in the OST is used to classify the images according to different sets of attributes (e.g., optical elements, calibration sources, etc.). The classification is achieved by applying user-definable rules, and the result of the classification (e.g., an optical path) is saved in the OST. EMMI (Melnick et al. 1992), one of the instruments mounted at the NTT, allows a wide range of observing modes, from wide-field imaging to high-dispersion spectroscopy, including long-slit and multi-object spectroscopy. Because of its complexity, EMMI has been chosen for the first implementation of the DO in an on-line environment.

2.3. Association of Exposures

The association of scientific frames with suitable calibration images is achieved by selecting and ranking all calibration frames which for a given scientific exposure match a set of user defined selection criteria. One may expand the search by submitting a "second choice" set of criteria when not enough frames match the original one. For instance, a search may be expressed in natural language as: "Find for each scientific frame two dark exposures that have been observed within the same night, and for which the mean detector temperature did not vary by more than 1 degree. If unsuccessful, look for dark exposures observed within the same observing run, and for which the mean detector temperature did not vary by more than 2 degrees." Selected frames can be ranked by applying weights to the attributes invoked in the same process. For example, one may give more importance to the time difference than to the detector temperature difference.

3. On-line Implementation of the DO at the NTT

The DO has been conceived as a general purpose tool, whereas the implementation at the NTT provides very specific services. In particular, it knows about

FITS keywords that characterize EMMI files. This customization was achieved by providing a set of NTT specific configuration tables which contain definitions of different exposure types (e.g., SCI, FFDOME, FFSKY, and BIAS) and instrument modes (e.g., blue imaging, red medium-dispersion spectroscopy). Each time a new file is delivered by the acquisition system, the exposure is classified according to a predefined set of rules, and the result is appended to the OST. Because it is essential that observers can interact efficiently with the tools offered to them in an on-line environment, a versatile Graphical User Interface has been fitted to the DO. It is shown and further explained in Figure 1.

4. Further Applications

After the adaptation of the association part of the DO to the particular requirements of EMMI, the basis for automatic on-line data reduction will be available. The first MIDAS package to be interfaced to it will be the CCD package. Observers and archival researchers will receive their data together with the corresponding OST. At the NTT, an observatory-supplied calibration database will be maintained. A dedicated OST will provide the overview of that database and enable observers to identify additional calibration observations which could be applicable to their own data. It is expected that at the Very Large Telescope (VLT) the Data Organizer will play a more central role in the on-line data flow.

References

ESO-IPG 1993, in MIDAS Users Guide, ESO Operating Manual, No. 1 (Garching, ESO)

Peron, M., Ochsenbein, F., & Grosbøl, P. 1992, in Astronomy from Large Databases II, ESO Conference and Workshop Proceedings, No. 43, eds. A. Heck & F. Murtagh (Garching, ESO), p. 433

Peron, M., Albrecht, M. A., & Grosbøl, P. 1994, in Handling and Archiving Data from Ground-Based Telescopes, ESO Conference and Workshop Proceedings, No. 50, eds. M. A. Albrecht & F. Pasian (Garching, ESO), p. 57

Melnick, J., Dekker, H., D'Odorico, S., & Giraud, E. 1994, in EMMI & SUSI, ESO Operating Manual, No. 15, Version 2.0

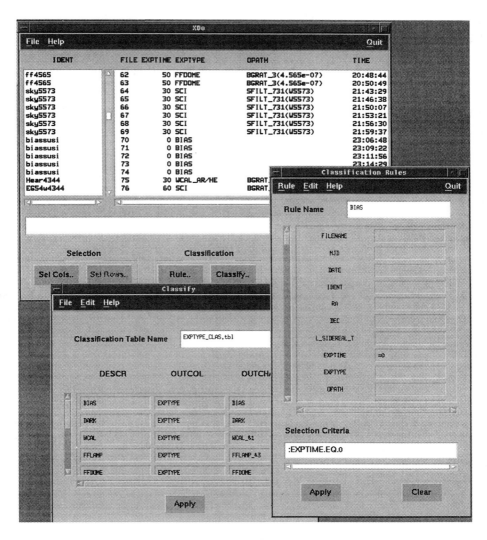

Figure 1. The Graphical User Interface to the Data Organizer as installed at the NTT. The main window in the upper left corner shows (part of) the OST which is the database on which all operations supported by the DO are performed. A table-like widget ("Classification rules") can be activated by a pushbutton to edit and modify classification rules. The rule "BIAS" for classifying BIAS exposures is being edited in the figure. Another table-like widget ("Classify") can be used for entering classification rules to be applied as well as the character string which in the OST will identify the selected frames.

Astronomical Data Analysis Software and Systems IV
ASP Conference Series, Vol. 77, 1995
R. A. Shaw, H. E. Payne, and J. J. E. Hayes, eds.

IMPORT/EXPORT: Image Conversion Tools for IRAF

M. Fitzpatrick

IRAF Group, NOAO[1], PO Box 26732, Tucson, AZ 85726

Abstract. We present a set of IRAF[2] tasks for converting foreign image formats to and from the native IRAF format. Traditionally, this has been done by writing individual one-to-one conversion utilities, either as single tasks or as procedures within a multiple-format conversion tool. This approach often results in the unnecessary duplication of code for formats that are actually very similar.

1. Introduction

A survey of the various image formats in use today was done to find the factors common to most formats. It was then possible to write a description of each format into a database, using parameters that describe the layout of the pixels, and expressions that determine the values of these parameters on an image by image basis. Using this database allows one piece of code to read multiple formats. In cases where a format does not fit the general raster model but should be supported because it is particularly useful or popular, or where more detailed processing is required, specialized code can be written. The database approach also means that a new image format can usually be added *by the user* without modifying the existing code.

When exporting IRAF images to other formats the only built-in knowledge of the format is the structure of the image header. Since the database only uses a fraction of what may be in an image header, it cannot be used to define completely the output file. Still, the code for writing the image data is general in nature, using format parameters set internally. This usually results in only a page or two of specialized code for a specific format, most of which is for writing the header. A user-defined file of header information can be prepended automatically. Since the task parameters define the pixel layout fully, there is some minimal capability for producing formats not fully integrated into the task. Table 1 shows the list of currently supported formats.

At the heart of the conversion tasks is the vector expression evaluator. In IMPORT this is not only used for computing the converted pixels, but also to evaluate the database expressions that define the image. A wide variety of

---

[1] National Optical Astronomy Observatories, operated by the Association of Universities for Research in Astronomy, Inc. (AURA) under cooperative agreement with the National Science Foundation

[2] Image Reduction and Analysis Facility, distributed by the National Optical Astronomy Observatories

Table 1. Currently supported or recognized image formats.

Image Format	IMPORT Task	EXPORT Task
Raw Binary	supported generic model	supported generic model
ASCII Text		all task options
AVS X Image	24-bit RGB plus alpha channel	...
Encapsulated PostScript	...	8-bit grayscale, 24-bit color
EXPORT task raw binary	all types	...
FITS	simple images only	...
GIF	8-bit CLT	8-bit CLT
IRAF OIF image	all IRAF data types	all data types
PDS	8-bit PDS3 images only	...
PGM Format	raw file only	raw file only
PPM Format	24-bit raw file only	24-bit raw file only
SGI RGB image	8-bit no CLT, 24-bit RGB	8-bit w/opt CLT, 24-bit RGB
Sun Rasterfile	8-bit w/opt CLT, RGB/RGBA	8-bit w/opt CLT, RGB/RGBA
Sun Taac	8-bit no CLT or 24-bit RGB	...
JPL Vicar	byte, short, int or floating point	...
X10 Window Dump	8-bit w/opt CLT	...
X11 Window Dump	8-bit w/opt CLT, RGB/RGBA	8-bit w/opt CLT, RGB/RGBA
CMU WM Raster	recognition only	...
Fuzzy Bitmap	recognition only	...
JPEG	recognition only	...
NCSA HDF	recognition only	...
Utah RLE Toolkit	recognition only	...
TIFF	recognition only	...

standard math functions is available, as well as specialized functions for reading header values, combining or separating image colors, or defining the layout of images. Expression operands may include user-defined image *tags* that refer to a line of pixels in the image, image header parameters, or scalar values. Operators permit concatenation and replication of pixels, and boolean operations used to select pixels, as well as the standard arithmetic operators.

2. The IMPORT Task

The *IMPORT* task can also be used simply to print information about a file, or list the pixels. By default, the database of formats will be used to identify the file before processing; task parameters can be used to override certain database values or to define the format completely. When a text listing or image is the requested output type, an expression parameter can be used to process the pixels before writing the result. Expressions typically convert RGB values to grayscale, apply a gamma correction, flip an image, or reverse the channel order.

2.1. The Generic Data Model

With a few exceptions, the following model features apply well to a large number of commonly used image formats: (1) pixels are even multiples of 8 bits, short integers may be signed or unsigned, floating point may be native or IEEE, (2) any padding is also a multiple of 8 bits, (3) no compression is used in storing the data, (4) multiple bands of pixels may either be stored sequentially, interleaved line-by-line, or pixel-by-pixel, (5) padding may come before or after the image, before or after each line in the data, or before or after groups of pixels, (6) byte-order of integers may be defined, and may require byte-swapping on a particular host, and (7) pixels are not required to be all the same type.

The task can read any format meeting the above guidelines without changing the code. New formats may be added by appending their own format database or by specifying the task parameters directly.

2.2. The Format Database

The format database is composed of records of *keyword = expression* pairs. Expressions often use the specialized I/O functions to read the image header at a specific offset, e.g., to get a magic number that identifies the format or to get an image dimension at a known byte offset. The records are examined sequentially until an expression that identifies the format is evaluated as true. Each expression in the record is then evaluated to set the keyword value, the keywords mirror the task parameter and define, for instance, the amount of header to skip before reading pixels. The expressions are evaluated for each image individually, so there is no requirement that each image in a list be the same size or, indeed, the same format.

3. The EXPORT Task

The *EXPORT* task can easily convert an IRAF image to any of the supported formats. Again, the generic data model is used as a basis for the image conversion code, but there is additional support for formats that meet special needs. Aside from raw binary data, the task can also write the pixels as ASCII text, for input to other programs or for debugging. Support for Encapsulated PostScript allows for direct conversion of images for hardcopy output or for inclusion in documents (such as this one). GIF output allows for data to be processed by common X utilities or to be presented on the WWW. Lastly, images may be written out as new IRAF images if the user wants to combine several images into a single file for further processing (e.g., annotating the image using the *TVMARK* task) before writing it out in some final format.

Where possible, the output may be color images—something not supported directly by IRAF image model itself. By combining three images which may represent the three color bands, an output file may be written as an RGB color

Figure 1. An example mosaic produced by the EXPORT task showing three BVR images of the Trifid Nebula. The images were converted directly from IRAF format to Encapsulated PostScript using the EXPORT task.

image. Since formats such as GIF do not support RGB data, an expression function is provided that will automatically compute an 8-bit colormap from three images using a Median Cut Algorithm and Floyd-Steinberg dithering. This colormap can be output to any file that supports it. When only one image is available, but color output is desired, an artificial colormap can be specified. There are currently eleven colormap selections built in to the task; a user-specified file may also be used.

The *EXPORT* task also provides several ways of scaling the dynamic range of the IRAF image to the allowed range of the output format. These include: (1) the *zscale* algorithm used by the *DISPLAY* task with default values to give an optimal scaling, (2) user-specified *z1* and *z2* values, (3) colormap scaling using display brightness and contrast values, (4) a linear transformation function, and (5) any arithmetic expression or function (e.g., sqrt(), log()). Automatic datatype conversion will be done unless something specific is requested.

Perhaps the most useful feature of the task is the ability to mosaic any number of input images into a single output image. Expression functions can be used to place images side-by-side or from top-to-bottom. In Figure 1, for example, three IRAF images were written directly to the EPS file included in this paper. Each image was intensity scaled independently. The expression syntax specified the image layout as well as the spacing between the images. In such mosaics, images do not have to be the same size; the size of the final image will be determined automatically. Writing an image mosaic as a new IRAF image allows the user, for instance, to mark object stars and output the marked image as PostScript.

Convert: Bridging the Scientific Data Format Chasm

D. G. Jennings[1] and W. D. Pence

NASA Goddard Space Flight Center, Greenbelt MD 20771

M. Folk

National Center for Supercomputing Applications, Champaign IL 61820

Abstract. The *Convert* project is a newly funded NASA endeavor whose goal is to provide the scientific community with software tools for data format conversion. Among its goals are the development of utilities that transform data in non-standard formats into selected standard formats and allow for the inter-convertibility of data between those same standard formats. This paper discusses several aspects of the *Convert* project including current software tools, planned software tools, standard data formats under consideration and the efforts to devise transformation mappings between these formats.

1. Introduction

There currently exists within the scientific community a plethora of formats used to store and analyze data. To help unify the data into understandable and transportable formats, we have undertaken an effort known as the *Convert* project. The goals of *Convert* are twofold. First, *Convert* will provide tools to convert data in non-standard data formats into selected standard data formats. These tools will initially perform conversion of data into FITS (Wells et al. 1981), although plans to accommodate HDF (NCSA 1993a) conversion are under consideration. Secondly, *Convert* will allow for the inter-convertibility of data between FITS, HDF, and eventually netCDF (Unidata 1991) formats.

The first goal of *Convert* benefits astrophysical science since it will, for the first time, provide the astronomical community with a set of general purpose tools to transform instrument and mission-specific data products into FITS format. Data from many older astronomy missions (e.g., *EXOSAT, SAS-2, Vela*) does not reside in FITS and therefore requires post-mission FITS conversion in order to remain useful to future analysis efforts. Even data from some current astrophysics missions (e.g., *GRO, ULYSSES*) requires special processing to convert it from instrument specific formats into FITS.

The second goal of *Convert* should allow astronomers to convert their FITS data into formats used by other disciplines. Besides being useful for collaborative efforts between astronomers and scientists from non-astronomical fields, this ability will make it possible for astronomers to utilize the software tools

---

[1]Employed by Hughes STX Corporation

(e.g., data visualization, data management, data analysis) being developed in other fields. As funding sources for science, especially astronomy, continue to shrink, the sharing of resources between disciplines becomes both attractive and necessary.

The following sections elaborate on the work underway by the *Convert* project participants. Section two gives background on the software tools that form the basis of the *Convert* software package, as well as the planned enhancements to these tools. Section three describes the invertible transformation being developed between FITS and HDF. Finally, section four provides a summary of *Convert* project activity and supplies reasons why this activity should be of interest to the astronomical community.

2. Software Tools

Convert shall make use of three pre-existing, public-domain software packages: FITSIO (Pence 1994), ToFU (Jennings 1993) and the HDF application programming interface library (NCSA 1993b). The FITSIO subroutine library, in wide use in NASA-sponsored astrophysics missions, provides an easy-to-use and reliable, low-level I/O programming interface to FITS files. Layered on top of FITSIO is the ToFU (To FITS Utilities) subroutine library. ToFU consists of routines that facilitate the conversion of data into FITS format. Lastly, the HDF application programming interface (API) consists of both high- and low-level routines arranged into six separate modules, with each module corresponding to a supported HDF data object.

Updated and modified versions of ToFU and the HDF API will constitute build 1 of the *Convert* software package. Work is currently underway (and should be complete by the printing of this paper) on porting the ToFU library to C and greatly enhancing its usability. Once complete, ToFU will provide *Convert* with its advertised FITS conversion capabilities. We also intend to integrate the FITSIO subroutine library into the HDF application interface. Thus, the HDF library will be able to use FITSIO to read and write FITS files, just as it currently does with HDF and netCDF files. The integration of FITSIO into the HDF API forms the basis of *Convert's* data format inter-convertibility capabilities, since the HDF API will then be able to read in FITS, HDF, or netCDF files and write out HDF or FITS (or eventually netCDF) files.

An interesting consequence of FITSIO incorporation into the HDF API is that FITS and HDF will then share a common software interface. In this sense, there will be no operational difference between the two formats. Applications may input or output FITS (HDF) formatted files as easily as they input or output HDF (FITS) formatted files. Additionally, the FITSIO/HDF API merger provides FITS with a standard and widely distributed application interface of its own: something that the FITS community has long needed.

3. Inter-Convertibility between FITS and HDF

Before the HDF API can make use of the FITSIO library to read and write FITS formatted files, there must be an invertible transformation defined between FITS and HDF data objects. Without such a transformation, HDF interfaces will not understand how to interpret the contents of FITS primary arrays, images, ASCII or binary tables. Even though FITS and HDF implement their data structures differently, they both make use of the same basic set of abstract data types. (Note, however, that netCDF and CDF use the concept of named N-dimensional variables, which does not conform well to either the HDF or FITS models. This makes the inter-convertibility between FITS and CDF/netCDF a more complex problem than the inter-convertibility between FITS and HDF. Thus, the issue of FITS to CDF/netCDF mappings will be left for future work). The mapping between FITS and HDF is, with a few notable exceptions, a relatively straightforward exercise. The following list demonstrates preliminary mappings between FITS and HDF data objects:

- FITS Primary Array and Image Extensions -> HDF SDS Objects
- FITS ASCII and Binary Table Extensions -> HDF Vdata Objects
- HDF 8-bit and 24-bit RIS Objects -> FITS Image Extensions
- HDF Palettes and SDS Objects -> FITS Image Extensions
- HDF Annotation Objects -> FITS ASCII Table Extensions
- HDF Vdata Objects -> FITS Binary Table Extensions

For a more detailed comparison of the FITS and HDF formats, see Jennings & McGlynn (1993).

The only unresolved issues in the FITS to HDF mapping are the lack of support within HDF for certain FITS data types (i.e., 4-byte complex, 8-byte complex, bit, boolean) and the need for field-associated attributes (i.e., keyword = keyvalue pairs) in HDF Vdata objects. The HDF development group at NCSA intends to solve both problems by modifying the HDF data format to accommodate new data types and Vdata field attributes.

The most significant unresolved issue in the HDF to FITS mapping is the lack of a robust FITS hierarchical grouping structure. HDF often associates its data objects into hierarchical groupings and then operates on the groups as if they were a single entity. Therefore, the FITS to HDF mapping requires that FITS adopt an HDF-like hierarchical grouping structure so that FITS may preserve HDF data object associations during HDF to FITS transformations. To this end, the authors have begun work on a proposal to augment FITS with its own hierarchical grouping convention. Even though the HDF to FITS mapping motivates this effort, the authors will endeavor to develop a grouping structure that is general enough for all FITS applications to use.

4. Summary

It could be argued that the field of astronomy focuses most of its efforts on internal concerns, neglecting to stay abreast of work being done in other disciplines. This attitude is ultimately self-limiting, because in terms of money and resources astronomy is just one small part of the world scientific community. The earth science, space science, and atmospheric science disciplines invest significant portions of their (often greater) resources into developing software to visualize, manage, and analyze data. If astronomers wish to leverage their own resources and take advantage of the infrastructure built by other disciplines, then astronomical data must be convertible to the formats used by those infrastructures.

Even within astronomy, research teams do not always concern themselves with producing data products in standard formats. This creates situations in which data becomes increasingly less understandable over time and more difficult to analyze with standard analysis packages. To ensure that the astronomical community has a cost effective means to transform data into standard formats and between standard formats, the *Convert* project has begun work on two software tool sets. The first, ToFU, converts generic data streams into FITS format. Secondly, the HDF application programming interface will allow the inter-convertibility of FITS and HDF formatted data sets. Both software packages will make use of the FITSIO subroutine library to perform low level FITS file manipulations.

Before true inter-convertibility between FITS and HDF can be achieved, an invertible mapping between the two formats is needed. This mapping is reasonably straightforward because of the high level of abstract conformability between FITS and HDF. Only minor changes need occur to the FITS and HDF formats to make the mapping possible. Amongst these changes are additions of new data types to HDF, support for field attributes in HDF Vdata objects, and the creation of a hierarchical grouping convention for FITS. The work necessary to modify both data formats is underway.

Acknowledgments. We gratefully acknowledge the support of the NASA Applied Information Systems Research Program (AISRP), under which this effort is funded.

References

Jennings, D. J., et al. 1993, BAAS, 25, 962

Jennings, D. J., & McGlynn, T. A. 1993, in Astronomical Data Analysis Software and Systems III, ASP Conf. Ser., Vol. 61, eds. D. R. Crabtree, R. J. Hanisch, & J. Barnes (San Francisco, ASP), p. 526

NCSA 1993a, NCSA HDF Specifications and Developers Guide, Version 3.2 (Champaign, NCSA)

NCSA 1993b, Getting Started with HDF, Version 3.2 (Champaign, NCSA)

Pence, W. D. 1994, this volume, p. 245

Unidata 1991, NetCDF User's Guide: An Interface for Data Access (Boulder, Unidata Program Center)

Wells, D. C., Greisen, E. W., & Harten, R. H. 1981, A&AS, 44, 363

Representations of Celestial Coordinates in FITS

E. W. Greisen

National Radio Astronomy Observatory

M. Calabretta

Australia Telescope National Facility

Abstract. The initial descriptions of the FITS format provided a simplified method for describing the physical coordinate values of the image pixels, but deliberately did not specify any of the detailed conventions required to convey the complexities of actual image projections. Building on conventions in wide use within astronomy, this paper proposes changes to the simple methods for describing coordinates and proposes detailed conventions for describing most of the methods by which spherical coordinates may be projected onto a two-dimensional plane. Simple methods for converting from the existing coordinate conventions are described. This paper does not attempt to address the politically sensitive questions of frequency/velocity coordinates, nor does it address various other types of coordinates, such as time.

1. Introduction

The initial paper describing the Flexible Image Transport System, or FITS format, (Wells, Greisen, & Harten 1981) proposed keywords to describe the physical coordinates of the image. They were CRPIXn for the reference pixel location on pixel axis n, CRVALn for the coordinate value at that pixel, CDELTn for the increment at that pixel in the coordinate value, and CTYPEn for the type of coordinate. Coordinate rotation—of an unspecified nature—was allowed, and a few possible values for CTYPEn were proposed. The original authors chose to defer discussion of the technical details of coordinate specification until the basic FITS format was accepted generally and until a deeper understanding of image coordinate specification and computation could be obtained.

The time for that discussion is now. While participating in the development of the AIPS software package of the National Radio Astronomy Observatory, Greisen (1983) developed FITS-like syntax and semantics to define both velocity and celestial coordinates. The latter have been widely used for interchanging imagery from a number of instruments at widely differing spectral domains and are fundamental to the present proposal. Greisen defined the reference pixel for celestial coordinates to be the tangent point of the projection. He specified that the first four characters of CTYPEn should be used to give the type of celestial coordinate while the next four characters specified the type of projection (e.g., DEC--TAN). Greisen (1983) gave the mathematics for four projections: orthographic (SIN), gnomic (TAN), zenithal equidistant (ARC), and a special coordinate used by East-West radio interferometers (NCP). In a second paper, Greisen (1986) added specifications for the stereographic (STG), sinusoidal (GLS), Hammer-Aitoff (AIT), and Mercator (MER) projections. The current proposal (Greisen and Calabretta, 1994) extends these earlier, widely tested proposals to clarify the logical process by which celestial coordinates are computed, and to specify a very wide range of possible projections. In addition, the current proposal specifies a method to define skew, offset rotations, and even rotations of axes of different physical type into each other. It also specifies a method to describe the units of the coordinates and to provide a second coordinate description for an axis.

2. Coordinate Computation

In the current proposal, we regard the conversion from simple pixel counts to a full coordinate description as a multi-step process containing one optional and four required steps. These steps are indicated conceptually in Figure 1. The first and optional step is used to correct the actual image pixel numbers into those which would have been recorded by an ideal instrument. The corrections in this "pixel regularization table" are expected to be rather small, so that they may be ignored except in high precision computations. In the second step, for all types of coordinates, the vector of reference pixels is subtracted from the vector of pixel numbers and the result multiplied by a pixel conversion (PC$iiijjj$) matrix to convert from pixel numbers to offsets from the reference pixel along physical axes but still in pixel units. The third step is a multiplication by a diagonal matrix (CDELTi) to convert to relative coordinate in physical units.

The fourth step in the process of finding the true coordinates depends on the type of axis given in CTYPEn. For simple linear axes, the true coordinate is found by adding the offset found above to the reference pixel value given by CRVALn. Otherwise, some function of the offset(s), the CRVALn, and, perhaps, other parameters must be established by convention and agreement. For celestial coordinates, the proposed fourth step involves converting the linear offsets into longitudes and latitudes in the "native coordinate system" for the specified type of projection. These are rotated, in the fifth step, by the usual spherical formulæ to longitudes and latitudes in the desired standard coordinates (e.g., Equatorial, Galactic, etc.) The native coordinate system is, for azimuthal and conical projections, one which has its north pole at the reference pixel. For cylindrical and conventional projections, the native coordinate system has its origin at the reference pixel. The rotation from native to standard coordinates is illustrated in Figure 2. The keyword LONGPOLE is proposed to specify the native longitude of the north pole of the standard system. The default value for LONGPOLE is to be 180 degrees to support current usage. Extra keywords PROJPj are defined to provide additional parametric information needed by some of the projections.

3. Other Proposed Conventions

The original FITS paper (Wells, Greisen, & Harten 1981) naively assumed that the units along each axis could be implied simply by the contents of the CTYPEn keyword and that they would be in the basic SI units. Outside of celestial coor-

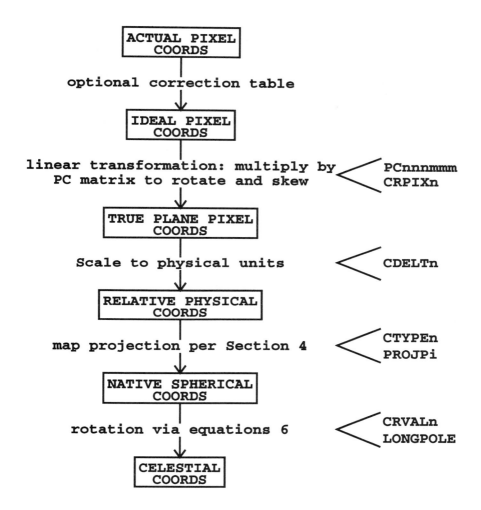

Figure 1. Conversion of pixel to celestial spherical coordinates

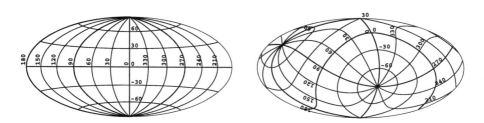

Figure 2. Conversion of native (left) to standard (right) spherical coordinates for the Hammer-Aitoff projection

dinates, both of these assumptions have apparently failed in practice. Therefore we propose that a new character-valued keyword CUNITn be added to describe the units used for coordinates on axis n. For celestial angular coordinates, following the proposed projection conventions, these units will be degrees ('deg '). Additional discussion and agreements will be needed to determine how one will represent other coordinate types.

In some cases, the axes of an image may be described as having more than one coordinate. An example of this would be the frequency, velocity, and wavelength along a spectral axis (only *one* of which, of course, could be linear). To allow up to 8 additional descriptions of each axis, we propose the addition of the follow optional, but now reserved, keywords.

CmVALn	coordinate value at reference pixel
CmPIXn	reference pixel array location
CmELTn	coordinate increment at reference pixel
CmYPEn	axis type (8 characters)
CmNITn	units of CmVALn and CmELTn (character valued)

where $m = 2, 3, \ldots, 9$ for the second through ninth alternate axis coordinate and $n = 1, 2, \ldots, 999$ for axis 1 through 999.

To improve the use of these coordinates for astrometric purposes, three new keywords are proposed. EQUINOX replaces EPOCH for the epoch of the mean equator and equinox in years. MJD-OBS gives the modified Julian date of observation in days and RADECSYS gives the frame of reference of equatorial coordinates as FK4, FK4-NO-E, FK5, GAPPT.

References

Greisen, E. W. 1983, AIPS Memo No. 27 (Charlottesville, National Radio Astronomy Observatory)

Greisen, E. W. 1986, AIPS Memo No. 46 (Charlottesville, National Radio Astronomy Observatory)

Greisen E. W., Calabretta, M. 1994, in preparation[1]

Wells, D. C., Greisen, E. W., & Harten, R. H. 1981, A&AS, 44, 363

---

[1] http://fits.cv.nrao.edu/documents/wcs/wcs.html

A Generic Data Exchange Scheme Between FITS Format and C Structures

W. Peng and T. Nicinski

Fermi National Accelerator Laboratory, PO Box 500, Batavia, IL 60510

Abstract. A flexible and efficient scheme allowing arbitrary FITS Binary and ASCII Tables to be converted to arbitrary C structures at run-time is presented. This scheme has been successfully implemented and used with Shiva[1] (Survey Human Interface and Visualization Environment), a package developed by Fermilab for the analysis of Sloan Digital Sky Survey data.

1. Introduction

The Sloan Digital Sky Survey (SDSS), for which Fermilab has been actively developing software and hardware, uses the Flexible Image Transport System (FITS) (NOST 1993) as the standard exchange format for survey data. Portions of the data are presented in FITS Binary and ASCII Tables. Accessing such arbitrary data from C structures, without knowing the FITS Table layout, can be difficult.

We have developed a versatile scheme that allows data transfer between FITS Tables and C data structures. This generic scheme uses two supporting structures: a TBLCOL to contain an arbitrary FITS Binary or ASCII Table (or both), and a translation table that maps TBLCOL to a user-specified C structure. FITS Tables are read into a TBLCOL structure. With a translation table filled in at run-time, C structures can be filled with data from TBLCOLs, and vice versa. This functionality is incorporated into Shiva, a package developed at Fermilab for analyzing SDSS data. The reading (writing) of arbitrary FITS Tables into (from) TBLCOLs and the translation of TBLCOL data to C structures are performed from the Shiva command line at run-time, without any compile-time knowledge of the FITS Tables and the C structures.

All primitive C data types, including characters, integers, floating point numbers, and strings, as well as arrays and structures of these types, are supported. Indirect data can also be accessed (through pointers).

2. TBLCOL Format[2]

Under Shiva, FITS Binary and ASCII Tables are read into and written from TBLCOLs. The TBLCOL format is flexible enough to accommodate any tabular

---

[1] http://www-sdss.fnal.gov:8000/shiva/doc/www/shiva.home.html

[2] http://www-sdss.fnal.gov:8000/shiva/doc/www/shTblHome.html

data. Once data is in a TBLCOL, the originating FITS Table type is irrelevant. This makes it possible to read in a FITS ASCII Table and then write it out as a Binary Table, and vice versa (as long as the resulting FITS Table is legal).

The TBLCOL format uses three major structures to achieve its goal of supporting arbitrary tables: TBLCOL, ARRAY, and TBLFLD.

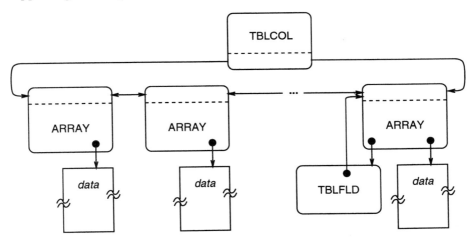

Figure 1.   TBLCOL Format Components

As a FITS Table is read in, each field is placed into an ARRAY. The TBLCOL structure simply heads the list of ARRAYs. An ARRAY points off to the FITS Table data, where each ARRAY element corresponds to the field data from a FITS Table row. The TBLFLD structure is optional, containing information about a field such as its name (akin to the FITS TTYPEn keyword), scaling and zeroing (FITS TSCALn and TZEROn keywords), etc. This organization allows quick and easy retrieval of data in a column/field oriented way. It also allows a FITS Table to be read into memory without any *a priori* knowledge of the FITS file contents or Table structure.

The ARRAY structure supports FITS Binary Tables having fields that are multidimensional arrays in themselves. The data type is *not* restricted only to primitives. Structures, and arrays of structures, can also be stored in the ARRAY and accessed properly. However, such use is not recommended if the TBLCOL is intended to be written out as a FITS Binary or ASCII Table (FITS does not permit such structures).

3.   Translation Table[3]

TBLCOLs allow users to read in arbitrary FITS Binary or ASCII Tables. But, access to TBLCOL data is only efficient if it is processed on a field by field basis (it is relatively expensive to "bounce" to another field). The use of translation tables to move some or all data from a TBLCOL to a C structure can be used to

---

[3] http://www-sdss.fnal.gov:8000/shiva/doc/www/shSchema.html

circumvent this inefficiency. Users build translation tables at run-time, instructing how **TBLCOL** fields and C structure members are related. A translation table is a collection of textual entries of the form:

EntryType FldName C_MemName C_MemType OptInfo

where *EntryType* can be either "name" or "cont" and *OptInfo* contains optional dimension information. Each entry represents a mapping that associates a **TBLCOL** field, *FldName*, to a C structure member, *C_MemName*.

Data copying routines use this mapping, along with a C structure's *schema*[4] to properly copy between the **TBLCOL** and C structure (or vice versa). Type casting, checking structure member and **TBLCOL** field sizes, allocating memory, and traversing pointers are done transparently during the copy.

3.1. Primitive Data Types and Fixed-size Arrays

For primitive data types (such as characters, integers, and floating point numbers), the relation is a straightforward one-to-one mapping. The translation routines simply copy the data directly between a **TBLCOL** field and a C structure (with any appropriate type conversions). For example,

```
name RA_IN_DEG   ra   double
```

indicates that data from the **TBLCOL** field RA_IN_DEG be copied as a double precision floating point number to the ra member in a C structure, or vice versa.

Fixed-size arrays of primitive data types are handled in a similar fashion. Their size is already embedded in the C structure declaration and reflected in the structure's schema (see Section 4).

3.2. Dynamically Allocated Arrays

Non-trivial C structures can have, for example, arrays of C primitive types whose memory is allocated at run-time. When transferring data from **TBLCOL**, memory must be allocated properly for the receiving C structure. The size of this transfer is obtained from additional information in a translation table entry. For instance, a size of "5x10" indicates that the C structure member is a 2-dimensional array (5 by 10). When transferring data to **TBLCOL**, the **TBLCOL** field should have the appropriate space.

3.3. Indirect Data

In practice, C structures have pointers to different memory areas. FITS does not support pointer data types in Tables. The translation tables take this into account through multiline entries. (a main **name** line followed by one or more continuation lines). Each line can have independent dimension information, imitating the process of traversing memory links to the ultimate data.

For example, consider the following two C structures

```
typedef struct {                typedef struct {
         ⋮                               ⋮
```

---

[4] A *schema* describes a C structure at run-time. Applications can understand a C structure *without* needing to be compiled with the structure declaration. See Section 4 for more information.

```
        REGION *reg;                    char *name;
            ⋮                               ⋮
      } MY_STRUCT;                     } REGION;
```

The translation to match a FITS Table field, REG_NAME, could be

```
      name    REG_NAME    reg     struct
      cont    reg         name    string   -dimen=10
```

which states that, reg in MY_STRUCT is a pointer to a REGION object. A mapping between REG_NAME and reg→name is established. When transferring data to TBLCOL, the data pointed to by reg→name are copied. Likewise, when transferring from TBLCOL, two memory allocations are done (one each for reg and name) to ensure problem-free copying from REG_NAME to reg→name.

4. Schemas and Concluding Remarks

A *schema* is a run-time description of a C structure. It permits applications to understand a C structure *without* having been compiled with the structure declaration. During compilation, the Shiva environment parses C header files to generate schema for structures. Information about a structure, such as the member names, their sizes, offsets, dimensions, etc., are retained and are available at run-time. Currently, there are about 50 C structures used in Shiva. With the translation tables, passing data between arbitrary FITS Tables and these structures is possible. Applications built on top of Shiva also enjoy this capability.

Acknowledgments. This research is sponsored by DOE Contract number DE-AC02-76CHO3000.

References

NASA Office of Standards and Technology 1993, Definition of the Flexible Image Transport System (FITS) (Greenbelt, NASA/OSSA)

A Proposed Convention for Writing FITS Data Tapes: DRAFT 0

ROSAT/ASCA/XTE Development Team
Astrophysics Data Facility, NASA Goddard Space Flight Center, Code 631, Greenbelt, Md. 20771

Abstract. Even with today's advances in networking, file system capacities and CD technology it is often necessary to transport and store scientific data sets on magnetic tape. The FITS data format standard contains guidelines on how to write FITS files to magnetic tape but does not address the problem of indexing or organizing tape files. Currently available magnetic tape media can store multiple gigabytes of information on a single tape, which translates into thousands of FITS files per tape. Thus, the lack of a standard tape indexing and organizing scheme can, in many instances, become a serious problem.

Faced with the above dilemma, the Astrophysics Data Facility at Goddard Space Flight Center has developed a simple in-house convention for indexing the contents of FITS data tapes that allows software to quickly and easily inventory tape contents. This paper describes the convention used by our organization. We propose that this convention be adopted into the FITS standard as the way to index and organize the contents of magnetic tape media.

1. Introduction

Originally defined as an exchange format for data on nine-track tape (NOST, 1993), the Flexible Image Transport System (Wells et al. 1981) has in recent years expanded into a very general and useful logical data format. The current FITS format accommodates data exchange and archival storage on a wide range of media, and is also used as a working native data format (Pence et al. 1992) for many new software applications.

While the main emphasis of FITS has moved away from magnetic tape-based data storage and transport there are, perhaps surprisingly, new and growing needs for its use as a data tape format. Data volumes from many current and planned astronomy missions preclude the use of electronic data distribution. Examples of such missions are the *ASCA* X-ray Observatory Satellite which produces $\sim$ 500 MB (200–1000 files) of FITS formatted data per observation, and the soon to be launched *XTE* (X-ray Timing Explorer) satellite with an estimated data size of $\sim$3 GB (3000–20000 files) files per observation. Storing and distributing data on CD-ROM media can be a good alternative to electronic distribution in some cases, however, this technology costs an order of magnitude more to use than magnetic tape media and the storage capacity is many times less (600 MB per CD-ROM vs. 8–16 GB per 4mm tape).

Storing and transporting data sets on magnetic tape solves the problem of large data volume, but it does not provide FITS readers with the information needed to construct a catalog of the tape contents (short of reading every file from the tape). Keeping tape content information (e.g., file names, file sizes, file order) tends to be less of a problem when the number of FITS files per tape remains small, since this information can be easily stored external to the tape. However, when the numbers of files on a single tape grow into the hundreds or thousands, it becomes desirable for the tape to be *self-describing* just as a single FITS file is considered self-describing.

The FITS standard provides simple guidelines on how to write FITS formatted data to magnetic tape (Grosbøl & Wells 1994), but it does not address the issue of tape content indexing and cataloging. This paper presents a convention for writing self-describing magnetic tapes that contain FITS formatted data files. By using this convention, FITS readers may quickly access a catalog of the tape contents and determine the names, sizes, positions and meanings of every data file contained on a tape.

2. Tape Structure

In addition to the recommended guidelines for tape block sizes (up to 28800 bytes per block in increments of 2880), file separators (single tape marks) and logical tape labels (ANSI standard labels or no labels) this convention requires that the first file written to every FITS formatted data tape be a catalog of the tape contents. The catalog file is itself a FITS formatted file with a null primary array and an ASCII table (Harten et al. 1988) as its first extension. The second and subsequent tape files may be written in any order as long as this corresponds to the recorded order in the tape catalog file.

3. Tape Catalog File

All information pertaining to tape content resides in an ASCII table, which must be the first extension of the tape catalog file. Other FITS extensions may follow this ASCII table but the primary array shall be empty (null). This ensures that a dump of the tape catalog file produces readable output, at least up to the end of the first extension.

The ASCII table containing tape content information is composed of four table columns and one row for each FITS file on the tape. The four column entries provide the following information about each FITS file on tape: (1) original file name, (2) file size, (3) a brief description of file contents, and (4) the file's tape position number (first file on tape = tape catalog file = 1).

The order of the columns within the extension is not important and any additional columns describing tape contents are allowed. However, the four required columns must have associated TTYPE keywords with the following values:

- TTYPEnnn= 'filename', TTYPE keyword value for column holding original names of files on tape.

- TTYPEnnn='filesize', TTYPE keyword value for column containing sizes of files on tape. Recommended file size unit is bytes, kilobytes or megabytes.

Proposed Convention for Writing FITS Data Tapes 243

- TTYPEnnn='descrip', TTYPE keyword value for column describing contents of files on tape.

- TTYPEnnn='filenum', TTYPE keyword value for column holding positional values of files on tape.

4. Example Tape Catalog File

The following is an example of a tape catalog file currently being used for guest observer distribution tapes for the ASCA X-ray Observatory.

```
SIMPLE  =                    T / file does conform to FITS standard
BITPIX  =                   16 / number of bits per data pixel
NAXIS   =                    0 / number of data axes
EXTEND  =                    T / FITS dataset may contain extensions
FNAME   = 'ad13000000_050_tape.cat' / Original file name
SEQNUM  =             13000000 / Sequential number from ODB
PROCVER = 'P4.0.0  '           / Processing Configuration number
SEQPNUM =                  050 / Number of times sequence processed
USPINUM =                 5000
END

XTENSION= 'TABLE   '           / ASCII table extension
BITPIX  =                    8 / 8-bit ASCII characters
NAXIS   =                    2 / 2-dimensional ASCII table
NAXIS1  =                  135 / width of table in characters
NAXIS2  =                  201 / number of rows in table
PCOUNT  =                    0 / no group parameters (required)
GCOUNT  =                    1 / one data group (required)
TFIELDS =                    4 / number of fields in each row
TTYPE1  = 'filenum '           / label for field   1
TBCOL1  =                    1 / beginning column of field   1
TFORM1  = 'I4      '           / Fortran-77 format of field
TTYPE2  = 'filename'           / label for field   2
TBCOL2  =                    6 / beginning column of field   2
TFORM2  = 'A57     '           / Fortran-77 format of field
TTYPE3  = 'filesize'           / label for field   3
TBCOL3  =                   64 / beginning column of field   3
TFORM3  = 'I7      '           / Fortran-77 format of field
TUNIT3  = 'kilobytes'          / physical unit of field
TTYPE4  = 'descrip '           / label for field   4
TBCOL4  =                   72 / beginning column of field   4
TFORM4  = 'A64     '           / Fortran-77 format of field
HISTORY   This FITS file was created by the FCREATE task.
SEQNUM  =             13000000 / Sequential number from ODB
END
```

5. Summary

The above simple scheme for producing self-describing FITS data tapes allows both humans and software to understand the contents of a tape without unloading and examining every file. The true utility of this feature can be realized when one considers data tapes containing gigabytes of data and hundreds (or thousands) of individual FITS files.

If adopted as a standard FITS convention, this method of writing data tapes will allow users to pseudo-randomly access files or groups of files from a tape, knowing in advance the disk space necessary to hold them. It will also ensure that the contents of a tape is documented, since the catalog becomes part of the data set and FITS readers will always know where to find the information.

References

NASA Office of Standards and Technology 1993, Definition of the Flexible Image Transport System (FITS) (Greenbelt, NASA/OSSA)

Wells, D. C., Greisen, E. W., & Harten R. H. 1981, A&AS, 44, 363

Pence W., Blackburn J. K., & Greene E. 1992, in Astronomical Data Analysis Software and Systems II, ASP Conf. Ser., Vol. 52, eds. R. J. Hanisch, R. J. V. Brissenden, & J. Barnes (San Francisco, ASP), p. 541

Grosbøl, P., & Wells D. 1994, Blocking of Fixed-block Sequential Media (Greenbelt, NOST Office GSFC), available via anonymous ftp at nssdca.gsfc.nasa.gov

Harten, R. H., Grosbøl, P., Greisen, E. W., & Wells, D. C. 1988, A&AS, 73, 365

FITSIO Subroutine Library Update

W. D. Pence

HEASARC, NASA/GSFC, Greenbelt, MD 20771

Abstract. The FITSIO Subroutine Library has been available for 3 years, and has matured into a stable, full-featured interface for reading and writing FITS format files. The main features of the FITSIO interface are discussed, as well as some ideas for future enhancements.

1. Introduction

The FITSIO subroutine library for reading and writing FITS files is now three years old and is used by many projects as the standard programming interface for accessing FITS format data files. The FITSIO library was previously described at the first ADASS conference (Pence 1992) but many enhancements have been made since then, and they are briefly described here.

2. Main Features of the FITSIO Interface Library

The FITSIO library provides a high-level, machine-independent interface for reading and writing FITS files. Programs that use FITS files and FITSIO for data input and output, can readily be ported to just about any type of computer with no modification to the application source code. Data analysis systems that adopt FITS for their on-line data analysis format have the advantage that the FITS data files are completely portable to any machine without any conversion to the machine's native number storage formats.

FITSIO insulates the programmer from most of the internal structure and format of the FITS files by providing simple subroutines to read and write keyword values, and to read and write data values in an image or table. The FITSIO interface keeps track of where the information is stored in the FITS file, and ensures that the FITS file adheres to all the FITS format rules.

FITSIO also provides the ability to dynamically increase or decrease the size of the FITS file, if necessary, when keywords are added or deleted in the FITS header, or rows are added or deleted in a FITS table. This capability makes is possible to use FITS files in dynamic applications, such as databases or observation logs, where the size and contents of the FITS file needs to be frequently modified.

All currently recognized types of FITS files are supported by FITSIO, including "random grouped data" and IMAGE, ASCII TABLE, and BINTABLE extensions. The support for binary tables includes both fixed length and variable length vector columns.

The core of the FITSIO subroutine library is written in ANSI standard FORTRAN-77 and consists of about 30,000 lines of source code in more than 350 subroutines. There is a small set of machine-specific subroutines that must be ported to each new type of machine; currently FITSIO is supported on virtually all commonly used machines, including SUN workstations, DECstations, VAX/VMS, Alpha/VMS, and Alpha/OSF-1 systems, Cray supercomputers, IBM mainframe computers, and IBM and Macintosh personal computers.

For the convenience of C programmers, there is also a complete set of C wrapper routines called CFITSIO that layer on top of the FITSIO routines, and perform the necessary translations between the C and FORTRAN calling sequences. A version of FITSIO that is fully integrated into the IRAF data analysis system and uses the IRAF VOS I/O calls, rather than FORTRAN data I/O, is also available, and may be called from tasks written in either FORTRAN or SPP.

There are two documents that provide information on using the FITSIO library. One is the "FITSIO User's Guide," which describes in detail how to build and use the FITSIO subroutines. The other is the "FITSIO Cookbook," which gives annotated listings of many real programs that read or write various types of FITS files. Programmers are free to use and modify these programs for their own projects.

The FITSIO software and documentation are freely available from the HEASARC (High Energy Astrophysics Science Archive Research Center) at Goddard Space Flight Center, via anonymous ftp. Currently the software resides on the *legacy.gsfc.nasa.gov* computer in the */software/fitsio* and the */software/fitsio/cfitsio* directories.

3. Future Plans

In the future, FITSIO will be maintained to keep pace with any new standards or conventions that are adopted by the FITS community. One area of particular interest is image compression; it is hoped that at some point FITSIO will transparently support one or more compression algorithms, to reduce the disk space currently required for large data archives. Other possible new capabilities that may be added to FITSIO are support for celestial coordinate system units conversions, the verification of FITS files using a checksum algorithm, and accessing FITS files across computer networks, perhaps using a client-server architecture.

References

Pence, W. D. 1992, in Astronomical Data Analysis Software and Systems I, ASP Conf. Ser., Vol. 25, eds. D. M. Worrall, C. Biemesderfer, & J. Barnes (San Francisco, ASP), p. 22

FITS Checksum Verification in the NOAO Archive

R. Seaman

National Optical Astronomy Observatories[1], *P. O. Box 26732, Tucson, Arizona 85726*

Abstract. There is no standard procedure for verifying the integrity of FITS data files. While a FITS file may be subjected to the same checksum or digital signature calculation as any other data file, the resulting sum or signature must normally be carried separately from the FITS file since writing the value into the header will change the checksum.

A simple method for embedding an ASCII coded 32 bit 1's complement checksum within a FITS header (or any ASCII text) is described that is quick to compute and has desirable features such as: the checksum of each FITS file or extension is set to zero; the checksum may be accumulated in any order; and the checksum is easily updated with simple arithmetic. On-line verification of tapes for the NOAO/IRAF *Save the Bits* archive is discussed as an example.

1. Introduction

There is no standard way to verify FITS files. Various checksums may be calculated for FITS as for other data, but the results must be kept separate from the FITS file since writing the value into the header will change the checksum.

There is a tradeoff between the error detection capability of an algorithm and its speed. The overhead of a digital signature or a cyclic redundancy check (CRC) may be prohibitive for multimegabyte files, and a CRC, tuned to be sensitive to the bursty nature of communication line noise, may not represent the best model for FITS bit errors.

A simple method of embedding an ASCII coded 32 bit 1's complement checksum within a FITS header is described. A 1's complement checksum (as used by TCP/IP) is preferable to a 2's complement checksum (as used by the UNIX `sum` command, for example), since overflow bits are permuted back into the sum and therefore all bit positions are sampled evenly. A 32 bit sum is as easy to calculate as a 16 bit sum because of this symmetry, providing greater sensitivity to errors. A binary to ASCII conversion (analogous to `uuencode`) allows writing the checksum, an unsigned integer, into a string valued FITS header keyword, such that the ASCII bytes sum four at a time. This method has several desirable features:

---

[1]NOAO is operated by AURA, Inc. under contract to the National Science Foundation.

- The checksum of each FITS file is forced to zero by writing the complement of the calculated checksum into the header. Verifying a particular file requires only that the checksum computes to zero.

- Since 1's complement addition is commutative and associative, the checksum may be accumulated in any order.

- If a FITS header is changed, the checksum is updated with simple arithmetic. Only the checksum of keywords that change need be recalculated. A simple rearrangement of keywords leaves the checksum unchanged.

- The checksum of the data records is written into a separate header keyword and is not recomputed unless the data are modified.

- The checksum for individual FITS extensions is separately preserved. Extensions may be added and removed at will from a larger FITS file without disturbing the checksum.

2. Algorithm

The 1's complement checksum is fast and simple to compute. A third of the following C code implementation handles odd length input records—a case that does not apply to FITS. Just zero sum32 and step through the FITS records:

```
checksum (buf, length, sum32)
char *buf;
int length;                    /* < 2^18, or carry can overflow */
unsigned int *sum32;
{
        unsigned short *sbuf;
        unsigned int hi, lo, hicarry, locarry;
        int len, remain, i;

        sbuf = (unsigned short *) buf;
        len = 2*(length / 4);  /* make sure it's even */
        remain = length % 4;   /* add odd bytes below */

        hi = (*sum32 >> 16);
        lo = (*sum32 << 16) >> 16;
        for (i=0; i < len; i+=2) {
            hi += sbuf[i];
            lo += sbuf[i+1];
        }
        (remain >= 1) ? hi += buf[2*len] * 0x100;
        (remain >= 2) ? hi += buf[2*len+1];
        (remain == 3) ? lo += buf[2*len+2] * 0x100;

        hicarry = hi >> 16;    /* fold carry bits in */
        locarry = lo >> 16;
        while (hicarry || locarry) {
            hi = (hi & 0xFFFF) + locarry;
            lo = (lo & 0xFFFF) + hicarry;
            hicarry = hi >> 16;
            locarry = lo >> 16;
        }
        *sum32 = (hi << 16) + lo;
}
```

Encoding the unsigned integer checksum into an ASCII string is simply a matter of dividing each initial byte into four bytes—this permits each quarter of the original 8-bit byte to fit within the range of the ASCII alpha-numerics, including an offset from ASCII zero (hex 0x30).

```
unsigned exclude[13] = { 0x3a, 0x3b, 0x3c, 0x3d, 0x3e, 0x3f, 0x40,
                         0x5b, 0x5c, 0x5d, 0x5e, 0x5f, 0x60 };

int offset = 0x30;                    /* ASCII 0 (zero) */

char_encode (value, ascii)
unsigned int value;
char *ascii;
{
    int byte, quotient, remainder, ch[4], check, i, j, k;

    for (i=0; i < 4; i++) {
        byte = (value << 8*i) >> 24;  /* each byte becomes four */
        quotient = byte / 4 + offset;
        remainder = byte % 4;
        for (j=0; j < 4; j++)
            ch[j] = quotient;
        ch[0] += remainder;

        for (check=1; check;)          /* avoid ASCII punctuation */
            for (check=0, k=0; k < 13; k++)
                for (j=0; j < 4; j+=2)
                    if (ch[j]==exclude[k] || ch[j+1]==exclude[k]) {
                        ch[j]++;
                        ch[j+1]--;
                        check++;
                    }

        for (j=0; j < 4; j++)          /* assign the bytes */
            ascii[4*j+i] = ch[j];
    }
    ascii[16] = 0;
}
```

The basic idea is the same as used by the Internet checksum (Braden et al. 1988; Mallory & Kullberg 1990). See Stevens (1994) for an overview, and Zweig & Partridge (1990) for alternatives. An integer is embedded within each data packet (FITS header) which forces the checksum of the entire packet (FITS HDU) to zero. To find this integer, zero the checksum field in the packet and accumulate the checksum—the necessary value is just the complement (additive inverse) of the checksum.

In this case, the equivalent of zeroing the checksum field is to set the 16 character string value of the CHECKSUM keyword to all ASCII 0s (hex 0x30). The checksum is accumulated and complemented in the same fashion. The ASCII encoded complement of the checksum is written into the header replacing the ASCII 0s, which are in effect subtracted back out of the encoding to restore the original value. The checksum and its complement sum to zero. (Actually they sum to *negative* zero, all 1's—1's complement addition has two identity elements.)

Note that the checksum field must be integer aligned, whether the checksum is being stored as an integer or an encoded string. In either case, this requirement only applies byte-by-byte. To begin the string at an arbitrary odd byte offset, just permute the bytes. Note also that the same zeroing effect could be gained by embedding the complemented value in a comment as well as in a keyword.

3. Verification in the NOAO Archive

The NOAO/IRAF *Save the Bits* archive is described in Seaman (1994). Images from several telescopes on Kitt Peak are multiplexed onto tape as large FITS image extension files. As each image is processed, the checksum of the resulting FITS extension is forced to zero by writing its complement into the header:

```
          XTENSION= 'IMAGE    '           / FITS image extension
              ...                ...              ...
          RECID    = 'kp09m.940909.082728' / archive ID for observation
          RECNO    =              318747  / NOAO archive sequence number
          CHECKSUM= ' cHjjc9ghcEghc9gh '  / ASCII 1's complement checksum
          DATASUM = ' 5ZNF4XME4XME4XME '  / checksum of data records
          END
```

As the tape files are assembled from the individual extension files, the checksum for the primary FITS header is zeroed. This zeroes the checksum for the entire multiple image file since each extension's checksum is the additive identity. After each tape (actually a duplicate pair) fills up, the archive takes the drive off-line and verifies the checksums.

The checksum of the data records is saved separately in the DATASUM keyword. This simplifies updating the checksum during subsequent header operations, as when an image is later extracted from the archive. Simple arithmetic suffices to recalculate the checksum no matter where in a file changes occur.

Other checksum schemes are possible (Peterson & Weldon 1972). Checksums, CRCs, and digital signatures such as MD5 (Rivest 1992) are all examples of hash functions. Many possible images will hash to the same checksum—how many depends on the number of bits in the image versus the number of bits in the sum. The utility of a checksum to detect errors (but not forgeries) depends on whether it evenly samples the likely errors. The 1's complement checksum is a good, quick way to do this.

References

Braden, R. T., Borman, D. A., & Partridge, C. 1988 (September), "Computing the Internet Checksum", Internet RFC 1071

Mallory, T. & Kullberg, A. 1990 (January), "Incremental Updating of the Internet Checksum", Internet RFC 1141

Peterson, W. W., & Weldon Jr., E. J. 1972, Error-Correcting Codes, Second Edition (Cambridge, Mass., MIT Press)

Rivest, R. 1992 (April), "The MD5 Message Digest Algorithm", Internet RFC 1321 (see also RFC 1319 and RFC 1320)

Seaman, R. 1994, in Astronomical Data Analysis Software and Systems III, ASP Conf. Ser., Vol. 61, eds. D. R. Crabtree, R. J. Hanisch, & J. Barnes (San Francisco, ASP), p. 119

Stevens, W. R. 1994, TCP/IP Illustrated Vol. 1 (Reading, Mass., Addison-Wesley)

Zweig, J., & Partridge, C. 1990 (March), TCP Alternate Checksum Options, Internet RFC 1146

Part 6. Object Detection and Classification

Automated Classification of a Large Database of Stellar Spectra

R. K. Gulati and R. Gupta

IUCAA, Post Bag 4, Ganeshkhind, Pune 411 007, India

P. Gothoskar and S. Khobragade

NCRA, TIFR Center, P.O. Box 3, Pune 411 007, India

Abstract. An Artificial Neural Network (ANN) is a versatile tool which has been used both in academic research and industrial applications. In astronomy, this technique has been used for a variety of applications, such as telescope adaptive optics, classifying galaxies, and separating stars from galaxies. The classification of a large database of stellar spectra, which would be a Herculean task for human classifiers if done visually, is an ideal problem for the ANN technique, which can handle such problems without manual intervention. Recently, increased computational power, combined with improvement in the ANN techniques, has provided an efficient way to perform automatic classification.

We have implemented ANN to classify stellar spectra from large spectral databases. We present here the Multilayer Back Propagation Network (MBPN), which is used to classify stellar spectra obtained in the optical and ultraviolet regions. The performance of MBPN shows that the ANN is capable of classifying ultraviolet stellar spectra to an accuracy of about one spectral subclass for most of the cases. The scope of this technique is expected to be expanded with the availability of large homogeneous digitized stellar spectral databases.

1. Introduction

Progress in ground based and space instrumentation has brought us to a new era of spectroscopy, where a large quantity of good quality stellar spectra has started becoming available through well organized data centers. In order to analyze these spectra and extract useful physical information about stars and stellar systems, we need to develop fast and accurate methods. One way to analyze these spectra is to classify them in terms of common visible properties.

Spectral classification, which conventionally has been done by human classifiers (Houk 1983; Houk & Smith-Moore 1988), involves large, time-consuming efforts. We now require automation of the classification process. The main advantages of automated over human classification is not only the speed with which it can be done, but also accuracy, detection of variability, the elimination of personal error, and the possibility of classification of higher dimensionality.

We (Gulati et al. 1994) have initiated a project to implement automated classification schemes to digitized databases of optical and UV spectra by using conventional metric distance minimization methods and Artificial Neural Networks. We have been using the Multilayer Back Propagation Network (MBPN). Similar efforts have also been employed by another group to classify the stellar spectra of high-dispersion objective prism plates using a neural network scheme (von Hippel et al. 1994). ANN has also been applied to classify a near infra-red database (Weaver 1994).

Here we present the performance of the ANN scheme on libraries of optical and ultraviolet stellar spectra by comparing classifications determined by ANN with those of human classifiers (i.e., catalog classifications).

2. Input and Pre-processing of Optical & UV Data

The optical data were taken from Silva (Silva & Cornell 1992) and Jacoby (Jacoby et al. 1984) libraries. A set of 55 spectra selected from the former library was treated as the template database, and the test database was a set of 158 spectra from the latter library. Both sets were brought to a uniform wavelength range of 3510–6800 Å with 5 Å sampling and 11 Å resolution, and normalized to a value of 100 at 5450Å. Instead of using the full spectral information, a set of 161 wavelength positions was used to monitor the fluxes which are diagnostic of the spectral classes as given by human experts (Jaschek and Jaschek 1990). Catalog classifications of the spectra were taken from the respective libraries. The spectra covered stars of solar metallicity, types O–M, and luminosity classes I–V. Each spectro-luminosity class was coded with a number $x = (1000 \times A1 + 100 \times A2 + (1.5 + 2 \times A3))$, where A1 was the main spectral type of the star (i.e., O to M types coded from 0.0 to 9.5), A2 was the sub-spectral type (coded from 0.0 to 9.5) and A3 the luminosity class (i.e., classes I to V coded as 0 to 4). For example, a B2I star and a G9.5V star would be coded as 2201.5 and 5959.5, respectively.

The input database for the UV data was the *IUE* Low Resolution Spectra (Heck et al. 1984). A set of 128 spectra spanning 75 spectro-luminosity classes was selected as the template and another set of 83 spectra was used as the test set. The catalog classification was taken from this catalog, where like MK classification, the UV classification is given as O, B, F, etc., as main classes, subclasses ranging from 0.0 to 9.5, and luminosity classes represented as s, g, and d for super-giants, giants, and dwarfs, respectively. The wavelength range of the UV spectra is 1213–3201 Å with 2 Å sampling and 6 Å resolution. The spectra were monitored at 35 wavelength positions which are the diagnostic of these spectral classes as given in Table 1 of the *IUE* catalog (Heck et al. 1984). Here, too, the spectral coding was done by using a number $x = (1000 \times A1 + 100 \times A2 + A3)$, where A1 was the main spectral type of the star (i.e., O to F types coded as 1 to 4), A2 was the sub-spectral type of the star (coded from 0.0 to 9.5), and A3 the luminosity class of the star (i.e., 2, 5, or 8 for s, g, or d). For example, stars dB2.5, gO9.5 and sF7 (ultraviolet classes) were coded as 2258, 1955, and 4702, respectively.

3. The ANN Architectures

We used the standard feedforward supervised neural network, known as "multilayer backpropagation network (MBPN)" (Rumelhart et al. 1986), for classifying

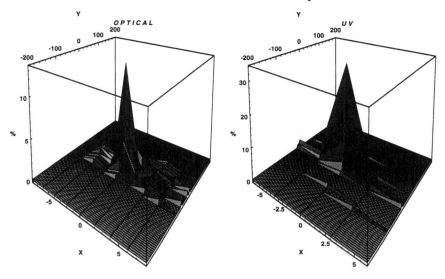

Figure 1. 3D plots for classification errors in luminosity (x-axis) and spectral type (y-axis) vs. the % of total number of test spectra for Optical and UV data.

the databases of optical and ultraviolet stellar spectra into different classes of stars. As mentioned earlier, the number of output classes was 55 in the optical case and 75 in the UV case. The input data points for optical and UV data classifiers were different, so the ANN classifiers were selected with different architectures for optical and UV data. The optical classifier was found to be optimal with the configuration 161:64:64:55 and the UV classifier was configured as 35:71:75. The configuration numbers show the input size, hidden nodes, and, at the end, the output nodes, respectively. Once the training was over, the networks could classify the large databases of stellar spectra within a minute, without any human intervention.

4. Performance

The performance of the ANN technique can be judged from Figure 1, which shows the 3D plots of classification errors in luminosity and spectral type on x and y axes and the percentage of total test sample along the z axis, respectively, for optical and UV data. In these plots an ideal classification would appear as a single peak of 100% value in the center of the (x, y) plane, signifying that all spectra are classified correctly with no errors in either luminosity or spectral type. One sub-spectral type error means 100 units error along the y-axis of these 3D plots. Statistical parameters, such as linear correlation coefficients and standard deviations, were computed on the scatter plots (for details see Gulati et al. 1994), and it was found that the classification error for optical was about two subclasses and for UV it is about one subclass, barring a few stars which clearly show more than 100 units of error. These stars require further detailed studies.

5. Conclusions and Future Steps

We do not see any gross mis-classification with the automated ANN scheme and the schemes are quite efficient for large databases. However, we feel that a more homogeneous and complete database is required for the ANN training to perform better. The implementation of ANN on a parallel computer would significantly reduce the training time. In the future, we plan to use a library of synthetic spectra based on stellar atmosphere models (Gulati et al., 1993) to tag information on the stellar physical parameters.

Acknowledgments. R. K. Gulati wishes to acknowledge the generous financial support from the organizing committees of ADASS IV, which allowed him to present this paper at the conference.

References

Gulati, R. K., Malagini, M. L., & Morossi, C. 1993, ApJ, 413, 166
Gulati, R. K., Gupta, R., Gothoskar, P., & Khobragade, S. 1994, ApJ, 426, 340
Heck, A., Egret, D., Jaschek, M., & Jaschek, C. 1984, in IUE Low-Resolution Spectra: A Reference Atlas–Part I, Normal Stars, ESA SP-1052
Houk, N. 1983, in The MK Process and Stellar Classification, ed. R. F. Garrison, (Toronto, David Dunlop Observatory), p. 85
Houk, N., & Smith-Moore, M. 1988, in University of Michigan Catalogue of Two-Dimensional Spectral Types for the HD Stars, 1988, Vol. 4
Jacoby, G. H., Hunter, D. A., & Christian, C. A. 1984, ApJS, 56, 257
Jaschek, C., & Jaschek, M. 1990, The Classification of Stars, (Cambridge, Cambridge Univ. Press)
Rumelhart, D. E., Hinton, G. E., & Williams, R. J. 1986, Nature, 323, 533
Silva, D. R., & Cornell, M. E. 1992, ApJS, 81, 865
von Hippel, T., Storrie-Lombardi, L. J., Storrie-Lombardi, M. C. & Irwin, M. J. 1994, MNRAS, 269, 97
Weaver, Wm. B. 1994, in The MK Process at 50 Years: A Powerful tool for Astrophysical Insight, ASP Conf. Series, Vol. 60, eds. C. J. Corbally, R. O. Gray, and R. F. Garrison (San Francisco, ASP), p. 303

Classification of Objects in the Guide Star Catalog

O. Yu. Malkov, O. M. Smirnov

*Institute of Astronomy of the Russian Academy of Sciences,
48 Pyatnitskaya Str., Moscow 109017 Russia*

Abstract. A status report of the The GSC Object Classifier (GOC) project is presented.

1. Introduction

As was reported at the ADASS III meeting (Malkov and Smirnov 1994a), software for retrieving data from the GSC (GUIDARES) and mapping of object regions has been developed. The GSC contains multiple entries for some objects due to plate overlap. It classifies objects as stars, non-stars, galaxies, blends and potential artifacts; the latter three classifications are rare, and only exist in GSC version 1.1, to which they were added manually. It should also be noted that many objects, especially near plate edges, are misclassified in the catalog.

The original GUIDARES had only two ways of dealing with multiple-entry objects (MEOs); compute a weighted average of the data, or report all entries for the object. We are working on a third way of looking at the GSC: analyzing the nature of the object, using data from the GSC, to come up with a "correct" classification. We do not have the intent nor the means to revise the GSC; rather, the object of the GOC project is to come up with extensions to the GUIDARES software that will be able to read the GSC and produce correct classifications within any region specified by the user. To implement GOC, we shall examine a set of object parameters, among which are; the brightness of an object, its coordinates (equatorial, galactic, and ecliptic), plate quality, distance from plate center, and others. Multiplate analysis will be done as well. We plan to establish probabilities for appearances of stars, galaxies, close binaries, "rapid" objects (like minor planets), as well as blends, defects and artifacts, as functions of these parameters. Artificial intelligence methods will be explored as well. The goal is to produce an automatic classifier that can overcome the original catalog's misclassifications by making use of all the other available information.

2. GSC classifications vs. GOC

GSC 1.1 contains the following classifications:

0 —star,

1 —galaxy,

2 —blend or incorrectly resolved blend,

3 —non-star,

5 —potential artifact.

We are designing GOC to recognize the following objects: stars, galaxies, close binaries, variable stars, objects in overcrowded stellar fields, Solar System objects, plate defects, and artifacts.

3. Implementation

The GOC will perform classification with the help of three separate modules. The first is a rule-based expert system that either produces definite classifications or fails. The second is a probability engine, which computes probabilities of the object belonging to one of the above classes, and the third is an artificial intelligence (AI) classifier that uses the test vote method to see if the object is similar to a class of predefined training objects. A fourth module will act as an arbitrator between the three classifiers to produce final results.

3.1. The basics: a Preliminary Analysis

To perform a preliminary analysis in the expert system, we will compare three values: (1) the number of GSC entries for the object (E), (2) the number of entries with the stellar (0) classification (S), and (3) the number of plates that overlap this point on the sky (P). Some obvious rules are, for example, $P = S = E \Rightarrow$ the object is a star; $E > 1 \Rightarrow$ the object exists on the sky (i.e., is not a plate defect) and has a "constant" position and brightness (i.e., is not a rapid object). We regard original GSC classifications of type 1, 2, and 5 as unconditionally correct (since they were added manually).

3.2. Carrying on: Multi-plate Analysis

For a multi-plate analysis, the entries originating from higher-quality plates, or those positioned closer to plate center (i.e., not distorted) are considered to be more "trustworthy" than those on low-quality plates and/or near plate edges. GOC will use this information when analyzing the different entries. Near plate edges, the original object images become distorted and often lead to incorrect classifications. The effective edge of plate is the threshold beyond which information from the plate should no longer be considered valid (Malkov & Smirnov 1994b).

3.3. Estimation of Probabilities

Can it be a galaxy? The probability of the object being a galaxy (independent of how the GSC classifies it) is a function of brightness and galactic latitude. GOC will combine this with the E-S-P analysis to produce final probability estimates.

Can it be a rapid (Solar System) object? The probability is a function of brightness, ecliptic latitude, date of plate exposure, and result of the E-S-P analysis (see above).

3.4. Internal Databases of Special Objects

The GOC will include a small database of special objects that can severely affect the original GSC classifications, for example, bright stars and clusters, or artifacts, which are usually expected in the vicinity of bright stars. The database can be used to quickly establish their locations (since searching the GSC itself can take much longer), and overcrowded regions in the vicinities of star clusters, which again can be established with the help of the database.

3.5. Artificial Intelligence Methods

A lot of artifacts are close to bright stars, and form line- and arc-like patterns. The distance from the nearest bright star, magnitude, and orientation relative to neighboring objects can indicate the nature of the object. AI methods will be used to analyze this information.

Close binaries are fully resolved on some plates, and have a single "non-star" or "stellar" classification on others. By using a catalogue of close binaries in multiparametric GSC-analysis, we hope to find an implicit dependence between separation, magnitude difference, plate quality, and some other parameters. The test vote method will be used to find this dependence and apply it to GSC objects.

4. Current Status and Future Plans

At the moment some principles of GOC are fully realized, others are taking shape, and others are still under discussion. When ready, GOC will be built into GUIDARES as an optional tool.

Acknowledgments. This presentation was made possible by financial support from ST ScI. Part of this work is funded by International Science Foundation Grant number R3Z000.

References

Malkov, O. Yu., & Smirnov, O. M. 1994a, in Astronomical Data Analysis Software and Systems III, ASP Conf. Ser., Vol. 61, eds. D. R. Crabtree, R. J. Hanisch, & J. Barnes (San Francisco, ASP), p. 183

Malkov, O. Yu., & Smirnov, O. M. 1994b, in Astronomical Data Analysis Software and Systems III, ASP Conf. Ser., Vol. 61, eds. D. R. Crabtree, R. J. Hanisch, & J. Barnes (San Francisco, ASP), p. 187

Object Detection Using Multi-Resolution Analysis

F. Murtagh[1]

ST-ECF, ESO, Karl-Schwarzschild-Str. 2, D-85748 Garching

W. Zeilinger

Department of Astronomy, University of Vienna, Türkenschanzstr. 17, A-1180 Vienna

J.-L. Starck

CEA, DSM/DAPNIA, F-91191 Gif-sur-Yvette Cedex

A. Bijaoui

OCA, BP 229, F-06304 Nice Cedex 4

Abstract. What we look for in an image is scale-dependent. Multi-resolution allows for image analysis in terms of different scales. Scale-space filtering, quadtrees, pyramid representations, and the wavelet transform have all been used to yield multi-resolution views of an image. One approach to the demarcation of significant structure, at different resolutions, is to determine statistically significant parts of each scale. A multi-resolution support data structure may be built up in this way. We can go further and incorporate a priori knowledge into the construction of this multi-resolution support. Mathematical morphology offers a convenient framework for removing detector artifacts, objects which are too small in extent to be of interest, etc. This approach is used for the detection (and subsequent analysis) of globular cluster systems around elliptical galaxies, and for image characterization and object trawling in large image databases.

1. Introduction

Content-based image retrieval can be tackled by using text annotations which characterize image content. Such an approach has been prototyped[2] for *Hubble Space Telescope* (*HST*) WF/PC-1 data. In this paper we describe the use of multi-resolution analysis for finding and making an inventory of objects in an image. For the multi-resolution transform, we used the pyramidal median image transform: see Starck, Murtagh, & Louys (1995) for further details.

---

[1] Affiliated with the Astrophysics Division, Space Sciences Department, ESA

[2] http://ecf.hq.eso.org/~fmurtagh/hst-navigate.html

The use of Minkowski morphological operators allows prior knowledge relating to the objects of interest to be introduced. The structuring element is related to the morphology of the objects of interest in the image. The choice adopted is point-symmetric; and its size is related to potentially relevant objects. Image dilations expand the object in a locally adaptive manner. The dilated object area may help in providing extra faint or background pixels later when the object is characterized using ellipticity, moments, or other properties. Closings, or erosions with a suitable structuring element followed by dilations, may be used to remove linear features that are not usually of interest in astronomical imagery (e.g., extended cosmic ray impacts, bleeding and charge-overflow effects in a CCD detector, and intersections of mosaiced images).

2. Globular Clusters Surrounding Elliptical Galaxies

Earlier work by Meurs et al. (1994) aimed at finding faint, edge-on galaxies in WF/PC images. For each object found, properties such as the number of pixels in the object, peak-to-minimum intensity difference, a coefficient characterizing the azimuthal profile, and the ellipticity of the principal axis, were used to allow discrimination between potentially relevant objects on the one hand, and faint stars or detector faults on the other. This work is currently being extended with the study of globular cluster systems. NGC 4636 was discussed by Kissler et al. (1993), and characterized as a rich globular cluster system in a normal elliptical galaxy. An extracted 512 × 512 sub-image from an ESO New Technology Telescope (NTT) image is shown in Figure 1. A multi-resolution support of this image was obtained. This multi-resolution support image with values 0, 2, 4, and 8 represents, in one image, the four resolution levels examined. Figure 2 shows the result of transforming the support image in Figure 1 to a boolean image: any pixel with value greater than 2 was assigned a value of 1, followed by two openings. The object "islands" are labeled, and an associated report file produced principally with Gaussian profile fits in X and in Y. This information can be used to discriminate between objects of interest and those which are not.

3. Object Detection in Large Image Databases

The applicability of this approach to object detection in large image collections has been investigated. Rather than using original WF/PC data, the associated preview images provide two advantages. Firstly they are available on-line. Secondly they are smaller in size: typically about 1 MB per image when compressed, compared to 10.5 MB in original form. On the negative side, the compression scheme used in these preview images produces artifacts, and in particular the image texture may be considerably affected. Mosaicing the WF/PC quadrants also leaves the intersections bare.

A selection of WF/PC images was chosen with dense stellar fields, sparse faint object fields, and filamentary structure. These images were subjected to the following operations to allow for the great differences in the image properties, due to varying exposure lengths, the different nature of objects viewed, and presence of detector artifacts:

1. Multi-resolution support construction.

Figure 1. Part of ESO NTT image of NGC 4636.

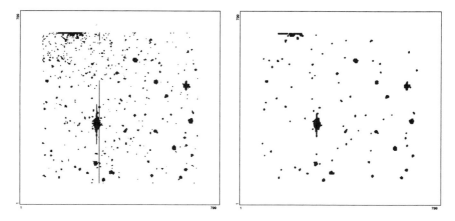

Figure 2. Booleanized multi-resolution support before and after 2 closings.

2. Level 4 in the multi-resolution support corresponds to large gradient local peaks—if it and level 3 accounted for more than 20% of all pixels, then level 4 alone was retained; if levels 4 and 3 together accounted for less than 20% of the image's pixels, then they were combined into a single boolean image. These rules were employed to specify what could be considered as significant objects in a given image.

3. A set of up to 3 openings with point-symmetric structuring elements were applied to remove remaining detector artifacts and cosmic ray hits.

4. Finally the contiguous regions were labeled and elementary properties were determined (object extent, defined by numbers of pixels; peak-to-minimum intensity within the object area, etc.).

This approach works well. Some remaining problems include the following: closely located objects are gathered into contiguous object "islands" in the final boolean image; and the presence of faint filamentary structure in the image in not explicitly handled when using structuring elements and a multi-resolution transform kernel which presuppose small, point-symmetric objects.

4. Conclusion

In this work, we ask ourselves whether it is feasible to consider automated image characterization through object detection and description. Traditional object inventory packages include FOCAS and MIDAS/INVENTORY, and are semi-interactive. Can a multi-resolution approach improve generality of treatment, and accuracy of results? Such established object inventory packages have been limited to relative peaks in flux intensity and/or intensity gradient. The multi-resolution transform incorporates these, and adds other potentially valuable image properties. Included among these properties are: object resolution scale (indirectly linked to gradient information); automatically ignoring the image background (through differencing the images which make up the multi-resolution transform); and facilitating incorporation of user-specified image processing rules using, for example, Minkowski image operators.

References

Kissler, M., Richtler, T., Held, E. V., Grebel, E. K., Wagner, S., & Cappaccioli, M. 1993, ESO Messenger, 32

Meurs, E. J. A., Murtagh, F., & Adorf, H.-M. 1994, IAU General Assembly

Starck, J.-L., Murtagh, F., & Bijaoui, A. 1995, this volume, p. 279

Starck, J.-L., Murtagh, F., & Louys, M. 1995, this volume, p. 268

Unsupervised Catalog Classification

F. Murtagh[1]

Space Telescope – European Coordinating Facility, European Southern Observatory, Karl-Schwarzschild-Str. 2, D-85748 Garching, Germany

Abstract. Automatic classification of large catalogs may be carried out for survey objectives, or to elucidate the catalog's contents in a minimally-restricted way. A cohesive mathematical framework is of benefit, since otherwise the boundaries between hand-crafted analysis tools will themselves require close monitoring at all times. We investigate the use of the Kohonen self-organizing feature map (SOFM) method for simultaneous clustering and dimensionality-reduction. We find a difficulty with the summarizing properties of this method, and propose a mathematically-coherent enhancement to the SOFM method to overcome it. This involves use of an agglomerative contiguity-constrained clustering method on the SOFM output. An application to the *IRAS* Point Source Catalog is presented.

1. The Kohonen SOFM Method

The Kohonen self-organizing feature map (SOFM) method may be described in these terms (Kohonen 1988; Kohonen 1990; Kohonen et al. 1992; Murtagh & Hernández-Pajares 1994; Hernández-Pajares et al. 1994): each item in a multi-dimensional input data set is assigned to a cluster center; the cluster centers are themselves ordered by their proximities; and the cluster centers are arranged in some specific output representational structure, often a regularly-spaced grid. The output representational grid of cluster centers is structured through the imposition of neighborhood relations. Different variants on this algorithm are possible. A description of our exact, coded implementation may be found in Murtagh & Hernández-Pajares (1994). The regularly-spaced grid leads to a close relationship with a range of widely-used dimensionality-reduction methods (Sammon nonlinear multidimensional scaling, principal components analysis, etc.). The given dimensionality m is mapped into a discretised 2-dimensional space.

A practical problem arises with this SOFM method: if a two- or three-cluster solution is potentially of interest, it is difficult to specify an output representational grid with just this number of cluster centers. One would be more advised to use a traditional k-means partitioning method. Alternatively, the thought comes to mind to further process the cluster centers, perhaps in a way which is motivated by hierarchical cluster analysis. This is close to the graphical proposals of Ultsch (1993a, 1993b).

---

[1] Affiliated to Astrophysics Division, Space Science Department, European Space Agency

2. Contiguity-Constrained Clustering

Cluster centers are formed by the SOFM method in such a way that similarly-valued cluster centers are closely located on the representational grid. This reinforces the need for any subsequent clustering of these cluster centers to take such contiguity information into account. Contiguity-constrained clustering has been reviewed by Murtagh (1985a), Gordon (1981), and Murtagh (1994). Split-and-merge techniques may well be warranted for segmenting large images. We however seek to segment a grid, at each element of which we have a multidimensional vector, which is ordinarily not of very large dimensions (say, if square, 50×50 or 100×100). On computational grounds, a pure agglomerative strategy is eminently feasible.

A contiguity-constrained enhancement on such algorithms is to only allow objects x and y to be agglomerated if in addition we have: there is some $q \in x$ and some $q' \in y$ such that q and q' are contiguous. The definition of contiguity on the regular grid is immediate if we require, for example, that the 8 possible neighbors of a given grid intersection form contiguous neighbors.

A number of theoretical and practical issues arise in the context of such algorithms. Such issues include the following which are discussed in Murtagh (1994):

1. Criteria other than the centroid one could be of value. In particular the variance criterion is often favored for forming homogeneous, compact (and hyperspherical) groups.

2. A hierarchical agglomerative method may suffer from inversions or reversals in the sequence of criterion values. When this happens, or when it can be avoided, is criterion-dependent.

3. Nearest-neighbor algorithmic implementations offer computational savings for unconstrained agglomerative methods. Such approaches may also be feasible for contiguity-constrained methods. We can specify the conditions under which such alternative algorithmic implementations should be used.

4. Parallel implementations are possible with array operations supported by image processing and other packages. An efficient IDL array-operator based implementation was used for the example discussed below.

3. Application

The *IRAS* Point Source Catalog (*IRAS* PSC) contains measurements on 245,897 objects. Coarse selection criteria, based on flux values or combinations of them, have been used for separating stars from non-stellar objects. Among non-stellar objects, in some studies, classes such as the three following ones have been distinguished: thin Galaxy plane; a "cirrus" region of objects surrounding this, due to dust; and the extragalactic sky. When using different normalizations of the input data, and different agglomerative criteria, we used this prior information (see Prusti et al. 1992; Boller et al. 1992; Meurs & Harmon 1988) to select one result over others.

No preliminary selections were made: all 245,897 objects were used. A set of four variables was derived from the four flux values (F_{12}, F_{25}, F_{60}, F_{100}), in line with the studies cited above: $c_{12} = \log(F_{12}/F_{25})$, $c_{23} = \log(F_{25}/F_{60})$, $c_{34} = \log(F_{60}/F_{100})$, and $\log F_{100}$. The first three of these define intrinsic colors. The SOFM method was applied to this 245897 × 4 array, mapping it onto a regular 50 × 50 representational grid. This output can be described as 2500 clusters (defined by a minimum distance criterion, similar to k-means methods); with the cluster centers themselves located on the discretized plane in such a way that proximity reflects similarity.

Normalization of the input data (necessitated, if for no other reason, by the fourth input being quite differently-valued compared to the first, second and third), was carried out in two ways: range-normalizing, by subtracting the minimum, and dividing by maximum minus minimum (this approach is favored in Milligan & Cooper 1988); or by carrying out a preliminary correlation-based principal components analysis (PCA). Three principal component values were used as input, for each of the 245,897 objects. Such normalizing is commonly practiced but is viewed negatively e.g., in Chang (1983), since mapping into a principal component space may destroy the classificatory relationships which are in fact sought. However, we found that this provided what we considered as the best end-result.

In all cases, the SOFM result was obtained following 10 epochs, i.e., all objects were successively cycled through 10 times. Typical computation times were 9 hours on a loaded SPARCstation 10. All other operations described in this paper (contiguity-constrained clustering, display, etc.) required orders of magnitude less time.

The numbers of objects assigned to each of the 50 × 50 cluster centers were not taken into account when carrying out the segmentation of this grid. We chose 8 clusters as a compromise between having some information about a desired set of 3 or 4 clusters (cf. discussion above regarding stars, thin plane, cirrus and extragalactic classes); and convenience in representing a somewhat greater number of clusters. Results obtained are further discussed in the more complete version accompanying this paper: see next section for access details.

4. Conclusions

The combined SOFM-CCC approach provides a convenient framework for clustering. We have noted how the CCC approach is the most appropriate for SOFM output. To aid in interpretation of such a method, the use of tracer objects has been found to be very advantageous (Hernández-Pajares et al. 1994).

Large astronomical catalogs are common-place, and the *IRAS* PSC is not untypical. Unsupervised classification has often taken the form of "hand-crafted" approaches, e.g., through pairwise plots of variables and the visually-based specification of discrimination rules. Our aim is to have reasonably standard, automated approaches for handling such data. We do not wish to have major "fault lines" running through our methodology, but seek instead an integrated, cohesive framework.

Open issues are (1) further study of the input data normalization issues; and (2) the enhancement of the method described here to handle input data which have associated quality/error values and/or which are censored.

References

Boller, Th., Meurs, E. J. A., & Adorf, H.-M. 1992, A&A, 259, 101

Chang, W. C. 1983, Appl. Stat., 32, 267

Gordon, A. D. 1981, Classification: Methods for the Exploratory Analysis of Multivariate Data (London, Chapman and Hall)

Hernández-Pajares, M., Floris, J., & Murtagh, F. 1994, Vistas in Astronomy, in press

Kohonen, T. 1988, in Proc. Connectionism in Perspective: Tutorial (Zurich, University of Zurich)

Kohonen, T. 1990, Proc. IEEE, 78, 1464

Kohonen, T., Kangas, J., & Laaksonen, J. 1992, SOM_PAK Version 1.0. Available by anonymous ftp from *cochlea.hut.fi* (130.233.168.48)

Meurs, E. J. A., & Harmon, R. T. 1988, A&A, 206, 53

Milligan, G. W., & Cooper, M. C. 1988, Journal of Classification 5, 181

Murtagh, F. 1985a, The Computer Journal, 28, 82

Murtagh, F. 1985b, Multidimensional Clustering Algorithms (Würzburg, Physica-Verlag)

Murtagh, F. 1994, in Partitioning Data Sets eds. I. J. Cox, P. Hansen and B. Julesz (New York, AMS), in press.

Murtagh, F. 1994, Pattern Recognition Letters, submitted

Murtagh, F., & Hernández-Pajares, M. 1994, Journal of Classification, in press

Research Systems, Inc. 1992, Interactive Data Language, Version 3.0 (Boulder, RSI)

Prusti, T., Adorf, H.-M, & Meurs, E. J. A. 1992, A&A, 261, 685

Ultsch, A. 1993a, in Information and Classification, eds. O. Opitz, B. Lausen, and R. Klar (Berlin, Springer-Verlag), p. 301

Ultsch, A. 1993b, in Information and Classification eds. O. Opitz, B. Lausen, and R. Klar (Berlin, Springer-Verlag), p. 307

Astronomical Image Compression Using the Pyramidal Median Transform

J.-L. Starck

CEA, DSM/DAPNIA, CEA-Saclay, F-91191 Gif-sur-Yvette Cedex, France

F. Murtagh[1]

ST-ECF, ESO, Karl-Schwarzschild-Str. 2, D-85748 Garching, Germany

M. Louys

LSIT, ENSP, 7 rue de l'Université, F-67084 Strasbourg Cedex, France

Abstract. We describe image compression based on the pyramidal median transform, introduced as an alternative to the use of a wavelet transform. The use of a multiresolution support allows "protection" of astronomical objects in the image. Non-support parts of the images are subjected to noise suppression. Experimental applications are presented.

1. Introduction

Image compression is required for preview functionality in large image databases (e.g., *HST* archive), for linking image and catalog information in interactive sky atlases (e.g., Aladin), and for image data transmission, where more global views are communicated to the user, followed by more detail if desired. We describe an approach to astronomical image compression through noise removal. Noise is determined on the basis of the image's assumed stochastic properties. This approach is quite similar to the wavelet transform-based hcompress approach. We begin by explaining why transforms other than the wavelet transform are important for astronomical image compression.

2. Wavelet Transform and Image Compression

Practical problems related to the use of the wavelet transform include:

Negative Values. By definition, the wavelet coefficient mean is null. Every time we have a positive structure at a scale, we have negative values surrounding it. These negative values often create artifacts during the restoration process, or complicate the analysis. For instance, if we threshold small values (noise, non-significant structures, etc.) in the wavelet transform,

---

[1] Affiliated to Astrophys. Div., Space Sci. Dept., ESA

and then reconstruct the image at full resolution, the structure's flux will be modified.

Point Objects. We often have bright point objects in astronomical images (stars, cosmic ray hits, etc.), and the convolution of a Dirac function with the wavelet transform is equal to the wavelet transform. So at each scale, and at each point source, we will have the wavelet. Cosmic rays can pollute all the scales of the wavelet transform.

This leads us to develop other multiresolution tools which we now present.

3. Pyramidal Median Transform

3.1. Multiresolution Median Transform

The median transform is nonlinear, and offers advantages for robust smoothing (i.e., the effects of outlier pixel values are mitigated). The multiresolution median transform consists of a series of smoothings of the input image, with successively broader kernels. Each successive smoothing provides a new resolution scale.

The multiresolution coefficient values constructed by differencing images at successive resolution scales are not necessarily of zero mean, and so the potential artifact-creation difficulties related to this aspect of wavelet transforms do not arise. For integer input image values, this transform can be carried out in integer arithmetic only, which may lead to computational savings.

3.2. Pyramidal Median Transform

Computational requirements of the multiresolution median transform are high, and these can be reduced by decimation: one pixel out of two is retained at each scale. In the Pyramidal Median Transform (PMT), the kernel or mask used to obtain the succession of resolution scales remains the same at each level. The image itself, to which this kernel is applied, becomes smaller. While this algorithm aids computationally, the reconstruction formula for the input image is no longer immediate. Instead, an algorithm based on B-spline interpolation can be used for reconstruction.

An iterative scheme can be proposed for reconstructing an image, based on pyramidal multi-median transform coefficients. Alternatively, the PMT algorithm, itself, can be enhanced to allow for better estimates of coefficient values, yielding an Iterative Pyramidal Median Transform.

4. PMT and Image Compression

The principle of the method is to select the information we want to keep, by using the PMT, and to code this information without any loss. Thus the first phase searches for the minimum set of quantized multiresolution coefficients which produce an image of "high quality." The quality is evidently subjective, and we will define by this term an image for which (1) there is no visual artifact in the decompressed image, and (2) the residual (original image minus decompressed image) does not contain any structure. Lost information cannot be recovered, so if we do not accept any loss, we have to compress what we take as noise, too, and the compression ratio will be low (only 3 or 4).

The method employed involves the following sequence of operations:

1. Determination of the multiresolution support.

2. Determination of the quantized multiresolution coefficients which gives the filtered image.

3. Coding of each resolution level using the Huang-Bijaoui method (Huang & Bijaoui 1991). This consists of quadtree-coding each image, followed by Huffman-coding the quadtree representation. There is no information lost during this phase.

4. Compression of the noise, if this is desired.

5. Decompression consists of reconstituting the noise-filtered image (plus the compressed noise, if this was specified).

Note that we can reconstruct an image at a given resolution without having to decode the entire compressed file.

5. Example

A simulated *HST* WF/PC (pre-refurbishment) stellar field image described in Hanisch (1993) was used. The image dimensions were 256 × 256. Here we used the aberrated, noisy image. With default options for the approach described in this article (pcomp/pdecomp; i.e., 5 iterations, 3σ thresholding, 4 multiresolution scales, and no conservation of the noise) a compressed image with 6643 bytes was obtained from the original image of 268,800 bytes. This is a compressed image equal to 2.5% of the original. Even with I2 storage of the input image, we have compression to about 5%. The total intensity dropped from 412,998 to 412,980 in compressing and decompressing, i.e., a loss rate of 0.0044%.

Using the known coordinate positions of the 470 stars in this image, we obtained the intensities at these positions in the reconstructed image and compared magnitudes in the reconstructed image with magnitudes in the input image. There was reasonable fidelity over about 8 magnitudes. For fainter objects, the noise filtering causes greater difficulty, as one would expect.

6. Conclusion

The approach described here works well in practice. Further experiments are described in the full paper. Work comparing the approach described in this paper with other well-known astronomical image compression procedures is continuing.

References

Hanisch, R., ed. 1993, Restoration—Newsletter of ST ScI's Image Restoration Project (Baltimore, Space Telescope Science Institute)

Huang, L., & Bijaoui, A. 1991, Experimental Astronomy, 1, 311

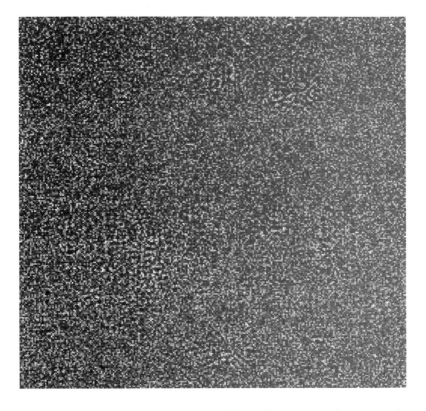

Figure 1. Difference between original and decompressed image, using a stellar field.

Astronomical Data Analysis Software and Systems IV
ASP Conference Series, Vol. 77, 1995
R. A. Shaw, H. E. Payne, and J. J. E. Hayes, eds.

Clustering Analysis Algorithms and Their Applications to Digital POSS-II Catalogs

R. R. de Carvalho[1], S. G. Djorgovski, and N. Weir

California Institute of Technology, MS 105-24, Pasadena, CA 91125

U. Fayyad, K. Cherkauer[2], J. Roden, and A. Gray

Jet Propulsion Laboratory, MS 525-3600, Pasadena, CA 91109

Abstract. We report on the preliminary results of experiments using a Bayesian cluster method to cluster objects present in photographic images of the POSS-II. Our goal is to explore the power of unsupervised learning techniques to classify objects meaningfully, and perhaps to discover previously unrecognized object categories in digital sky surveys. Our primary finding is that the program we used, *AutoClass*, was able to form several sensible categories from a few simple attributes of the object images, separating the data into four recognizable and astronomically meaningful classes: stars, galaxies with bright central cores, galaxies without bright cores, and stars with a visible "fuzz" around them. Also, in an independent experiment we found out that the two types of galaxies have distinct color distributions (the more concentrated class being redder, as indeed expected if they are predominantly early Hubble types), although no color information was given to *AutoClass*. This illustrates the power of unsupervised classification techniques to discriminate between astronomically distinct types of objects on the basis of data alone. We believe that the application of such algorithms to large-scale astronomical sky surveys can aid in cataloging the detected objects, and may even have the potential to discover new categories of objects.

1. Introduction

The last two decades have witnessed the cataloging of the northern and southern hemispheres through the use of high-quality photographic plates combined with CCD frames (van Altena 1993). These digital sky surveys amount to ∼5–6 TB worth of data, resulting in catalogs of many millions—or even billions—of objects. This richness of information requires new, efficient tools to explore the resulting data spaces (Weir et al. 1993a).

A crucial point in constructing scientifically useful object catalogs is the star/galaxy separation. Various supervised classification schemes can be used

---

[1]On Leave of Absence from Observatório Nacional/Cnpq, Rio de Janeiro, CEP 20921, Brazil

[2]University of Wisconsin, Madison, WI 53706

to produce consistent results in this task (Valdes 1982; Beard et al. 1990; Odewahn et al. 1992; Weir et al. 1995). However, a more difficult problem is systematically and objectively to provide at least rough morphological types for the galaxies detected, without visual inspection of the plates or scans—which is impractical for obvious reasons. We have thus started to explore new clustering analysis and unsupervised classification techniques for this task. Our goal is to try to separate astronomically meaningful morphological types on the basis of the data themselves, rather than some preconceived scheme.

Thus we investigate the possibility of finding natural (data-based) partitions of the attribute spaces which show high correlations between the plate-measured attribute space, and the CCD-based attribute space, or a high degree of separation between expected classes such as stars versus galaxies, spirals versus ellipticals, or galaxies of different concentrations. These partitions of the data may be used for investigations of unusual regions of the attribute space, and may even lead to a discovery of the previously unknown objects or classes of objects.

2. Data and Methodology

We use the data from the digitized version of the Second Palomar Observatory Sky Survey (POSS-II). For brief descriptions of the survey, see (Djorgovski et al. 1994; Reid and Djorgovski 1993; Weir et al. 1993b; Weir et al. 1994; Weir 1995). We have used data from 3 fields from POSS-II, numbers 380 (J-Band), 442 (J-Band), and 679 (J and F Bands).

The following attributes were used for the analysis: (1) resolution scale, (2) resolution fraction (these two are described in Valdes 1982), (3) ellipticity, (4) normalized core magnitude, (5) normalized area, (6) first intensity moment, and (7) the S parameter introduced by Collins et al. (1989). We have used only objects classified as galaxies and stars by using the Decision Tree technique (Weir et al. 1995). It is important to emphasize that we are not intentionally using legitimate attributes like colors, mean surface brightness and concentration index, which are available in our catalogs, because at this point they can help us understand the association between the classes which come out from the experiment and the large scale distribution of galaxies. Also, the classification is not given to the algorithm but is only used to judge its performance.

AutoClass (Cheeseman et al. 1988) is an unsupervised learning algorithm that fits user-specified probability distribution models to a set of examples represented as feature vectors. Classes are represented probabilistically as particular parameterizations of the models. In these experiments, we used multidimensional Gaussian models. *AutoClass* uses Bayesian techniques to estimate the parameter values of each class. It also tries to find the most probable number of classes by comparing the likelihoods of the fits for different numbers of classes. Objects are then assigned probabilistic memberships in the output classes.

The Gaussians used to model the classes can range from noncovariant (i.e., axis-aligned) to fully covariant based on prior knowledge the user may have of the attributes and problem. In our experiments here, we used only models that had no covariance. We ran some simple tests using synthetic data to verify that *AutoClass*'s behavior was reasonable in each of these cases.

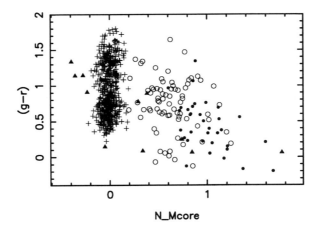

Figure 1. The $(g-r)$ color versus normalized core magnitude, for the four types of objects found by *AutoClass*: galaxies without a bright core (open circles), galaxies with a bright core (solid circles), stars (crosses), and stars with fuzz (solid triangles).

3. Discussion

In our first experiment we used data from the fields 380 and 442. *AutoClass* was able to find four natural classes of objects in the data space. These four classes were, by visual inspection, identified with stars, galaxies with a bright core, galaxies without a bright core, and stars with fuzz around them. Thus, the object classes found by *AutoClass* are astronomically meaningful, even though the program itself does not know about stars, galaxies, and such! These results were obtained using data in a given bin of magnitude ($17^m < r < 18^m$), although the same trends were found for a bin one magnitude fainter. The results are robust and repeatable from field to field.

By inspecting the so-called confusion matrix, we found that each cluster identified by *AutoClass* corresponds to the type of the objects, as classified by the Decision Tree (a supervised classification approach). The Decision Tree was trained to recognize only two classes of objects, stars and galaxies, and no attempt was made to make any morphological distinctions among the galaxies.

Another experiment was done using another field in two colors (442 J and F), both in order to check the previous finding, and also to explore a little more deeply the meaning of these classes. Again, *AutoClass* found the same four significant classes in the data space, which confirms the robustness of the method.

Figure 1 displays the $(g - r)$ color versus the normalized core magnitude (one of the attributes used in the experiment). As can be seen, the two morphologically distinct classes of galaxies, represented by solid and open circles, populate different regions of the data space, and have systematically different colors, *even though AutoClass was not given the color information*. In this figure we display stars as crosses and stars with fuzz around as solid triangles.

The confusion matrix for such experiments indicates that stars and galaxies, as classified by Decision Tree, are well separated in different classes. Galaxies are distributed in two classes, representing redder and bluer systems, respectively— presumably the early and late Hubble types, respectively.

We are now exploring our database from POSS-II in a systematic way using such techniques to map the large scale structure (clustering in the physical space) in an unbiased fashion. One project is to objectively define and discover clusters and groups of galaxies, which can then be used for a variety of follow-up studies.

A full paper will be presented in near future describing in detail the application of *AutoClass* to POSS-II and similar data.

References

Beard, S. M., MacGillivray, H. T., & Thanisch, P. F., 1990. MNRAS, 247, 311

Cheeseman, P., et al. 1988, in Proc. Fifth Machine Learning Workshop, ed. J. Laird (San Mateo, Calif., M. Kauffmann), p. 54

Collins, C. A., Heydon-Dumbleton, N. H., & MacGillivray, H. T. 1989. MNRAS, 236, 7p

Djorgovski, S., Weir, N., & Fayyad, U. 1994, in Astronomical Data Analysis Software and Systems III, ASP Conf. Ser., Vol. 61, eds. D. R. Crabtree, R. J. Hanisch, & J. Barnes (San Francisco, ASP), p. 195

Odewahn, S. C., Stockwell, E. B., Pennington, R. L., Humphreys, R. M., & Zumach, W. A. 1992. AJ, 103, 318

Reid, I. N., & Djorgovski, S. 1993, in Sky Surveys: Protostars to Protogalaxies, ASP Conf. Ser., Vol. 43, ed. B. T. Soifer (San Francisco, ASP), p. 125

Valdes, F. 1982, in Instrumentation in Astronomy, IV, ed. D. L. Crawford, SPIE Proc., 331, 465

van Altena, W. F. 1993, in Astronomy from Wide-field Imaging, IAU Symp. 161, ed. H. T. MacGillivray et al. (Dordrecht, Kluwer), p. 193

Weir, N., Djorgovski, S., Fayyad, U., Smith, J. D., & Roden, J. 1993a. in Astronomy from Wide-field Imaging, IAU Symp. 161, ed. H. T. MacGillivray et al. (Dordrecht, Kluwer), p. 205

Weir, N., Fayyad, U., Djorgovski, S., Roden, J., & Rouquette, N. 1993b, in Astronomical Data Analysis Software and Systems II, ASP Conf. Ser., Vol. 52, eds. R. J. Hanisch, R. J. V. Brissenden, & J. Barnes (San Francisco, ASP), p. 39

Weir, N., Djorgovski, S., Fayyad, U., Smith, J. D., & Roden, J. 1994, in Astronomy From Wide-Field Imaging, IAU Symp. 161, ed. H. T. MacGillivray et al. (Dordrecht, Kluwer), p. 205

Weir, N. 1995, Ph.D. Thesis, California Institute of Technology.

Weir, N., Djorgovski, S., & Fayyad, U. 1995. AJ, in press

Part 7. Image Restoration and Analysis

Multiresolution and Astronomical Image Processing

J.-L. Starck

CEA, DSM/DAPNIA, CE-Saclay, F-91191 Gif-sur-Yvette Cedex, France

F. Murtagh[1]

Space Telescope–European Coordinating Facility, European Southern Observatory, Karl-Schwarzschild-Str. 2, D-85748 Garching, Germany

A. Bijaoui

Observatoire de la Côte d'Azur, B.P. 229, F-06304 Nice Cedex 4, France

Abstract. We present several wavelet transform algorithms and their applications in astronomical image processing (restoration, object detection, compression, etc.).

1. The Discrete Wavelet Transform

1.1. Mallat's Transform

Extensive literature exists on the wavelet transform and its application (Chui 1992; Daubechies 1992; Meyer 1989). A discrete wavelet transform approach can be obtained from multiresolution analysis (Mallat 1989). Multiresolution analysis results from the embedded subsets generated by interpolations at different scales. A function $f(x)$ is projected at each step j onto the subset V_j. This projection is defined by the scalar product $c_j(k)$ of $f(x)$ with the scaling function $\phi(x)$ which is dilated and translated:

$$c_j(k) = <f(x), 2^{-j}\phi(2^{-j}x - k)> . \qquad (1)$$

$\phi(x)$ is a scaling function which has the property

$$\frac{1}{2}\phi\left(\frac{x}{2}\right) = \sum_n h(n)\phi(x-n). \qquad (2)$$

Equation 2 permits the set $c_{j+1}(k)$ to be computed directly from $c_j(k)$. If we start from the set $c_0(k)$, we compute all the sets $c_j(k)$, with $j > 0$, without directly computing any other scalar product:

$$c_{j+1}(k) = \sum_n h(n - 2k)c_j(n). \qquad (3)$$

---

[1] Affiliated to Astrophysics Division, Space Science Department, European Space Agency

At each step, the number of scalar products is divided by 2; the signal is smoothed and information is lost. The remaining information can be restored using the complementary subspace W_{j+1} of V_{j+1} in V_j. This subspace can be generated by a suitable wavelet function $\psi(x)$ with translation and dilation:

$$\frac{1}{2}\psi\left(\frac{x}{2}\right) = \sum_n g(n)\phi(x-n). \tag{4}$$

We compute the scalar products $<f(x), 2^{-(j+1)}\psi(2^{-(j+1)}x - k)>$ with

$$w_{j+1}(k) = \sum_n g(n-2k)c_j(n). \tag{5}$$

With this analysis, we have built the first part of a filter bank. In order to restore the original data, Mallat uses the properties of orthogonal wavelets, but the theory has been generalized to a large class of filters (Cohen, Daubechies, & Feauveau 1992). Filters $\tilde{h}$ and $\tilde{g}$, the conjugates of h and g, have been introduced (Daubechies 1992), and the restoration is performed with

$$c_j(k) = 2\sum_l [c_{j+1}(l)\tilde{h}(k+2l) + w_{j+1}(l)\tilde{g}(k+2l)]. \tag{6}$$

In order to get an exact restoration, two conditions are required for the conjugate filters: (1) the De-aliasing condition:

$$\hat{h}(\nu + \frac{1}{2})\hat{\tilde{h}}(\nu) + \hat{g}(\nu + \frac{1}{2})\hat{\tilde{g}}(\nu) = 0, \tag{7}$$

and (2) exact restoration:

$$\hat{h}(\nu)\hat{\tilde{h}}(\nu) + \hat{g}(\nu)\hat{\tilde{g}}(\nu) = 1. \tag{8}$$

Many sets of filters have been proposed, especially for coding. It has been shown (Daubechies 1992) that the choice of these filters must be guided by the regularity of the scaling and the wavelet functions. The complexity is proportional to N. The algorithm provides a pyramid of N elements. The 2D algorithm is based on separable variables leading to x and y directions being prioritized. This implies a non-isotropic analysis, which often runs counter to physical patterns.

1.2. Feauveau's Transform

Feauveau (1990) introduced the quiconx analysis according to Adelson (Adelson, Simoncelli, & Hingorani 1987). This analysis is not dyadic and allows an image decomposition with a resolution factor equal to $\sqrt{2}$. By this method, we have one wavelet image at each scale, and not three as with the previous method.

1.3. Wavelet Transform with the FFT

If $\phi(\nu)$ has a cut-off frequency, the discrete filter $h(n)$ used to compute the coefficients at a given resolution from the previous one (Starck & Bijaoui 1994), is

$\hat{h}(\nu) = \hat{\phi}(2\nu)/\hat{\phi}(\nu)$. Therefore, between two resolutions, the cut-off frequency is divided by 2, allowing us to reduce the number of samples. The information lost in the filtering is given by the discrete filter $g(n)$ defined by $\hat{g}(\nu) = \hat{\psi}(2\nu)/\hat{\phi}(\nu)$. The signal is reconstructed with the two filters $\tilde{h}$, $\tilde{g}$:

$$\hat{\tilde{h}}(\nu) = \frac{\hat{h}^*(\nu)}{\mid \hat{h}(\nu) \mid^2 + \mid \hat{g}(\nu) \mid^2} \text{ and } \hat{\tilde{g}}(\nu) = \frac{\hat{g}^*(\nu)}{\mid \hat{h}(\nu) \mid^2 + \mid \hat{g}(\nu) \mid^2}. \tag{9}$$

The described algorithm is easily adapted to image processing. As opposed to classical multiresolution analysis, it is a decomposition with isotropic scaling and wavelet functions. The frequency band is also reduced by a factor 2 at each step. Applying the sampling theorem, we can build a pyramid of $N + N/2 + \ldots + 1 = 2N$ elements. For image analysis the number of elements is $4N^2/3$. The increase in storage is small.

1.4. À Trous Algorithm

The "à trous" algorithm is a powerful algorithm for the following reasons: (1) the computational requirement is reasonable, (2) the algorithm is easy to program, (3) in two dimensions, the transform is practically isotropic, (4) we can use compact scaling functions, (5) the reconstruction algorithm is trivial, (6) the transform is known at each pixel, allowing detection without any error, and without interpolation, (7) we can follow the evolution of the transform from one scale to the next, and (8) invariance under translation is completely verified.

Details of the algorithm are given in Bijaoui, Starck, & Murtagh (1994) and Shensa (1992). The wavelet transform of an image by this algorithm produces, at each scale j, a set $\{w_j\}$ which we will call a wavelet plane throughout the following discussion. This has the same number of pixels as the image. The original image c_0 can be expressed as the sum over wavelet planes and the smoothed array c_p

$$c_0(x, y) = c_p(x, y) + \sum_{j=1}^{p} w_j(x, y) \tag{10}$$

1.5. Pyramidal Algorithm

The pyramidal algorithm (Starck 1993) is derived from the "à trous" algorithm. At each scale, w_{j+1} is obtained by computing the difference between c_j and $F.c_j$ (where F denotes a linear filtering). But c_{j+1} is equal to $U.F.c_j$ (U being the undersampling or decimating operator), and not equal to $F.c_j$ (we first undersample $F.c_j$ by keeping one pixel out of two). This transform produces data similar to those obtained by the Burt & Adelson (1983) pyramidal Laplacian algorithm, but it is done with one wavelet. Burt's algorithm needs three wavelets and is therefore not isotropic (Bijaoui 1991). The number of elements is $4N^2/3$, as in the case of the transform using the FFT, but this method has the advantage to allow all computation to be carried out in direct space. An iterative algorithm is necessary for an exact restoration.

1.6. Problems Related to the Wavelet Transform

Anisotropic Wavelet. The 2-dimensional extension of Mallat's algorithm leads to a wavelet transform with three wavelet functions (three wavelet coefficient sub-images at each scale). An isotropic wavelet seems more appropriate, specially in astronomical imaging where objects are often isotropic (e.g., stars).

Invariance by Translation. Mallat's and Feauveau's methods provide a remarkable framework to code a signal, and especially an image, with a pyramidal set of values. But, contrary to the continuous wavelet transform, these analyses are not covariant under translation. At a given scale, we derive a decimated number of wavelet coefficients. We cannot restore the intermediate values without using the approximation at this scale and the wavelet coefficients at smaller scales. Since the multiresolution analysis is based on scaling functions without cut-off frequency, the application of the Shannon interpolation theorem is not possible. The interpolation of the wavelet coefficients can only be done after reconstruction and shift. This is unimportant for a signal coding that does not modify the data, but is not the same if we want to analyze or restore an image.

Scale Separation. If the image I we want to analyze is the convolution product of an object O by a point spread function (PSF) ($I = P * O$), we have

$$\hat{W}^{(I)}(a, u, v) = \sqrt{a}\hat{\psi}^*(au, av)\hat{I}(u, v), \qquad (11)$$

where $W^{(z)}$ are the wavelet coefficients of z, and a is the scale parameter. We deduce that

$$\hat{W}^{(I)}(a, u, v) = \sqrt{a}\hat{\psi}^*(au, av)\hat{P}(u, v)\hat{O}(u, v) \qquad (12)$$
$$= \hat{O}(u, v)\hat{W}^{(P)}(a, u, v). \qquad (13)$$

We can directly analyze the object from the wavelet coefficients of the image. But due to decimation effects in Mallat's, Feauveau's, and pyramidal methods, this equation fails. The wavelet transform using the FFT (which decimates using Shannon's theorem) and the "à trous" algorithm (which does not decimate) are the only ones which respect the scale separation property.

Negative Values. By definition, the wavelet coefficient mean is null. Every time we have a positive structure at a scale, we have negative values surrounding it. These negative values often create artifacts during the restoration process, or complicate the analysis. For instance, if we threshold small values (noise, insignificant structures, etc.) in the wavelet transform, and if we reconstruct the image at the full resolution, the structure's flux will be modified. Furthermore, if an object is very high, the negative values will be important, too, and will lead to false structure detections.

Point Objects. We often have bright point objects in astronomical imaging (stars, cosmic ray hits, etc.) The convolution of a Dirac function by the wavelet function is equal to the wavelet function, so at each scale, and at each point source, we will have the wavelet. Cosmic rays can pollute all the scales of the wavelet transform.

1.7. Conclusion

There is no ideal wavelet transform algorithm—the selection will depend on the application. The two last problems cannot be solved by the wavelet transform, and they lead to other multiresolution tools which we now present.

2. Other Multiresolution Approaches

The search for new multiresolution tools was motivated by problems related to the wavelet transform. We would prefer that a point structure (represented in one pixel in the image) is present only at the first scale, and that a positive structure in the image not create negative values in the multiresolution space. We will see how such an algorithm can be found, using morphological filters such as the median filter.

2.1. Multiresolution from the Median

By modifying the "à trous" algorithm, we easily obtain the desired transform. The algorithm becomes:

1. We define a mask M_t with a size t ($t = 2*s+1$) (for instance a square mask with a size 3×3, implying that $t = 3$).

2. We initialize j to 0, and we start from data c_0.

3. med being the filtering median function, we calculate $c_{j+1} = med(M_t, f_j)$ and median coefficients at scale j by: $w_{j+1} = c_j - c_{j+1}$

4. We double the mask size: $s = 2*s$ and $j = j+1$.

5. If j is less than the number of scales we want, return to 3.

The reconstruction is carried out by a simple addition of all the scales:

$$c_0 = c_p + \sum_j w_j \quad (14)$$

Such an algorithm has several advantages: (1) the transform can be carried out with integer values, (2) energy in structures at each scale is real, and is not modified by the treatment, (3) structure contours are better respected, (4) there are no negative values around positive structures, and (5) the algorithm can easily be modified to work at intermediate scales (in order to have a non-dyadic analysis). We just multiply the size s at the step 4 by a coefficient different from 2 (if we want half resolution, we multiply by 1.5).

However this algorithm has a big disadvantage: computation time. The mask increases by two at each scale, and we have to sort a number of values which increases considerably. To solve this problem, we introduce decimation.

2.2. Pyramidal Multi-Median Transform

We reduce the number of samples (pixels) by keeping one pixel out of every two (in each dimension) at each scale. The algorithm is:

1. We define a mask M_t with a size t ($t = 2*s+1$).

2. We initialize j to 0, and we start from data c_0.

3. Calculate $m_{j+1} = med(M_t, f_j)$ and median coefficients at scale j as $w_{j+1} = c_j - m_{j+1}$

4. We calculate c_{j+1} from m_{j+1} by decimating.

5. Set $j = j + 1$. If j is less than the number of scales we want, return to 3.

In this transform, the mask size is constant, and there is no time computation problem anymore. However, the reconstruction is not exact, and an iterative procedure has to be introduced during the transformation or at the reconstruction. If we choose the iterative transform, we have:

1. Initialize i to 0, R^i to the image, and multiresolution coefficients $w_j^{(I)}$ to 0.

2. Compute the pyramidal multi-median transform of $R^{(i)}$; we have $w_j^{(R)}$.

3. $w_j^{(I)} = w_j^{(I)} + w_j^{(R)}$.

4. Reconstruct $\tilde{I}$ from $w_j^{(I)}$. B-spline interpolation is used.

5. Compute $R^{i+1} = I - \tilde{I}$.

6. Set $i = i + 1$, and return to 3.

R^{i+1} in step 6 tends towards a null image (cf. Van Cittert 1931). We have found that 4 or 5 iterations are sufficient.

2.3. Conclusion

Such a transform does not replace the wavelet transform, but complements it. When images contain cosmic ray hits, for instance, the use of the pyramidal multi-median transform avoids cosmic ray hits affecting all scales. Other morphological tools can be used to perform a similar transform, such as opening (N erosions followed by N dilations). However, results with the median filter were superior.

3. Significant Coefficients

Images generally contain noise (Gaussian or Poisson, usually) and hence the wavelet coefficients are noisy, too. In most applications, it is necessary to know whether a coefficient is due to signal or to noise. The wavelet transform yields a set of resolution-related views of the input image. A wavelet image plane at level j has coefficients given by $w_j(x, y)$. If we obtain the distribution of the coefficient $w_j(x, y)$ for each plane, based on the noise, we can introduce a statistical significance test for this coefficient.

Given stationary Gaussian noise, it suffices to compare $w_j(x, y)$ to $k\sigma_j$, where σ_j is the standard deviation of wavelet plane j, and k is often chosen as 3. If $w_j(x, y)$ is small it is not significant, and could be due to noise. If $w_j(x, y)$ is large, it is significant. If the noise in the data I is Poisson, the transform $T(I(x, y)) = 2\sqrt{I(x, y) + 3/8}$ acts as if the data arose from the Gaussian white noise model with unit standard deviation (Anscombe 1948). Taking the wavelet transform of $T(I)$, $w_j^{(I)}(x, y)$ will be significant if $w_j^{(T(I))}(x, y)$ is above a given threshold. (The superscript on the wavelet coefficients indicates the image to which the wavelet transform was applied.)

The appropriate value of σ_j in the succession of wavelet planes is assessed from the standard deviation of the noise σ_I in the original image and from study of the noise in the wavelet space. Details of this study can be found in Starck & Murtagh (1994).

4. Applications

4.1. Multiresolution Support

We will say that a multiresolution support (Starck, Bijaoui, & Murtagh 1994) of an image describes in a logical or boolean way whether an image I contains information at a given scale j and at a given position (x, y). If $M^{(I)}(j, x, y) = 1$ (or *true*), then I contains information at scale j and at the position (x, y). M depends on several parameters: (1) the input image, (2) the algorithm used for the multiresolution decomposition, (3) the noise, and (4) all additional constraints we want the support to satisfy.

Such a support results from the data, the treatment (noise estimation, etc.), and from knowledge on our part of the objects contained in the data (size of objects, linearity, etc.). In the most general case, *a priori* information is not available to us.

The multiresolution support of an image is computed in several steps:

1. Compute the wavelet (or other multiresolution) transform of the image.

2. Binarization of each scale leads to the multiresolution support. We have:

$$M(j, x, y) = \begin{cases} 1 & \text{if } w_j(x,y) \text{ is significant} \\ 0 & \text{if } w_j(x,y) \text{ is not significant} \end{cases} \quad (15)$$

3. *A priori* knowledge can be introduced by modifying the support.

The last step depends on the knowledge we have of our images. For instance, if we know there is no interesting object smaller or larger than a given size in our image, we can suppress, in the support, anything which is due to that kind of object. This can often be done conveniently by the use of mathematical morphology. Or we can add star positions from a catalog in order to improve a restoration. In the most general setting, we have no information to add to the multiresolution support.

The multiresolution support indicates where the information is, and allows all information we have about the image (data, noise, catalog information, etc.) to be integrated. It can then be used to visualize or to further process the data. The fact that we can add constraints and laws to the support makes this a very convenient data structure for introducing information into the treatment.

4.2. Visualization

The visualization of the wavelet transform allows us to have better knowledge of the structure of the image. The "*à trous*" algorithm is particularly efficient, because each scale has the same size as the original image, and can be visualized, processed, and compared very easily. See Starck (1993) and Bijaoui, Starck, & Murtagh (1994) which use various ways to visualize a wavelet transform.

4.3. Detection

In general in astronomy, it is not the image which is of interest but rather the objects in the image! The multiresolution support provides all of the information needed to demarcate and then label the objects. Varying background is automatically catered for by the multiresolution transform (equation 14 or its equivalent in the case of a pyramidal transform). This is a very real advantage, since alternative approaches must devote considerable processing attention to adaptive or other methods in order to allow for variable backgrounds or superimposed objects.

Multiresolution provides much information on the scale of the objects. Thus foreground objects (which may not be of interest, e.g., cosmic ray hits) as well as background objects (e.g., faint galaxies) can be found—assuming an appropriate wavelet (or transform kernel) and processing chain—at different resolution scales. It is not difficult to determine all objects, in this way, and then to retain only those which are of immediate interest. Examples of the use of the multiresolution approach to object detection can be found in Murtagh, Zeilinger, Starck, & Bijaoui (1994).

4.4. Quality Criteria

Objective comparison of two images is often necessary, for instance when we want to evaluate the restoration quality. Very few quantitative parameters can be extracted. The correlation between the original image and the restored one provides a classical criterion. Another way to compare two pictures is to determine the mean-square error. These criteria, however, are not sufficient. They give no information about the resulting resolution. A comprehensive criterion must take into account the resolution. We can compute for each dyadic scale the correlation coefficient and the quadratic error between the wavelet transforms of the original and the restored images. This allows us to compare, for each resolution, the restoration quality (Starck & Bijaoui 1994).

4.5. Filtering

An easy way to filter an image is to keep only the significant wavelet coefficients and to reconstruct an image (Starck & Bijaoui 1994; Donoho 1992). We have shown (Starck, Bijaoui, & Murtagh 1994) that, with few iterations and by using the multiresolution support, we can have excellent results.

4.6. Deconvolution

The object-image relation is often given by:

$$I = O * P + N \tag{16}$$

O is the observed object, I is the obtained image, P is the point spread function (PSF) of the imaging system, and N is an additive noise. We want to determine O knowing I and P, the main difficulties being the existence of: (1) a cut-off frequency of the PSF, and (2) the additive noise (e.g., Cornwell 1988). The multiresolution approach (Starck & Murtagh 1994) allows the noise to be controlled during the deconvolution process, and leads to a regularization of this ill-posed problem. The Poisson noise case and the use of a multiresolution support framework have been treated in (Murtagh, Starck, & Bijaoui 1994).

4.7. Interferometric Deconvolution

In interferometric imaging, measurements are carried out in the Fourier space but the (u,v) plane is not completely covered. The image ("dirty map"), is obtained by a simple inverse Fourier transform of the data, and the PSF ("dirty beam"), by an inverse Fourier transform of the (u,v) plane coverage. The presence of secondary lobes in the dirty beam creates very important artifacts in the dirty map and a deconvolution is necessary. By applying the CLEAN method (Högbom 1974) at each scale of the wavelet transform using the FFT, we can localize significant structures. An iterative reconstruction algorithm allows solutions to be found which satisfy the positivity constraint, and the fidelity to measurements constraint (i.e., at each measured $V_m(u,v) \pm \Delta_m(u,v)$, we require that the solution O satisfies $\mid \hat{O}(u,v) - V_m(u,v) \mid < \Delta_m(u,v)$. More details can be found in Starck & Bijaoui (1994) and Starck, Bijaoui, Lopez, & Perrier (1994).

4.8. Compression

Several studies have used the orthogonal wavelet transform to compress astronomical data (Press 1991; White 1994). Using the multiresolution support should improve the quality of the decompressed image. Results presented in Starck, Murtagh, & Louys (1994) show how good results can be.

5. Conclusion

The wavelet transform, and more generally the multiresolution transform, provides a powerful and versatile framework for astronomical image processing. A data structure is created which allows such objectives as the following to be handled: visualization, object detection, filtering, deconvolution, and compression. Studies in all of these areas have been cited here, and in many cases the results obtained are considerably better than traditional approaches.

References

Adelson, E. H., Simoncelli, E., & Hingorani, R. 1987, SPIE Visual Communication and Image Processing II, 845, 50

Anscombe, F. J. 1948, Biometrika, 15, 246

Bijaoui, A. 1991, Ondelettes et Paquets d'Ondes, Inria, Rocquencourt

Bijaoui, A., Starck, J. L., & Murtagh, F. 1994, Traitement du Signal, 3, 11

Burt, P. J., & Adelson A. E. 1983, IEEE Trans. on Communications, 31, 532

Cornwell, T. J. 1988, Proc. NATO Advanced Study Institute on Diffraction-Limited Imaging with Very Large Telescopes, Cargèse, 273

Chui, C. H. 1992, Wavelet Analysis and its Application (New York, Academic Press)

Cohen, A., Daubechies, I., & Feauveau, J. C. 1992, Comm. Pur. Appl. Math., 45, 485

Daubechies, I. 1992, Ten Lectures on Wavelets, Philadelphia.

Donoho, D. L. 1992, Proc. Progress in Wavelet Analysis and Applications (Toulouse, Ed. Frontières)

Feauveau, J. C. 1990, Thesis, University Paris Sud.

Högbom, J. A. 1974, A&AS, 15, 417

Mallat, S. 1989, IEEE Trans. on Pattern Analysis and Machine Intelligence, 11, 574

Meyer, Y. 1989, Wavelets, ed. J. M. Combes et al., (Berlin, Springer Verlag), 21

Murtagh, F., Starck, J. L., & Bijaoui, A. 1994, A&A, submitted

Murtagh, F., Zeilinger, W., , Starck, J. L., & Bijaoui, A. 1994, this volume, p. 260

Press, W. H. 1991, in Astronomical Data Analysis Software and Systems I, ASP Conf. Ser., Vol. 25, eds. D. M. Worrall, C. Biemesderfer, & J. Barnes (San Francisco, ASP), p. 25

Shensa, M. J. 1992, Proc. IEEE Transactions on Signal Processing, 40, 2464

Starck, J. L. 1993, The Wavelet Transform, MIDAS Manual (Garching, ESO)

Starck, J. L., & Bijaoui, A. 1994, Signal Processing, 35, 195

Starck, J. L., Bijaoui, A., Lopez, A., & Perrier, C. 1994, A&A, 283, 349

Starck, J. L., & Bijaoui, A. 1994, J. Opt. Soc. Am. A, 11, No. 4

Starck, J. L., & Murtagh, F. 1994, A&A, 288, 343

Starck, J. L., Bijaoui, A., & Murtagh, F., 1994, CVIP: Graphical Models and Image Processing, submitted

Starck, J. L., Murtagh, F., & Louys, M., 1994, this volume, p. 268

Van Cittert, P. H. 1931, Physik, Z., 69, 298

White, R. L. 1993, in Space and Earth Science Data Compression Workshop, ed. James C. Tilton, NASA Conference Publication 3183, p. 117

Astronomical Data Analysis Software and Systems IV
ASP Conference Series, Vol. 77, 1995
R. A. Shaw, H. E. Payne, and J. J. E. Hayes, eds.

Improvements in Filter Design for Removing Galactic "Cirrus" from IRAS Images

J. P. Basart, L. X. He

Department of Electrical and Computer Engineering, Iowa State University, Ames, IA 50011

P. N. Appleton

Department of Physics and Astronomy, Iowa State University, Ames, IA 50011

J. A. Pedelty

Space Data and Computing Division, NASA Goddard Space Flight Center, Greenbelt, MD 20771

Abstract. Design improvements continue to be made in the filter developed using mathematical morphology principles for the Infrared Astronomical Satellite (*IRAS*) images. The purpose of this filter is to eliminate Galactic cirrus emission from extragalactic fields. Current improvements are based on dividing the structure in the *IRAS* images into several classes, and then filtering the information in the image class by class. The current procedure gives significantly improved results compared to those of the first version of the filter.

1. Introduction

"Cirrus" emission in *IRAS* images at 60 μm and, especially, at 100 μm, has hampered the study of extended extragalactic structures. The problem is made more difficult by the fact that the cirrus emission can exhibit a wide spread in far-IR color temperatures, ranging from 20 K to 35 K. Our approach for reducing the IR cirrus in *IRAS* images is based upon an image processing procedure called *mathematical morphology* (Serra 1982). Results from the first version of our filter are discussed in Basart, Siqueira, & Appleton (1992), Appleton, Siqueira, & Basart (1993), and Pedelty, Appleton, & Basart (1994). Previous filtered results of the M81 group showed considerable improvement over the original image. Not all undesired features were filtered out, however, so filter design continued. The second version of the filter, presented here, produces significantly better results than the first filter.

2. Filter Design

We briefly review the basic procedure underlying both filters. The process starts by performing an opening operation (Appleton et al. 1993) on the filter, with

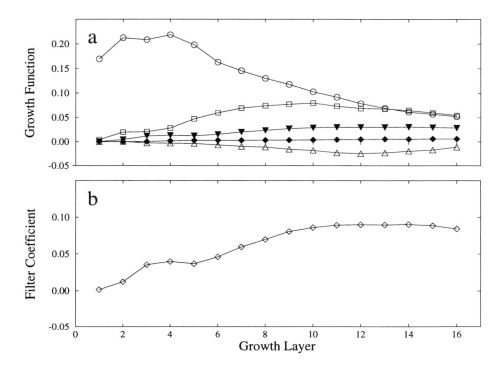

Figure 1. The top plot (a) shows an example of the growth functions for various structural types as explained in the text. The bottom plot (b) shows the filter coefficients created from the growth curve and used to filter the image shown in Figure 2b.

a Gaussian shaped structuring element. This opening operation eliminates all structural information smaller than the structuring element. We then open the original image again with a slightly larger structuring element, and subtract the first result from the second. This difference produces an image with a narrow range of structural sizes: those that lie between the sizes of the two structuring elements. We then open the original image with a third structuring element, whose size is a little larger than the second structuring element. We subtract this third opening from the second opening, giving another image with a different range of structural sizes—larger sizes than those of the first differenced image. We continue this process until we get a set of 16 images whose range of structural sizes varies from seven pixels to thirty nine pixels. We call the resulting plot of intensity vs. structural size at one pixel a "growth curve."

The growth curve contains information central to the filtering operation. In version one of the filter, the growth curves for many cirrus pixels not containing galaxies were averaged together, and the results used as the filter curve. With an appropriate normalization, the filter curve was applied to each pixel in the image to remove the Galactic cirrus. This filter had a high degree of success because the growth information of the galaxies differed from that of the cirrus. However, not all information averaged together to make the filter curve was homogeneous, causing artifacts to be introduced into the final image. This difficulty led to the development of version two of the filter. In this extended approach a classification procedure was introduced. The growth curve was treated as feature information in 16 dimensional space—one dimension for each difference in openings. After placing all opening differences into the feature space, a clustering operation was performed to determine groupings. Five clusters were allowed. The growth curves for each of the five structural types are shown in Figure 1a. Four of the curves are somewhat similar while the top curve is much different. Viewing the central portion of the graph, the identification of the structure from the top curve to the bottom curve is: (1) small bright objects, (2) regions around the small bright objects, (3) small cloud structure, (4) large cloud structure, and (5) corrupted structure caused by the boundaries of the image. It is apparent from the curve that filter performance could be improved over that of version one by selectively filtering by structure type.

3. Results

Version two of the filter uses a filter curve based upon type 3 structure, as identified above. A filter curve was created by normalizing the growth curve by the area under this curve. The resulting filter curve is shown in Figure 1b. This curve, with appropriate re-normalization, was applied to all pixels in the image.

Figure 2a shows an example of *IRAS* field I363B4H0 before filtering, and Figure 2b shows the results after filtering. The original image (Figure 2a) is very heavily contaminated with 100 μm IR cirrus emission. The image shown is a portion of an *IRAS* field which contains a variety of non-cirrus structure ranging from Galactic nebulae to galaxies. Small objects, such as galaxies, are difficult, if not possible, to detect. Even more ambiguous is the tenuous structure on the periphery of galaxies. The purpose of the filter is to minimize the presence of Galactic IR cirrus in the image in order to make the extragalactic IR emission more visible.

The filtered image contains considerably less Galactic cirrus than the unfiltered image. The remaining structure in the filtered image is primarily from non-diffuse objects. About 30% of the point-like sources in the image are galaxies and the remaining objects are extended structures. Many of the extended structures appear to be highly correlated with unusually bright and sharply defined reflection nebulosity. The latter conclusion was drawn by comparing the filtered image with an optical image.

Preliminary testing for flux integrity during the filtering process has been completed. Throughout the image, flux retention is better than 1% in consistency. In absolute terms, the filtered flux is within a few percent of what is assumed to be a true flux of an object. Overall, the filter has been extremely effective at extracting unusual and interesting sources. We are currently following up our results with wide-field CCD observations of these objects.

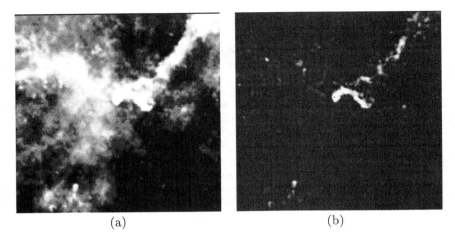

(a) (b)

Figure 2. The left figure (a) is a portion of the *IRAS* field I363B4H0 observed at 100 μm. The right figure (b) is the result of filtering out much of the IR cirrus structure using a filter based upon mathematical morphology.

References

Appleton, P. N., Siqueira, P. R., & Basart, J. P. 1993, AJ, 106, 1664

Basart, J. P., Siqueira, P. R., & Appleton, P. N. 1992, in Astronomical Data Analysis Software and Systems I, ASP Conf. Ser., Vol. 25, eds. D. M. Worrall, C. Biemesderfer, & J. Barnes (San Francisco, ASP), p. 283

Pedelty, J. A., Appleton, P. N., & Basart, J. P. 1994, in Astronomical Data Analysis Software and Systems III, ASP Conf. Ser., Vol. 61, eds. D. R. Crabtree, R. J. Hanisch, & J. Barnes (San Francisco, ASP), p. 308

Serra, J., 1982, Image Analysis and Mathematical Morphology (London, Academic Press)

Star Finding and PSF Determination using Image Restoration

R. N. Hook and L. B. Lucy[1]

Space Telescope–European Coordinating Facility, European Southern Observatory, Karl-Schwarzschild-Str. 2, D-85748 Garching bei München, Germany

Abstract. Crowded-field stellar photometry consists of three main phases: locating the point sources, determining the point spread function (PSF) and measuring the point source brightnesses. In earlier work (Lucy 1994; Hook & Lucy 1994) we have described a two-channel restoration method which provides photometric fidelity and addresses the last of these items in a restoration context. Here we describe two further enhancements. First, an experimental method for locating stellar images by enhancing the sharp cores of stars using a multi-channel entropy *minimization* technique is described. In addition an extended version of the two-channel method is given which allows PSFs to be extracted from designated point-sources in images during restoration. A *default* PSF may be used to regularize the result. A simple but accurate model for the form of ground-based PSFs (Saglia et al. 1993) has been implemented and seems a suitable choice of default PSF for ground-based images. Examples using a typical ground-based CCD image of a star cluster are given.

These methods are available in preliminary implementations running within both the IRAF and MIDAS data analysis packages.

1. Introduction

In earlier papers (Lucy 1994; Hook & Lucy 1994) we described a two-channel restoration method in which one channel contains point-sources and the other contains a smooth background to represent the sky or an extended object. The second channel is regularized by the addition of an entropy term to the expression being maximized, but the first is treated as a simple likelihood maximization. This method has many advantages including the suppression of the artifacts often seen around bright stars in conventional single channel restorations and photometric fidelity without the bias found in most restoration techniques (e.g., Cohen 1991).

The technique of multiple channels with different regularization is here extended in two directions of relevance to crowded field photometry. First an experimental three-channel method is described in which one channel has a *negative* coefficient for the entropy term. This has the opposite effect to the normal smoothing which follows from positive entropy coefficients and enhances point

---

[1] Affiliated with the Astrophysics Division, Space Science Department, European Space Agency

Figure 1. An example of star finding: see text for details.

sources, ultimately to the point where they become δ-functions (single pixels or sub-pixels). Secondly, ways of extracting a PSF from an image containing designated point-sources are described as well as different ways in which such a PSF may be regularized. An implementation of a recently suggested theoretical form for the PSF of ground-based telescopes is available and is suitable for use as a *default* PSF.

2. Star Finding

Maximum likelihood restorations often develop noise "spikes" when large numbers of iterations are used and regularization methods, often based on maximizing an entropy expression, are used to impose smoothness. However, point sources in images are really δ-functions on the sky and hence are the opposite of smooth. This suggests a method for finding point sources using an entropy *minimization* rather than the normal maximization used to impose smoothness. In this case an objective function of the following form is maximized:

$$Q = \sum_i \tilde{\Phi}_i ln \Phi_i + \alpha S + \beta T, \qquad (1)$$

where $\tilde{\Phi}$ is the observed intensity distribution, Φ is the current estimate and the summation is performed over all pixels in the image. The first term is the likelihood, and S and T are entropy-type expressions. The second term has positive α and acts as a standard regularization to enforce smoothness on the background channel. The third term has $\beta < 0$ and leads to a minimization of the entropy of a "points" channel. Both entropy expressions are evaluated relative to floating priors, the second term relative to a highly smoothed version of the total estimated intensity in the image and the third relative to a slightly smoothed version of the current estimate for the points channel. It is necessary to choose the parameters α and β, as well as the degree of smoothing applied when creating the floating defaults, so that even faint points are successfully found but noise clumps are not detected. An experimental implementation has been coded as a program called **stars**. Unlike normal star finding methods which seek local maxima this method is global and works on the entire image.

Figure 1 (*left*) shows a typical deep CCD image of a star cluster. This frame is part of an I–band image of the cluster M71 taken by F. G. Jensen

(Aarhus) using the Nordic Optical Telescope, and used with his permission. Figure 1 (*center*) shows the points (low entropy) channel produced by applying this method to Figure 1 (*left*), and Figure 1 (*right*) shows the smooth (high entropy) background channel obtained simultaneously. All the stars visible to the eye in the input have been found and there are few spurious detections. The background smooth image shows small artifacts caused by the displacements of the stars from the centers of pixels. This map of the star positions may then be used directly as input to the PLUCY two-channel code to obtain *unbiased* magnitudes for the designated point-sources.

3. PSF Determination

Successful image restoration and photometry both require a good knowledge of the PSF. Many restoration methods can be generalized to allow simultaneous "blind iterative restoration" in which the PSF is obtained simultaneously with the restored image. Such methods are generally thought to be unreliable and are little used. However, when the additional information of designated point-sources is added much greater robustness and reliability is achieved. Such simultaneous PSF determination can easily be added to the PLUCY two-channel code and a preliminary example of its use on *HST* data is given in Hook & Lucy (1994).

Such results tend to retain noise features from the data frames, particularly around bright stars, and are clearly not optimal. It would be advantageous to use extra information about the PSF expected and also to provide regularization to produce a resultant PSF which is smooth. The code has now been updated to include such regularization and found to be effective. We now need a suitable choice of form for the regularizing, default PSF.

Several models for ground-based PSFs have been proposed and used. These are typically simple analytic functions (such as the Moffat function or multiple Gaussians) which may be conveniently fitted to stars but do not have any physical basis. However, recently Saglia et al. (1993) have investigated another form for such PSFs which can be derived from the theory of atmospheric seeing and they show that this form fits observed PSFs as well as, if not better than, the more traditional ad hoc forms. This new form has only two parameters (one of which is simply the size of the star images as defined by the FWHM) but has the minor disadvantage of not being analytic, being instead the Fourier transform of a simple exponential function. This form for the PSF has been implemented as an IRAF compatible task called "seeing" and seems an excellent choice for a default PSF.

Figure 2 shows the steps in the derivation of a good PSF from the same M71 image used above. First the FWHM of bright (but unsaturated) stars in the frame is measured. This value is then used to produce a default PSF of the Saglia et al. (1993) form (*upper-right*). This in turn is used as the default PSF with the PLUCY code to finally give the required best estimate for the PSF (*lower-left*). A simple circular disc is used as the first approximation for the PSF (*upper-left*). A clear elongation of the images has been well modelled and the result is smooth and free of noise features. The PSF found by DAOPHOT, using the same input data, is given at the lower-right for comparison.

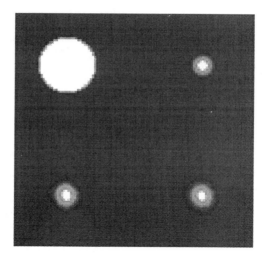

Figure 2. PSF Determination—see text for details.

4. Conclusions

We have extended earlier work on multi-channel, regularized image restoration to produce experimental codes which allow both the mapping of point-source positions in an image and the estimation of the PSF during a subsequent photometric image restoration. A recently suggested form for ground-based PSFs has been implemented and found to be a suitable default. Tests have successfully been made using a typical deep, ground-based CCD frame. All codes have been implemented using the F77/VOS interface to IRAF and are available on request.

References

Cohen, J. G. 1991, AJ, 101, 734

Hook, R. N. & Lucy, L. B. 1994, in The Restoration of HST Images and Spectra II, ed. R. J. Hanisch & R. L. White (Baltimore, Space Telescope Science Institute), p. 86

Lucy, L. B. 1994, in The Restoration of HST Images and Spectra II, ed. R. J. Hanisch & R. L. White (Baltimore, Space Telescope Science Institute), p. 79

Saglia, R. P., et al. 1993, MNRAS, 264, 961

Registering, PSF-Matching and Intensity-Matching Images in IRAF

A. C. Phillips

Lick Observatory, University of California, Santa Cruz, CA, 95064

L. E. Davis

National Optical Astronomy Observatories, P. O. Box 26732, Tucson, AZ 85726

Abstract. We have developed a set of tasks in the IRAF environment for registering, matching the point-spread functions, and matching the intensity scales of two or more images. This software can be applied to a wide variety of astronomical problems including searches for transient events such as extragalactic novae and supernovae, continuum-subtraction for emission line imaging, and the examination of small scale color gradients. We briefly describe the algorithms and illustrate the software with images obtained under different seeing conditions.

1. Introduction

A variety of astronomical programs involve comparing similar image data of the same field. Examples include searches for extragalactic novae, supernovae and variable stars; continuum-subtraction for narrow-band emission-line images; and production of color or polarization maps, to name a few of the more obvious cases. For these types of observations, it is usually essential that all observational differences be eliminated before comparison. While instrumental signatures can be removed accurately, varying conditions (e.g., seeing, transparency) are more problematic. This is particularly crucial where the objects of interest are either extended or are found in regions of strong background gradients.

The process of "matching" an image to a reference image requires: (1) applying a geometric transformation to spatially register an image to the reference image; (2) convolving the registered image with an appropriate kernel to degrade the point-spread function to match that of the reference image; and (3) scaling and offsetting the intensity of the convolved image to match the intensity and background sky level of the reference image.

We have developed a set of IRAF tasks to compute and apply the necessary transformations for image matching. In this paper we briefly describe the algorithms and tasks and illustrate the software with a pair of images obtained under different seeing conditions. We conclude by describing the current status and future development plans of the software.

2. Spatial Registration

The current software performs spatial registration by supplying the registration task with a reference and input image list, and a list of features common to both images. Two spatial registration tasks are available: a "polynomial warping" procedure for general cases, and a new cross-correlation task for the special case that only translations are required.

The polynomial warping task is an IRAF script which combines existing IRAF object selecting, centering, and geometric transformation tasks. This task is required for the general case that pixel scales and/or orientation differ between the input and reference images. The user supplies this task with a set of star positions in the reference image, and the positions of two stars in both the reference and input images. The task computes an initial transformation using the two specified stars, refines it using the full reference star list, and applies the computed transformation to the input image.

If the input and reference images have identical pixel scales and orientation, they can be registered using the cross-correlation task. The user supplies this task with a list of rectangular image regions, each containing high signal-to-noise features suitable for cross-correlation. The task computes the shift from the cross-correlation function and applies it to the input image.

3. Point-Spread Function Matching

Point-spread function (PSF) variations have many sources: seeing changes, guiding errors and focus variations are the most common. Removing the PSF differences is easy in principle but difficult in practice. If r is the reference image, i is the (spatially-registered) input image, and R and I are their Fourier transforms, then
$$r = i \star k,$$
and the Convolution Theorem gives us
$$R = I \times K,$$
where k is some convolution kernel describing the difference in seeing, etc., and K is its Fourier transform. k also contains scaling information for differences in exposure and transparency. The unknown kernel k is then given simply by
$$k = FT\{R/I\}.$$

In practice, unfortunately, the high-frequency components of the images are dominated by noise and so the ratio in Fourier space is poorly behaved. We have found that satisfactory results can obtained by replacing the high-frequency components with a Gaussian model fit to the lower-frequency (higher signal-to-noise) components. This approximation, as well as many other considerations, are discussed in Phillips (1993b).

In principle, the kernel k can be used to deconvolve an image to match a better-seeing image, but in practice, noise is again a severe problem, and deconvolution is a costly procedure. Degrading the better-seeing image is usually acceptable. The convolution smooths the noise as well as the object signal, so only contrast—not signal-to-noise—is lost.

Two PSF-matching options are available in our IRAF task. Both require the reference image to have lower resolution than the input image. In the first, the user must supply the task with the reference and input image and a list of high signal-to-noise point sources. In the second, the user must supply pre-computed PSFs for the reference and input images. (Pre-computed PSFs may be derived from other IRAF software such as the DAOPHOT package or be computed by the user.) In both cases, the task computes the required kernel and performs the convolution. The PSF-matching task also includes options for removing the background around the input point sources, as well as control over the replacement of the high-frequency components with the Gaussian model.

4. Intensity Matching

Many image matching problems require the final images to have the same intensity scale. The current software requires the input and reference images to be spatially-registered and PSF-matched. The user specifies a rectangular data region which contains both object and (ideally) sky data. The task uses linear least squares techniques (including a user-supplied noise model for both images, and automatic deviant pixel rejection) to compute the required scale factor and zero point offset, and applies the transformation.

5. Results: An Illustration

Figure 1 shows two B-band images of the Phoenix dwarf galaxy at different epochs (top). Also shown are the difference image after registration and intensity matching only (bottom left), and the same with PSF-matching (right). In the fully-matched case, about a dozen Cepheids variables are easily seen. Note also the high-proper motion star just below and left of the center. Additional illustrations may be found elsewhere (Phillips 1993a,b; Margon et al. 1992; Ruiz et al. 1987).

6. Current Status and Future Plans

A prototype version of this software is currently available from A. Phillips (*phillips@lick.ucsc.edu*) on a use at your own risk basis. Prospective users should be aware that the software is unsupported and minimally documented, although the author welcomes comments and suggestions for improvement.

The cross-correlation task for spatial registration will be included in the next release of IRAF, and is already available to users as add-on software. A new version of the PSF-matching task is nearing completion and will be available as add-on software shortly after this meeting. Contact L. Davis (*davis@noao.edu*) for a status report.

Future plans include modifying the cross-correlation task to handle images with small scale and orientation differences and adding the intensity scaling task to IRAF itself.

Acknowledgments. ACP wishes to thank NOAO for providing him support as a Visiting Specialist during the summer of 1988, when the PSF-matching

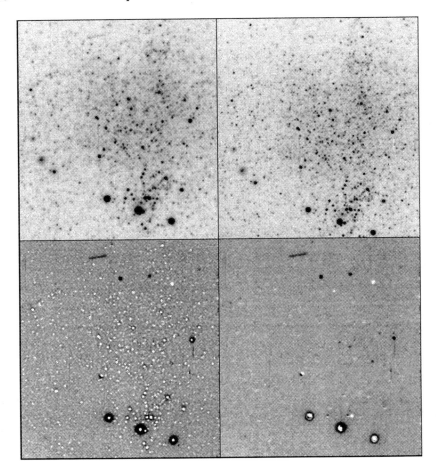

Figure 1. Searching for Cepheid variables in the Phoenix dwarf galaxy. See text. (CTIO 4-m images courtesy of Nelson Calwell.)

algorithm and software were developed. He also thanks the IRAF group for their hospitality and patient instruction during that time.

References

Margon, B., Phillips, A. C., Jacoby, G. H., & Ciardullo, R. 1992, AJ, 103, 924
Phillips, A. C. 1993a, AJ, 105, 486
Phillips, A. C. 1993b, PhD Dissertation, University of Washington, Seattle
Ruiz, M. T., Blanco, V., Maza, J., Heathcote, S., Phillips, A., Kawara, K., Anguita, C., Hamuy, M., & Gómez, A. 1987, ApJ, 316, L21

Restoration of HST WFPC2 Images in Gyro-Hold Mode

J. Mo and R. J. Hanisch

Space Telescope Science Institute[1], 3700 San Martin Drive, Baltimore, MD 21218

Abstract. In gyro-hold tracking mode, *HST* is stabilized only with its gyros, and the Fine Guidance Sensors are not used to maintain pointing. With gyro drift rates of order $0''\!.002$ per second of time, a WFPC 2 Planetary Camera image with a typical exposure time of 100 s can be blurred by as much as five pixels. Image restoration techniques developed for use on aberrated *HST* images have been adapted for use to remove this motion blur. Experiments have been done on WFPC 2 images of NGC 330 and NGC 422 (having exposure times of 80 and 100 s in gyro-hold mode) with complete success.

1. Introduction

Since the first servicing mission in 1993 December, *HST*'s optical performance has been corrected to nearly the original design specifications. As a result, however, spacecraft stability and the guiding mode used by the Fine Guidance Sensors (FGSs) have become much more significant factors in determining overall image quality. The FGSs now have only two guiding modes: fine lock and gyro-hold. In order to utilize gaps in the *HST* observing program that would otherwise be unused, about 200–300 "snapshot" observations are placed on the program in each observing cycle. In order to minimize spacecraft overhead, snapshot observations are taken without guiding, using gyro pointing control only (e.g., Bond 1994). Gyro-hold pointing drifts at a rate ah high as 1.4 ± 0.7 mas s^{-1}. This motion causes a significant blur in WFPC 2 images, even for exposures as short as 100 s. Given our experience in restoring the aberrated images from WFPC 1, we decided to explore the possibilities of removing the motion blur from images taken in gyro-hold pointing mode with WFPC 2.

Two WFPC 2 snapshot observations of NGC 422 and NGC 330 (kindly provided by M. Shara) have been restored successfully using the image restoration techniques which have been implemented in the STSDAS package.

2. Motion Blurred WFPC 2 Images

An NGC 422 WFPC 2 WC2 image and an NGC 330 WFPC 2 PC1 image were taken in mode FSGLOCK = GYROS in 1994 January. A synopsis of these images

---

[1] Operated by AURA, Inc., for NASA

is given in Table 1. In the table, the column "Image Section" indicates the pixel coordinates of the detector in the sequence ($x_{min} : x_{max}, y_{min} : y_{max}$). The restorations have each been done on a 256 × 256 pixel subimage.

Table 1. WFPC 2 Images

Object	Camera	Filter	Exp Time	Date	Image Section
NGC 422	WF2	F450W	100 s	25/01/94	(103:358,494:749)
NGC 330	PC1	F450W	80 s	27/01/94	(499:754,99:354)

3. Image Restoration

The restoration of the motion blurred images is carried out in three stages: (1) standard restoration using a model PSF, (2) determination of the motion-blur function, and (3) final restoration. The initial model PSFs were computed using Tiny TIM Version 4.0. For each observation, the PSF position coincides with the peak position of a bright star.

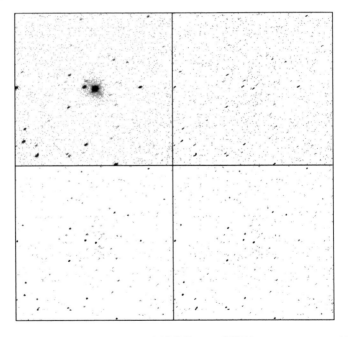

Figure 1. WFPC 2 image of NGC 422 (WF2, exposure = 100 s). Top left: original observation. Top right: standard Tiny TIM deconvolution. Bottom left: MEM restoration (final). Bottom right: Lucy restoration (final).

In order to generate the motion-blur function, a standard Tiny TIM PSF restoration is made for each observation. The blurring function is determined by averaging the images of several bright stars in the initial restored frame. The final PSF is constructed by convolving the motion blur function with the Tiny TIM PSFs.

The restorations were done using the maximum entropy method (MEM) (Wu 1992, 1994) as implemented in STSDAS. The restored image adopted is the ME solution after approximately 100 iterations. The Richardson-Lucy method has also been tested in this experiment using the task *lucy* implemented in STSDAS (White 1993; Stobie et al. 1994). The motion-blurred images of NGC 422 and NGC 330, and their restorations are illustrated in Figs. 1 and 2, respectively.

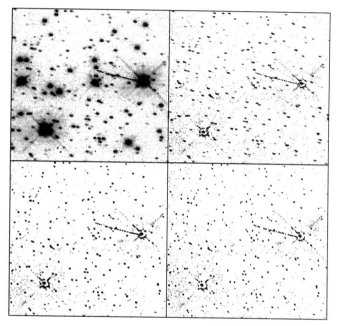

Figure 2. NGC 330 WFPC 2 (PC1, exposure = 80 s). Top left: original observation. Top right: standard Tiny TIM deconvolution. Bottom left: MEM restoration (final). Bottom right: Lucy restoration (final).

4. Discussion and Conclusions

The point spread function for data taken in gyro-hold mode is the convolution of the normal PSF with the motion blurring function. Because our PSF models are now quite accurate, deconvolution of a motion-blurred image with a model PSF yields an image in which the remaining shapes, and features associated with point sources, define the motion-blur component. This residual also encompasses

Figure 3. Motion-blurred PSF. Left: PSF of NGC 422 WFPC 2 (WF2, exposure = 100 s). Right: NGC 330 WFPC 2 (PC1, exposure = 80 s).

any additional mismatches between the model PSF and the observed PSF in the absence of motion blur.

WFPC 2 is undersampled, especially in the UV in WF mode. In order to obtain a better representation of the motion blurring function, several (partially) restored star images are averaged together. The success of the method relies upon having several bright but unsaturated star images in the field of view.

Our experiments indicate that gyro drift is not uniform linear motion, but rather must be described by a two-dimensional function. Indeed, the final PSF obtained for the NGC 422 image shows a multiply-peaked structure, indicating that within the exposure time the telescope dwelled longer at certain locations than in others. A surface plot of this PSF and the final PSF of the NGC 330 image are shown in Figure 3. Both the MEM and Richardson-Lucy algorithms implemented in STSDAS were successful in restoring the image with a multiply-peaked PSF, and gave virtually indistinguishable results.

The standard image restoration techniques developed for use on aberrated *HST* images can be adapted for removing the motion blur from snapshot images taken under gyro-hold pointing mode. This should allow observers to obtain optimal spatial resolution and dynamic range on many of these data sets.

Acknowledgments. Support from ST ScI's Image Restoration Project, funded by NASA as a contract augmentation to NAS5–26555, is gratefully acknowledged. The authors thank M. Shara for providing his *HST* WFPC2 observations prior to publication.

References

Bond, H. E., ed. 1994, Hubble Space Telescope Cycle 5 Call for Proposals (Baltimore, Space Telescope Science Institute)

Stobie, E. B., Hanisch, R. J., & White, R. L. 1994, in Astronomical Data Analysis Software and Systems III, ASP Conf. Ser., Vol. 61, eds. D. R. Crabtree, R. J. Hanisch, & J. Barnes (San Francisco, ASP), p. 296

White, R. L. 1993, in Newsletter of ST ScI Image Restoration Project, ed. R. J. Hanisch (Baltimore, Space Telescope Science Institute), p. 11

Wu, N. 1992, in Astronomical Data Analysis Software and Systems II, ASP Conf. Ser., Vol. 52, eds. R. J. Hanisch, R. J. V. Brissenden, & J. Barnes (San Francisco, ASP), p. 520

Wu, N. 1994, this volume, p. 305

Astronomical Data Analysis Software and Systems IV
ASP Conference Series, Vol. 77, 1995
R. A. Shaw, H. E. Payne, and J. J. E. Hayes, eds.

MEM Task for Image Restoration in IRAF

N. Wu

Space Telescope Science Institute, 3700 San Martin Drive, Baltimore, MD 21218

Abstract. This presentation is devoted to the MEM (Maximum Entropy Method) task, version D, for image restoration in IRAF. Described in detail are its enhanced functions of Poisson noise only case handling and subpixelization technique; improved algorithms for maximization including the preconditioned conjugate method, accurate Newton method, the accurate one-dimensional search, and the model updating technique. The task's limitations and possible development are also discussed.

1. Introduction

The IRAF mem0 package, version B, was released to the public in October 1992, and was upgraded to version C in December 1993. Version C resides in the anonymous ftp site at NOAO (*iraf.noao.edu*, directory ~/*contrib/*). Now, the most important MEM task of the mem0 package has been upgraded to version D; the other four tasks remain unchanged. This task is included in the subpackage stsdas.analysis.restore with the name mem. Its executable code on SUN/UNIX has also been put into the ftp site at NOAO with the task name dmem. The two relevant files are dmem.notes and dmem.tar.Z.

The task mem, version D, features the following sophisticated functions and techniques:

1. Poisson noise only case handling.
2. Subpixelization technique.
3. Model updating technique.
4. The preconditioned conjugate method and accurate Newton method.
5. The accurate one-dimensional search in maximization.

They are described in detail in the following section.

2. Algorithm and Programming

2.1. The Case of Poisson Noise Only

CCD images from the Wide Field and Planetary Camera (WF/PC) of *Hubble Space Telescope* (*HST*) have both readout noise, of Gaussian type, and signal noise, of Poisson type. In this case the total noise variance is calculated using the parameters *noise*, in electrons, for the former, and *adu* (analogue-to-digital unit conversion constant or gain), in electrons/DN, for the latter. Then this variance is used in the Gaussian likelihood function or, equivalently, in the expression of χ^2.

On the other hand, images from the photon counting detectors of the Faint Object Camera (FOC) of *HST* contain only signal noise of Poisson type. Not only should the parameter *noise* be set to zero, but, more importantly, the Poisson likelihood function must be used in calculation. Failure to do so will lead to wrong and unacceptable results.

Earlier versions of the task mem could not handle correctly the case of Poisson noise only. In version D, setting the parameter *poisson* to yes selects the correct expression for the likelihood function, automatically sets the parameter *noise* to zero, and takes other actions especially designed for the case of Poisson noise only.

2.2. Subpixelization Technique

It has been shown that the subpixelization (subsampling) technique may improve resolution in restored images, or at least result in more pleasant appearances for objects (Weir & Djorgovski 1990).

Subpixelization means to restore an image on a grid finer than that on which the input data image is defined. In this process one "normal size" pixel of the input image is "split" to several "subpixels" of the restored image. This mechanism must be built into the program, but cannot be accomplished simply by replicating each pixel of the input image before restoration.

Subpixelization is activated by setting the parameter *nsub* to a value greater than one. In this case all input images, including the point spread function (PSF) but excluding the data image, must be subpixelized by the user by a factor of *nsub* in each dimension. The required core memory and computational time are approximately proportional to $nsub^2$.

2.3. Model Updating Technique

This technique was described in detail in Wu (1994). The MEM program uses a double iteration scheme: the values of the Lagrange multipliers α and β, respectively for the data constraint and the total power constraint, are revised in the outer iteration, while the inner iteration is for finding the ME solution for the particular α and β of each outer iteration.

The basic idea behind the model updating technique is to use the ME image converged for particular α and β in each outer iteration as the model to start the next iteration. In this way in maximization, the approximate solution to the linear equations used in the preconditioned conjugate method is more accurate, or the accurate solution to the linear equations required in the accurate Newton method is easier to be found. Therefore, the total number of iterations is considerably reduced and much computational time is saved. The restored image also has improved photometric linearity.

2.4. Preconditioned Conjugate and Accurate Newton Methods

In previous versions of the task mem, only the zeroth-order approximate Newton method of maximization was available. Hence the name of the package: mem0. Zeroth-order approximation means that in the solution of a large set of linear equations (or equivalently the inversion of a large matrix) non-diagonal elements are ignored, under the assumption that the diagonal ones dominate. In this way, solving the equations becomes a simple operation. However, this simple method may result in very slow convergence.

Now, in version D, much more sophisticated methods are used to calculate the change in the iteration, i.e., to determine the search direction in maximization. The first method is the preconditioned conjugate method (as opposed to the "standard" conjugate method commonly used), or the conjugate method based on the approximate Newton method described in the above. More specifically, in each outer iteration for particular α and β, the approximate Newton method is used to calculate the search direction for the first inner iteration. Thereafter, in each inner iteration, a direction is calculated using the approximate Newton method. The component orthogonal or conjugate to the search direction used in the previous inner iteration is calculated, and used as the new search direction for the maximum point. By careful programming, the core memory requirement and the number of FFTs are the same as in the approximate Newton method.

The second method is the accurate Newton method (as opposed to the approximate Newton method). Here the accurate solution to a set of linear equations of large size is calculated by iteration, each of which requires two convolutions, i.e., four FFTs.

The accurate Newton method is the most efficient in the sense that the fewest total number of (inner) iterations is needed because the search direction is determined accurately. However, many FFTs may well be required in each (inner) iteration to solve the linear equations. In contrast, the preconditioned conjugate method requires a greater total number of (inner) iterations, but no extra FFTs in each (inner) iteration are needed to calculate the conjugate direction.

By default, in the case of Poisson noise only ($poisson$=yes) the accurate Newton method is used, otherwise ($poisson$=no) the preconditioned conjugate method is used. This is the best choice. In the case of Poisson noise only, the Hessian of the objective function, which is a measure of the curvature of the image space, changes rapidly from (inner) iteration to iteration. Consequently, the preconditioned conjugate method, as a method taking advantage of memory in the iteration, is not effective in determining the search direction.

2.5. Accurate One-Dimensional Search in Maximization

After determining the search direction, an optimal step (length) in this direction should be calculated. In earlier versions of the task **mem**, quadratic extrapolation and cubic interpolation are used for this purpose. They are both approximate methods for calculating the optimal step. Now, in version D, the accurate one-dimensional (1-D) search is available for interpolation. Specifically, the approximate maximum point found by cubic interpolation is used as the initial guess to start a search for the accurate maximum point, using the Newton method in a single variable. In such a way, the maximum point is found with little extra effort but much higher accuracy.

The (approximate) quadratic extrapolation remains. In most cases, especially at the late stage of iteration, interpolation but not extrapolation is desirable in the 1-D search.

2.6. Other Revisions

Apart from employing the methods described in the previous subsections to enhance the task's function and to speed up convergence, having a good user interface is also important. Every effort has been made to create a user friendly interface. The number of positional parameters is kept to a minimum. The default values of hidden parameters are carefully chosen. Automatic schemes to adjust some variables and to deal with some predicted ill-conditioned cases are built in the program. Options and parameters used only for testing are hidden from the user. The diagnostic messages are informative and grouped logically, and can be output at three different levels of verboseness to meet the user's needs. The most detailed level ($message=3$) is primarily for debugging purposes. The help file is well written and should be read before the first attempt to run the task.

3. Concluding Remarks

The current version of the task mem has its limitations: Like other MEM programs, it can only handle the case of space-invariant PSF. It cannot be used for multi-channel and multi-data set restoration like the package MEM/MemSys5 (Weir 1991). The algorithm used in the accurate Newton method to find the accurate solution of a large set of linear equations should be improved, e.g., using the preconditioned conjugate method. Finally, criteria for convergence should be investigated, especially in the case of Poisson noise only, and more reasonable ones should be adopted. For this version of the task, the user's judgment is very important in obtaining satisfactory results.

References

Weir, N. & Djorgovski, S. 1990, in The Restoration of HST Images and Spectra, ed. R. L. White & R. J. Allen (Baltimore, Space Telescope Science Institute), p. 31

Weir, N. 1991, in Proceedings of the 3rd ESO/ST-ECF Data Analysis Workshop, ed. P. J. Grosbøl & R. H. Warmels (Garching, ST-ECF), p. 115

Wu, N. 1994, in The Restoration of HST Images and Spectra II, ed. R. J. Hanisch & R. L. White (Baltimore, Space Telescope Science Institute), p. 58

Part 8. Statistical Analysis

Statistical Consulting Center for Astronomy

E. D. Feigelson

Department of Astronomy & Astrophysics, Penn State University, University Park PA 16802

M. G. Akritas, J. L. Rosenberger

Department of Statistics, Penn State University, University Park PA 16802

Abstract. We announce the formation of a Statistical Consulting Center for Astronomy (SCCA), designed to provide prompt high-quality advice on statistical and methodological issues to the astronomical community. Questions should be sent by e-mail to *scca@stat.psu.edu*. Questions and answers can be examined on the World Wide Web[1].

1. Introduction

The astronomer extracting scientifically useful information from astronomical data often encounters complex and subtle problems. Statistical techniques such as least-squares model fitting, Kolmogorov-Smirnov two-sample test and χ^2 goodness-of-fit test can be applied to many simple situations, but are inadequate for other problems. A few examples of such data analysis problems are: satellite surveys with flux limits and nondetections; discrimination between stars and galaxies in digitized optical surveys; detection of weak sources in photon-counting detectors with variable backgrounds; characterization of quasi-periodic or stochastically variable objects; identification of filaments and voids in anisotropically clustered galaxies; analysis of the Lyman-α forest in quasar spectra; repeated application of calibration regressions in the cosmic distance scale; and error analysis in all of these situations.

The field of mathematical statistics and its many areas of application (biometrics, econometrics, chemometrics, geostatistics, quality control, etc.) have made huge advances in recent decades. Mathematics libraries have dozens of journals and hundreds of monographs on specialized problems in statistics that are rarely if ever read by the astronomer. The problem encountered by an astronomer has often been addressed, and perhaps clearly resolved, by statisticians working in other fields. In other cases, the astronomical problem is methodologically unique, and its treatment might challenge a top statistician specializing in the relevant field.

We have created the Statistical Consulting Center for Astronomy (SCCA) to help bridge the wide gap between the astronomical and statistical communi-

---

[1] http://www.stat.psu.edu/scca/homepage.html

ties. Through the SCCA, astronomers can ask a team of statisticians questions about the data analysis problems they are facing today. If a good solution is readily known, the SCCA will respond rapidly with an answer and guidance into the appropriate statistical literature. If the problem is particularly tricky or important, the SCCA will seek out top quality statisticians to consult with, and possibly collaborate with, the astronomer.

The need for improved statistical treatment of astronomical data is clear. A scan of the *Astronomy & Astrophysics Abstracts* indicates that 100–200 papers are published annually are principally concerned with methodological issues in the astronomical literature, and dozens of additional observational papers have discussions of statistical issues. Statistical issues arising in astronomical data analysis have been presented at a growing number of conferences (e.g., Jaschek & Murtagh 1990; Feigelson & Babu 1992; Subbarao 1995; various ADASS and European workshop proceedings). Yet except for the 1991 Penn State conference, there has been little involvement of the academic and professional statistical community in addressing the problems arising in astronomy.

2. Operation of the SCCA

The SCCA is a team of Penn State faculty with interest and expertise in statistical problems arising in astronomical research. The Center has contacts with experts in the international statistical community. The goals of the Center are to: (1) address the immediate statistical needs of astronomers by providing prompt high-quality statistical advice, (2) make publicly available questions and answers for the benefit of the wider astronomical community, and (3) encourage interdisciplinary collaboration between the fields of statistics and astronomy.

Any individual in the astronomical community can submit a question to the SCCA: a graduate student preparing a dissertation; a scientist confronted with a tricky data set, preparing or revising a paper for publication; a scientist preparing software for an instrument or a data analysis software system; or a scientist organizing a major observational program. Incoming questions are reviewed by the members of the team and colleagues in the Department of Statistics at Penn State. Many questions will be answered in-house, but particularly complex and important problems will be sent to top-ranked experts worldwide. The turn-around time for answering straightforward problems should be no more than three weeks. Summaries of questions and answers will be made publicly available through the Internet/WWW and publications.

The operation of the SCCA is partially supported by the NASA Astrophysics Data Program starting in fall 1994. Initially, consulting can be free of charge to U.S. astronomers. However, we strongly encourage questioners to pay a nominal fee for the service. This will ensure the continuance of the Center into the future, and the availability of top-quality external consultants.

When a question for the SCCA arises, astronomers should send e-mail to *scca@stat.psu.edu* or FAX the Center at (814) 863-7114. Questions and answers will be available by anonymous ftp at *ftp.stat.psu.edu* (cd to the *pub/scca* directory) and on the World Wide Web at the SCCA Homepage.

3. Some Early Questions & Answers

Q: Can a partial correlation coefficient be applied to data with upper limits?
A: One can construct a partial correlation coefficient for censored data using (say) the generalized Kendall's τ bivariate coefficient implemented in the ASURV package (LaValley et al. 1992), but no tests of significance are available. In fact, no significance testing method is available for the partial Kendall's τ even with uncensored data (Hettmansperger 1984, p. 208). For uncensored data, we recommend instead either multiple regression (Murtagh & Heck 1987) or Pearson's linear partial correlation coefficient (Anderson 1984). Unfortunately, the extension of multivariate analysis to censored data has proved to be quite difficult and there are no available methods. Thus, no fully satisfactory answer to your question exists, but an expert in the field has promised to work on developing a method for testing the hypothesis that the partial Kendall's τ is zero.

Q: How can one assess the likelihood and amplitude of variability of an X-ray source from *ROSAT* observations consisting of 20 disjoint good time intervals? 100-1000 total counts are collected, which is a bit low for the χ^2 test.
A: If you can confidently assume that the underlying distribution of counts follows a Poisson distribution, we recommend the likelihood ratio test. You want to test that $\lambda_1 = \cdots = \lambda_{20} = \lambda$, where λ_i times the exposure time t_i gives the expected counts $E(X_i)$ in the i-th interval. The likelihood ratio statistic is

$$L_R = 2 \sum_{i=1}^{n} X_i \log\left(\frac{X_i}{\hat{\lambda} t_i}\right). \tag{1}$$

If the hypothesis of no variability is true, then L_R has a χ^2 distribution with 19 degrees of freedom. Thus, the null hypothesis is rejected at significance level $\alpha = 0.01$ if $L_R > 36.2$. If the hypothesis of constancy is rejected, the amplitude of variability can be examined from the estimated parameters $\hat{\lambda}_i = X_i/t_i$. The likelihood ratio test is presented in Hogg & Tanis (1993), and its use in astronomy under the Poisson hypothesis is discussed by Cash (1979).

Q: Consider two clusters of galaxies, one with $N_1 = 80$ galaxies with 40% spirals and the other with $N_2 = 120$ galaxies with 70% spirals. Is the spiral fraction difference significant?
A: Let $\hat{p}_1 = X_1/N_1$, and $\hat{p}_2 = X_2/N_2$ be the two proportions. We can suggest two test statistics for determining if the proportions are significantly different (Arnold 1990; Miller, Freund, & Johnson 1990):

$$T_1 = \frac{\hat{p}_1 - \hat{p}_2}{[\frac{\hat{p}_1(1-\hat{p}_1)}{N_1} + \frac{\hat{p}_2(1-\hat{p}_2)}{N_2}]^{1/2}} \tag{2}$$

$$T_2 = \frac{\hat{p}_1 - \hat{p}_2}{[\hat{p}(1-\hat{p})(N_1^{-1} + N_2^{-1})]^{1/2}} \tag{3}$$

where $\hat{p} = (X_1 + X_2)/(N_1 + N_2)$. Under the null hypothesis, both statistics have a normal distribution with mean zero and variance one. T_2 is equivalent to Pearson's χ^2 and is more commonly used, though its applicability is limited to testing the null hypothesis. T_1 is a Wald-type statistic and can be used to give confidence intervals for the true difference $p_1 - p_2$. For the problem at hand, the two proportions would be declared significantly different at significance level $\alpha = 0.01$ if $|T_2| > 2.58$. Miller, Freund, & Johnson (1990) also consider a k-sample version of this statistic.

Q: I am teaching a graduate course on astronomical techniques, and would like to include a short section on Bayesian analysis. Can you suggest a general reference?

A: An excellent review of Bayesian inference in astronomy is given by Loredo (1992) and further applications are discussed in Ripley (1992). Background references might include Lindley (1965) and Howson & Urbach (1993).

Acknowledgments. The SCCA is partially funded by NASA grant NAS5-32669.

References

Anderson, T. 1984, An Introduction to Multivariate Statistical Analysis (New York, Wiley)

Arnold, S. 1990, Mathematical Statistics (Englewood Cliffs, Prentice-Hall), p. 386

Cash, W. 1979, ApJ, 228, 939

Feigelson, E. D., & Babu, G. J., eds. 1992, Statistical Challenges in Modern Astronomy (New York, Springer-Verlag)

Hettmansperger 1984, Statistical Inference Based on Ranks (New York, Wiley)

Hogg, R., & Tanis, E. 1993, Probability and Statistical Inference (Macmillan)

Howson, C., & Urbach, P. 1993, Scientific Reasoning: The Bayesian Approach (Chicago, Open Court)

Jaschek, C., & Murtagh, F. (eds.) 1990, Errors, Bias and Uncertainties in Astronomy (Cambridge, Cambridge Univ. Press)

LaValley, M., Isobe, T., & Feigelson, E. 1992, BAAS (Software Report), 24, 839

Lindley, D. 1965, Introduction to Probability and Statistics from a Bayesian Viewpoint, 2 vols., (Cambridge, Cambridge Univ. Press)

Loredo, T. 1992, in Statistical Challenges in Modern Astronomy, eds. E. D. Feigelson & G. J. Babu (New York, Springer-Verlag), 275

Miller, I., Freund, J., & Johnson, R. 1990, Probability and Statistics for Engineers (Englewood Cliffs, Prentice-Hall), p. 282

Murtagh, F., & Heck, A. 1987, Multivariate Data Analysis (Dordrecht, Kluwer)

Ripley, B. D. 1992, in Statistical Challenges in Modern Astronomy, eds. E. D. Feigelson & G. J. Babu (New York, Springer-Verlag), p. 329

Subbarao, T., ed. 1995, Applications of Time Series Analysis to Astronomy and Meteorology (New York, Chapman-Hall)

Stochastic Relaxation as a Tool for Bayesian Modeling of Astronomical Images

I. C. Busko

Astrophysics Division, National Space Research Institute, CP 515, CEP 12201-970, S. J. dos Campos, SP, Brazil

Abstract. Sampling techniques are used to explore the Bayesian posterior density in imaging problems. Besides the familiar MAP estimators, such techniques can easily provide more detailed information on the posterior, such as moments and marginal densities. From these, error bars and confidence levels can be assessed, and hypothesis testing performed. The algorithms are implemented in IRAF, and are being tested on both simulated and actual *ROSAT* images.

1. Introduction

The image formation process introduces uncertainties in pixel values, due to both deterministic (Point Response Function, PRF) and random (noise) effects. The traditional approach to this problem involves some sort of inversion technique, and these are usually unable to use optimally all of the information available both *a priori* and in the data. An alternative approach is to adopt a data modeling perspective of the imaging problem. The work reported here aims at developing Bayesian-inference signal modeling methods based on sampling techniques. These methods are able to provide optimal estimators for pixel values, but, more importantly, can also provide consistently their error bars, and confidence levels for detected structures in the image. Hypothesis testing capabilities can also be built-in in these methods, as well as the capability of handling complicated, non-analytic imaging models.

2. Bayesian Inference and Sampling

The goal of Bayesian inference in imaging problems is to compute the posterior probability density given by Bayes' rule

$$\mathcal{P}(\mathbf{F}|\mathbf{G},\mathcal{H}) = \mathcal{P}(\mathbf{G}|\mathbf{F},\mathcal{H}) \frac{\mathcal{P}(\mathbf{F},\mathcal{H})}{\mathcal{P}(\mathbf{G},\mathcal{H})}$$

where $\mathbf{G}$ is the data vector (observation), $\mathbf{F}$ is the parameter vector to fit, and $\mathcal{H}$ is the hypothesis space containing the model being fitted.

MAP (Maximum A Posteriori) estimation is the simplest Bayesian inference operation; it corresponds to finding the posterior's maximum (mode). It is relatively easy to perform since there is no need to compute the normalization term $\mathcal{P}(\mathbf{G},\mathcal{H})$. Several MAP methods exist, adapted to a variety of problems, usually seeking the fastest way to the solution through the use of efficient maximization algorithms.

If one wants more information on the posterior besides its mode, sampling methods can be of help (Gelfand & Smith 1990), since they offer a very general way of handling intractable probability densities. Stochastic Relaxation (SR) is such a method, with foundations in statistical physics. It was devised to study equilibrium properties of large systems of identical "particles". When combined with an "annealing schedule," SR can be used as a maximization tool as well (Geman & Geman 1984; Kirkpatrick et al. 1983). It is robust (does not get stuck in secondary maxima), intrinsically parallel, and very easy to code, in the sense that the algorithm does not depend on the details of the imaging problem. Thus, it can easily take into account such complicated imaging situations as signal-dependent noise, variable/disjoint PRFs, and non-analytic prior densities $\mathcal{P}(\mathbf{F}, \mathcal{H})$. Thus, it is an excellent testbed for experimenting with such imaging situations.

In this work an IRAF task that implements SR was created and is being tested with both artificial and actual ROSAT HRI data. The task can be used either as a MAP searcher or as a Gibbs sampler for the posterior density (Geman & Geman 1984), and includes both Maximum Likelihood (ML) and Maximum Entropy (ME) prior densities. An on-axis PRF was computed by task rosprf in the xray package and used in all experiments.

3. Results

Figure 1 depicts MAP solutions obtained by the SR algorithm with both ML and ME priors, using as input a simulated ROSAT HRI observation. The Richardson-Lucy (R-L) algorithm is used here as comparison, since it is the Expectation Maximization implementation of ML for Poisson data.

Figure 2 depicts values assumed by the likelihood $\mathcal{P}(\mathbf{G}|\mathbf{F}, \mathcal{H})$ and prior $\mathcal{P}(\mathbf{F}, \mathcal{H})$ densities along the iteration sequence for each algorithm. The likelihood can be expressed in a zero-to-one probability scale by the bootstrap sampling technique of Bi & Boerner (1994). In this way, it is a measure of how well a given estimator "fits the data." Notice how the likelihood increases monotonically with iteration number, while probability for both ML and R-L reaches a maximum and decreases afterwards. This behavior, which is also seen with actual ROSAT data, was used to define a "best" estimator, which is the last one along the iteration sequence for which the likelihood, as expressed in a probability scale, is ~ 1. In other words, the best estimator is the most restored one which still fits the data with high probability. In this experiment, the best R-L estimator appears at iteration ~ 15, best ML at iteration ~ 90, and best ME at convergence. This ME estimator is depicted in Panel 4 of Figure 1. Figure 3 depicts the best ML and R-L estimators.

When operated as a Gibbs sampler, SR output can be used along with the ergodicity theorem of Geman & Geman (1984) to compute moments of the posterior density. Figure 3 depicts the first moment for the ML prior. This is the posterior mean, which should be compared with its mode in Panel 3 of Figure 1. Panel 8 in Figure 3 depicts iso-confidence level contours for the same prior. These were derived from both the first and second moments of the posterior. They show that the extended structures detected on the estimator are backed

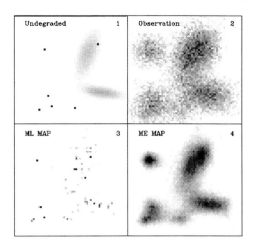

Figure 1. **Panel 1:** artificial sky field. **Panel 2:** 10,000 photon simulated *ROSAT* HRI observation. **Panel 3:** ML MAP estimator. The same result is obtained by either SR or R-L. **Panel 4:** ME MAP estimator, obtained by SR.

up by the data at the 1–2 σ level. Stars are backed up by the data at the 3 σ level and above.

4. Future Work

Two main areas of future work are envisaged. The first is the inclusion of new prior types, such as Maximum Sparsity (Jeffs & Elsmore 1991) and Markov Random Field, and adaptation of the likelihood computation to specific problems (e.g., coded mask). The second is the implementation of a different sampling scheme, that would automatically and explicitly take care of the posterior normalization.

References

Bi, H., & Boerner, G. 1994, A&A, 108, 409

Gelfand, A. E., & Smith, A. F. M. 1990, J. Amer. Statist. Assoc., 85, 398

Geman, S., & Geman, D. 1984, IEEE Transactions on Pattern Analysis and Machine Intelligence, PAMI-6, 721

Jeffs, B. D., & Elsmore, D. 1991, International Conference on Acoustics, Speech and Signal Processing, ICASSP91-4, 2937

Kirkpatrick, S., Gelatt, C. D. Jr., & Vecchi, M. P. 1983, Science, 220, 671

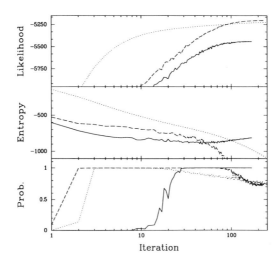

Figure 2. Likelihood (log), entropy and probability for *ROSAT* HRI simulated data, along the iteration sequence of each algorithm. Solid: SR with entropy prior. Dashed: SR with Maximum Likelihood prior. Dotted: Richardson-Lucy.

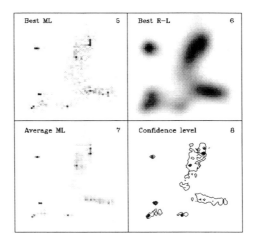

Figure 3. **Panel 5:** "Best" ML estimator obtained by SR. **Panel 6:** Best R-L estimator. **Panel 7:** Averaged Gibbs sampler output (150 samples) with ML prior. **Panel 8:** Iso-confidence curves for Panel 7 estimator, ranging from 1 to 7 σ in steps of 1.

Spatial Models and Spatial Statistics for Astronomical Data

L. Pásztor

MTA TAKI, H-1022 Budapest Herman Ottó út 15, Hungary

L. V. Tóth

Dept. of Astr., Eötvös Univ., H-1083 Budapest Ludovika tér 2, Hungary

Abstract. A statistical model is a convenient conceptual representation of an observed phenomenon. A statistical model represents the observations in term of random variables which can then be used for description, estimation, interpretation, and prediction based on Probability Theory. Interest in statistical methodology is increasing rapidly in the astronomical community. New questions arising from old and new technologies require new statistical models, and many of the new problems are spatial in nature. Spatial statistics is still a young discipline, the application of which is not yet widespread in the astronomical community.

1. The Formalization of the General Spatial Model

Consider $\{Z(t) : t \in T\}$; where $T \subseteq \Re^d$. Here T is the index set, $Z(.)$ is the spatial process, $\{z(t) : t \in T\}$ is a realization of the process. In the present paper we give a brief overview on the most important spatial statistical models, to illustrate the range of problems that can be addressed and the wide applicability of spatial statistical models in astronomy.

2. Point Processes

A usual spatial point process is defined as $Z(t) = 1; \forall t \in T$ (i.e., the index set is the points of $\Re^d$) or $Z(A) = N(A)$ [the number of points within A]; $A \subseteq T$ (i.e., the index set is the units of $\Re^d$), where both $Z(.)$ and T are random. First- and second-order properties of a spatial point process are the intensity function: $\lambda(t) = lim_{|dt| \to 0} \{E[N(dt)]/ \mid dt \mid\}$; and the second-order intensity function: $\lambda_2(t, u) = lim_{|dt|,|du| \to 0} \{E[N(dt)N(du)]/ \mid dt \mid\mid du \mid\}$. Spatial point processes are the mathematical models producing point patterns as their realization. A number of processes are available for modeling the patterns that arise in nature:

Complete Spatial Randomness process (CSR; the white noise of spatial point processes) ≡ homogeneous Poisson process (HPP). The number of points for $\forall A$ has a Poisson distribution with mean $\lambda \mid A \mid$; counts in disjoint sets are independent.

Processes with tendency to produce aggregated patterns. For an inhomogeneous Poisson process (IPP) the number of points for $\forall A$ has a Poisson distribution with mean $\int_A \lambda(t)dt$. Counts in disjoint sets are independent. For a Cox process (CP; doubly stochastic point process) $\Lambda(t)$ is a non-negative valued stochastic process. Conditional on $\Lambda(x) = \lambda(x)$, the events form an IPP with intensity function $\lambda(x)$. For a Poisson cluster process (PCP; Neyman-Scott process) parent events form an IPP. Each parent produces a random number of offspring, realized independently according to a discrete probability distribution. The position of the offspring relative to their parents are independently distributed according to a d-dimensional density function. The final process is composed of the superposition of offspring only. Multi-generation process is the generalization of PCP, where offspring are parents of the next generation.

Processes with tendency to produce regular patterns. For a simple inhibition processes (SIP; hard core processes) no two events may be located within a minimum permissible distance, d, of each other. (Matern models, Matern-Stoyan, Matern-Bartlett, simple sequential inhibition models are examples.) The Markov point process (MPP) is a more flexible framework for modeling inhibition processes. $\lambda(u \mid t_1, t_2, ..., t_i, ..; t_i \in A \backslash \{u\}) = \lambda(u \mid t_1, t_2, ..., t_i, ..; t_i \in b(u,d)\backslash\{u\})$, where $b(u,d)$ is the closed ball of radius d centered at u (Strauss process, Pair-potential Markov point process, Gibbs process).

Multivariate spatial point processes. Defined as $Z(t) = 1, 2, 3, ..., m$; $\forall t \in T$ (i.e., the index set is the points of $\Re^d$) or $Z(A) = N_i(A)$ [number of i points within A]; $A \subseteq T$ (i.e., the index set is the units of $\Re^d$), where both $Z_i(.)$ and T are random. The m univariate spatial point processes are the components of the multivariate process, which is thus characterized by m intensity functions and $m(m+1)/2$ second-order intensity functions. The terminology reflects the components of the process (e.g., bivariate Cox process).

Examples of applicability in astronomy include: (1) revealing regularity in the spatial distribution of point-like objects, (2) identification of important scales in the spatial distribution of point-like objects, (3) stellar statistics (deriving distributions, testing of predicted distribution functions, identification of clusters and associations of stars, search for wide binaries and multiple systems), and (4) cosmological problems (testing of predicted distribution functions, identification of galaxy clusters, voids, etc.).

3. Theory of Regionalized Variables (Geostatistics)

The spatial index t varies continuously throughout a fixed subset T of a d-dimensional Euclidean space. Term "regionalized" was introduced in order to emphasize the continuous spatial nature of the index set T. The prefix "geo" reflects the fact that the theory's roots are in geographical and geological applications. Random processes are usually characterized by their moment measures. In geostatistics, "semivariogram" plays a crucial role. If $var(Z(t_1) - Z(t_2)) = 2\gamma(t_1 - t_2)$ for $\forall t_1, t_2 \in T$; γ is called semivariogram. If $E(Z(t)) = m$ for $\forall t \in T$

and $\gamma(.)$ exist, $Z(.)$ is intrinsically stationary. Semivariogram is conditional negative-definite. If $Z(.)$ is second-order stationary $2\gamma(d) = 2(C(0) - C(d))$. Linear, spherical, and exponential models are simple isotropic (semi)variogram.

The most important application of the (semi)variogram is "kriging," a stochastic spatial interpolation method which depends on the second-order properties of the process. The principal aim of kriging is to provide accurate spatial predictions from observed data. Kriging techniques are all related and refined versions of the weighted moving average originally used by Krige (1951) and based on the simple linear model: $Z(t) = \sum_{i=1}^{n} w_i Z(t_i)$, where $\sum_{i=1}^{n} w_i = 1$. Kriging provides optimum prediction in a sense of minimizing mean-squared prediction error, and also provides the estimation. A useful decomposition is $Z(.) \equiv \mu(.) + W(.) + \eta(.) + \epsilon(.)$, where $\mu(.) \equiv E(Z(.))$ is the large-scale variation, $W(.)$ is the smooth small-scale variation, $\eta(.)$ is the micro-scale variation, $\epsilon(.)$ is the measurement error. These models are widely applied in geosciences.

A number of astronomical applications of the method come to mind: (1) the creation of contour and/or surface maps in the case of incompletely sampled maps in extended radio surveys, (2) testing for completeness in sampling (whether the expected structure is revealed as spiral or filamentary), (3) testing whether resolution is achieved (in the cores of galaxies), (4) the creation of maps with resolution higher than the physical resolution of the observation (interpolations arising from the co-addition of separate sky coverage by $IRAS$ or ISO), and (5) interpolations to reach a higher virtual resolution for comparisons (e.g., $IRAS$ 12 and 100 μm images).

4. Further Cases of the General Model

Spatial model on lattices. The index set T is a countable collection of regularly or irregularly scattered spatial sites and these sites are supplemented with a neighborhood structure. Neighborhood structure is generally modeled either by the connectivity matrix (C is an $n \times n$ matrix, $c_{ij} = 1$ if sites i and j are juxtaposed, $c_{ij} = 0$ if not; n is the number of sites) or by a graph-theoretic formalism (the sites become vertices, which are connected with edges for contiguous objects). Examples for realizations of lattice processes in 2-D are spot maps, mosaics, and digital images. The most important application of lattice models is statistical modeling of spatial images, which is widespread in astronomical image processing (restoration, segmentation, classification, reconstruction, etc.).

Fuzzy sets theory. The elements of T are random sets. The premise of the approach that all the data are imprecise, even after they have been observed.

Multivariate spatial statistics. $Z(t)$ is multidimensional. An example of multivariate spatial statistics is provided by multiband image processing. A generalization of the univariate spatial statistical methods is provided by cokriging, where spatial prediction of a variable is carried out with the aid of another.

Examples of applicability to astronomy include: (1) 2-D classification of objects by their shape on images (e.g., star, galaxy identification on CCD or photographic images), (2) cloud identification from coordinate-velocity "data cubes" (e.g., radio spectroscopic observations), and (3) any advanced image processing technique, like maximum entropy or deconvolution (e.g., maximum correlation method in "HIRES" *IRAS* data processing at IPAC).

Acknowledgments. This research was partially supported by the Hungarian State Research Found (Grant No. OTKA-F 4239). L. Pásztor is grateful to ADASS and the Hungarian State Research Found for the travel grants.

Basic References on the Astronomical Applications of Spatial Statistics

Bahcall, J. N., & Soneira, R. M. 1981, ApJ, 246, 122

Bahcall, J. N., Jones, B. F., & Ratnatunga, K. U. 1986, ApJ, 60, 939

Bucciarelli, B., Lattanzi, M. G., & Taff, L. G. 1993, ApJS, 84, 91

Cliff, A. D., & Ord, J. K., 1973, Spatial Autocorrelation (London, Pion)

Cressie, N. A. C. 1991, Statistics for Spatial Data (New York, Wiley)

Diggle, P. J. 1983, Statistical Analysis of Spatial Point Patterns (London, Academic Press)

Getis, A., & Boots, B. 1978, Models of Spatial Processes (Cambridge, Cambridge University Press)

Huang, J. S., & Shieh, W. R. 1990, Pattern Recognition, 23, 147

Journel, A. G., & Huijbregts, Ch. J. 1978, Mining Geostatistics (London, Academic Press)

Matheron, G. 1965, La Theorie des Variables Regionalisées et ses Applications (Paris, Masson)

Molina, R., Olmo, A., Perea, J., & Ripley, B. D. 1992, AJ, 103, 666

Pásztor, L., Tóth, L. V., & Balázs, L. G. 1993, A&A, 268, 108

Pásztor, L. 1993, in Astronomical Data Analysis Software and Systems II, ASP Conf. Ser., Vol. 52, eds. R. J. Hanisch, R. J. V. Brissenden, & J. Barnes (San Francisco, ASP), p. 7

Pásztor, L. 1994, in Astronomical Data Analysis Software and Systems III, ASP Conf. Ser., Vol. 61, eds. D. R. Crabtree, R. J. Hanisch, & J. Barnes (San Francisco, ASP), p. 253

Peebles, P. J. E. 1973, ApJ, 185, 413

Prusti, T., Adorf, H-M., & Meurs, E. J. A. 1991, ESO Sci. Preprint 797

Ripley, B. D. 1981, Spatial Statistics (New York, Wiley)

Ripley, B. D. 1988, Statistical Inference for Spatial Processes (Cambridge, Cambridge University Press)

Upton, G., & Fingleton, B. 1985, Spatial Data Analysis by Example, Vols I–II (New York, Wiley)

Wilson, C. D. 1991, AJ, 101, 1663

Spatial Structure of NGC 6822: An Example for Statistical Modeling of Astronomical Data

L. Pásztor

MTA TAKI, H-1022 Budapest, Herman Ottó út 15, Hungary

C. Gallart, A. Aparicio, J. M. Vílchez

IAC, Vía Láctea a/n 38200 La Laguna, Tenerife, Spain

Abstract. The spatial distribution of stars in the Local Group dwarf irregular galaxy NGC 6822 has been studied, using recent positions and deep photometry for about 15,000 stars in the galaxy. Based on photometrical data, OB stars and red stars could be studied separately. Spatial statistical tools have been applied to the analysis of the spatial structure. The primary aim of the analysis was to find associations among OB stars of the sample in statistical way; that is to quantify the grouping tendency visible in the images of the dwarf galaxy with the aid of merely statistical models. In the present paper the technical aspects of the analysis are discussed.

1. Introduction

The Local Group dwarf irregular galaxy NGC 6822 is characterized by small dimensions and structural simplicity; its distance is about 500 kpc. The surface distribution of stars in the direction of NGC 6822 (total: 22,958 objects) came from recent position and deep photometric observations (Gallart et al. 1994 and references therein). After removing the foreground contamination (estimated by the aid of a representative comparison field located near to the galaxy) 15,343 objects retained in the sample.

Blue stars [$(B - V) < 0.5$ and $(V - R) < 0.35$; a total of 1,631 objects] and red stars [$(V - R) > 0.65$; a total of 8,998 objects] in the filtered sample were separated in $(B - V) - (V - R)$ parameter space which was accompanied by a separation also in $2D$ geometrical space (a similar effect was published for NGC 3109 by Bresolin et al. 1993).

Our approach to group identification has been based on merely statistical tools, as opposed to recent works on similar efforts finding associations in nearby galaxies, like Bresolin et al. (1993 or Wilson (1991, 1992). The main characteristic of the present method is the subsequent refinement of point process models fitted to the sample.

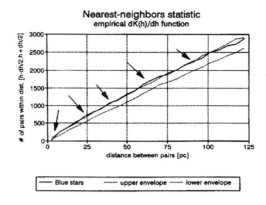

Figure 1. Identification of significant scales by NNS

2. Model Fitting, Step 1

Rejection of CSR, (complete spatial randomness; for details on the following spatial statistical models see Pásztor and Tóth 1995) was based on the results of the NNS (nearest-neighbors statistic). Principles of NNS can be summarized as follows. Consider every pair of objects whose separation is less than a predefined limit: The number of pairs whose distance is between $h - dh/2$ and $h + dh/2$ versus their separation is a well defined function, and widely used in point pattern analysis. This function is the derivative of another important point process function, the K function which is related to second-order properties of point processes (for details see Cressie 1991). For HPP (homogeneous Poisson point process), the function is linear: upward deviation indicates aggregation, while a deviation downward is due to some regularity in the point pattern. The significance of deviation from CSR results from comparisons to 100 Monte Carlo simulations of HPP, so the significance of deviation from CSR is $p < 0.01$ for every scale.

3. Model Fitting, Step 2

Refinement of the model was carried out by taking into consideration the apparent large-scale structure present in the sample. The newer 100 Monte Carlo simulations were generated as realizations of IPP (inhomogeneous Poisson point process) with fixed (and identical with that of the real sample) marginal distributions (Pásztor et al. 1993). Result of NNS (Figure.1) shows significant ($p < 0.01$) clustering at around the scales of 25, 40, 65, and 90 pc, and additionally vacancies at scales smaller than 5 pc. This latter is a resukt of an SIP (simple inhibition point process), and is probably due to the limit in resolution of the observations.

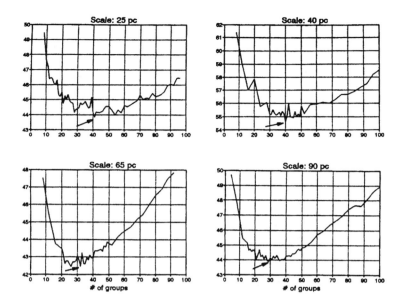

Figure 2.   Model selection by the aid of $CAIC$

4. Model Fitting, Step 3

All of the objects which dominant the significant clustering at a given scale are thought to be members of groups with a characteristic size comparable with the scale value. However the number, shape, and location of these groups is *a priori* unknown. A sequence of non-hierarchical clustering models was carried out, providing partitions of the sample as well as results of PCP (Poisson cluster process) models. In choosing an optimum and minimal model over the set of these competing models, an information theoretic criterion, the $CAIC$ (a more strictly inforced version of Akaike's Information Criterion) (Eisenblätter and Bozdogan 1988), was used (Figure 2).

5. Results

A final partition of the blue stars into groups (associations) on the level characterized by characteristics scale of 25, 40, 65, and 90 pc can be seen on Figure 3. Circles with radii of the scale values should not be interpreted as anything but models of fuzzy sets with radii varying around these values. Th centers scattered around the centers of the resultant circles, and shapes are approximate.

Acknowledgments.   L. Pásztor was partially supported by the Hungarian State Research Found (Grant No. OTKA-F 4239). L. Pásztor is grateful to the Organizers of the conference and the Hungarian State Research Found for the travel grants.

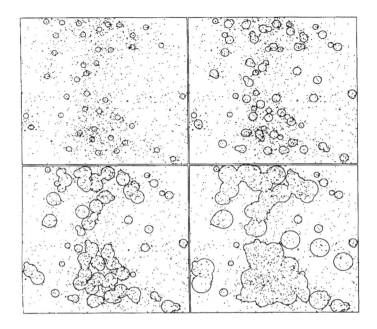

Figure 3. Clustering of OB stars on scales of 25, 40, 65 and 90 pc

References

Bresolin, F., Capaccioli, M., & Piotti, G. 1993, AJ, 105, 1779
Cressie, N. A. C. 1991, Statistics for Spatial Data (New York, Wiley)
Eisenblätter, D., & Bozdogan, H. 1988, in Classification and Related Methods of Data Analysis, ed. H. H. Bock (Amsterdam, North-Holland), p. 91
Gallart, C., Aparacio, A., Chiosi, C., Bertelli, G., & Vílchez, J. M. 1994, ApJ, 425, L9
Pásztor, L., Tóth, L. V., & Balázs, L. G. 1993, A&A, 268, 108
Pásztor, L., Tóth, L. V. 1995, this volume, p. 319
Wilson, C. D. 1991, AJ, 101, 1663
Wilson, C. D. 1992, ApJ, 384, L29

Cheating Poisson: A Biased Method for Detecting Faint Sources in All-Sky Survey Data

J. W. Lewis

Center for EUV Astrophysics, 2150 Kittredge St., University of California, Berkeley, CA 94720-5030

Abstract. One approach to compiling a catalog of point sources from all-sky survey data is to apply a source detection algorithm to the entire data set and include in the catalog any location whose significance exceeds some minimum value. The detection threshold is generally chosen to keep the expected number of spurious detections below some more-or-less arbitrary figure; in low signal-to-noise ratio data, such as the *Extreme Ultraviolet Explorer (EUVE)* survey skymaps, even a small change in the detection threshold can result in an explosion of spurious detections, destroying the usefulness of the catalog.

This result does not, however, imply that real sources below the limiting catalog threshold cannot be reliably detected. If one has some prior knowledge of where the real sources are likely to be found, it is possible to "cheat Poisson" and include these sub-threshold sources without introducing significant numbers of spurious detections. This paper describes the theoretical and practical aspects of the biased search technique as applied to *EUVE* all-sky survey skymaps.

1. Introduction

Consider the problem of producing a catalog of point sources from a data set dominated by background noise where many sources will have low signal-to-noise ratios. The goal is to include as many sources as possible without introducing a large number of spurious detections. It may be the case that a threshold strict enough to reduce the spurious detections to an acceptable number may exclude large numbers of faint, yet interesting sources. This loss is the unfortunate price one must pay to produce an unbiased catalog.

In some situations, however, certain types of bias may be acceptable. For example, the *Extreme Ultraviolet Explorer (EUVE)* has conducted an all-sky survey and is now being used to obtain deep, pointed exposures of interesting targets. A guest observer interested in a specific, perhaps rare, class of objects may wish to use the survey data to determine which objects of that class might be good candidates for pointed observations, even if the potential targets were too faint to be included in an unbiased survey catalog. The prior information that an object of the correct type is known to exist near the position of a marginal detection can increase our confidence that the detection is not spurious, and that scheduling an observation of that target will not be a waste of precious instrument time.

2. Unbiased Approach with Uniform Significance Threshold

In the unbiased approach, we apply the detection algorithm of choice to compute the significance at each point on the sky. Every significance value corresponds to a probability that the detection is a false alarm caused by random background variations. (The significance is usually expressed as a χ^2 score or number of standard deviations, but for this purpose it is more convenient to work with raw probabilities.) A uniform threshold is applied to the significance list to determine which detections are to be included in the catalog.

The number of spurious detections in the catalog will be a random variable, approximating a Poisson distribution with expectation pN_{eff}, where p is the threshold false alarm probability, and N_{eff} is the effective number of independent trials. The value of N_{eff} will depend strongly on the size and shape of the instrument point-spread function (PSF), the pixel size (for binned data), and the amount of sky covered by the survey. A crude estimate of N_{eff} is given by

$$N_{\text{eff}} = \frac{A_{\text{sky}}}{A_{\text{psf}}}, \qquad (1)$$

where A_{sky} is the sky area surveyed, and A_{psf} is some measure of the PSF area; but this estimate is highly dependent on the shape of the PSF. (Consider two points separated by less than one PSF diameter; their significance will be somewhat correlated because of overlapping PSFs, but the amount of correlation will depend on how peaked the PSF is.)

It may be easier to estimate N_{eff} empirically via Monte Carlo methods, e.g., generating a random, background-only data set and applying the detection algorithm to assess the false alarm rate (Lewis 1993). Simulation results indicate that N_{eff} is approximately 10^8 for the shortest wavelength *EUVE* survey coverage and PSF. Regardless of the PSF shape, N_{eff} will generally be proportional to the area of sky surveyed.

3. Biased Catalog Search

The disadvantage of the unbiased approach is that for large sky coverage and small PSF area, one must use a rather strict detection threshold to prevent catalog contamination from excessive numbers of spurious detections. For the first *EUVE* catalog (Bowyer et al. 1994), the detection thresholds were in the neighborhood from approximately 5.5 σ to 6 σ, which excluded many interesting sources.

Suppose the source search were restricted to those areas immediately surrounding a small (relative to N_{eff} for an all-sky unbiased survey) set of objects that we expect, a priori, to detect in the all-sky data. If the search radius around each catalog location is on the order of one PSF radius, the effective number of trials will be close to the size of the input catalog (assuming the points are well separated). This constraint can reduce N_{eff} by several orders of magnitude, allowing a corresponding relaxation in the threshold probability to achieve the same expected number of spurious detections. By using an input catalog of a few thousand objects, detection thresholds from approximately 3 σ to 4 σ become feasible, allowing a substantial increase in the number of objects detected without a severe penalty in spurious detections.

4. A Hybrid Method: Multiple-Threshold, Partially Biased Search

The biased catalog search suffers from the obvious problem of inheriting all biases present in the input catalog and will never result in unexpected detections (which are, in a sense, the most interesting kind). We can combine the best features of both approaches by using the following hybrid approach.

As in the unbiased case, we apply the detection algorithm to the all-sky data set and apply a strict, uniform significance threshold, T_1. Instead of immediately discarding detections failing the significance test, we apply a second, more liberal threshold, T_2, to the leftover detections. Any of these marginal detections corresponding to previously cataloged objects are added to the final catalog.

The existence of a cataloged object near a marginal detection is prior information that effectively increases our confidence that the detection is not spurious. The significance boost can be expressed in terms of the input catalog size and the positional tolerance in the matching process. We assume that the input catalog sources are in an approximately uniform distribution over the entire sky, and that they are almost always separated by at least one search radius. If $N_{\rm cat}$ is the size of the input catalog, and $A_{\rm search}$ is the area within one search radius of a detection, the probability q that a random point on the sky will be within one search radius of a cataloged object is given by

$$q = \frac{N_{\rm cat} A_{\rm search}}{A_{\rm sky}}. \qquad (2)$$

We presume the existence of a detection at a given point with false alarm probability p, and the existence of a cataloged object near that point with coincidence probability q, are independent events. Therefore the joint false alarm probability p' is simply

$$p' = pq. \qquad (3)$$

Since $p' < p$, we have lowered the false alarm probability by finding a nearby cataloged object. It is obviously advantageous to have q as small as possible. Any objects in the input catalog unlikely to be detected in the survey data should be pruned to reduce the number of potential coincidences. For example, we used several on-line catalogs such as SIMBAD and NED in an attempt to identify newly detected *EUVE* sources. Many of our on-line catalog "hits" turned out to be *IRAS* sources and faint galaxies in directions of high hydrogen column density (Bowyer et al. 1994) and, therefore, highly unlikely to be detected in the extreme ultraviolet (EUV) bandpasses. After pruning these implausible objects, we observed a coincidence rate q of about 0.03 in a sample of 100 random points using a search radius of 3 arcmin.

The expected number of spurious detections in the biased component of the hybrid catalog is qM, where M is the count of marginal detections between the two thresholds T_1 and T_2.

5. World Wide Web Resource: EUVE Survey Skymap Source Detection and Flux Service

Researchers interested in applying these concepts to *EUVE* all-sky survey data are invited to use CEA's in-house software via our on-line source detection server.[1] This service allows the user to supply a list of coordinates and receive a list of detection significance, flux, best-fit position, and other relevant data by e-mail, usually within a few hours of submitting the request. A skymap image server[2] is also available to allow users to obtain images of skymap regions of interest. A great deal of general *EUVE* sky survey documentation[3] is also available to assist users in interpreting the results.

6. Systematic Errors and Other Caveats

When dealing with marginal detections, it is important to keep in mind that the discussion in this paper only addresses spurious detections arising from random background fluctuations. A possibility always exists that at very low significance thresholds, any detection algorithm may respond more to deviations from the underlying background model than to the putative source itself. We have found that analysis of large ensembles of randomly placed test points is a useful tool to assess the presence and severity of systematic deviations from any claimed statistical properties of the significance reported by the detection software. In some cases, it may be advisable to use perturbed versions of the input catalog (e.g., adding 1° of ecliptic latitude to each object's coordinates) if one suspects spurious detections are correlated with known problematic skymap features. Finally, it is always a good idea to visually inspect the skymap at each claimed detection to rule out significance errors from diffuse skymap features, exposure edges, or strong background gradients.

Acknowledgments. We thank the principal investigator, Stuart Bowyer, and the *EUVE* science team for their advice and support. This research has been supported by NASA contract NAS5-30180.

References

Bowyer, S., Lieu, R., Lampton, M., Lewis, J., Wu, X., Drake, J. J., & Malina, R. F. 1994, ApJS, 93, 569

Lewis, J. 1993, Journal of the British Interplanetary Society, 46, 346

---

[1] http://www.cea.berkeley.edu/Archive/Survey/fluxform.html

[2] http://www.cea.berkeley.edu/Archive/Survey/mapform.html

[3] http://www.cea.berkeley.edu/Archive/Survey/Survey.html

Bias-Free Parameter Estimation with Few Counts, by Iterative Chi-Squared Minimization

K. Kearns, F. Primini, and D. Alexander

Smithsonian Astrophysical Observatory, 60 Garden St., Cambridge, MA 02138

Abstract. We present a modified χ^2 fitting technique, useful for fitting models to binned data with few counts per bin. We demonstrate through numerical simulations that model parameters estimated with our technique are essentially bias-free, even when the average number of counts per bin is ~1. This is in contrast to the results from traditional χ^2 techniques, which exhibit significant biases in such cases (see, for example, Nousek & Shue 1989; Cash 1979). Moreover, our technique can explicitly handle bins with 0 counts, obviating the need to ignore such bins or rebin the data. We conclude with a discussion of the problem of estimating goodness-of-fit in the limit of few counts using our modified χ^2 statistic.

1. Introduction

When fitting models to data with few counts, two of the most common methods used are the standard χ^2 method and the C statistic. Use of the χ^2 method requires that one avoid bins with 0 counts by either ignoring them or rebinning, and produces significantly biased results for data with few counts. The C-statistic gives unbiased results but is difficult to interpret in terms of goodness-of-fit. Neither approach is ideal, though each is useful in some cases. The Iterative Weighting Technique which we investigate here both addresses the deficiencies inherent in using the standard method for data with few counts, and provides a goodness-of-fit parameter which is indistinguishable from the standard χ^2 parameter for many datasets.

2. Iterative Weighting

Iterative Weighting (IW) is an example of the class of weighted least-squares estimators described by Wheaton et al. (1994), in which χ^2 is expressed as a weighted sum of squared deviations,

$$\chi^2 = \sum_i W_i [O_i - M_i(p_1, p_2 ...)]^2,$$

where O_i are the observed counts in bin i, $M_i(p_1, p_2 ...)$ are the counts predicted by the model M with parameters $(p_1, p_2 ...)$, and the weights W_i are the inverses of the true variances σ_i^2. As Wheaton et al. (1994) point out, the approximation $W_i \simeq O_i^{-1}$ leads to significant biases in the best-fit parameters, due to the strong anti-correlation between W_i and $[O_i - M_i]^2$. Similar biases are encountered if the approximation $W_i \simeq M_i^{-1}$ is used (Nousek & Shue 1989). The IW technique avoids such biases by estimating W_i through successive iterations, where for each iteration, j, $W_i^j \simeq [M_i(p_1^{*,\ j-1}, p_2^{*,\ j-1}...)]^{-1}$, and the best-fit parameters $p_1^{*,\ j}, p_2^{*,\ j}...$ are determined by minimization of

$$\chi^{2,\ j} = \sum_i \frac{[O_i - M_i(p_1^j, p_2^j...)]^2}{M_i(p_1^{*,\ j-1}, p_2^{*,\ j-1}...)}.$$

For the first iteration, all weights are set to 1. In our sample, we find that the minimum χ^2 values and best-fit parameters converge after about 6 iterations.

3. Data Simulation

To demonstrate the IW technique, we repeat the simple numerical experiment of Nousek & Shue (1989). For a range of total counts, N, from 25 to 1000, we generate an ideal power-law spectrum such that:

$$\overline{n_i} = N_o \int_{E_1+(i-1)\Delta E}^{E_1+i\Delta E} E^{-\gamma} dE, \quad N = N_o \int_{0.095}^{0.845} E^{-\gamma} dE$$

for $i = 1 \rightarrow 15$, $\Delta E = (0.845 - 0.095)/15$, and $\gamma = 2.0$ For each ideal spectrum, we simulate 1000 sample spectra $\{n_i\}$, where $\{n_i\}$ are random deviates drawn from Poisson distributions with means $= \overline{n_i}$. We then determine best-fit model parameters N_o^{calc} and γ^{calc} for each simulated spectrum, using IW and Powell's method for function minimization (Press 1988). For each N, we then compute the average N_o^{calc}/N_o and γ^{calc}/γ; compile the distributions of minimum χ^2 for comparison with the theoretical distribution; and compute the percentage of simulations for which the χ^2_{min} and $\Delta\chi^2$ contours include N_o and γ, for comparison with the expected percentages.

4. Results

In Table 1 we compare the biases (as measured by the ratios of average best-fit parameter values to true values) in 1000 IW runs with those found for traditional χ^2 and the C statistic by Nousek & Shue in 250 runs. We find that the IW biases are comparable to those encountered using the C statistic for all N. These results are displayed in Figure 1. In Figure 2 we compare both differential and cumulative theoretical χ^2 distributions with our observed distributions. We apply a KS-test to the cumulative distributions and find that at N=25 the match is poor, but by N=100 the two distributions are in good agreement.

The percentage of simulations for which the $\chi^2_{min} + \Delta\chi^2$ contours include N_o and γ, for $\Delta\chi^2$ values appropriate to various joint two-parameter confidence levels, is shown in table 2. For most N, the measured and expected confidence levels are in good agreement.

N	Traditional χ^2			Iterative Weighting			C-Statistic		
	N_{calc}/N_o	γ_{calc}/γ	%cnvg	N_{calc}/N_o	γ_{calc}/γ	%cnvg	N_{calc}/N_o	γ_{calc}/γ	%cnvg
25	0.709	1.152	96	1.145	1.003	98	1.269	0.958	86
50	0.647	1.134	100	1.055	1.008	99.6	1.079	0.998	100
75	0.636	1.130	100	1.025	1.009	100	1.078	0.995	100
100	0.673	1.109	100	1.008	1.008	100	1.053	0.996	100
150	0.707	1.094	100	1.025	1.001	100	1.015	1.005	100
250	0.767	1.072	100	1.019	1.000	100	1.019	1.000	100
500	0.863	1.040	100	1.007	1.000	100	0.997	1.004	100
1000	0.937	1.017	100	1.005	1.000	100	1.001	0.999	100

Table 1. Comparison of three fitting techniques.

	Minimum χ^2 plus ... (for 2 parameters of interest)					
N	2.30 (68.3%)	4.61 (90%)	6.17 (95.4%)	9.21 (99%)	11.80 (99.73%)	18.40 (99.99%)
25	69.8	87.0	92.1	96.8	98.4	99.9
50	68.9	88.1	93.7	97.2	98.2	98.3
75	67.7	87.8	93.5	98.4	99.3	99.8
100	68.1	89.1	94.1	98.2	99.0	99.8
150	67.0	87.1	93.6	97.7	99.3	100
250	68.0	90.3	95.6	99.2	99.8	100
500	69.1	90.1	95.5	98.9	99.6	99.9
1000	69.6	88.9	95.0	99.0	99.7	100

Table 2. Estimating Confidence Limits: Percentage of Best Fits Within Various χ^2 Boundaries, from a total of 1000 spectral fits.

5. Conclusions

We find that unbiased parameter estimates by χ^2 minimization are possible for binned data with few or no counts in some bins, provided the χ^2 calculation is modified slightly. Except for very small N, this modified χ^2 statistic is distributed according to the theoretical χ^2 distribution. Goodness-of-fit can therefore be assessed using traditional techniques. Further, this χ^2 statistic can be used to estimate confidence levels from standard $\chi^2_{min} + \Delta\chi^2$ boundaries.

Acknowledgments. This work is partially supported by NASA contract NAS5-30934.

References

Cash, W. 1979 ApJ, 228, 939

Nousek, J. A., & Shue, D. R. 1989, ApJ, 342, 1207

Press, W. H., Teukolsky, S. A., Vetterling, W. T., & Flannery, B. P. 1986, Numerical Recipes (New York, Cambridge University Press)

Wheaton, W. A. et al. 1995, ApJ, submitted

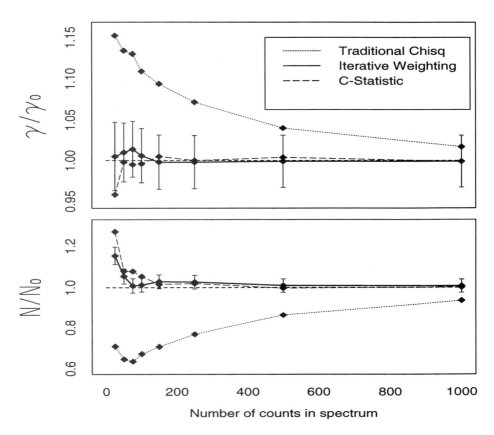

Figure 1. Bias in best-fit parameters for three fitting techniques.

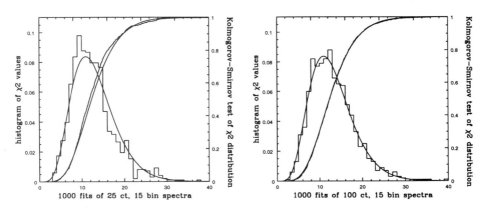

Figure 2. Comparison of theoretical χ^2 distribution with observed distribution for IW by KS-test, with overlaid histograms.

A Method for Minimizing Background Offset Errors When Creating Mosaics

M. W. Regan, R. A. Gruendl

Department of Astronomy, University of Maryland, College Park, MD 20742

Abstract. When creating a mosaic from a set of individual frames, the correct background offset for each frame must be determined for proper matching. In this paper we describe a method for constructing a mosaic from a set of individual frames by applying a χ^2 minimization technique. This technique uses the offset between all overlapping pairs of frames to determine the best background offset for each frame. This technique was developed for use with infrared arrays; it has been used with the existing Simultaneous Quad Infrared Imaging Device (SQIID) software within IRAF, but applies to any image that is created from several individual frames. We also describe several techniques that allow objective criteria to be used to identify individual frames to be excluded from a mosaic. We present examples of our technique using two computer programs and compare the results with simpler methods. This technique is extremely powerful for examining low surface brightness objects.

1. Introduction

Due to the high background in the near-infrared, it is necessary to split an observation into several small individual integrations. These individual frames must then be reformed into a composite image. Since the sky background can change on minute time-scales, the individual frames will have different background intensities. Usually a sky frame is formed and subtracted from all frames, but since the intensity of the sky can change between the object frame and the sky frame, each frame has a different zero-point.

The normal method used to create a mosaic is to determine the background offset in a region of overlap between two frames, and bootstrap, adding one frame at a time. The problem with this method is that an error in one of the relative offset determinations can propagate, leading to background intensities that do not match in regions where one frame overlaps with more than one other frame. A better way to match the backgrounds between individual frames is to use all the information that exists in the data. In general, each frame overlaps with several other frames, and a relative offset between each pair of overlapping frames can be determined.

In this paper we discuss a method for determining the correct background offset for each frame by generating a least-squares solution that minimizes the errors in the offsets for all frames.

2. Least Squares Solution

If i and j are the indices of two frames that overlap, then the model equation is

$$D(i,j) = O(i) - O(j), \tag{1}$$

where $D(i,j)$ is the measured difference in background intensities and $O(i)$ and $O(j)$ are the true background offsets of frames i and j, respectively. We wish to minimize the deviation between the observed difference and that expected. The quantity to be minimized in the least-squares solution is

$$\alpha = \sum_{i=1}^{m} \sum_{j=i+1}^{m} [D(i,j) - O(i) + O(j)]^2 w(i,j), \tag{2}$$

where $w(i,j)$ is set to one if frames i and j overlap, and is set to zero otherwise. Assuming that every frame overlaps with at least one other frame, there are m unknowns. The $m-1$ normal equations for the m parameters can be written as

$$\frac{\partial \alpha}{\partial O(i)} = \sum_{j=i+1}^{m} D(i,j)w(i,j) - O(i)w(i,j) + O(j)w(i,j). \tag{3}$$

Since there are only $m-1$ equations for the m unknowns, one of the offsets must be fixed before the equations can be solved. By setting $O(1)$ to zero, the set of equations becomes solvable. The final set of offsets has an arbitrary zero-point, which must be determined by other means. The best way to determine the correct zero-point is to include a set of overlapping frames that extend far enough from an object of interest to be sure that there is no low surface brightness emission. This region will determine the overall zero-point.

3. Implementation

The algorithm is currently implemented in two computer programs: (1) a FORTRAN routine that reads all the individual frames into memory, determines the background offset difference between each pair of overlapping frames, and writes this to a file, and (2) a C routine that reads the file containing the background offset differences and performs the χ^2 minimization. This routine also reads in a file that lists the frames to be excluded from the fit. Typically these are frames with some known defect (see §4). This program creates output that can be used by the SQIID routine **nircombine** to create a mosaic.

4. Recognition of Bad Frames

Determining which frames to exclude from the final mosaic is critical, since one bad frame can create background offset errors that propagate throughout the mosaic. The most common method—trial and error—is very subjective and can take a lot of time. A side benefit of our technique is that we determine the residual for each of the frames, which can be used as an objective basis for rejecting frames whose residual is greater than 3σ above the average residual.

Minimizing Background Offset Errors When Creating Mosaics

Frame	Initial offset	Initial uncer.		Final offset	Final uncer.
1	0.0	0.67		0.0	0.97
2	30.9	6.09	poor	32.2	0.41
3	-9.1	0.51		-8.8	0.19
4	20.5	0.97		20.0	0.61
5	49.3	2.21	poor	48.2	0.20
6	-45.6	0.58		-44.9	0.09
7	**-681.0**	**13.63**	**BAD**		
8	-29.8	4.03	poor	-28.0	0.19
9	17.2	0.79		17.1	0.46
10	52.1	1.05		52.4	0.62
11	-23.7	1.05		-23.7	0.59
13	-4.8	0.32		-5.4	0.14
14	-25.9	8.48	poor	-28.9	0.19
15	8.9	4.16	poor	8.9	0.14
16	4.6	0.61		5.7	0.12
17	-58.2	1.05		-57.6	0.70
18	-6.1	0.92		-7.7	0.11
19	-52.4	2.42	poor	-54.7	0.08
20	-0.4	3.05	poor	-2.9	0.01
21	-9.4	0.17		-12.0	0.02
22	-35.7	0.91		-36.9	0.49
23	4.6	0.83		3.6	0.28
Average		2.39			0.33

Table 1. Resultant Fit and Subsequent Improvement of Offsets

We used an iterative process to remove bad frames. On each iteration the frame with the highest residual above 3σ is removed and a new solution is calculated. This process is repeated until no frames have residuals above 3σ.

While it is clear that this process is applicable when sky emission variations occur, as in the near-infrared, a similar technique can be applied to determine and correct for the effects of variable atmospheric extinction. Variable extinction can have two causes: a poorly determined airmass correction, or the presence of high clouds. We have developed a similar set of routines that perform a χ^2 minimization of the gain variation between frames. The goal of this routine is to look for frames that show a very low gain relative to the other frames in the mosaic. If there is enough redundant coverage, the most conservative thing to do is to remove those frames that show excessive extinction. If there is not enough redundant coverage, then one corrects the frames by multiplying them by the inverse of the amount of extinction.

5. An Example

As an example of how to use this χ^2 minimization technique, and how to reject bad frames, Figure 1a shows a SQIID image of IC 348 made with the standard bootstrap approach. The background offsets mismatch at several places in the figure. The χ^2 minimization technique produces the initial solution shown in Table 1. It is clear from Table 1 that frame 7 is defective; we removed

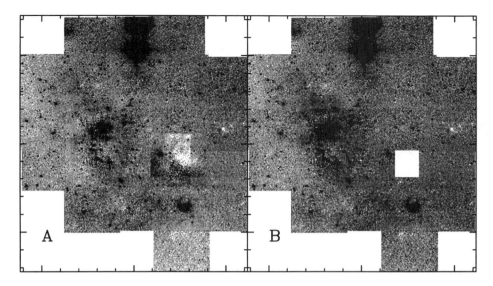

Figure 1. H-band mosaic of the cluster IC 348 taken with SQIID on the Kitt Peak 1.3 m telescope. Frame A (*left*) shows a mosaic made with the bootstrap method of 23 frames laid out in a grid. There are no two frames that share more than $\sim 1.''5$ (75 pixels) overlap. It is apparent that errors in the calculated background offset cause significant mismatches from frame to frame. Frame B (*right*) shows the same mosaic with the χ^2 minimization run and the rejection criteria applied. Low surface brightness emission features can be identified with some degree of confidence now that the variations have been reduced.

it from the mosaic and re-determined the background offsets. These results are shown in the last two columns of Table 1. It is apparent that the residuals for all the frames with poor fits have been reduced. The final mosaic (Figure 1b) is much improved and low-level emission features near the zero-point level are now visible and believable.

A copy of the source code that implements this algorithm is available by request from the authors.

Acknowledgments. We would like to thank Elizabeth Lada for letting us use her images of IC 348. We would also like to thank Stuart Vogel for his comments and direction throughout this work.

Part 9. Simulation

CCDs at ESO: A Systematic Testing Program

T. M. C. Abbott

European Southern Observatory, Casilla 19001, Santiago 19, Chile

R. H. Warmels

European Southern Observatory, Karl-Schwarzschild-Straße 2, D 85748, Germany

Abstract. ESO currently offers a stable of 12 CCDs for use by visiting astronomers. It is incumbent upon ESO to ensure that these devices perform according to their advertised specifications (Abbott 1994). We describe a systematic, regular testing program for CCDs which is now being applied at La Silla. These tests are designed to expose failures which may not have catastrophic effects but which may compromise observations. The results of these tests are stored in an archive, accessible to visiting astronomers, and will be subject to trend analysis. The test are integrated in the CCD reduction package of the Munich Image Data Analysis System (ESO-MIDAS).

1. Introduction

At the time of writing we at ESO, La Silla offer 12 CCDs for use by visiting astronomers. These CCDs range in quality from a venerable RCA with read noise of 32 electrons per pixel to the most recent, a thinned Tektronix 2048^2 pixel device. Supporting all of these CCDs poses some unusual problems. ESO serves a very broad community, and the astronomers who use our CCDs range in ability from those who are quite new to the field to those with many years of experience in the use of modern, state-of-the-art detectors. We must be aware at all times of the current status of our CCDs so that even the most exacting visiting astronomers can be satisfied that their data is of uniformly high quality and that they are completely informed of any problems or limitations. Likewise, we must work to protect the less experienced astronomers by convincing ourselves that our CCDs are providing data of sufficient quality to ensure the success of a broad spectrum of observing programs. It is, therefore, not sufficient that we trust our CCDs to remain in the state determined when they are commissioned, nor that we depend on visiting astronomers to identify problems as they arise. Instead, we must make a concerted effort to regularly investigate the quality of the data delivered, whether or not any problems are known. To that end, we have instigated a systematic program of standard CCD tests at ESO, La Silla.

2. The Test Data

We currently test one CCD each week, and thus each CCD is tested every 3 months. These tests are not intended to be as thorough as might be performed in a specialized CCD lab; instead, they should expose as many problems as possible with minimal technical intervention and under the simple setups available with the CCD on the telescope. We can then investigate any problems that we identify with more sophisticated methods, or, if we judge the CCD to be functioning satisfactorily, the test results provide a baseline of its performance.

For each test, we collect the following data: (1) nine bias frames, (2) sixteen pairs of flat fields (both of each pair have the same integration time) using a stable light source and with exposure levels ranging from just above bias to digital saturation, (3) nine low-count-level (of order a few hundred electrons per pixel) flat-fields with stable light source, (4) one flat-field exposure obtained with 64 rapid shutter cycles, (5) three 30-minute dark images, and (6) the time taken to read out and display an image. All images include bias overscan regions in both dimensions, cover the entire light-sensitive, unbinned area of the CCD and are collected under the same circumstances as normal observing.

The light source used to obtain the flat fields may be either an LED or a beta light. Beta lights consist of a fluorescent screen stimulated by β decay from a small bulb of tritium. Since these present a possible radiation hazard and are prone to variation with temperature ($\sim -0.3\%$ per °C (Florentin, private communication)) we are in the process of replacing them with compact light sources consisting of a battery-powered LED regulated by feedback from a photodiode. Like the beta lights, these are small enough to fit within a normal filter wheel in most La Silla instruments and exhibit a flux/temperature dependence of $\sim 0.2\%$ per °C (we expect to improve on this in future versions).

3. The Results

The information we expect to obtain from each test data set is as follows:

1. A 16-point transfer curve (Janesick et al. 1987, Figure 1a) generated for any window onto the images obtained.

2. Two 16-point linearity curves. We find that the linearity curves are most useful when expressed as count rate versus true exposure time (Figure 1b). We determine the mechanical shutter delay either by linear extrapolation of the normal linearity curve (observed counts versus exposure time), thus assuming the response of the CCD is linear, or by adjusting the exposure times such that the count rate curve is closest to a straight line, thus allowing for a first-order nonlinearity in the response of the CCD. We obtain the 16 pairs of frames in two groups of eight—the first with increasing exposure times and the second with decreasing exposure times, interleaved with those of the first group. In this way, we can reject trends observed in the CCD response that are probably caused by the effect of temperature variations on the light source. The linearity curves may be generated for any window onto the images obtained.

3. A map of hot pixels in bias frames (obtained from a median stack of the nine raw bias frames).

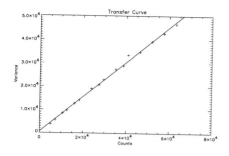 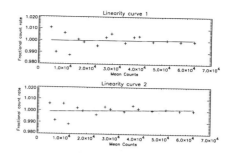

Figure 1. **a** (left): Sample transfer curve (TK#36). The abscissa is the mean counts per pixel in a 200^2 pixel region centered on the CCD. The ordinate is the variance of the same region in the image that results from the difference of two images of the same mean counts. **b** (right): Sample linearity curves expressed as count rate versus mean counts in an image using the same light source throughout (TK#36). The straight lines are linear fit to the data. The exposure times have been corrected for a shutter delay of 1.4 seconds.

4. A map of traps and other defects (obtained from a median stack of the nine low count level frames).

5. An estimate of bulk CTE in the horizontal and vertical directions (by the EPER method (Janesick et al. 1987)).

6. The amplitudes and frequencies of interference signals (from a Fourier analysis of raw bias frames).

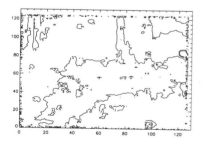 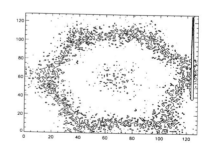

Figure 2. **a** (left): Sample dark current map (TK#36). Contours are labeled in electrons/pixel/hour. **b** (right): Sample shutter delay map (TK#25). The contours are at 0.016 seconds and 0.024 seconds. Note the hexagonal shape caused by the iris shutter.

7. The current values of all bias and clock voltages (normally measured by the CCD controller and recorded in image headers).

8. A map of dark current across the CCD (Figure 2a).

9. A map of the shutter pattern on the CCD (e.g., a star-shaped pattern in the case of an iris shutter (Figure 2b), obtained by analysis of the image made with 64 shutter cycles).

4. Implementation and Documentation

The test program has been integrated in the CCD reduction package of the Munich Data Analysis System (ESO-MIDAS) (ESO 1993), and will be available in the 94NOV release. Therefore, in addition to the already available pipe-line and interactive processing tools for CCD direct imaging data, the CCD package will also offer standard tools for testing the detector quality at ESO and at other institutes.

We issue a full report on the condition of a CCD each time a new CCD test data set is collected and reduced. We are in the process of developing an on-line test data archive to store the raw and reduced test data (ESO 1994) and a World Wide Web interface for browsing these data. These software systems are accessible via the Internet at the ESO Home Page[1]. We use the data obtained for normally functioning CCDs to define baselines for their performances. Trends in these data expose possible slowly developing problems and thus allow realistic preventive maintenance, reducing the probability of catastrophic failures. The most recent test data set combined with the history of a CCD's behavior provides the astronomer with an indication of the current performance and reliability of the device.

Acknowledgments. S. Deiries of ESO, Garching designed and built the stable LED light source. We are grateful to the ESO, La Silla Astronomy Department and CCD group for their cooperation in collecting the test data necessary for the success of this project.

References

Abbott, T. M. C. 1994, ESO CCD Catalogue

Janesick, J. R., Elliot, T, Collins, S., Blouke, M. M., & Freeman, J. 1987, Optical Engineering, 26, 69

ESO 1993, Document MID-MAN-ESO-11000-0002/0003/0004, ESO-MIDAS User Manual, Volumes A, B, and C (Garching, ESO)

ESO 1994, Document OSDH-SPEC-ESO-00000-0002/2.0, EMMI/SUSI Calibration Plan for an On-Line Calibration Database (Garching, ESO)

---

[1] http://www.hq.eso.org/eso-homepage.html

Modeling Scattered Light in the HST Faint Object Spectrograph

H. Bushouse

Space Telescope Science Institute, 3700 San Martin Drive, Baltimore, MD 21218

M. Rosa and Th. Mueller

Space Telescope–European Coordinating Facility, European Southern Observatory, Karl-Schwarzschild-Str. 2, D-85748 Garching, Germany

Abstract. We describe the software tool **bspec** which models the dispersion and diffraction of light in the *HST* Faint Object Spectrograph. The **bspec** program is available in both the MIDAS and IRAF environments.

1. Introduction

The *HST* Faint Object Spectrograph (FOS) uses blazed, ruled gratings and detectors that are sensitive over wide wavelength ranges. Therefore the FOS is subject to scattered light which has its origin in the diffraction patterns of the gratings and the entrance apertures, as well as the micro-roughness of the ruled gratings. This becomes a significant problem when red stars are observed at short wavelengths where the spectrum is often dominated by scattered red photons.

The analysis of laboratory and in-flight FOS data indicates that the instrument is very close to the performance anticipated from ideal optical surfaces. Therefore, the contamination of observations by scattered light can be predicted with reasonable accuracy.

The program **bspec** has been developed to model the dispersion and diffraction of light in the FOS with an accuracy sufficient for the estimation of scattered light contamination in observed FOS data. The program was developed and is maintained in the MIDAS environment and it was recently ported to the IRAF environment. **bspec** takes an input spectral distribution and disperses it into the most significant spectral orders—using the equations of blazed gratings and the grating parameters for the FOS—and convolves this multi-order spectrum with a model of the instrumental line spread function (LSF). It is light from the wing of order zero that constitutes a significant portion of the scattered light level seen in the blue wavelength end of first-order FOS spectra.

The spectra produced by **bspec**, which includes the intrinsic spectrum of the source as well as the predicted intrinsic plus scattered light spectrum, can be compared with FOS observations in order to determine the relative amount of scattered light contained in observed spectra.

2. The Details

The **bspec** program takes as input a spectral distribution of count rates which is dispersed into the most significant spectral orders (e.g., -5 to $+5$) using the equations of blazed gratings and the known parameters of the FOS gratings. Light from the wing of order zero constitutes a significant component of the scattered light in the blue part of the first order spectra imaged onto the detector. The amount of zero order light is determined in **bspec** as the residual flux not being distributed into higher orders. By comparison with more rigorous models, we find it sufficient to include orders up through 5. Even for stellar spectra as late as M5, the fractional improvement of including higher orders is below 10^{-4}.

The spectral shape in each order is the product of the input spectrum and the blaze function for a given order. The resultant multi-order spectrum is convolved with a model of the line spread function (LSF) which represents the effects of diffraction at the entrance aperture, the collimator, the grating and the detector faceplate, and includes a flat component to simulate micro-roughness and dust particle scatter.

In order to ensure that all significant light is collected and redistributed by both the grating equations and the convolution with the LSF, the computation is performed over a range of diffracted angles much larger than that seen by the actual detector. The red and blue FOS detectors both cover the range -1.47 to $+1.47$ degrees from the grating normal. The computations in **bspec** typically cover the range -10 to $+35$ degrees in diffracted angle, which includes all orders from zero to five.

3. Using BSPEC

In practice, correcting observational data for scattered light must be done before the count rate spectra are transformed to absolute fluxes. The **bspec** program is therefore intended to produce output data that are in units of counts per pixel, which then requires that the input data also be in this form. Calibrated spectra must be prepared for use in **bspec** by scaling by the sensitivity and transmission of all the *HST* and FOS optical components, except for the grating blaze function which is computed and applied within **bspec**. In particular the conversion from flux per unit wavelength to counts per pixel has to be made before using in **bspec**.

To make this job easier a program called **countspec** has been developed in IRAF/STSDAS which will convert a flux calibrated spectrum of any object into a count rate spectrum. The **countspec** program uses the known throughputs and sensitivities of the *HST* and FOS optical components—as contained in the *HST* Calibration Data Base System (CDBS)—to perform this conversion.

The **bspec** program, in both the MIDAS and IRAF environments, uses tables for all input and output data. The **countspec** program (only available in IRAF/STSDAS) also uses tables. The input and output tables for **countspec** and the input table to **bspec** are simple two-column tables of wavelengths and fluxes (or counts). The output table produced by **bspec** contains several columns of data including wavelengths and dispersed counts, both with and without scattered light, and the grating blaze function.

4. Example Calculations

For illustration let the target spectrum be the model atmosphere for the the solar-like star 16 Cyg B, which we wish to compare with an observed spectrum of this star obtained using the FOS blue detector, the G190H grating, and the 1″.0 round aperture.

First, we prepare the model spectrum—which is in absolute flux units—using the **countspec** task to convert the model data to a count rate spectrum. The calibrated model data are contained in table cyg16b.tab and the count rate spectrum will be written to table counts.tab. We run **countspec** as follows:

cl> countspec cyg16b fos,1.0,g190h,blue counts.tab

The second argument, "fos,1.0,g190h,blue", specifies the desired instrument observing mode.

Second, we run **bspec**, using the table counts.tab as input, to compute a predicted scattered light spectrum. The output from **bspec** will be written to table scatter.tab. It is only necessary to specify the input and output table names and the detector and grating names. Appropriate default values for the grating and LSF parameters will be chosen based on the selected detector/grating combination. We run **bspec** as follows:

cl> bspec counts.tab scatter.tab blue g190h

5. Results

Figure 1 shows the results of **bspec** computations for the star 16 Cyg B as observed with the FOS blue detector and G190H grating. The lower curve in Figure 1 shows what would be an "ideal" spectrum as observed by an unphysical instrument that relates wavelengths one-to-one with diffracted angles. The upper curve (offset vertically by a factor of 10) shows the "model observations" computed by **bspec**, i.e., the intrinsic spectrum dispersed by the blazed grating and further convolved with the scattering imposed by the entrance aperture, the ruled surface of the grating and a minute amount of dust on the optical surfaces. Orders -5 to $+5$ were computed but only orders 0 to 2 are shown in the figure. Note that the real detector covers only the wavelength range indicated by the horizontal bar near 2000 Å.

The shape of the zero order peak in Figure 1 reflects the actual LSF. The far wings of this LSF carry light from the peak of the original spectral distribution into regions where the target spectrum, filtered by the total throughput of the optical elements and the detector efficiency, produces few intrinsic counts. In addition, this LSF moves photons from the zero order peak into the adjacent parts of the first order seen by the detector.

Figure 2 shows an actual observed FOS spectrum of 16 Cyg B overlaid with the "ideal" and the "model" observations from Figure 1. The observed spectrum is offset vertically for clarity. For a solar-like spectrum, the scattered light component ranges between 1–99% of the observed signal in the FOS blue G190H mode.

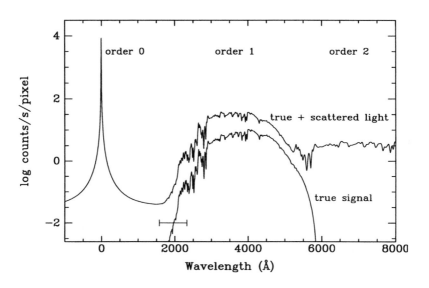

Figure 1. FOS blue G190H count rate spectra for a G5V model atmosphere.

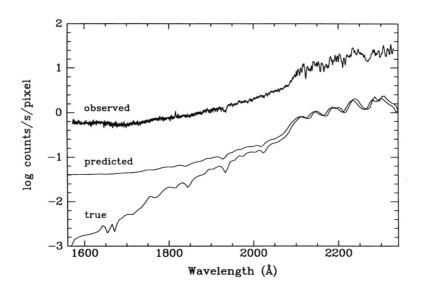

Figure 2. FOS blue G190H data for the G5V star 16 Cyg B.

Simulation of HST PSFs using Tiny Tim

J. Krist

Space Telescope Science Institute, 3700 San Martin Dr., Baltimore, MD 21218

Abstract. Tiny Tim has been used for generating *Hubble Space Telescope (HST)* point-spread functions (PSFs) for deconvolution, algorithm testing, proposal planning, and optical studies. The most recent version, V4.0, includes new mirror zonal error maps and revised aberrations, significantly improving the PSF models.

1. Introduction

Tiny Tim is a stand-alone program for simulating *HST* PSFs as viewed with the on-board imaging cameras (WF/PC-1, WFPC2, FOC, COSTAR/FOC). PSFs generated with **Tiny Tim** have been used extensively for deconvolution, algorithm testing, proposal planning, and optical studies. The most recent version, V4.0, is significantly better than earlier releases, producing more accurate PSFs due to better maps of the *HST* mirror zonal errors and improved aberration values. The software package is available via anonymous ftp at *ftp.stsci.edu* in the *software/tinytim* directory.

2. What is Included in Tiny Tim

The *HST* focus changes slowly over time due to shrinkage of the telescope truss from desorption. The secondary mirror has occasionally been moved away from the primary to compensate. The desorption changes have been characterized by measuring observed PSFs. **Tiny Tim** computes the focus for an aberrated camera with adjustments for desorption and mirror moves, based on a user-specified date of observation. Since desorption has nearly stopped, it is no longer included in the focus values for the corrected cameras (WFPC2 and COSTAR/FOC).

Each camera has its own set of aberrations (astigmatism, coma, spherical, etc.) which have been determined using phase retrieval methods. The measured aberrations for FOC f/96, WF/PC-1 PC6, WFPC2, and COSTAR/FOC are used by **Tiny Tim**. The PC6 values are used for the other WF/PC-1 channels, which may result in some PSF mis-matches. The measured focus offsets among the WFPC2 cameras are also included.

Circular zones in the *HST* mirrors resulting from the polishing process have important effects on the PSFs. In the aberrated PSFs, they affected the diffraction ring structures, and in the corrected ones, they cause scattering which results in a low level halo extending out to about $3''$. Maps of these errors were initially obtained from pre-launch interferograms by Perkin-Elmer and were used by **Tiny Tim** (up to V3.0). Recent maps obtained from phase retrieval of on-orbit WFPC2 data are much better and significantly improve the PSF models. These improvements were introduced in V3.0; they are illustrated in Figures 1 and 2.

The WF/PC-1 and WFPC2 instruments contain Cassegrain repeater optics with their own secondary mirrors and spiders. Since these obscurations are in the same plane as those from the telescope, they appear to shift with respect to the telescope's depending on field position. This leads to position-dependent PSFs. These effects were significant in WF/PC-1, but are less so in WFPC2. The shifts are determined by **Tiny Tim** based on user-specified object positions. Multiple positions can be simulated in a single run by providing a list of coordinates.

The *HST* PSF varies significantly with wavelength. In narrow-band filters, where the PSF does not change much over the bandpass, the diffraction rings are sharp. However, in wide-band filters the changes in the PSF result in blurring of the diffraction structures. These effects are accounted for in the software by adding together PSFs from different wavelengths with weights appropriate for a given filter. By default, **Tiny Tim** will create a PSF integrated onto detector-sized pixels. It can, however, creates PSFs at finer samplings. Such PSFs are useful in deconvolution of undersampled data (like WF/PC-1 or WFPC2) and in peak-centering analysis. Note that in WFPC2 there is a sub-pixel variation which is not included (except as a general pixel-level scattering function applied only to normally sampled PSFs).

3. What is Not Included in Tiny Tim

Despite including all of the above parameters, **Tiny Tim** does not account for all important factors. This is typically because they are not well characterized, or vary on short timescales.

The focus of the telescope is known to change slightly over the period of an orbit due to thermal effects ("breathing"). Also, it is known to change sometimes depending on where the telescope is pointed. While a model for breathing has been developed, it relies on thermal data for a given observation which is not readily available. The pointing-dependent effect, which is about three times worse than breathing, has not been characterized. Note that these effects are not limited to changes in focus but also in coma and astigmatism.

Tiny Tim does not include large angle scattering. The scattering at angles larger than about $3''$ from the core was below expected levels in WF/PC-1 (the limited FOC dynamic range does not allow such scatter to be measured in that camera). In WFPC2, however, it is about an order of magnitude greater and includes streaks which radiate from the star. This is due to scattering by the electrode structure of the front illuminated CCDs (WF/PC-1 was back illuminated). It is not possible to model this effect (which is typically seen only in highly saturated star images). This will lead to possibly significant errors at large angles from the core in the models.

The WF/PC-1, WFPC2, and COSTAR/FOC instruments have as-designed field dependent astigmatism. However, the changes over the field of view are practically unmeasurable, and thus are not included in the models. Geometric distortions in WF/PC-1 and WFPC2 have been measured but are not included.

Simulation of HST PSFs using Tiny Tim

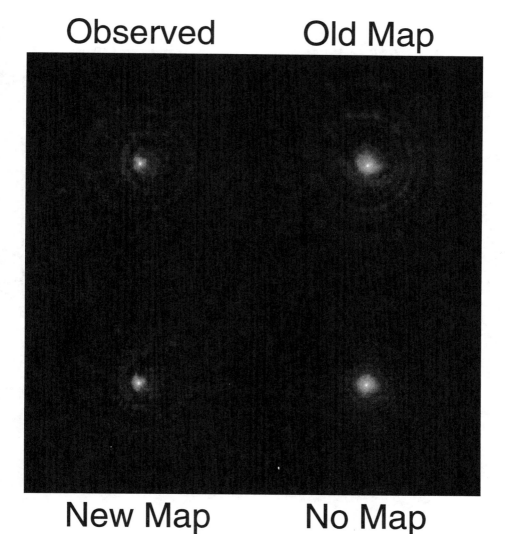

Figure 1. FOC f/96 observed and **Tiny Tim** model PSFs (pre-COSTAR) at 253 nm. The PSF generated using the old mirror map is typical of those produced by Tiny Tim previous to V3.0. The PSF using the new mirror map was created using V4.0b.

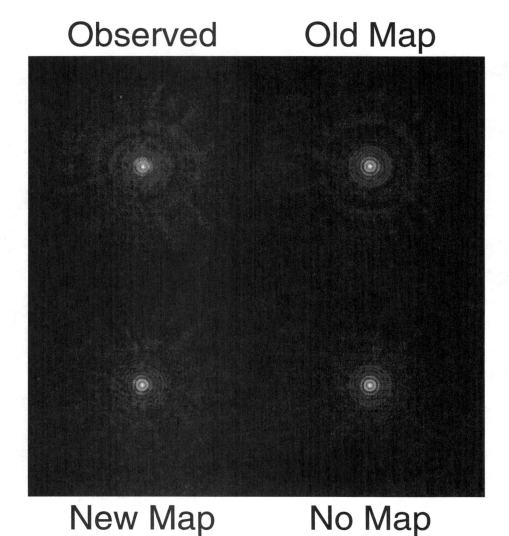

Figure 2. FOC f/96 observed and **Tiny Tim** model PSFs (pre-COSTAR) at 486 nm.

QPSIM: An IRAF/PROS Tool for Source Simulation

K. R. Manning, J. DePonte, and F. Primini

Smithsonian Astrophysical Observatory, 60 Garden St., Cambridge, MA 02138

Abstract. The generation of simulated sources whose physical properties are well understood is useful in assessing the functionality of point source analysis software. **Qpsim** is an IRAF/PROS task which will generate events from sources of given intensity, shape, and position and either overlay them on a flat background in a new QPOE file or inject them into an existing QPOE file. We present an overview of the capabilities of **qpsim** and some examples of its use.

1. Introduction

Qpsim is a task which will simulate X-ray events for a *ROSAT* High Resolution Imager (HRI) observation in QPOE format. The task may be used to generate a random flat background as well as sources of a specific intensity, shape and position. The simulated data can be a useful tool in calibrating point source analysis software.

2. Describing the Data

An STSDAS table file is used to describe the source and background data to be generated. Each row in the table represents information for a single source with the background treated as a special case. An example is shown in Table 1.

(row)	x pixels	y pixels	itype	intensity count s^{-1}|counts	prf_type	prf_param oaa|sigma
1	2000	4000	rate	0.50	gauss_oaa	18.00
2	5000	2500	counts	200.00	roshri	20.00
3	3000	3000	counts	150.00	gauss_sig	15.00
4	0	0	rate	5.00	bkgd	INDEF

Table 1. Example of a **qpsim** source table

The desired intensity may be specified as either "counts" or "rate." In the latter case, the user must provide a value for the livetime in order to calculate the source counts. The livetime value may be entered via a task parameter or extracted from a reference QPOE file.

The background data is designated by "bkgd" in the "prf_type" column. Two point-response functions (PRFs) are available for source data, either the *ROSAT* HRI PRF (described by David et al. 1993) or a Gaussian function. The *ROSAT* HRI PRF, designated "roshri", is described by an off-axis angle value. The Gaussian may be described by either a sigma value ("gauss_sig") or by an off-axis angle ("gauss_oaa") where the off-axis dependency of sigma is deduced from a 50% power radius assuming a Gaussian distribution (David et al. 1993, p. 17). Off-axis angle values are given in units of arc minutes, and sigma values in units of pixels.

Off-axis angle values are treated independently from source position, giving the user control over source shape. This allows the generation of identical sources at many positions in the field. Such fields can increase the efficiency of methods used to study simulated data and to calibrate source detection software. If an off-axis angle value of "INDEF" is given, the value will then be calculated from the source position.

All source positions are constrained to lie within the field size limits of an unrolled image. For the *ROSAT* HRI, this is taken to be the central region defined by the pixel range [2048:6144]. The exception to this is the background specification for which the table values of X, Y and "prf_param" are ignored.

3. Algorithm for Generating Events

Qpsim loops over the rows in the source table, generating the data for each source independently. The background "source" is treated as a special case, but the source data is generated with a general routine which invokes the appropriate PRF function. In this way, other PRF functions can be easily added to the program.

The IRAF system function "urand" is used to to select random, real numbers between 0.0 and 1.0. The X, Y coordinates of the background events are generated independently through a simple translation and scaling of random numbers:

$$ran\_num = urand(seed) \tag{1}$$
$$X = ran\_num * X_{max} \tag{2}$$

If the resulting number is not within the field limits, a new random number is generated and the process repeats until a valid coordinate value is found.

Azimuthally averaged versions of the PRFs are used to generate source events. The distance of each event from the source center is determined by the solution of the integral equation:

$$\int_0^R PRF(r)\, 2\pi r\, dr = ran\_num \tag{3}$$

for R. An angle is then randomly selected, and the projected X, Y coordinates are calculated, so that we have:

$$\Theta = 2\pi * urand(seed) \tag{4}$$
$$X = SRC_X + R * cos(\Theta) \tag{5}$$
$$Y = SRC_Y + R * sin(\Theta) \tag{6}$$

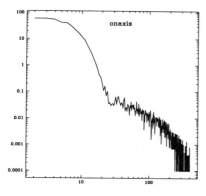 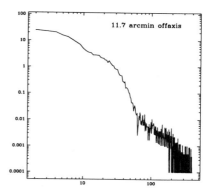

Figure 1. Radial Profiles of Simulated HRI sources (Radius (pixels) vs. counts/pixel). On the left is a source generated with the on-axis PRF, on the right, a source generated at 11.′7 off-axis.

where SRC_X, SRC_Y in equations (5) and (6) are the coordinates of the source center. This method ensures that the events are uniformly distributed over azimuth. The coordinates of the source events are then screened to fall within the field limits, and the process is repeated until a valid X, Y pair is determined.

The simulated data are output as an ASCII list representing events from all of the sources specified in the source table. The list is then converted to QPOE format with the IRAF/PROS tasks **qpcreate** and **qpappend**; at this point the user may select a real observation to which the events are appended.

4. Radial Profiles of Simulated Sources

Radial profiles of simulated sources were generated to verify that the data exhibits the expected spatial properties. Profiles of Gaussian and HRI sources at different off-axis angles were created. Figure 1 shows the radial profiles of two HRI sources of equal intensity. The source on the left was generated on-axis and the source on the right at 11.′7 off-axis.

The *ROSAT* HRI PRF is described as the combination of two Gaussian functions and an exponential function (David et al. 1993). In the on-axis figure, the two Gaussian functions are not easily distinguishable, and quickly decrease to the exponential tail. In the off-axis figure, the two Gaussian functions are discernible and they decrease more gradually to the exponential factor.

5. Projected Uses of the QPSIM

The ability to generate random background and sources of known position, shape and intensity can useful in calibrating source detection software. Some examples of what may be accomplished with data generated by **qpsim** are: assess the ability of the software to calculate source properties, such as position, source counts and signal to noise ratio; assess the significance of assuming a particular

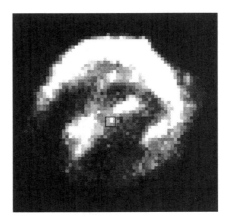

Figure 2. A *ROSAT* HRI observation of the Kepler supernova remnant with a point source injected into the center of the shell (shown in the box region).

source model when identifying a detection (for example, the use of a Gaussian model when applied to HRI data); determine the software's ability to discriminate sources of equal intensity in close proximity.

In addition, simulated data can be injected into a real observation in order to determine physical properties in a region of interest. For example, a point source can be added to a supernova remnant to determine the capability of analysis software to detect a pulsar in the midst of the remnant. An example of such a simulation is shown in Figure 2.

6. Extendibility of QPSIM

Currently, QSIM will only simulate data for the *ROSAT* HRI. However, the program has been designed to be easily extended to other instruments provided an azimuthally averaged version of the instrument Point Response Function is available. The simulation of source data in event rather than image format allows the extension of **qpsim** to simulate other events attributes, such as event time or pulse height.

Acknowledgments. This work is partially supported by NASA contracts to the *ROSAT* Science Data Center (NAS5–30934) and *Einstein* (NAS8–30751).

References

David, L. P., Harnden, F. R., Jr., Kearns, K. E., & Zombeck, M. V. 1993, The ROSAT High Resolution Imager (HRI) (Boston, U.S. ROSAT Science Data Center/SAO)

The SAO AXAF Simulation System

D. Jerius, M. Freeman, T. Gaetz, J. P. Hughes, and W. Podgorski

Smithsonian Astrophysical Observatory, 60 Garden St., Cambridge, MA 02138

Abstract. As part of our efforts to support the *AXAF* program, the SAO *AXAF* Mission Support Team has developed a software suite to simulate the *AXAF* telescope. The software traces the fate of photons through the telescope, from the X-ray source through apertures, baffles, the telescope optics, and finally to the photons' ultimate interactions with the focal plane detectors. We model relevant physical processes, including geometrical reflection, scattering due to surface microroughness, distortions of the optics due to the mirror mounts, attenuation through baffles, etc. The software is composed of programs and scripts, each specialized to a given task, which communicate through UNIX pipes. Software tasks are centered about functional components of the telescope (e.g., apertures, mirrors, detectors) and provide a comfortable and flexible paradigm for performing simulations. The use of separate programs and the UNIX pipe facility allows great flexibility in building different configurations of the telescope and distilling diagnostics from the photon stream through the telescope. We are able to transparently use symmetric multi-processing (e.g., SPARCStation 10s and SGI Challenges) and can easily use sequential multi-processing (via workstation clusters). Some of the tasks are amenable to parallel processing and have been implemented using the MPI standard.

1. Introduction

As part of SAO's support of the design and calibration of the Advanced X-ray Astrophysical Facility (*AXAF*), we have undertaken the task of simulating the optical performance of the telescope. We seek to create a high fidelity model which takes into account the major factors which will impact the optics' X-ray performance. These include distortions of the optics due to mechanical and thermal stresses in the mounting assembly, manufacturing figure errors in the optics, and scattering due to mirror surface microroughness and surface contamination.

We have a far-ranging set of investigations which we are, and will be pursuing with this system. These investigations range from studies of requirements for calibration, to studies of the impact of detector characteristics upon the observation of astrophysical sources. Among the issues which we are currently investigating are: (1) the ability of the telescope, as designed, to achieve its performance goals; (2) the performance of the mirror assembly during testing at the Marshall Space Flight Center X-Ray Calibration Facility; (3) the development of models for the on-orbit science performance of the telescope based upon ground calibration data; (4) the performance of flight instruments; and (5) the efficacy of ground and on-orbit calibration procedures.

2. System Design

The diverse set of studies which we must perform and the goals of the *AXAF* calibration effort require of our software system a high degree of flexibility and efficiency. Because of limited human resources, we have striven for the former, while keeping open our options for the latter.

AXAF is a complicated system, and the mirror assembly and assorted structures comprise a highly complex set of obstructions, apertures, and reflective surfaces. In order to understand how the various components affect each other, and to diagnose problems (both with our software and the hardware designs), we must be able to insert or remove routines which model the apertures, baffles, collimators, and detectors, or perform diagnostics, at any point in the photons' path. At the same time, we must retain enough efficiency so that we can trace $\sim 10^7$ rays through the system on a routine basis with a reasonable turn-around time. To cap it off, we need to be able to encapsulate existing software which may not follow our final paradigm. For example, our surface reflection module is based upon a mature FORTRAN program written in the days of mainframes and minimal core memory.

One possible approach is to develop a single, monolithic, program, which follows a ray from generation to collection. This avoids interim output (on the order of hundreds of MB) and is quite efficient. We rejected this approach early on, however, because it would require a complicated scheme to control the various system components and allow us the flexibility we desire. From a software development point of view, it forces a recompilation and re-verification each time a new component is added. In an environment composed of scientists, engineers, and programmers with different levels of programming sophistication, ensuring the integrity of such a system would be a nightmare.

Instead, we chose to model each optical or structural component as a separate program, based on the UNIX filter paradigm. A program reads a ray from its standard input stream, manipulates it and writes it to its standard output stream. The rays are encoded as binary data in dynamically extensible records, allowing routines to add more information to the ray packet, as necessary. The programs use a common user interface (based upon the IRAF parameter interface) with parameter storage. As programs are completely independent from each other, validation and verification are quite straightforward. Programs communicate with each other via UNIX pipes, which are easily accessible via the standard UNIX shells. We provide higher level control by embedding the commands in UNIX shell or Perl scripts. We have created a Perl library to easily interface our Perl scripts to the parameter interface, presenting a uniform environment to the user.

In some instances efficiency has won over flexibility, and we have combined several "atomic" operations into one program. One example of this is the program which simulates apertures. At the front end of the mirror assembly is a set of thermal baffle plates, which act to control the thermal environment inside the mirror assembly. It consists of some nine circular plates, with annular apertures to let the photons pass through to the mirrors. Rather than model each plate with a separate program, our aperture program can deal with any number of plates. In addition, we have designed and are prototyping an aperture program which uses an embedded language and a flexible script-driven approach to assemble complex apertures from simple building blocks.

3. Components

As an illustration of how the individual programs work together to model the telescope, we present a short synopsis of the major modules, in the order of their affect upon an incoming photon.

3.1. Ray Generation

The ray generator spews forth rays from a variety of sources, constraining the ray directions to fully illuminate the entrance aperture of the telescope. Sources may be placed at any location in the telescope's field of view, and may have any of several simple geometrical shapes. They may also be diffuse, simulating all-sky background radiation or extended X-ray sources.

3.2. Apertures

The aperture module, as noted above, is used to model both optical and thermal baffles, support structures, and the like. Surfaces may be partially transparent (via input of an absorption depth), and may be composed of struts and annular plates. Complicated apertures will be possible with our new hierarchical aperture module, which can use overlapping geometrical shapes and "cookie-cutters".

3.3. Surface Reflection

The surface reflection program is based upon a NASA routine (the "Optical Surface Analysis Code", or OSAC). OSAC can handle deformed optics, with deformations approximated by Fourier (azimuthal) and Legendre (axial) polynomials. We have built upon the deformed optics module (which we have renamed as SAOsac), integrating it into our filter-based environment, and have extended it to use bi-cubic splines as well polynomials. Deformation in the optics is caused by loads at the mirror support points, due either to thermal or mechanical stresses and (in the case of ground calibration) a one-g gravitational field. The deformations are modeled by a highly detailed finite-element model of the mirror assembly. An IDL program (TRANSFIT, Freeman 1993) was written which takes output from the finite-element modeling code (ANSYS) and interactively fits the distortions in the optics.

3.4. Scattering due to Surface Microroughness and Contamination

After a ray's reflection from an optic has been determined, high frequency mirror roughness (surface microroughness) is modeled by scattering the ray according to a statistical model of the surface roughness. The model is based upon the measured point-spread distribution (PSD) of high-frequency surface height fluctuations of the optical surfaces. A similar process is used to model scattering due to surface contamination.

3.5. Ray Accumulation

There is a bevy of detectors which will be used both on orbit and at ground calibration, including gas proportional counters, CCD's, solid-state spectrometers, and microchannel plate detectors. We have developed models of these detectors to provide us with end-to-end predictions of the telescope performance. In one case (modeling scans of the ray distribution through circular apertures) we were able to parallelize the aperture-detector combination (using the MPI standard), allowing us to gang several workstations together (Nguyen & Hillberg 1995).

4. Conclusion

This system permits an essentially unlimited numbers of rays and all of the flexibility that one could desire. At any point between programs we can siphon off rays for diagnostics or analysis. Manipulating configurations is simple and intuitive. Writing detector simulators or diagnostic programs is straightforward. As an added advantage, when running on multiprocessor computers with symmetric processing (e.g., Solaris 2 or IRIX 5), the system will automatically switch processes to unused CPU's. For CPU dominated calculations, we can chain multiple machines together, each operating on one part of our process chain and passing the result on to the next machine.

The paradigm has proven highly successful, enabling us to provide high precision results without a complicated setup procedure to construct the particular configuration of the telescope that we desire. The efficiency (even allowing for the overhead of the inter-process communications) is reasonable, and on modern workstations has not a proven a barrier to our need for rapid turn-around.

References

Freeman, M., 1993 TRANSFIT - Finite Element Analysis Data Fitting Software, Interim Report SAO-AXAF-DR-93-052 (Cambridge, SAO)

Nguyen, D., & Hillberg, B. 1995, this volume, p. 361

Simulations of Pinhole Imaging for AXAF: Distributed Processing Using the MPI Standard

D. Nguyen and B. Hillberg

Smithsonian Astrophysical Observatory, 60 Garden St., Cambridge, MA 02138

Abstract. The pinhole simulation program is a computationally and memory intensive task. A conventional sequential approach would limit the size and complexity of such a problem when investigated in the framework of the SAO *AXAF* simulation system. A parallel version was developed instead, to enable distributed processing on a cluster of workstations. The program makes use of the Message Passing Interface (MPI) standard for parallel processing, implemented as an API (Application Programming Interface) to the Local Area Multicomputer (LAM) programming environment developed at the Ohio Supercomputer Center.

1. Introduction

As part of our efforts to support the *AXAF* program, the SAO *AXAF* Mission Support Team has developed a software suite to simulate *AXAF* images generated by the flight mirror assembly (Jerius 1994). One of the tasks of this system is to simulate pinhole imaging of the X-ray source.

2. Pinhole Simulation in Sequential Mode

The task of the pinhole program is to tabulate the weight of the photons detected through a pinhole. The weight of a photon represents the probability of finding the photon at a given position. Photons are generated with weights equal to 1 at the source. Weights are reduced at every reflection point along their paths toward the detector. The condition for a photon to successfully pass through a pinhole is:

$$(x - x_o)^2 + (y - y_o)^2 < r_o^2 \qquad (1)$$

where (x,y) is the photon position in the plane of the pinhole, relative to the pinhole of radius r_o centered at (x_o, y_o). In order to simulate a two dimensional scan at the focal plane for pinholes of various radii, the pinholes are laid out on a cubic lattice. The pinholes on a rectangular grid simulate a two dimensional scan, the stack of rectangular grids represents the pinholes of various radii.

A naive approach in writing the pinhole program would require

$$O(N * X * Y * Z) \qquad (2)$$

operations to finish. N is the number of incident photons, X and Y are the number of grid sites along the x and y axes of the rectangular grid, and Z is the number of radii to be calculated. An efficient program should minimize the number of times equation (1) has to be executed. The layout of the pinholes on a cubic lattice is used for this purpose, since in this case the photon stream only needs to be read once. Further reduction in execution time can be achieved by realizing that the pinholes of different radii are concentric. In the current implementation, the program requires

$$O(N * Overlap^2 * log_2(Z) * (log_2(X) + log_2(Y))) \qquad (3)$$

operations. N, X, Y, and Z are as defined above, and $Overlap$ is the number of overlapping pinholes in x and y directions. The drawback of the cubic lattice layout approach is that the three dimensional array that holds the photon weights can be prohibitively large for some class of problems; e.g., a 4000 by 4000 grid size would require 128 MB of memory.

Verification of the software was done as follows: a spatially uniform distribution of photons was generated with the weight of each photon set to unity. The photons were traced to the plane of the pinholes. The number of photons which make it through a pinhole must be equal to the density of the incident photon beam times the area of the pinhole.

3. Parallel Processing

These efficiency improvements achieve optimum speed for the pinhole simulation program running on a given machine. This, however, is still not satisfactory given the large volume of data that needs to be simulated. The next step was to consider parallel processing across workstations on a local area network (LAN).

Distributed multicomputing on a network-connected workstations provides a cost-effective environment for high performance scientific computing. Software packages exist that support parallel processing on workstation clusters by managing the communications and data traffic in a way transparent to the application. These packages provide sets of Application Programming Interfaces (API) for various languages so that their functions can be called from an application. Examples of such packages are Express, PARMACS, PVM, and P4. Although these packages have very similar functionalities, their APIs are very different. The result of non-standard APIs is that third party software becomes specific to a given package, and cannot be used with other packages.

4. The MPI Standard

A standardization process for a message passing system was initiated at the "Workshop on Standards for Message Passing in a Distributed Memory Environment," held in Williamsburg, Virginia in 1992 November. The Message Passing Interface (MPI) forum consists of researchers from government laboratories, universities, and industry, along with vendors of concurrent computers. The MPI standard was derived from the best features of its predecessors, rather than adopting one of the existing systems. The MPI standard includes: Point-to-point communication, Collective operations, Process groups, Communication context, Process Topologies, Bindings for FORTRAN 77 and C, Environment Management, and Inquiry and Profiling Interfaces.

The pinhole program makes use of the MPI standard for parallel processing implemented as an API to the Local Area Multicomputer (LAM) programming environment developed at the Ohio Supercomputer Center (Burns 1989). LAM is a distributed memory MIMD programming and operating environment for heterogeneous UNIX computers on a network.

5. Pinhole Simulation in Parallel Mode

The pinhole program can be completely parallelized by using data decomposition. More than one instance of the program can run, each on a different machine, analyzing a different subset of the data. One process (the master) initiates all the other processes (the slaves), and, at the end, collects all the results. In the master-slave computing paradigm, each slave communicates with the master, but there is no inter-slave communication.

The expected speedup from the sequential process for p number of slaves is (Almasi & Gottlieb, 1994):

$$speedup = \frac{p * r_1}{(r_1 + p^2)} \quad (4)$$

$$r_1 = \frac{T_{sequential}}{T_{communication1}} \quad (5)$$

where $T_{sequential}$ is the time needed on one machine and $T_{communication1}$ is the time needed to communicate with one slave. For a given r_1 the maximum speedup occurs at $p = \sqrt{r_1}$. The maximum speedup is therefore $0.5\sqrt{r_1}$.

The pinhole program was modified and executed on a system of one master and ten slave processes, each running on a different workstation. The results were consistent with equations (4) and (5), i.e., r_1 was measured to be about 100, so for ten slaves a factor of five in speed was gained.

6. Summary

MPI enables the writing of portable high-performance libraries for distributed-memory machines, easing the burden for application programmers.

References

Almasi, G. & Gottlieb, A. 1994, Highly Parallel Computing, 2nd ed., (Redwood City, Benjamin/Cummings)

Burns, G. 1989, in Proceedings of the Fourth Conference on HyperCubes, Concurrent Computers, and Applications (Los Altos, Golden Gate Enterprises)

Jerius, D., Freeman, M., Gaetz, T., Hughes, J. P., & Podgorski, W. 1995, this volume, p. 357

Message Passing Interface Forum 1995, MPI: A message passing Interface Standard, Technical Report (Knoxville, University of Tennessee), in press

Part 10. Software Systems

FTOOLS: A FITS Data Processing and Analysis Software Package

J. K. Blackburn[1,2]

NASA Goddard Space Flight Center, Code 664.0, Greenbelt, MD 20771

Abstract. FTOOLS, a highly modular collection of over 110 utilities for processing and analyzing data in the FITS (Flexible Image Transport System) format, has been developed in support of the HEASARC (High Energy Astrophysics Science Archive Research Center) at NASA's Goddard Space Flight Center. Each utility performs a single simple task such as presentation of file contents, extraction of specific rows or columns, appending or merging tables, binning values in a column or selecting subsets of rows based on a boolean expression. Individual utilities can easily be chained together in scripts to achieve more complex operations such as the generation and displaying of spectra or light curves. The collection of utilities provides both generic processing and analysis utilities and utilities specific to high energy astrophysics data sets used for the *ASCA*, *ROSAT*, *GRO*, and *XTE* missions. A core set of FTOOLS providing support for generic FITS data processing, FITS image analysis and timing analysis can easily be split out of the full software package for users not needing the high energy astrophysics mission utilities. The FTOOLS software package is designed to be both compatible with IRAF and completely stand alone in a UNIX or VMS environment. The user interface is controlled by standard IRAF parameter files. The package is self documenting through the IRAF help facility and a stand alone help task. Software is written in ANSI C and FORTRAN to provide portability across most computer systems. The data format dependencies between hardware platforms are isolated through the FITSIO library package.

1. Introduction

The FTOOLS software project began in late 1991 (Pence 1992) as part of a goal to standardize high energy astrophysics data sets to the FITS (Wells et al. 1981) format at NASA's HEASARC. The earliest releases of FTOOLS consisted of a collection of tools for creating, viewing, and manipulating data sets using the FITS format. Today, the FTOOLS software has grown into a collection of subpackages supporting not only generic manipulation of FITS formatted data, but also tools for data analysis of high energy astrophysics missions using the FITS standard such as *ASCA*, *GRO*, *ROSAT*, and now *XTE*.

---

[1] Hughes STX Corporation, 4400 Forbes Blvd., Lanham, MD 20706

[2] HEASARC

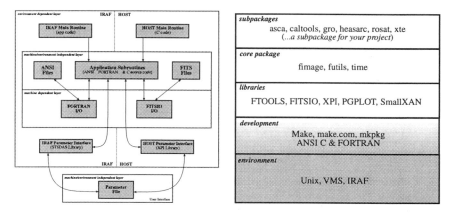

Figure 1. FTOOLS Task and Package Design.

1.1. Portability

In the design of the FTOOLS, portability was given the highest consideration. The software specifications called for ANSI FORTRAN and C as the base languages. To bind the differences in these languages between supported architectures, the C macro package CFORTRAN develop by Burkhard Burows and distributed by CERN is used. The FTOOLS design also required a common user interface in the three most popular environments used by the astronomical community today, UNIX, VMS and IRAF. Under UNIX and VMS, the FTOOLS are referred to as stand-alone or "Host" to distinguish this environment from IRAF. However, the user interface to the FTOOLS is identical in all environments. To achieve this cross environment support, a user interface based on the IRAF parameter file was adapted. This particular interface provides for command line assignment of any or all parameters, range checking, defaults and user prompts. By isolating the interface to the data and parameter files files to standardized subroutines common to all environments, the differences in making FTOOLS for UNIX, VMS and IRAF are resolved by linking to the appropriate libraries such as IRAF or the stand-alone XPI parameter interface. This is illustrated to the left in Figure 1.

1.2. Open Development

The FTOOLS software package is layered into a collection of subpackages supporting a generic (or core) set of tools and a collection of subpackages for specific missions in high energy astrophysics. These subpackages are layered over the necessary libraries and environments that the FTOOLS build under. This is illustrated to the right in Figure 1. By having this layering, additional subpackages supporting new missions or even the particular needs of an individual user or project can easily be plugged into the FTOOLS package with minimal edits to underlying layers.

1.3. On-line Help

On-line help is available for each of the tasks distributed with the FTOOLS. Under IRAF, help is obtained just as with any other IRAF task. Under the stand-alone Host FTOOLS, a specialized FTOOL named fhelp is used to access the on-line help database. In both cases, simply follow the appropriate help command with the name of the FTOOLS task. The on-line help presented provides a usage, description, list of options (through the parameters) and examples.

2. Example

As an example of the use of the FTOOLS, consider a FITS data set of photon events from an supernova remnant made by an X-ray imaging telescope which has within its field of view a calibration source. First the structure of the FITS dataset needs to be determined. This is accomplished with the **fstruct** task. Using the Host FTOOLS this would look like

```
ftools.gsfc.nasa.gov kent[1] % fstruct raw.evt
  No. Type EXTNAME BITPIX Dimensions(columns) PCOUNT GCOUNT

  0 PRIMARY 32 0 0 1
  1 BINTABLE EVENTS 8 30(12) 55753 0 1
  2 BINTABLE STDGTI 8 16(2) 139 0 1
```

From this it is determined that the photon events are located in the first extension, which is a FITS binary table extension with 55753 rows (events). Now determine the format of the table using the **flcol** task to list the names of the columns

```
ftools.gsfc.nasa.gov kent[2] % flcol raw.evt+1
___Column_Names_____Formats_____Units___
        TIME 1D s
        X 1I pixel
        Y 1I pixel
        PHA 1I channel
        PI 1I channel
        RISE_TIME 1I
```

There are several columns in this data set. Using the X and Y columns, basic statistics can be learned about these columns with the **fstatistic** task or an image can be built up from the photon list using the **f2dhisto**. Using the PHA column and the **fhisto** task a quick-look spectrum can be made and viewed with **fplot** (see spectrum to the left in Figure 2).

Now through the use of these FTOOLS tasks one determines that the supernova remnant is centered at (114,129) in the pixel space of the image with a radius of 40 pixels. Using the **fselect** task to select out only the photons associated with the supernova and the background and then the **fhisto** task on the resulting supernova events, the spectrum on the right in Figure 2 having no calibration peak is binned and then viewed.

```
ftools.gsfc.nasa.gov kent[3] % fselect raw.evt+1 snova.evt
```

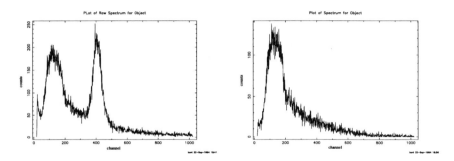

Figure 2. Raw spectrum on left showing supernova remnant with calibration source and the spatially filtered spectrum to right.

```
Name of output FITS file[] snova.evt
Selection Expression[] sqrt((X-114)**2 + (Y-129)**2) < 40
ftools.gsfc.nasa.gov kent[4] % fhisto snova.evt+1 spec.fits PHA 1
```

The FTOOLS distribution comes complete with a *User's Guide* which has a short tutorial going into more detail about using FTOOLS. Other references discussing FTOOLS are found in Blackburn & Pence (1994). The IRAF documentation are also available to learn more on the FTOOLS software package.

3. Distribution

The FTOOLS software package is available to the public with access to the Internet. It is distributed as a compressed tar file through anonymous ftp from *legacy.gsfc.nasa.gov*. In the directory software/ftools/release you will find the a README file, release notes, source code, reference data for *ASCA* data analysis, and documentation covering the installation, usage, and development of the FTOOLS software. Further information on the FTOOLS can be found on the World Wide Web at the FTOOLS Home Page[3].

Acknowledgments. I am grateful to all the members of the FTOOLS group at the HEASARC for the effort they have made in bringing together under one paradigm, a significant and useful astronomical software package.

References

Pence, W. 1992, Legacy - The Journal of the HEASARC, 1, 14
Wells, D. C., Greisen, E. W., & Harten, R. H. 1981, A&AS, 44, 371
Blackburn, J. K., & Pence, W. 1994, Legacy - The Journal of the HEASARC, 4, 5

---

[3] http:://heasarc.gsfc.nasa.gov/0/docs/software/ftools/ftools_menu.html

Migrating the Starlink Network from VMS to Unix

C. Clayton

Rutherford Appleton Lab., Chilton, Didcot, Oxon OX11 0QX, U.K.

Abstract. The Starlink Project is a UK-wide astronomical computing service consisting of a network of computers used by UK astronomers at over 25 sites, a collection of software to calibrate and analyze astronomical data, and a team of people to give hardware, software, and administrative support.

In order to exploit the most cost-effective hardware and to maintain compatibility with the international community, Starlink is migrating from an entirely VAX/VMS based service to UNIX-based systems. This migration is almost complete, and this paper describes some of the solutions adopted for the wide variety of problems which were encountered.

Migration of the hardware platform is discussed first. Equipment which can be re-used under Unix is identified. System software and non-astronomical applications which are required to allow a smooth transition from VMS to Unix are considered next. While many VMS functions can be replaced with Unix equivalents, it has become apparent that there is a small number of key VMS applications which must be provided on the replacement Unix platform to avoid considerable disruption to users.

Various strategies for moving the users themselves from VMS to UNIX are considered and their relative merits compared. Fast migration routes are considered to be more effective as long as certain key applications and user aids are already in place. The porting of the Starlink Software Collection is discussed, as is the problem of migrating large quantities of private user code.

1. Introduction

In order to exploit the most cost-effective hardware and to maintain compatibility with the international community, the formerly VMS-only Starlink Project has switched to using Unix-based computers. A mixed Unix/VMS service, with its extra system management and software support costs, cannot be afforded and at present the existing VMS service is being run down.

A complete change to Unix will have occurred by April 1995. A central fall-back VMS service will, however, be retained for emergencies (for example, for applications where source code is not available). Starlink is currently operating DEC Alpha and Sun SPARC CPUs.

2. Hardware

2.1. Hardware Migration

The first Sun was installed at a Starlink site in March 1990 and hence the total elapsed migration time has been 5 years. However, the majority of the hardware purchasing has taken place in the last 3 years. During this period, we have had to run VMS and Unix in parallel. The VAXes at a given site have then been switched off when virtually all of the software which is used at that site has been ported to Unix and when there is sufficient Unix hardware at that site to support the resident user community.

Unfortunately, it has been necessary to divert funds from software effort in order to complete the move so swiftly. A high rate of change was deemed more cost effective; savings are then possible in the areas of maintenance (we can switch off older, expensive-to-maintain equipment earlier) and system management (less manpower required if the VMS and Unix systems are run in parallel for a shorter time).

2.2. Hardware Re-use

We have attempted to re-use hardware from our VMS system where it is **both** technically possible **and** cost-effective. Such equipment includes 1/2" tape drives, exabytes, CD-ROMs, SCSI hard disks, printers, and networking kit. Equipment which could not be moved cost effectively included VAX CPUs, non-SCSI disks, Honeywell image display cameras, and DEC LAT-only terminal servers. Further details are available from the author.

3. Operations Software and Migration Utilities

One of the most vital aspects of the migration was the provision of migration aids for users. These took the form of VMS-equivalent software for Unix and friendlier versions of certain "user-hostile" Unix utilities which come bundled with the operating system.

- **TCP/IP and NFS for VAXes** - This allowed users continued access to their old VMS files when they ventured onto the Unix machines. We used DEC's **UCX** product.

- **Mail interface** - Users disliked the non-X-windows Unix mail interfaces. Hence, we adopted the friendly **pine** system which has proven very popular, even with VMS diehards. We also provided Perl scripts to convert users' VMS MAIL folders.

- **Editors** - This was the single most contentious issue. Initially we tried to persuade users to adopt either **vi** or **emacs**, depending on the level of sophistication they required, on the grounds that these are the most standard editors and likely to be found on virtually any Unix system encountered. The users rebelled, claiming that **vi** was too unfriendly and that **emacs** was too complex. In the end we were forced to adopt two additional editors:

    - **jed** - a public domain editor which can be set up to emulate VMS EDT

- **nu/TPU** - a commercial version of VMS TPU for Unix

- **VMS BACKUP** - Many of our users have for years been archiving data to tape using VMS BACKUP. It was not possible or practical to re-write this large body of data using Unix tape utilities. Instead, we tried using a public domain VMS BACKUP reader for Unix called **vmsbackup**. However, this package, while useful, was not able to read a number of problem tapes, and we were eventually forced to purchase a commercial VMS BACKUP system for Unix called **Vbackup**.

- **Motif** - We purchased (at modest cost) Motif for our Suns so that all of our systems (VMS and all flavors of Unix) would have the same user interface.

- **tcsh** - We adopted the tcsh as the default user shell, mainly because the users liked the VMS-like command line recall system.

- **Template files** - We provided template files for new Unix users which included files to remap keyboards, allowing transparent use of the VAXes from the alien Unix keyboards, as well as templates for the **.login, .tcshrc, .xinitrc, .emacs** and **.Xdefaults** files.

- **VAXnotes** - The Starlink Project makes heavy use of the DEC VAXnotes electronic conferencing product. This is not available from DEC under Unix and we have yet to find a suitable replacement. Usenet news has some similar functionality but does not meet all of our requirements. It may be possible to implement something similar using WWW.

4. Training Staff and Users

Due to budgetary limitations, we cannot afford formal Unix training for all of our users. Instead we have adopted a strategy of sending system managers on Unix management courses, encouraging system managers and other local experts to give Unix seminars to local users, and encouraging users to attend other local Unix courses (e.g., ones held for new undergraduates at university computing centers). We also provide copies of the book *Unix for VMS users*, make WWW Unix guides readily available from our home pages, and encourage users to help each other. In order to transfer our large (1800) user population, we adopted the following broad strategies: (1) new users are given accounts on Unix machines *only*; (2) exciting new software is only made available on the Unix machines (this included Mosaic and Usenet news as well as new astronomical software products); (3) the VMS systems were made to look unattractive by a variety of methods, including transferring useful peripherals to the Unix systems (e.g. printers) and reducing the level of support available to VMS users; and finally, (4) switching off the VAXes.

It was found that users who ventured onto Unix, but returned to VAX/VMS whenever they encountered a problem, *never* made the transition to Unix until their VAXes were actually switched off. Those users who did not look back, and overcame each problem as it arose, typically completed the transition in about three weeks. Those users who returned to the VMS service temporarily seemed to forget most of what they had learned while on the Unix machines and hence each attempt to migrate was a painful as the previous. However, those who immersed themselves consolidated what they had learned, and very rapidly became proficient under Unix.

5. Porting the Starlink Software Collection

The Starlink Software Collection is based on a software environment (called ADAM) supported by a team of professional programmers, and hence the problem of porting to Unix was tractable. The software which was VAX-specific has been made portable, and it now runs on a number of flavors of Unix (SunOS 4.x, Solaris 2.x, Ultrix 4.x and OSF/1) as well as VMS. Ports to other Unix implementations are expected.

Users were given several years warning of the switch to Unix and were instructed to ensure that they ported their private code themselves. Unix porting advice has been available to users encountering difficulties. Much of the code has been standard FORTRAN 77 and has been moved to Unix with relative ease.

The change to Unix gave us the opportunity to kill off certain hard-to-maintain packages, which were originally poorly written, on the grounds that they were too difficult to port. These have been replaced with better engineered products. However, use of certain packages in this category were too entrenched and we were forced to port these despite the difficulty of the task.

6. Software Distribution

This is now done via anonymous ftp. For further details contact the Starlink Software Librarian (*ussc@star.rl.ac.uk*).

Astronomical Data Analysis Software and Systems IV
ASP Conference Series, Vol. 77, 1995
R. A. Shaw, H. E. Payne, and J. J. E. Hayes, eds.

Porting CGS4DR to Unix

P. N. Daly

Joint Astronomy Centre, 660 N. A'ohōkū Place, Hilo HI 96720.

Abstract. This paper discusses the port of the CGS4 Data Reduction system to Unix, and describes a novel solution to some non-portable code.

1. Introduction

The CGS4 Data Reduction system, CGS4DR, is a suite of tasks for automatically reducing data from the long-slit spectrometer on the 3.8m United Kingdom Infra-Red Telescope (UKIRT) on Mauna Kea, Hawai'i (Daly et al. 1994 and references therein). Delivered with the instrument in 1991, the core software consists of four ADAM tasks written in FORTRAN under VMS. Since that time, British astronomy has moved away from VMS and towards Unix. The ADAM Support Group was charged with porting the ADAM system, and a considerable fraction of that project is now complete. Although not required by UKIRT in the foreseeable future, since the data acquisition system will remain VAX-based for some time, CGS4DR is currently being ported to UNIX—it will provide a strict test of the final release of the ported ADAM system.

2. Portability Issues

During the period 1991–1994, the code was modified substantially to enhance the facilities available to the astronomer in reducing spectroscopic data in real-time at the telescope, while remaining true to the original design paradigm. The more obvious features of non-portable code, VMS run time library calls and the like, were removed, and the code was altered to conform to ADAM V2 instrumentation software standards (Kelly & Chipperfield 1992). Two minor, but annoying, problems were discovered: mixed data types within a common block, and common blocks with variable lists exceeding the maximum line length, even when compiled in "–e" mode. Reformatting the common block cured both problems. Code porting problems were few, and error-free compilation has been achieved. Local extensions to Figaro (Shortridge 1993), used by CGS4DR for accessing data structures and so forth, have also been ported.

3. Design Issues

A major design flaw of CGS4DR was the use of indexed sequential access method (ISAM) files. Such files are DEC FORTRAN compiler extensions not supported under many other operating systems. They were used to allow the data acquisition (DA) system and the data reduction (DR) system should communicate only via a time stamped queue. A second use of an ISAM file—maintaining an index of reduced observations—was added later.

3.1. QMAN The Generic Queue Manager

The (ISAM disk file) data reduction queue accepted time stamped commands and was periodically read by the DR system by task re-scheduling. The following data reduction sequence could have been generated by the DA system automatically adding observation numbers 2–3 to the "tail" of the queue and the observer inserting observation number 1 at the "head" of the queue:

```
000913:11:12:49.81    REDUCE 0940913_1
000913:11:12:49.82    END 0940913_1
940913:11:12:44.38    REDUCE 0940913_2
940913:11:12:44.47    END 0940913_2
940913:11:12:44.50    REDUCE 0940913_3
940913:11:12:44.51    END 0940913_3
```

CGS4DR *always* read the *oldest* command from the queue and the use of an ISAM file with a time stamp as the primary key was, at the time, an efficient solution to these requirements. Since this file was also on disk, it was easy to recover in the event of a software failure.

In porting this application, a new task was created to replace the ISAM file: QMAN, the generic queue manager (Daly 1994). QMAN is a standard ADAM I-task and maintains a list of 1000 character string entries *in memory*. Incoming strings are time stamped using an accurate modified Julian date (MJD) and the queue position may be specified as either "OLDEST" or "NEWEST." To QMAN, the above queue might look like this:

```
 49608.47223391203614  REDUCE 0940913_1
 49608.47223390046202  END 0940913_1
-49608.47206021990678  REDUCE 0940913_2
-49608.47206023148101  END 0940913_2
-49608.47206024305524  REDUCE 0940913_3
-49608.47206025462947  END 0940913_3
```

In this scheme, the "NEWEST" position has the maximum MJD whereas the "OLDEST" position has the latest MJD *negated* and so has the minimum value. The MJD is a monotonically increasing function, and provides a powerful and flexible means of maintaining a time stamped queue. With a suitable combination of reads and writes, QMAN may mimic either a FIFO or a LIFO buffer.

An important distinction now is that the Unix–CGS4DR software always reads the *newest* command from QMAN. Further, after each write by the DA system the queue is saved to disk and each read by the DR system is a destructive process that releases that slot in memory for subsequent commands. QMAN also features an optional (global) password protection scheme and a (local) lock manager to provide exclusive access to the database.

3.2. The Index of Reduced Observations

There is second use of an ISAM file in CGS4DR: to maintain an index of already reduced observations that may be used as calibration frames in later reduction sequences. When such a calibration frame is required, a search is made according to some defined criteria—data size, wavelength, oversampling parameters and so forth—and the frame closest in time to the present observation being reduced is selected. There are several alternatives for re-coding this functionality.

First, clearly, a separate invocation of the QMAN task could fulfill this role. Take the simplest example of a BIAS frame. CGS4DR requires only that the detector size should match and so a BIAS could be filed using the code fragment:

```
      CALL TASK_OBEY( 'QMAN', 'WRITE',
     :   'STRING="OBSN=ODIR:O940913_1 TYPE=BIAS QLTY=GOOD '/
     :   /'ROWS=256 COLS=256" QPOSITION="NEWEST"',
     :   OUTVAL, QMAN_PATH, QMAN_MESSID, STATUS )
```

When a BIAS is required, therefore, QMAN searches for the "NEWEST" observation (i.e., closest in time) using the READ command in SEARCH mode:

```
      CALL TASK_OBEY( 'QMAN', 'READ',
     :   'READ_MODE="SEARCH" SEARCH_MODE="NEWEST" '/
     :   /'DESTRUCTIVE=FALSE STRING="TYPE=BIAS QLTY=GOOD '/
     :   /'ROWS=256 COLS=256" QPOSITION="NEWEST"',
     :   OUTVAL, QMAN_PATH, QMAN_MESSID, STATUS )
```

This search is made *non-destructively* so that the same index is maintained throughout. At the end of the data reduction session, the complete index may be saved to an ASCII text file. To file an observation as BAD would simply require searching the database in *destructive* mode and ignoring the result of the read (which is always passed back). The BAD observation would, therefore, be erased from the database (and could be re-filed with "QLTY=BAD" so that all observations appear in the database as they do at present). The problem with invoking a second incarnation of QMAN in this way is that it takes more memory, uses up a task slot and the reduction task—an A-task in ADAM parlance—would then have to talk directly to another task which is bad form since that is the domain of I-tasks. A-tasks should simply "do something and return."

Using simple ASCII text files would overcome these problems. A separate file for each calibration type would be used; e.g., cgs4_940913.bias and so forth. Continuing with the BIAS example, the string written to the file would be similar to that above but with the addition of an encoded MJD as used by QMAN. Thus the difference in time between the observation currently being reduced and the calibration frame could be quickly evaluated. Such a text file for BIAS frames might contain the following lines:

```
RODIR:RO940913_1 BIAS GOOD 256x256 MJD=49608.47223390046202
RODIR:RO940913_2 BIAS GOOD 256x256 MJD=49608.47223391203614
```

The drawback here is the relative slowness of (sequential) disk file access although the size of these files would be small (typically, less than 50 frames of calibration data are taken per night). A further complication—not insurmountable—is the ability to (re-)file an observation as BAD when desired. The optimum solution to this problem has yet to be determined.

4. De-coupling the Software Tasks

VMS–CGS4DR consists of four tightly coupled tasks: a control task, a reduction task, a plotting task, and an engineering task. QMAN has essentially de-coupled the management of the data reduction queue from the control task. Might de-coupling benefit the other tasks? The plotting task is in the process of being de-coupled from the control task, making it standalone, and potentially useful to other applications, since it can access both Figaro DST and Starlink NDF format data files.

Under the VMS system, the control task, when requesting a sub-plot of an image, for example, would send a message such as:

```
IMAGE DATA="RODIR:RO940912_1" PORT=0       -
PLANE=DATA WHOLE=FALSE XSTART=20 XEND=40 -
YSTART=20 YEND=40 AUTOSCALE=FALSE HIGH=1000.0 LOW=0.0
```

Under Unix–CGS4DR, the plotting task now maintains its own noticeboard so the control task sends a far simpler message:

```
DISPLAY DATA="RODIR:RO940809_1" PORT=0
```

The plotting task then reads the display characteristics of port 0 from the noticeboard global section. This global section may be manipulated directly from the command line or user interface to tailor the display as desired.

5. User Interface and GUIs

At present, there is no clear way to interface Unix–ADAM tasks with a GUI although groups are investigating ICLmenus (the Starlink approach) and Tcl/Tk. The first release of Unix–CGS4DR, intended for delivery before the close of 1994, might well have a command line interface alone.

References

Daly, P. N., Bridger, A., & Krisciunas, K. 1994, in Astronomical Data Analysis Software and Systems III, ASP Conf. Ser., Vol. 61, eds. D. R. Crabtree, R. J. Hanisch, & J. Barnes (San Francisco, ASP), p. 457

Kelly, B. D., & Chipperfield A. J. 1992, "ADAM—Guide to Writing Instrumentation Tasks", SUN/134, Starlink Project (DRAL/Rutherford Appleton Laboratory)

Shortridge, K. 1993, in Astronomical Data Analysis Software and Systems II, ASP Conf. Ser., Vol. 52, eds. R. J. Hanisch, R. J. V. Brissenden, & J. Barnes (San Francisco, ASP), p. 219

Daly, P. N. 1994, "QMAN—A Generic Queue Manager", UON/16, UKIRT, (Hilo, Joint Astronomy Centre).

Astronomical Data Analysis Software and Systems IV
ASP Conference Series, Vol. 77, 1995
R. A. Shaw, H. E. Payne, and J. J. E. Hayes, eds.

The Array Limited Infrared Control Environment

P. N. Daly and A. Bridger

Joint Astronomy Centre, 660 N. A'ohōkū Place, Hilo HI 96720

D. A. Pickup and M. J. Paterson

Royal Observatory, Blackford Hill, Edinburgh EH9 3HJ, Scotland

Abstract. This paper describes the recently commissioned Array Limited Infrared Control Environment (ALICE), used to control a new 256^2 array imager, IRCAM3, on UKIRT. Future plans, including the delivery of a second ALICE to upgrade the UKIRT long-slit spectrometer, CGS4, are also discussed.

1. Introduction

Infrared astronomy was revolutionized in the 1980s by the advent of array detectors. One of the first such imaging cameras in regular operation was IRCAM (McLean et al. 1986), followed by the array spectrometer CGS4 (Mountain et al. 1990), both of which were commissioned onto the 3.8 m United Kingdom Infrared Telescope (UKIRT) on Mauna Kea, Hawaii. The control electronics of both these systems, however, were geared to driving the smaller SBRC 62 × 58 InSb array. What was needed was a flexible *and* extensible system capable of driving the new generation of arrays to the limits of their performance, hence the term *Array Limited Infrared Control Environment* (ALICE).

Built at the Royal Observatory in Edinburgh (ROE), ALICE forms the backbone of the program to upgrade both IRCAM and CGS4 to SBRC 256^2 InSb arrays. The ALICE design allows it to operate under a wide range of conditions such as high resolution low background near infrared spectroscopy, real-time shift-and-add image sharpening, and broad-band thermal imaging. To date, the first ALICE has been commissioned on the telescope and the new instrument, IRCAM3, saw first light on 1994 April 8 UT.

2. The Array Control System

The array control system (Pickup et al. 1993) applies well defined, and well controlled, bias and clock voltages to the array, and is also responsible for clocking the appropriate waveform to trigger read, reset, or read-reset operations. It also obtains time stamps from the UKIRT satellite clock and is able to control the secondary mirror in synchronization with the observations. All of these operations may be manipulated under software control from the user interface, thus giving the system considerable flexibility.

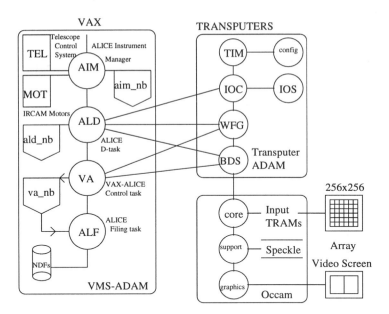

Figure 1. The ALICE Software Map.

The array control system software operates under the transputer ADAM environment (Kelly et al. 1993), and was originally written in parallel FORTRAN, although this will be replaced by a C version when the second ALICE is commissioned. The tasks comprise the time stamping process (TIM), the I/O controllers (IOC and IOS), and the waveform generator (WFG), as shown in Figure 1. The WFG is initialized with full frame read, reset, and read-reset waveforms—other waveforms, for sub-array reads and so forth, are downloaded as required.

The waveform generator hardware includes an Inmos T805-25 transputer with 16 MB of DRAM. The waveform memory has 16 banks of 256k × 32-bit field memory. Each of the 32 bits of the waveform handles one clock signal (three of which are reserved for system use). Moreover, any of the 16 banks can hold a single waveform sufficient to drive a full 256^2 array. Banks can also be read out seamlessly so that, if necessary, a waveform can be loaded across several banks. In this fashion, larger arrays should easily be accommodated as they become available.

3. The Bulk Data System

The bulk data system (BDS) supports the bi-directional transfer of data between ALICE and the host system at ~ 50 kB s^{-1}. Thus a full 256^2 4-bytes pixel^{-1} frame takes ~ 5 s to transfer. Four MB of on-board RAM are used to buffer the data to allow asynchronous acquisition of new data.

The bi-directional nature of the data transfer allows data to be passed up to the host computer and, for example, bad pixel masks, color tables, or background frames to be passed down to the transputer sub-system when required by the video display system. The BDS also handles the downloading of waveforms to the WFG when required to trigger different reads of the array and the transputer ADAM messaging traffic. Messages between the host computer (a VAX in the present application), are routed through a B300 TCP-linkbox (ethernet gateway).

4. The Data Acquisition System

The data acquisition system controls the analog conditioning of signals from the array, conversion to digital inputs for the INMOS link adaptors via ADCs (14-bit, 2 mega-samples s^{-1}), and for the collection and re-assembly of the full 256^2 image via a transputer network. Several frames may be coadded to form an integration before transmission to a host computer or local disk. There is also a video display system that updates at ~2 Hz, that can show either the current integration or both the current integration and the coadded image side by side, along with some simple image analysis (peak pixel in x, y, brightness, etc.).

ALICE makes extensive use of transputers (> 24), mainly T805s with 2 MB of memory each. The 16 input processors are hardwired as a pipeline on the motherboard so each processor, during normal operation, sees every 16th pixel from the array. The "core" process running on a single transputer acquires the data from the pipeline and re-assembles the data into full frames. A speckle option has recently been introduced using fiber-optic TRAMs that can "spurt" the data to a local (SCSI) disk directly for (very) fast acquisition. The transputer data acquisition sub-system is programmed in Occam and the bootable may be downloaded from either a Sun or a VAX.

Table 1. IRCAM3/ALICE Observing Modes

Observing Mode	80% Well Depth	Readout Area	Exp_{min}[a] (ms)	M_{sat}[b] J	K	L
Standard	80,000 e^-	256^2	120	8.6	7.8	7.4
(JHK)	(13,500 DN)	128^2	35	7.3	6.5	6.2
		64^2	12.5	6.2	5.4	5.0
Fast	80,000 e^-	256^2	72	8.1	7.3	6.9
(Shift and Add)	(13,500 DN)	128^2	20	6.7	5.9	5.5
		64^2	6.5	5.5	4.6	4.3
Deep Well	144,000 e^-	256^2	72	7.4	6.6	6.3
(nbL, L', nbM)	(24,000 DN)	128^2	20	6.1	5.2	4.9
		64^2	6.5	4.8	4.0	3.7

[a] Minimum exposure time in this mode.
[b] Saturation magnitude for Exp_{min} (0.″3 pixel^{-1})

5. Performance

Laboratory tests have shown that ALICE is capable of sustaining data rates for $256^2 \times 4$ bytes pixel^{-1} frames of $\sim$115 Hz for destructive reads and $\sim$85 Hz for non-destructive reads (Chapman et al. 1990). Real-time, post-detection image sharpening via a shift-and-add algorithm keyed on the brightest pixel in a sub-area, can be driven on the full array at $\sim$35 Hz. Unfortunately, the IRCAM3 array cannot be driven at these rates (Puxley et al. 1994). Table 1 summarizes the available ALICE/IRCAM3 observing modes. The ALICE/IRCAM3 observing regimes include STARE, ND_STARE, CHOP, ND_CHOP, and Shift-and-Add, presented to the observer via Starlink's SMS user interface.

6. Future Plans

The immediate plan is to install and commission the second ALICE as part of the CGS4 upgrade beginning in 1994 November. The major parts of this program are modifications to the CGS4 instrument (hardware) to handle the smaller pixel size in the larger array (30 μm as opposed to 76 μm), the introduction of algorithms into the automated data reduction system to compensate for curved slits and so forth, and the upgrade of the ALICE software to C. Another ALICE has been built for use with IRCAM2 on the 4.2 m William Herschel Telescope on La Palma in the Canary Islands. A second-generation ALICE is being built at the ROE to drive the mid-infrared imager/spectrometer (Michelle). Given the modularity of the ALICE design, new components such as faster ADCs or T9000 TRAMs should easily be accommodated, improving performance by a factor of $\sim$10 each.

References

McLean, I. S., Chuter, T. C., McCaughrean, M. J., & Rayner, J. T. 1986, in Instrumentation in Astronomy VI, Proc. SPIE, Vol. 627, ed. D. L. Crawford (Bellingham, SPIE), p. 430

Mountain, C. M., Robertson, D. J., Lee, T. J., & Wade, R. 1990, in Instrumentation in Astronomy VII, Proc. SPIE, Vol. 1235, ed. D. L. Crawford (Bellingham, SPIE), p. 25

Pickup, D. A., Sylvester, J., Paterson, M. J., Puxley, P. J., Beard, S. M., & Laird, D. C. 1993, in Infrared Detectors and Instrumentation, Proc. SPIE, Vol. 1946, ed. A. M. Fowler (Bellingham, SPIE), p. 558

Kelly, B. D., McNally, B. V., & Stewart, J. M. 1993, in Astronomical Data Analysis Software and Systems II, ASP Conf. Ser., Vol. 52, eds. R. J. Hanisch, R. J. V. Brissenden, & J. Barnes (San Francisco, ASP), p. 305

Chapman, A. R., Beard, S. M., Mountain, C. M., Pettie, D. G., Pickup, D. A., & Wade, R. 1990, in Instrumentation in Astronomy VII, Proc. SPIE, Vol. 1235, ed. D. L. Crawford (Bellingham, SPIE), p. 25

Puxley, P. J., Sylvester, J., Pickup, D. A., Paterson, M. J., Laird, D. C., & Atad, E. 1994, in Instrumentation in Astronomy VIII, Proc. SPIE, Vol. 2198, ed. D. L. Crawford (Bellingham, SPIE), p. 350

The SAX-LEGSPC Data Reduction and Analysis System: An Example of a Minimalist Approach

F. Favata, A. N. Parmar, U. Lammers, G. Vacanti, M. Busetta

Astrophysics Division, ESA/ESTEC, P.O. Box 299, 2200 AG Noordwijk, The Netherlands

J. J. Mathieu, P. Isherwood

Mathematics and Software Division, ESA/ESTEC, P.O. Box 299, 2200 AG Noordwijk, The Netherlands

Abstract. We present the data reduction and analysis system for the Low Energy Gas Scintillation Proportional Counter to be flown on the Italian-Dutch X-ray satellite *SAX*. The design philosophy of the system is presented, describing the design constraints and discussing the various choices made. Also, the problems encountered in following the chosen approach are described. In particular, the advantages and disadvantages of using FITS for all the steps in the data analysis chain are discussed.

1. Introduction

The Low Energy Gas Scintillation Proportional Counter (hereafter, LEGSPC) is one of the instruments flying on the Italian–Dutch X-ray mission *SAX* (Scarsi 1993), which will be launched in early 1996. The instrument is being provided from the Astrophysics Division of the European Space Agency at ESTEC, and is described in detail by Favata & Smith (1989), Parmar et al. (1990), and Erd & Bavdaz (1992). It is an imaging spectrometer, sitting behind a set of nested double cone mirrors that approximate Wolter type I.

We are currently implementing the data reduction and analysis system which will allow the observer to go from the raw telemetry tapes (or "final observation tapes," FOTs) being provided by the spacecraft operator (Telespazio), to the extraction of scientific results.

2. Design Choices

Given the characteristics of the detector and the limited resource level, we have made some radical design choices early on.

2.1. Detector Characteristics, Strong Points, and Limitations

The LEGSPC has a large energy coverage (almost two decades in energy) with good spectral resolution (actually better than current CCD detectors at the lowest energies, reaching about 28% full width at half maximum (FWHM) at the C 280 eV line), with relatively modest imaging capabilities (about 1 arcmin

FWHM on-axis, with rather extended wings). Additionally, the front window has a support structure in the form of a grid with an approximate 4 arcmin pitch, and the mirror point response function rapidly assumes an irregular shape with increasing off-axis angle. Therefore the best usage of the instrument will be for spectroscopy of on-axis point sources, and the complete data-analysis chain is tuned to this purpose, rather than, for example, the analysis of extended sources.

On the other hand, one of the nice features of the LEGSPC is that the raw data coming out of the detector are already rather clean, and only need a relatively small amount of processing before being scientifically useful. Two ^{55}Fe calibration sources are constantly shining in the field of view provide continuous monitoring of the gain, and therefore an easy to use reference point against which to monitor the performance of the detector.

2.2. Resource Limitations

The level of resources available internally to support the data analysis system set-up activities was limited to about 2.5 person-years per year, starting about 2 years before launch. This limit meant that nice-to-haves were excluded from the beginning and, given that we had to handle the data from the raw telemetry state down to final science, we had to concentrate on the truly essential tasks.

2.3. Baseline Design Requirements

Our baseline design approach has therefore been to set up an analysis system which would satisfy several criteria: The system must allow for cleaning of the LEGSPC photon lists, and remove all known instrumental effects (e.g., linearization of energy and x-y coordinates). It must allow for extraction of cleaned spectra of on-axis point sources for further scientific analysis. The system must also be optimized for in-house usage; its eventual integration with software for other *SAX* instruments, and eventual distribution and support to guest observers, must be provided by a team set up by the Italian Space Agency (ASI). Finally, although the system was written keeping portability in mind, it is being developed and run only on SunOS platforms, with no effort being made to provide multi-platform support.

3. Design Choices

To be able to complete the task within the limited amount of resources available we have decided to adopt a minimalist approach, i.e., to reuse as much available software as possible, and to concentrate on the instrument-specific stages (i.e., event cleaning and linearization, building of the detector response matrix). We left the generic tasks (image display, spectral extraction and fitting, etc.) to existing software.

3.1. File Formats: FITS and Others

First and foremost among the design choices was the selection of a file format. Driven by the aim of recycling existing software as much as possible we have chosen to adopt FITS, and in particular the use of binary tables for storing event lists. Given the large number of tools available for the reduction and analysis of *ASCA* data in FITS format, and given that FITS will most likely be adopted by forthcoming X-ray missions, we decided to adopt, as much as possible, a format adhering to the specifications for *ASCA* event lists and house-keeping (HK) files as produced by the HEASARC group at Goddard Space Flight Center. This immediately gave us the possibility of using the set of software tools produced by HEASARC known as FTOOLS. These include tools for performing elementary operations on FITS binary tables, tools for selecting events from a table based on arbitrary complex selection criteria, and building 1-D and 2-D histograms (images) from event lists, etc.

We have additionally decided to use FITS as the format for the calibration files, adopting an approach similar to the one used by HEASARC for the ASCA calibration database. We have used, for setting up the calibration database, the same software tools distributed by HEASARC (**caltools**), which allow to keep the indexing of calibration products and to retrieve automatically the correct calibration product for the observation being analyzed. As for the format of the calibration files, we have found that many of the HEASARC suggested formats were, while aiming to be very general, too complex or unnatural for our instrument, and have therefore decided to define our own formats which follow the data types at hand in a more natural way.

3.2. Disadvantages of FITS

While the advantages of using FITS for our data from the start of the chain should be evident from the previous section (e.g., the large body of software usable as-is, the ease of archiving and of sharing the data), we have found that this comes at the price of a significant amount of work in defining the files formats and keywords and in checking that the chosen format is actually compatible with the large body of software that we are going to use. As for the other supposed disadvantages of FITS, namely the inefficiencies both in terms of sheer file size and of I/O requirements, we have found that, with our typical data sizes (an observation file set being of order 10 MB in size) and with the speed of modern machines and the size of modern disks, these disadvantages are almost negligible, and that the choice of a more efficient but non-standard file format is therefore not justified.

4. The Reduction System

Starting from the raw telemetry tapes the data files go through the following steps:

1. conversion to FITS

2. non-interactive linearization

3. non-interactive selection of Good Time Intervals (GTI)

4. (optional) interactive selection of Good Time Intervals (GTI)

5. filtering of the data based on the GTI files produced above

6. interactive extraction of the spectrum of a source (PHA file), using XSELECT and SAOimage as image analysis tools

7. building of the response matrix for the extracted spectrum file

8. spectral fitting of the resulting spectrum using XSPEC

5. Calibration Data

To analyze the calibration data, thanks to the fact that the data produced during ground calibration by the electronic ground support equipment have essentially the same format as the flight data, we are using the same tools and the same approach as we will take for the flight data. This gives us the possibility of thoroughly testing the data analysis chain before flight.

6. Conclusions

The adoption of FITS and the extensive recycling of existing software has allowed us to put together a complete data analysis chain for the *SAX* LEGSPC detector using a relatively small amount of resources. This task would not have been possible, within the same resource constraint, by using a more traditional approach of re-designing everything from scratch in-house, even using fancy environments such as IDL.

References

Erd, C., & Bavdaz, M. 1992, in EUV, X-Ray, and Gamma-Ray Instrumentation in Astronomy III, SPIE Conference Proceedings, Vol. 1743 (Bellingham, SPIE), p. 133

Favata, F., & Smith, A. 1989, in EUV, X-ray and Gamma-Ray Instrumentation for Astronomy and Atomic Physics, SPIE Conference Proceedings, Vol. 1159 (Bellingham, SPIE), p. 488

Parmar, A. N., Smith, A., Bavdaz, M. 1990, in Observatories in Earth Orbit and Beyond, ed. Y. Kondo, (Dordrecht, Kluwer)

Scarsi, L. 1993, A&AS, 97, 371

The Data Analysis System For The PDS Detector On-board the SAX Satellite

D. Dal Fiume, F. Frontera[1], L. Nicastro, M. Orlandini, and M. Trifoglio

CNR/TESRE, via Gobetti,101, I-40129 Bologna, Italy

1. Introduction

PDS is the high energy (15–300 keV) instrument on board the *SAX* satellite (Scarsi 1993). The launch is currently foreseen for the end of 1995. PDS is a telescope composed of a square array of four phoswich (NaI(Tl)/CsI(Na)) units for a total geometric area of 800 cm^2. The detector is surrounded by active shields (CsI(Na)) on the lateral sides and by a thin plastic scintillator on the front side. The field of view is limited to 1.4^o by means of graded hexagonal mechanical collimators (Ta–Sn–Cu).

The telescope is operated via an analog processor that analyses the analog signals coming from the various subsystems composing the telescope. The result of this analysis is then processed by two digital processors (PDS Intelligent Terminal, hereafter PDS IT): one dedicated to data acquisition from the analog processor, and the other dedicated to handle data transfer and communications with the On Board Data Handling (OBDH). More details on the detector and on its performances can be found in Frontera et al. (1991).

Tests on the detector electronics have already begun on the LABEN premises (Vimodrone). Tests on the integrated detector started in 1994 October, and calibrations are scheduled to begin soon after. In-flight deep calibrations and performance verification will take place in the first few months after commissioning of the satellite. In the operative phase, routine calibrations will be performed on a typical rate of one per day and long term history files, including some quantities relevant for calibrations and long term trend analysis, will be stored by Telespazio at SDC (Scientific Data Center located on the Telespazio premises in Rome). In the first part of the mission, a relevant part of these data must be processed timely by our group in order to produce the first set of calibration data usable by *SAX* PDS end-users. The production of publically available calibration data should be gradually taken over by SDC, during the operative life of the satellite.

In this report we describe the data collection, analysis and archival system in development at TeSRE to host the data of PDS.

2. Data Acquisition During Ground Tests and Calibrations

During the ground calibrations, data from PDS will be acquired using LABEN ITE (Instrument Test Equipment), a complex data acquisition system composed by a probe of the data bus on *SAX* (BTB probe) and a VAXstation.

---

[1] Also Dipartimento di Fisica Università di Ferrara, Ferrara, Italy.

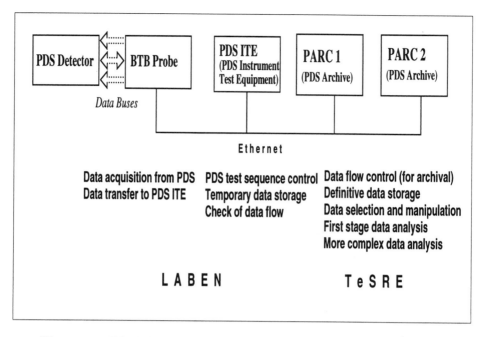

Figure 1. Block diagram of the data flow during the on-ground tests and calibrations.

Data comes from the PDS IT, which is composed of one analog processor and one digital processor. This latter is composed of two 80C86 microprocessors, on the SAX BUSes: the Response Bus (RB) and the Block Transfer Bus (BTB) (see Figure 1). The RB carries information on the health of the instrument (engineering HK, echos, responses to interrogations from IT to the central data handling system of SAX). The BTB transfers the scientific data in packets ready to be inserted in SAX telemetry. On SAX these buses connect all the ITs (one for each telescope) to the On Board Data Handling system (OBDH).

During the ground tests the RB and the BTB are probed by a dedicated instrument developed by LABEN called the bus probe. It retrieves output data from PDS coming on the RB and BTB and sends these data via TCP/IP to a VAXstation, where they are stored in files on hard disk. The LABEN ITE can control the data flow, with a dynamic display of the packets being transferred from PDS via bus probe. Data from LABEN ITE are then transferred to the PDS Data Analysis System, where the analysis and archiving are performed. These data are translated in native PDS format, exactly as they will be transmitted in SAX telemetry.

3. Data Acquisition, Event Handling, and Data Archiving

The front end to data coming from PDS is composed by some modules that: (1) reformat the input data, (2) handles events (as changes in the PDS configuration are made), (3) handles interprocess communication (as the scheduling of programs or information passes through the pipes), (4) archives the data in the PDS relational DataBase, and (5) allows for the display of data and performs data reductions/accumulations.

The **reformatter** is a module that transforms the data in PDS "native" format to the PDS data format of Final Observation Tapes (FOTs). This transformation will allow to have our archive data files with PDS packets in a single format for life of the PDS both pre-operational and operational. Actually, this reformatting is needed because the data distributed in FOTs are not a straight copy of telemetry data, but have undergone extensive reformatting, mainly consisting of the expansion of compressed data to an integer number of bytes. We will perform this reformatting, done during the operational phase at the Observation Control Center (OCC) operated by Telespazio, before archiving; recreating a data and file structure closely resembling that of the final data analysis of PDS. Prototypes of this reformatter were tested in 1994 October.

The **event handler** is a module that processes external events and notices, via message servers, the other applications running, of the type and effect of the noticed event. The event handler parses and analyzes the log-file produced by the "test sequence", a sequence of commands, macrocommands, actions, data produced by the LABEN ITE and that gives complete information on the operations performed on the PDS detector during the test session. The event handler is being developed.

The **message server** is a background process that handles requests to broadcast messages to the other running applications. The **process controller** is a deamon that handles inter-process communications and requests of process priorities. The message server and the process controller are now in prototype form and are being tested and completed.

4. The Insertion in the PDS Relational DataBase

The PDS Archival is based on the INGRES (t.m. of Ask Corporation) relational database. For each calibration record, we foresee archiving the following information: (1) the location of all the ancillary files related to a calibration, together with all the information needed by a user in order to analyze and reconstruct the calibration, (2) the results from the first level analysis (first processing of spectra and pseudo-images), and (3) the results from the second level analysis (PHA/ADC channel conversion, trend analysis, etc).

External tools to perform more sophisticated analysis can be used on data extracted from the database. The results of this analysis will be again archived in the database for further use. To this goal, we have built a Graphical User Interface (GUI) with Windows4GL scripts within the DataBase itself. The tasks that can be performed by means of the GUI are: (1) viewing database records, (2) creation, insertion and update database records (allowed only to authorized users—AU's), (3) the extraction of sets of records satisfying user conditions, (4) plotting one parameter vs. another (a correlation plot), and (5) fitting these correlation plots with linear, semilog or power law relations.

The View/Edit frame shows a list of all the records present in the database, that can be browsed with a scroll bar. For standard users, only the **View** and **Close** operations will be allowed. The **Create** and **Edit** tasks are accessible only for AUs. For AUs it is possible to perform three operations on the record: to insert the record in the database; to update the record, i.e., overwrite the record with the same ID Number, and delete the record.

Another operation possible from the View/Edit frame is the extraction of a subset of records that satisfy a search condition. The Selection frame allows the user to select particular values of the attributes and extract all the records in the database that satisfy the conditions imposed on the attributes. The subselections can be made in *cascade*. This means that the user can extract a subset from a subset, and so on, until the right sample is found.

By selecting the second task shown in the starting frame, we enter in the Plotting/Fitting frame. The user is asked to select the two attributes to be plotted, and to select the corresponding error columns, if present. The Fitting menu allows the user to access four kinds of fitting relations: linear, semilog (both in x and y), and log-log.

5. The Data Reduction and Display

IDL has been chosen as the graphical representation and manipulation tool for the *SAX* PDS data. We took advantage of the capability of IDL to build GUIs by writing user friendly, widget based, programs. In doing this, we assume that users have access to an X11 based System. IDL has seven basic widgets but several "compound widgets", and User Library routines that use the widgets are available. This allows most of the functionalities of our programs to be available just by "point and click".

Spectra, time series, and pseudo-images (counts vs. rise time vs. energy channel) are all accessible by a single, widget-based, program. The three types of data will have three different user interfaces with the relevant data shared using common blocks and temporary disk files. Default start-up settings are read-in from a user defined ASCII file. Other information exchange with the user uses both widget based windows and an IDL command window.

Thanks to the capability of IDL to spawn external processes via the SPAWN procedure, and to access external routines contained in sharable object libraries via the CALL_EXTERNAL function, once the data have been displayed, manipulated, and in some way selected, further analysis (e.g., model fitting) can be performed by invoking external tools. The results can then be returned to IDL for display and further processing.

References

Frontera, F., et al. 1991, Adv. Space Res., 11, 281
Scarsi, L. 1993, A&AS, 97, 371

Astronomical Data Analysis Software and Systems IV
ASP Conference Series, Vol. 77, 1995
R. A. Shaw, H. E. Payne, and J. J. E. Hayes, eds.

Use of Inheritance Techniques in Real-Time Systems under DRAMA

T. J. Farrell, K. Shortridge

Anglo-Australian Observatory, P.O. Box 296, Epping N.S.W. 2121 Australia

Abstract. In real-time instrument control systems such as the AAO's DRAMA system, where relatively complex individual tasks control their own parts of a system (spectrograph, camera, telescope, etc), there are often significant similarities between tasks. At AAO, all DRAMA tasks have a number of standard "actions" such as "INITIALISE", "RESET" and "EXIT", and all camera control tasks provide actions such as "WINDOW", "EXPOSE" etc. We have developed a mechanism which allows tasks to "inherit" code from a generic instrument task, a generic camera task, etc. This makes the initial coding easier and also enforces the standard behavior we require of our tasks.

Such techniques for inheriting the behavior of software objects have become popular over recent years, with most attention focusing on C++ and similar languages. DRAMA is written in C, but combines ideas from the X Windows Xt toolkit with its use of named actions in tasks to provide a successful, effective, inheritance mechanism.

1. DRAMA - A Quick Introduction

The DRAMA data acquisition environment is summarized elsewhere in these proceedings (Farrell, Bailey, & Shortridge 1995). For the purposes of this paper the important point is that DRAMA systems are built from individual tasks each responding to named "actions". These actions are invoked by name through messages sent from other tasks—either user interfaces or intermediate level control tasks.

2. Standards

DRAMA imposes no limits (other than an arbitrary 20 character length) on the names of actions. However, there are clear advantages to having standards imposed. For example, most tasks have an initialization action, but some authors will call it "INITIALIZE" and others "STARTUP". Some authors may specify "EXPOSE" to start a CCD exposure and others will use "RUN". This makes it hard to use an unfamiliar task; consistency is a virtue always emphasized by user interface guidelines for windows-based programs, and it applies here as well.

In 1988 the AAO began imposing well-defined standards on all its new tasks—which then were being written in the earlier ADAM system. All tasks were to obey a "Generic Instrumentation Task" standard. This specified, for example, the use of an action named "INITIALIZE" for initialization and "EXIT" to shut down a task. In addition, we took this technique one stage further and defined "Detector Tasks", "Data Handling Tasks", etc. Each of these standards "inherited" the "Generic Instrumentation Task" standard and imposed its own additions.

The benefits of such a scheme are greatly increased when you have to write complex control tasks—tasks which coordinate other tasks. A control task can now be designed to allow you to swap lower level tasks in and out as required— say changing the type of detector system used by changing the detector task, while leaving the data handling task unchanged. In addition, the control task's initialization and reset code is easier when all tasks behave in the same way. This scheme played a major part in the success of the AAO's "OBSERVER" CCD/Infrared System.

3. Improvements with DRAMA

Under ADAM, the major benefit to this scheme came purely from having consistent interfaces to the tasks. Coding the tasks themselves was not made significantly easier (except by some 'cutting and pasting' of existing code). The top level of an ADAM task has a fixed application routine, invoked for all actions, with a big, fixed, IF-ELSE-IF sequence that calls the appropriate action routine based on the action name passed to it. This rigid structure did little to ease the inheritance of existing code.

When we started to work on DRAMA, we imposed similar standards on the DRAMA tasks. It turned out that a few relatively simple changes in style made by DRAMA solved the inheritance problem in a very elegant way. DRAMA drops the requirement for a fixed application specific routine with a big IF-ELSE-IF sequence. Instead, the application sets up an association between action names and C routines. When an 'obey' message with a particular action name arrives, the appropriate C routine is invoked directly by a 'fixed part' of the task common to all DRAMA tasks. The association between actions and C routines is dynamic and may be changed by the task itself at any time. This gave rise to the following implementation of generic standards:

- Each defined standard is associated with a package of C routines.

- Each such package has an "Activation" routine.

- The activation routine of a package sets up the association between action names and C routines, defining default implementations of all actions defined by that standard.

- An application "inherits" standard packages by invoking the activation routine of each standard it is to meet.

- The application then sets up action name to C routine mappings for new actions specific to the application and for cases where it wishes to override the default implementation of standard actions.

For example, an AAO camera control task must obey the "Generic Instrumentation Task Specification" and the "Generic Camera Task Specification". In

DRAMA, such a task can pick up default implementations of all required actions by calling the activation routines of the packages associated with the standards.

4. Implementation of Packages

The easiest way to implement a Generic Package is to provide "null" routines for each action defined in the package. This does not gain you very much. Each application would have to override almost all Generic actions. The real power of such a scheme arises when it is possible to provide standard implementations of generic actions. For some generic actions, such as "INITIALIZE" this is probably not possible. However, for other actions it is obvious. An example is our "SIMULATE_LEVEL" action. This action is used to set the simulation mode of our tasks. All it has to do is to read its argument and check that it is valid. If so, it sets a global task parameter to that value. Most of our ADAM tasks ended up with their own versions of this action, often slightly modified versions of older implementations. In our DRAMA systems, there is only one such implementation, which is available when a task inherits the generic package.

Taking this idea further, many of our DRAMA programs are also written as packages that can be inherited. Consider a DRAMA program to run a simple Video Frame Grabber (VFG). By inheriting the "Generic Instrumentation task" and "Generic Camera Task" packages, almost all the required actions are defined and our program is quite simple. By inheriting the VFG package other DRAMA programs can add all the basic actions needed to control a VFG.

Although a similar effect could be obtained more conventionally by use of a VFG subroutine library, this scheme is much simpler. A subroutine library might provide the necessary action handlers, but these would still have to be associated explicitly with the named actions, and the various parameters needed would still have to set up for them. The inheritance scheme requires one simple subroutine call, with little more than a status argument, at the start of the program to give a task that has all the standard actions built-in.

A subroutine interface can of course be used to simplify the writing of actions that override the standard actions. In the case of the "Generic Camera Specification", a WINDOW action is defined to set the detector window size. A C interface to the standard action is provided so that an inheriting package may add to the semantics of the WINDOW action. For example, the basic specification does not use 'binning' (the on-chip summing of rows or columns of data). A inheriting task can add binning to the WINDOW action without rewriting the standard argument handling in the package. It redefines the WINDOW action to call the provided interface and then handles any additional action arguments itself.

5. Inheritance Techniques

We often describe our DRAMA tasks in an object-oriented way—a DRAMA task is a software object which receives and initiates messages. The package technique gives us a way of adding inheritance to this. The style is similar in some respects to that used by the X Windows Xt toolkit to inherit widgets. A significant difference is that in DRAMA existing handler routines are always overridden instead of being added to a list of routines to be executed. This is not a fundamental restriction and such a facility could be added. However, it was felt that in the real-time systems we are building it is better to avoid the overhead associated with searching such a list. We also gain here by having the behavior of the program clearly defined, a requirement of real-time systems.

6. Results

This approach has been used with considerable success in the AAO's Two degree Field (2dF) project. Using C, it has allowed DRAMA to provide many of the benefits normally expected from a well designed object-oriented scheme without the problems sometimes associated with purer O-O implementations.

References

Farrell, T. J., Bailey J. A., & Shortridge K. 1995, this volume, p. 113

Interfacing the Tk Toolkit to ADAM

D. L. Terrett

Rutherford Appleton Laboratory, Chilton, Didcot, Oxon OX11 0QX, U.K.

Abstract. This paper describes how Tcl and the Tk toolkit have been extended to send and receive Adam inter-task messages, and how it has been used to build graphically-oriented user interfaces for various data reduction tasks.

1. Introduction

The Starlink software environment (Adam) is used for telescope control (on the JCMT & UKIRT), instrument control and data acquisition (at the AAO, La Palma, UK Hawaii telescopes), and data reduction (throughout the Starlink network). In the real-time environment of telescope and instrument control, an Adam system consists of several independent tasks cooperating by exchanging messages using the Adam inter-task messaging systems. Usually, one or more of these tasks has the job of providing the user interface.

When used for data reduction, there are also several tasks in the system but usually only two—the user interface and one of the data processing tasks—are active at any one time. This is because, although the system allows several data processing tasks to run simultaneously, controlling such a system with a traditional command line interface is clumsy and confusing.

2. User Interfaces

The Adam software environment was designed to support a variety of user interfaces; there are three currently in use:

ICL A command line interface widely used for telescope and instrument control and for data reduction.

ADAMCL A command line interface, now considered obsolete, but still used by one of the UKIRT instruments. It only runs on VMS.

SMS A menu interface that runs on VT type terminals. It is used by several telescope and instrument control systems. It only runs on VMS but an X windows equivalent that will run on UNIX is being considered.

Adam data reduction tasks can also be run without any separate user interface (i.e., directly from the operating system shell) as each task has its own, built in, simple user interface.

A graphical user interface is widely seen as the best way of satisfying the demand that Adam systems should be easier to use, and Tcl and the Tk toolkit have been selected as the most promising tools for writing such an interface. Other approaches considered were:

- Write a GUI in C using an X toolkit such as Xt or Motif. This was rejected because of the time and effort required to get a prototype that would work well enough to be evaluated and to be useful for learning what styles of GUI would be most appropriate.
- Adapt an existing GUI such as AVS. Pilot experiments with AVS and XLips indicated that the behavior of Adam tasks does not match that of the GUI well enough for this to be successful.

Tcl and Tk (Ousterhout 1994) were selected because of the ease with which prototypes can be generated and the freedom this gives for exploring a wide range of interface styles in a reasonable time. Its status as freely available software, and its acceptance in a range of disciplines (including astronomy), are also attractive features. An Adam user interface has to communicate with Adam tasks. This means extending Tcl with commands for sending and receiving messages via the Adam inter-task messaging system.

3. The Message System

A typical Adam system consists of a user interface process plus one or more Adam tasks (processes); each task supports one or more commands (up to several hundred in the case of some data reduction tasks). The tasks are independent, and can all be processing a command at the same time.

The user interface uses the message system to control the task. There are four types of message that can be sent:

SET Sets the value of a parameter (parameter values are stored by the task).

GET Gets the value of a parameter.

OBEY Tells the task to execute a command. A typical data reduction task supports commands which, to the user, look like separate applications.

CANCEL Cancels the command currently being obeyed. This is only used in telescope and instrument control tasks.

SET, **GET** and **CANCEL** messages are acknowledged by the task with a single message; an **OBEY** can result in the exchange of many messages between the user interface and the task. These messages are used by the task to request new values for parameters and to display information to the user.

4. New Tcl Commands

The following Tcl commands have been created to allow Tcl to control Adam tasks:

adam_init Connects the Tcl application to the Adam message system.

adam_send Sends a message to a task.

adam_reply Sends a reply to message from a task.

adam_receive Reads an incoming message from the message queue.

adam_exit Closes down the message system.

If there are no messages in the message system queue when **adam_receive** is called, it will block until a message arrives. In a Tk application this will cause the user interface to "hang". For a GUI that is tightly bound to a single application, this is acceptable (in fact, it feels quite natural—the interface freezes while the application is busy, just like a conventional GUI application with the interface and application in a single process). However, for a user interface that controls several tasks, it is unacceptable.

The solution adopted (pending possible modifications to the message system) is to use two processes: one running the user interface and not communicating directly with tasks, and another, called the relay process (also written in Tcl), that does. The user interface sends messages to the relay task using the Adam message system, which forwards them to the Adam tasks. Replies from the Adam tasks go to the relay task, which forwards them to the user interface via a pipe. The relay task can safely call **adam_receive** since it does not have to respond to X events and the user interface never has to call **adam_receive** since all incoming messages are delivered via a pipe. The I/O on the pipe can be integrated with the X event loop with *Tcl_CreateFileHandler*.

5. GUI Architectures

When a GUI is being added to an existing application, the GUI has to handle prompts and information messages from the application; the number of messages and the order in which they arrive is unpredictable (except for very simple applications). Information messages can be difficult to display appropriately; for example, they may contain instructions to the user that only make sense in the context of a command line interface. Interactive graphical applications will bypass the user interface when doing the interaction and this results in an inconsistent interface style. Consequently, existing application will usually need some modification before a satisfactory GUI can be constructed.

An Adam application can be written specifically to be run by a GUI by ensuring that it never requests parameter values from the user interface and by formatting messages to be interpreted by the GUI rather than the user.

References

Ousterhout, John K. 1994, Tcl and the Tk Toolkit (Reading, Addison-Wesley)

The Stellar Dynamics Toolbox NEMO

P. Teuben

University of Maryland, College Park, MD 20742

Abstract. NEMO is a toolbox for users and programmers with which a wide range of experimental situations can be constructed, dynamically evolved, and analyzed. Primarily geared towards stellar dynamics, NEMO is by no means restricted to it. Great importance has been given to making the package extendible and to flexibly import and export data. NEMO has been implemented within, and integrated into, the UNIX environment. A unified command syntax provides users with a simple help facility, and provide programmers with a ready-made user interface. A general method for hierarchically structured binary data files is used to communicate between programs. We also describe a proposal to adopt FITS BINTABLEs as a basis for interchanging and archiving simulation results.

1. Introduction

With the advent of powerful workstations in the mid 1980s, the use of N-body simulations has become increasingly widespread. It was realized (e.g., Hut & Sussman 1986; Hut 1989) that this part of the astronomical community would benefit from a programming environment similar to what many observers have had for years in packages like AIPS and IRAF. Most researchers in the field have their own private packages—some of them with a large degree of sophistication. A first attempt to structure these efforts into a package was reported by Barnes et al. (1988), and became known as NEMO. In this paper we report on the current status of NEMO (see also Teuben 1994) and its future prospects.

2. System Components

To the user, NEMO appears as a collection of loosely coupled programs. The following components can be identified within NEMO:

A common user interface is shared by all programs; all programs are executed from the UNIX shell and accept a list of *keyword=value* pairs on the command line. There are **program** and **system** parameters, the latter shared by all programs. These system parameters control global properties such as error handling, debug output, graphics devices, etc. The user interface allows you to switch to a different look-and-feel user interface. Examples have been constructed in interfacing NEMO with the Miriad (an AIPS-like interface, see Sault et al. 1995) and KHOROS/CANTATA (a visual programming language) packages. In practice, however, we find shell scripts ("batch mode") the most powerful and self-documenting mode of operation.

A common file-structure is shared by all programs. Data, used to communicate between programs, is stored in files in a hierarchical structure of name- and type-tagged items, in binary format. Items can be single scalars, multi-dimensional arrays, and even arbitrary (C-type) structures. In addition, each program can have an input and output channel which is connected to a UNIX pipe instead of a disk-file. Since many programs transform one dataset into another, this allows for efficient chaining to build new tools.

On top of this file-structure, a number of packages have been defined. The most important one is *snapshot*, on which NEMO was originally based. Subsequently *images*, *tables*, and *orbits* were introduced, with a number of programs that convert data between them.

A common graphics interface is defined on the Applications Programming Interface (API) level. One simply needs to link with any graphics library (e.g., MONGO, PGPLOT, X11, PostScript, SunView, or GL) to create a working program. This hides the complexity of different graphics programs but is, of course, not as flexible since a fairly low common denominator has to be chosen for the API.

Programs can use a dynamic loader in order to compile, load, and run user-specific code. This supports very flexible and powerful analysis tools. For example, in orbit integration, a user-defined potential can be used to integrate and analyze orbits. And for snapshots, arbitrary user-defined expressions are used for many analysis and plotting programs.

3. Documentation

Extensive on-line help is present on three levels. First of all, each program comes with a dynamic inline help, which can remind the user of the keywords, their default values, and a small one line description of each keyword. It is part of the user interface for which the programmer is responsible. The advantage of this lowest help level is that it is much less likely to be out of date than the other levels, and does not necessarily need the entire NEMO environment to run.

The second form of help is a standard UNIX manual page created for each program. This is an example of how NEMO is integrated into the UNIX environment and where "standard" tools like xman and tkman can be immediately used for powerful (hypertext) browsing.

The third form of help is the extensive "Users and Programmers Guide." It contains numerous examples of how to use, program, and extend the toolbox. NEMO information is also available on the World Wide Web[1].

4. Future

After the on-going conversion of supporting Solaris 2.x, NEMO will have formal ANSI C and C++ support. Also, support for other packages, such as Starlab (Hut, Makino, & McMillan, private communication), and TIPSY (Katz & Quinn, private communication) is on-going.

---

[1] http://astro.umd.edu/nemo

5. Example

To illustrates some of the features in NEMO, the following C-shell script sets up an encounter between two 256 body Plummer models, evolves the encounter, and plots the evolution in a 3 by 3 panel:

```
mkplummer out=mod1 nbody=256
mkplummer mod2 1024/4
snapstack in1=mod1 in2=mod2 out=mod5 deltar=10,2,0 deltav=-0.6,0,0

hackcode1 in=mod5 out=mod5.out freqout=1 tstop=80 > mod5.log

snaptrim in=mod5.out out=- times=80 |\
    hackforce - - |\
    snapcenter - - 'weight=-phi*phi*phi' report=f |\
    unbind - - |\
    snapplot - xrange=0:9 yrange=-3:3 nxticks=5 nyticks=5 \
        xvar=r 'yvar=(x*vy-y*vx)/sqrt(x*x+y*y)' \
        times=0,10,20,30,40,50,60,70,80 nxy=3,3
```

Note that command line parameters do not have to be specified by name but can also be passed by order. UNIX pipes can be used to chain commands, in which case the files which denote the ends of the pipe must be denoted with a "-" symbol. Lastly, numeric values can generally be given as arbitrary expressions.

6. N-Body Data Interchange Format

Over time, NEMO has accumulated a number of routines that support import from, and export to, a large variety of ASCII and binary formats. A draft was written (Teuben 1994) where the FITS BINTABLE (Cotton, Tody, & Pence 1995) has been used to define a conceptual format with which different packages can exchange their simulation results. The FITS BINTABLE extension is well documented, and allows for efficient storage of binary tables. Each row represents a particle, and different columns represent different attributes of the particle (a column can be a vector), such as mass, position, velocity, potential, density, etc.

The BINTABLE format is merely a prescription for storing tabular data. Any FITS reader should be able to read (and/or skip) this NBODY data, but only specialized readers would know what to do with this data. There are a number of implementation details that still have to be decided. We note the following:

- A naming convention for the column names (the so-called TTYPE's) is required. This will be the most visible part of the implementation. Apart from registering a new extension name (e.g., NBODY) with the FITS standards office, the community needs to agree on things like the column names, units, coordinate systems, etc.

- Datasets with different kinds of particles which have different attributes. For example, SPH particles have extra attributes (such as temperature and density) which pure N-body particles formally don't have. It would not be very efficient to tag along such unneeded data attributes. The "variable length array" construct within the BINTABLE proposal could solve this problem, but this has to be worked out in more detail. A more obvious solution is to store them in separate FITS extensions. This is clearly a much simpler and easier to understand implementation, but requires FITS readers to look ahead.

- Some variables are global variables. There may be more efficient ways to store these values. For example, if all the masses (**MASS**) are the same, they could be stored in the header (which is, however, in ASCII). Another example is **TIME**. Although, in principle, each particle may have it's own variable time, a snapshot has, by definition, a single time shared by all particles. The format needs to allow for both.

- For certain complex situations, extra data structures may be needed to fully describe an experimental situation. For example, descriptions of hierarchical double and triple stars could be stored in an accompanying table. The names and descriptions of these extensions would need to be part of the standard, too, but they can always be added to the proposal at a later date. This is a very common technique in radio interferometry (VLBA data is stored with many associated tables).

Acknowledgments. NEMO was initially developed by Joshua Barnes, Piet Hut, and myself during the 1986–87 academic year in Princeton. I wish to thank them for the stimulating environment in those years.

References

Barnes, J. E., Hernquist, L., Hut, P., & Teuben, P. 1988, BAAS, 20, 706
Cotton, W. B., Tody, D., & Pence, W. 1995, Binary Table Extensions to FITS, in preparation
Hut, P. 1989, Celestial Mechanics, 45, 213
Hut, P., & Sussman, G. J. 1986, in The Use of Supercomputers in Stellar Dynamics, ed. P. Hut & S. McMillan (Berlin, Springer-Verlag), p. 193
Sault, R. J., Teuben, P. J., & Wright, M. C. H. 1995, this volume, p. 433
Teuben, P. 1994, PASJ, in press

Data Reduction Software for SWAS

Z. Wang

Smithsonian Astrophysical Observatory, 60 Garden St., Cambridge, MA 02138

Abstract. The Submillimeter Wave Astronomy Satellite (*SWAS*) is an unique instrument that will observe from space several important spectral lines between 538 μm and 615 μm in wavelength. In anticipation of its operation starting 1995, a set of software tools has been developed, mainly as a layered package in the IRAF environment, to allow efficient inspection, processing, archiving and retrieval of the scientific data. In this contribution, we discuss the basic design concept and a few special features of these tools.

1. Introduction

As part of NASA's Small Explorers program, preparation is currently underway for the *SWAS* mission: a 0.6 m radio telescope in earth orbit capable of carrying out pointed observations in the submillimeter wavelengths. Its primary goal is to perform spectroscopic observations of the ground state or a low-lying transition in several important atomic and molecular species that are either difficult or impossible to detect from the ground. The *SWAS* Science Operation Center, established at the Harvard-Smithsonian Center for Astrophysics, is in charge of planning the scientific observations, preliminary calibration and analysis, and archiving all science data from the mission (see also Kleiner 1995)

In essence, *SWAS* is a self-contained small single-dish telescope working at very high frequencies (Tolls et al. 1994). It has a pair of heterodyne receivers that can be independently tuned, and a single acousto-optical spectrometer (AOS) with 1400 1 MHz channels of Thomson linear CCD digital output, enabling simultaneous observations of different lines. Despite similarities with ground-based single-dish millimeter and submillimeter telescopes however, *SWAS* is unique in several important aspects of its basic design and operation that require a software reduction package specifically tailored to its needs. In a departure from conventional software design for radio telescopes, we have implemented the *SWAS* software as a layered package in the IRAF/VOS environment, making extensive use of the host's powerful features such as the command interface, image I/O, vector processing, line fitting, and graphics capabilities. We have also incorporated the STSDAS Tables facility to help manage the relatively large amount of spectral header information. The main parts of this software have been coded in SPP and are now undergoing preliminary tests.

2. Science Requirements for SWAS Software

During each of its 97 minute orbits, *SWAS* will normally point at three to five different sources and perform repeated integrations at on- and off-source positions. There will be spectroscopic chop observations for compact sources, standard nod observations for extended objects, as well as mapping mode observations. The sampling rate of the spectrometer is one full spectrum every 2 s. The actual on-off source data calibrations are performed on every 2 s spectrum in each of its two spectral bands individually (each band occupies one half of the 1,400 channels of the AOS). Moreover, because the Doppler effect resulting from the motion of the spacecraft itself needs to be corrected for the upper and lower sidebands in opposite directions, the single AOS can actually yield four different spectra in each integration.

The observations are naturally grouped into segments depending on orbit and target. Each segment consists of up to about 35 minutes of on-off source integrations, plus calibration spectra. Sources in need of longer integration times will be observed in multiple segments. This requires that the shifting (for Doppler correction) and co-adding of spectra be easy and flexible in the reduction software, not only for spectra within each segment, but also between segments.

3. Outline of SWAS Software Design

The *SWAS* data reduction software is designed to perform the following basic functions: (1) check data integrity, find time gaps and possible noise spikes, and flag bad data according to various diagnostics and ancillary ancillary (mostly instrument housekeep) information; (2) apply (ON-OFF/OFF-ZERO) and passband calibration to each of the 2 s spectra; (3) evaluate and apply wavelength calibration spectra, and derive and apply Doppler corrections to each individual sideband before co-adding; (4) perform similar processing to the broad-band (continuum) channels to establish flux calibration points; (5) recognize planet observations and perform beam centering and planet calibration calculations; (6) sort and partition the data in ways that can be easily handled by a system-wide database management tool; (7) provide a user-friendly means to sample, evaluate, and display the data at every stage of the reduction process; and (8) produce FITS data files, each containing co-added spectra of one or more segments.

Five new IRAF tasks have been implemented in a local package (currently named "*submm*") to perform the calibration in consecutive steps. It is intended to work as a data pipeline, with many adjustable parameters in each of the steps to accommodate actual data reduction needs. In normal processing however, a list of data files can also be run through these tasks in a batch mode with entirely preset parameters.

Although the raw data files are unaltered in the reduction, output data files will contain stamps recording the time of the reduction and name of the reducer. This allows different reduction procedures to be performed independently on the same dataset, and leaves open the possibility of re-applying all of the "standard" calibration procedures with different parameter sets at a latter time.

4. Several Special Features

Because of the large number of short-integration (2 s) spectra that *SWAS* is to transmit to the ground, and the fact that each of these spectra has its own set of attributes, we have designed the *SWAS* data format in a way significantly different from the conventional approach.

Both raw and calibrated spectra are written in 2-D *images*, loosely resembling an IRAF "multispec" spectral image file. However, only the information that is common to all spectra are written into the image header. Most variable parameters for each individual spectrum are instead stored in two "associated" table files. In other words, each row in the table file represents a set of attributes for a row in the corresponding spectral line image. This approach eliminates the burden of an excessively long header for the image file, and more importantly, allows structured, efficient, yet flexible operations to be performed on nearly all of the spectral parameters. Tests show that this scheme is working for our intended purposes and the processing speed is satisfactory.

Because spectra of different receiver bands and sidebands share much of the same information, the actual data storage in file is a set of 2-D images, with the third dimension being "band". Only a small number of columns in the header table file are band-specific. This avoids redundancy and improves the efficiency of data processing and storage. The associated image and table files are clearly identified by their naming convention, and are grouped according to observing segments.

We archive the raw and calibrated *SWAS* data in the same fashion. Changes to a different file format can be performed with an auxiliary task, but is necessary only to export the data reduction product to other external data reduction systems for analysis or comparison. The output data will be made available in the more conventional "header plus spectral data" fashion and written in FITS, using the IRAF multispec format.

In the calibration, there is also a problem of automatically selecting appropriate OFF and ZERO to pair with given ON and CAL spectra in order to minimize noise and artifacts. These selections can be complicated because of the continuous nature of *SWAS* data taking, and the different modes of observations (nodding, chopping, and mapping). We have adopted a simple "stacking" approach, in which the most "recent" available OFFs and ZEROs are pushed onto the stack for subsequent processing of the ONs and CALs. The advantage of this approach is that it can be applied to all observing modes with little modification. Several user-changeable parameters are built-in the processing schemes such that we may select the best way to perform the reduction depending on the exact nature of the data.

5. Some Practical Considerations

SWAS data will be relayed to the SOC on a daily basis. The estimated volume of science data alone would be 120 to 150 MB per day. This is not a huge volume by today's standard, but it does require a substantially automated pipeline system that can process bulk of the data in an unsupervised batch mode. Also, the possible need for re-processing raw data through the pipeline requires that the processing time to be relatively short. Our current design calls for a processing time of a few hours for 24 hours worth of *SWAS* data (assuming a dedicated Sparc-20 workstation with 64 MB memory).

The RF COMB spectra taken in-orbit are fitted for precise line center positions (accurate to 0.02 pixel). These positions are then used to check for possible sudden frequency changes (mode hops). A dispersion relation for each band is then derived from the COMB lines and is represented by four coefficients. Actual wavelength calibration is performed by interpolating the dispersion relation time-wise.

We are currently using both SPLOT and SPECPLOT in IRAF to display the data. Although these are adequate for showing the spectra in pixel coordinates, some modifications need to be made to incorporate the World Coordinate Systems of our data to display physical coordinates. We are also hoping to adopt the new GUI packages SPECTOOL or ASpect currently under development (e.g., Hulbert et al. 1995) for our use.

It is straightforward to convert the IRAF image and table files into FITS files for archival purposes. The current plan is to make these FITS files into CD ROM's as permanent archive and distribution media. A problem still being explored is how to search and retrieve part of the data from the archive. As IRAF itself does not have a complete database system, we tentatively decide to save the key parameters which are likely to be searched for in a commercial database management system (Sybase).

6. Outlook

Although there are a number of existing software packages for radio telescopes that all do an excellent job in performing versatile data reduction procedures, none provides the exact tools for the specific needs of *SWAS*. One the other hand, being a relatively small project, we can not afford to develop a reduction system from scratch. The IRAF environment provides a powerful, open platform upon which smaller packages like our own can be built, and they can take advantage of many of the existing tools from NOAO as well as facilities being developed elsewhere. More importantly, since IRAF itself is continuously being improved and enhanced, we are looking forward to being part of its large active user community and remaining updated in this ever-changing astronomical software environment.

Acknowledgments. My thanks to Eric Mandel of CfA, Phil Hodge of ST ScI, and Frank Valdes of NOAO for their help in software related questions. I also wish to thank the *SWAS* science team and *SWAS* SOC staff for their support.

References

Hulbert, S. J., Eisenhamer, J. D., Levay, Z. G., & Shaw, R. A. 1995, this volume, p. 121
Kleiner, S. 1995, page this volume, p. 136
Tolls et al. 1994, IEEE Proc. Vol. 2268

# A New PROS Task for Calculating HRI Source Intensities or Upper Limits

J. C. Chen, M. A. Conroy, J. DePonte, and F. A. Primini

Smithsonian Astrophysical Observatory, 60 Garden St., Cambridge, MA 02138

Abstract. The srcinten task provides PROS users with a tool to compute count rates for point sources in the *ROSAT* High Resolution Imager (HRI). Count rates are corrected for point response function, vignetting, and quantum efficiency effects. Corrected upper limits are cited at locations where the count rate falls below the source detection threshold.

1. Introduction

Approximate count rates for point sources observed with the *ROSAT* HRI can easily be calculated by using the IRAF/PROS **imcnts** task to accumulate X-ray events in a (usually) circular aperture about the source, with background derived from either a model background map or a region adjacent to the source. However, more accurate determinations require corrections for various instrumental effects, such as reduction in effective area for off-axis sources (vignetting), fraction of total source counts in the source aperture (encircled energy), and variations in detector quantum efficiency with event position on the detector. Although level-1 processing properly corrects for such effects, PROS users who wish to apply non-standard data screens, or who wish to study sources not detected in standard processing, are required to compute and apply such corrections by hand. The task is complicated by the fact that the correction factors include contributions from a range of off-axis angles or detector positions, due to the induced $\sim 3'$ satellite wobble present in most observations.

Srcinten is a new PROS task that computes and applies the above correction factors to the net counts in a user-specified aperture. If significance of the net counts is less than a user-specified threshold, an upper limit is calculated. As a pedagogical exercise, the task was designed to minimize additional coding and maximize the use of other existing tools in PROS and IRAF. It will be fully integrated into IRAF and available in the PROS **xray.xspatial** package in the next PROS release.

2. Calculation of Correction Factors

Both the vignetting and encircled energy corrections are primarily functions of source off-axis angle. The existing PROS task QPSPEC is used both to compute net counts in the source aperture and to compile an off-axis angle histogram, whose entries comprise the fraction, f_i, of counts in the aperture corresponding to off-axis angle θ_i.

For each angle θ_i, the vignetting correction $V(\theta_i)$ at 1 keV is evaluated using the function in Figure 1 (David et al. 1993). The average vignetting correction is then

$$<V> = \sum_i f_i V(\theta_i).$$

The encircled energy correction is simply the integral of the point response function (PRF) within the source aperture. Although in general the *ROSAT* HRI PRF exhibits significant azimuthal asymmetries for off-axis sources, the azimuthally-averaged function $P(r, \theta_i)$ may be used to estimate encircled energy (David et al. 1993). For each angle θ_i in the off-axis angle histogram, the encircled energy within aperture radius R is given by

$$E(R, \theta_i) = \int_0^R P(r, \theta_i) 2\pi r \, dr.$$

$E(R)$ for various off-axis angles θ is shown in Figure 1. The average encircled energy is given by

$$<E(R)> = \sum_i f_i E(R, \theta_i).$$

Relative quantum efficiency variations of approximately $\pm 10\%$ across the detector are present in the *ROSAT* HRI, and their correction requires knowledge of the event positions in detector coordinates (i.e., a coordinate system fixed to the detector). These coordinates are present in the event structure of QPOE files generated with PROS version 2.3, and may be added to earlier QPOE files using the PROS task **upqpoerdf**.

The PROS task **qpcopy** is used to generate an image of the events in the source aperture, blocked at a resolution equal to that of the calibration quantum efficiency map (QE). The average quantum efficiency correction is then

$$<QE> = \sum_{i,j} f_{i,j} QE_{i,j},$$

where $f_{i,j}$ is the fraction of total counts in binned detector pixel (i, j).

3. Structure Chart

Srcinten uses the IRAF CL script language and, as described above, takes advantage of a number of existing IRAF/PROS tools. The structure chart in Figure 2 shows all the tools used in **srcinten**, and a brief overview is given below.

qpspec extracts the total counts from the source and background regions and computes net counts and error. It also produces an off-axis angle histogram file in table format.

qpcopy copies one QPOE file to another through region filtering, to generate a source image in detector coordinates.

imcalc multiplies the source and QE images.

imcnts computes the total counts in the source image.

tabpar, keypar, tinfo get keywords and values from tables, needed in computing correction factors.

tcalc performs arithmetic operations on columns to compute corrections.

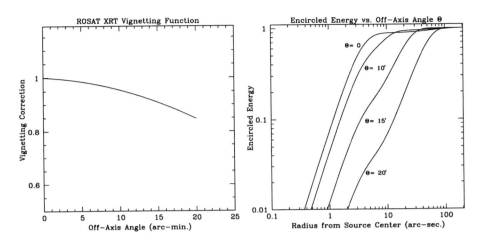

Figure 1. *ROSAT* XRT vignetting function at the energy 1 keV (*left*). *ROSAT* HRI encircled energy at various off-axis angles (*right*).

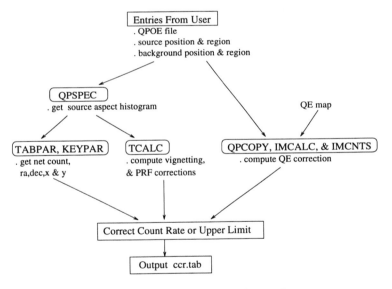

Figure 2.   Structure Chart for **srcinten**

4. Using srcinten in PROS

Before using **srcinten**, users are required to load the **xspectral**, **xspatial**, **ximages**, **images**, and TABLES packages. An example **srcinten** run is shown below, and the output table is displayed in Table 1. Both the count rate and upper limit columns are included in the output, but for sources above the user-specified significance threshold, only the count rate column is filled.

```
xs> srcinten
source qpoe file: xdata$rh110267n00.qp
source region descriptor: circle  2427.  2806.  82.
background qpoe file: xdata$rh110267n00.qp
bkgd region descriptor: annulus  2427.  2806.  82.  100.
output root name (_ccr.tab): src

-- Creating aspect histogram table from qpoe --
-- Computing vignetting correction factor --
-- Computing prf correction factor --
-- Computing qe correction factor --
-- SRCINTEN created Output table: src_ccr.tab --
```

ra	dec	x	y	radius	cntrate	cntraterr
deg	deg	pix	pix	asec	c/ks	c/ks
331.8	45.7	2427.	2806.	41.	83.1	10.7

uplrat	totcnt	bkgcnt	exptim	prf	qe	vign
c/ks	c	c	sec	none	none	none
INDEF	188.	43.	2124.4	0.96	0.93	0.92

Table 1. Output table src_ccr.tab

5. Conclusions

Although **srcinten** is a simple tool, it reproduces much of the functionality of its HRI level-1 processing counterpart and demonstrates the utility of assembling existing PROS tasks into CL scripts to address real analysis problems. Possible future enhancements include computing an energy-averaged vignetting function weighted by the source spectrum, allowing apertures of arbitrary shape, and correcting for spatially-varying exposure through an HRI exposure map.

Acknowledgments. This work is partially supported by NASA contract NAS5–30934.

References

David, L. P., Harden, F. R., Kearns, K. E., & Zombeck, M. V. 1993, The *ROSAT High Resolution Imager*

The PROS Big Picture: A High-Level Representation of a Software System

J. DePonte, J. Chen, K. R. Manning, D. Schmidt, and D. Van Stone

Smithsonian Astrophysical Observatory, 60 Garden St., Cambridge MA 02138

Abstract. We present a high-level representation of the packages and tasks within IRAF/PROS. The development of this overview is part of a comprehensive plan to document, and to improve consistency throughout, the IRAF/PROS analysis software. We anticipate that both current and future programmers will benefit from this document, and that the guidelines we have established will be applicable to other software projects.

1. Introduction

PROS is a multi-mission X-ray analysis software system designed to run under the Image Reduction and Analysis Facility (IRAF) (Tody 1986). The PROS software includes spatial, spectral, timing, data I/O, and plotting applications, and algorithms for performing general arithmetic operations with image data (Worrall et al. 1992). We have produced a high-level overview of PROS. The present paper describes this picture and the project that produced it.

2. Approach

A project to document a system requires a good definition for success. It can easily become unwieldy and misguided. The approach we took was based on four objectives. The first three were: a statement of mission, or what the project was intended to accomplish; a strategic plan to meet the objectives; and a set of short-term goals, to be met by means of scheduled task assignments. We also defined our long-term goals, a broader context for the mission.

The mission statement for the PROS overview specified four objectives: to communicate the content and organization of PROS to current and future programmers and projects; to make it easier to maintain the system and to evaluate enhancements; to identify code of high algorithmic content, for possible inclusion in libraries; and to identify temporary or prototype code for replacement or elimination.

The strategy was to do the project in stages, focusing at each stage on low-effort, high-payoff work that would produce useful deliverables. The incremental stages were the short-term goals: a project overview diagram (i.e., an organizational diagram of software modules); high-level diagram (i.e., a bubble diagram of each module in the system); and unifying element tables (i.e., tables of task vs. software attributes).

3. Implementation

To meet the long term goals within short term objectives, a hierarchical chart showing all packages and tasks was developed. The diagram was the basis for task assignments and progress evaluation (see Figure 1).

Next, we developed a high-level representation of the system. Each task was shown along with file I/O in a bubble diagram. Connectors were placed to show paths to other tasks within the same package. Standard analysis and design diagramming were applied, but standards were customized when appropriate for the application. For example, we used our own notation to represent inclusive/exclusive ORs and defined an icon to represent graphical output (see Figure 2).

Once diagrams were generated for all of the PROS tasks, the unifying elements of the system were identified. The elements (e.g., file I/O, code/data dependencies, maintenance of header keywords) were derived by analyzing the libraries and key software modules. We represented them as table columns and categorized the elements into four groups: I/O attributes (i.e., activities such as file I/O for each type of data file in the system); data attributes (e.g., the dependence of reference data on time); form attributes (e.g., whether the task is compiled code or a script, or whether it utilizes lookup tables); and derived attributes (e.g., whether the task maintains WCS or file header values, or performs error computation).

Next, a unifying element table was built for all tasks in the system. In the table, elements were represented as columns and tasks as rows. The parameter files, help files, and code, were used to analyze the modules and complete the tables. Table 1 shows an example of a table of I/O attributes of tasks in the PROS **xtiming** package.

The last phase focused on each identified attribute, documenting its usage in the code. Different projects could use different views for this phase of documenting their systems. For example, the structure charts of each module could be derived, or one aspect of the larger infrastructure could be identified and examined. In our preliminary work on this phase, we have taken the latter course, investigating the interrelations among library routines involved in implementing regions and masks.

4. Project Status

We have completed the project overview diagram, the bubble diagrams, and the attribute vs. task tables for all modules in PROS. We are currently analyzing code with identified attributes and documenting the usage throughout the system.

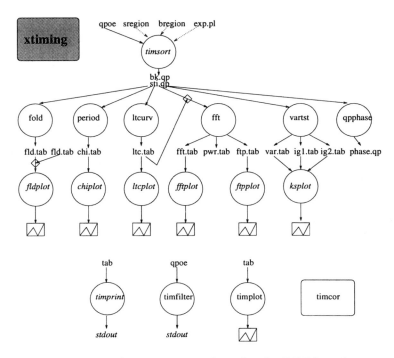

Figure 1. Task organization chart for the PROS package.

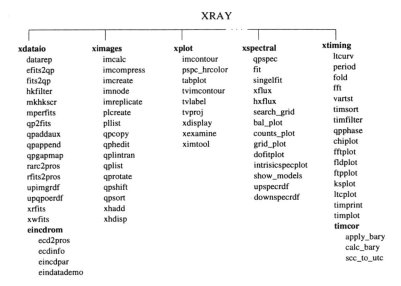

Figure 2. Bubble diagram for the PROS **xtiming** package.

Task	imio	qpio	plio	tabio	ASCII	bin	FITS	stdout	plt
ltcurv	–	x	–	x	–	–	–	–	–
period	–	x	–	x	–	–	–	–	–
fold	–	x	–	x	–	–	–	–	–
fft	–	x	–	x	–	–	–	–	–
vartst	–	x	–	x	x	–	–	–	–
timfilter	–	x	–	–	x	–	–	x	–
timplot	–	–	–	x	–	–	–	x	–
qpphase	–	x	–	–	–	–	–	–	–
ksplot	–	–	–	x	x	–	–	–	x

Table 1. Data I/O attributes table. The columns are I/O attributes. The rows are tasks in the **xtiming** package. Attributes that apply to a task are marked with an "x". Attributes that do not apply are marked with a "–".

5. Conclusion

The implementation decisions enabled us to produce useful products at each phase of the project. The high-level graphical representation of the system has already proved a useful tool in communicating PROS to others. The attribute vs. task tables identify library code, flag code that are trouble spots in preparing a software release, and aid in evaluating test procedures.

We anticipate that the procedures that we have established will be applicable to other large software projects. The effort we describe distinguishes each phase and can be customized to bring added understanding to a large software system.

Acknowledgments. This work is partially supported by NASA contracts to the *ROSAT* Science Data Center (NAS5–30934) and *Einstein* (NAS8–30751).

References

Tody, D. 1986, in Instrumentation in Astronomy VI, Part 2, SPIE, 627
Worrall, D. M., et al. 1992, in Data Analysis in Astronomy IV, eds. V. Di Gesù, L. Scarsi, R. Buccheri, P. Crane, M. C. Maccarone, & H. U. Zimmermann (New York, Plenum Press), p. 145

An IRAF-Independent Interface for Spatial-Region Descriptors

D. Schmidt

Smithsonian Astrophysical Observatory, 60 Garden St., Cambridge, MA 02138

Abstract. Regions for spatial masking are a valued part of PROS. To move toward open analysis software, we plan to make the PROS regions subsystem available from processes outside of IRAF. We also plan to make the region descriptor parsing system available for purposes beyond the creation of masks, such as an interface to an image display program, accepting a region descriptor and generating commands to display the region on an image.

Last year, we put in place support for arbitrary actions using region descriptors. We discuss this architecture and its implications for opening up the use of regions. We then discuss the challenge of giving the subsystem an IRAF-independent interface.

1. Introduction

Limiting attention to photons from specified source and background areas is a necessary part of astrophysical X-ray analysis. The IRAF/PROS X-ray analysis system provides the required masking capability through its regions subsystem, described by Mandel et al. (1993). The interface to the PROS regions subsystem is a language for constructing ASCII region descriptors. The language is based on a set of simple geometric shapes and a scheme for specifying complex shapes as combinations of the simple shapes. The regions subsystem supports a variety of coordinate systems, both celestial and pixel-based.

Anticipating two kinds of new requirements, we reconsidered the design of the PROS region descriptor processing subsystem. One class of new requirements would be for products other than spatial masks from region descriptors. The other new requirement would be to allow new tasks, within or outside of IRAF, to use the subsystem to generate masks or other products.

2. Functions of a Region Descriptor

The primary function of a PROS region descriptor is to create an IRAF PLIO pixel mask to select photons for analysis. Certain secondary functions were already present in PROS and SAOimage:

- to display the regions on an image;

- to document the regions used in an analysis;

- to extract specific regions and their parameters for special processing, such as validity checks;

- to transform coordinates into a different system (e.g., to move data between SAOimage and PROS, when using two different pixel-based systems).

The presence of these secondary functions in PROS, however, did not satisfy our goals of functional extensibility and open interfacing.

3. The Problems We Faced

Each of the secondary functions of region descriptor processing was implemented *ad hoc*, each in its own program with its own lexical analyzer and parser. Each parser defined the region descriptor language somewhat differently. Each parser had side effects, making it difficult to add or change functionality. There was no provision for defining new functions or novel combinations of existing functions, and the system was available only within IRAF.

4. Our Solution

We defined a set of requirements for a new implementation of the regions subsystem: (1) there should be a single language definition, and a single parser, (2) there should be a single entry to the descriptor-processing subsystem, (3) selection of functions and return of resultants should be through a simple, extensible control structure, (4) functions should be independent, so that any combination would be selectable in a single call to the parser, and (5) the interfacing of functions to the general language processing scheme should be modular, so that we could define recipes for adding and changing functionality.

We determined that the underlying metaphor of a *software CPU*, executing instructions compiled by the parser, remained a sound conceptual framework for the syntactic and semantic analysis of region descriptors. In this framework, we treat a descriptor like a program for a CPU that we emulate in software:

- *Lexical analysis* provides tokens to a compiler.

- *Compilation* produces an object program of software CPU instructions.

- *Execution* of the object program creates a mask or other resultant.

The implementation, however, needed a new layer of design to provide a general method of plugging in new functions; the program interface needed a uniform access protocol, so that we might open the subsystem to general tasks.

We retained xyacc as the language definition language. Using xyacc simplified generating a correct and reliable parser. The integral processing, with xyacc, from lexical analysis through compilation challenged flexibility that we wanted in using the lexical analysis and object program. We defined a uniform and extensible system of control flags to achieve the desired flexibility in using information at any stage of the linguistic analysis.

Figure 1 shows schematically how various applications use a uniform protocol to interact with the subsystem via a versatile *parsing control structure*,

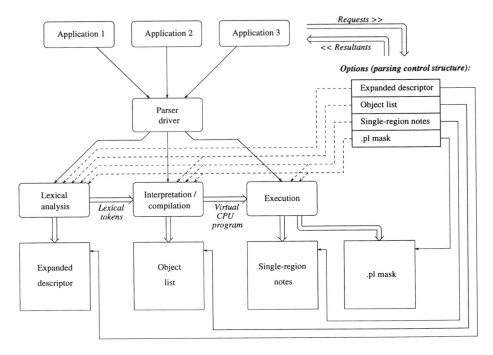

Figure 1. The PROS regions subsystem program interface.

and how the underlying subsystem is structured to enable functions to access, as needed, any of the three stages of the linguistic processing.

5. Using the New Design

This design prepared us for extending the regions subsystem's functionality and for making the subsystem available for use by programs in a more open environment. To make it easy for programmers to begin to exploit the new design, we created "recipes" for two ways of extending region descriptor functionality. Outlines of those procedures follow.

First, a recipe for processing a region descriptor: (1) call rg_open_parser(), to create a parsing control structure, (2) call one or more of the request setup routines, to create request/resultant structures and link them into the parsing control structure, (3) call the parser, rg_parse(), (4) retrieve the resultants through a set of macros provided, and (5) call rg_close_parser(), to release any request/resultant structures and the parsing control structure.

Second, a recipe for adding a new function: (1) define a request/resultant structure for the new option, and standard-format macros to access its fields, (2) create a slot for the new option in the parsing control structure, (3) create option setup and release routines, using standard templates, (4) integrate the option into the system of "request set query" routines and control flags, according to the processing required for the option, and (5) add the new functionality at one of four locations in the system, according to what form of the descriptor it uses. Full utilization of the new capabilities remains for future applications.

6. Conclusion

The redesign and re-implementation of the program interface to the IRAF/PROS regions subsystem has provided PROS programmers with a cleaner environment for developing new regions applications. It has been used successfully for more than a year. We anticipate greater benefit in the future, when applications outside of IRAF may make use of the new request/resultant protocol for much freer access to the functions of region descriptor processing.

Acknowledgments. This work was supported under NASA contract to the *ROSAT* Science Data Center (NAS5–30934).

References

Mandel, E., Roll, J., Schmidt, D., VanHilst, M., & Burg, R. 1993, in Astronomical Data Analysis Software and Systems II, ASP Conf. Ser., Vol. 52, eds. R. J. Hanisch, R. J. V. Brissenden, & J. Barnes (San Francisco, ASP), p. 430

Recreating Einstein Level One Processing Exposure Masks and Background Maps in IRAF

D. Van Stone, M. Garcia, J. McDowell

Center for Astrophysics, 60 Garden St., Cambridge, MA 02138

Abstract. This paper describes the main algorithms used by the Einstein Level One processing to create the exposure masks and the background maps for *Einstein* IPC images, and how these algorithms were recreated in the IRAF environment.

1. Introduction

Our goal was to recreate the algorithms used by the Level One Processing to create exposure masks and background maps for *Einstein* Image Proportional Counter (IPC) data (cf. sections 2.5 and 2.7 of Harnden et al. 1984). Although these images have not been archived on CD-ROMs, the aspect and original background image information have been. From this information, one can either recreate the original background images, or recalculate and create exposure masks and background maps from within IRAF. Special consideration is given to extend the background map algorithm to work on the *Einstein* unscreened IPC images[1].

2. Exposure Masks

The aspect information for an image contains a timeline of the aspect solution for the duration of the observation. The exposure mask is simply the sum of multiple images, where each image is the result of applying an aspect solution to the IPC geometry, weighted by that aspect's duration. This algorithm does not attempt to correct for vignetting.

2.1. Exposure Mask Algorithm

For each set of aspect data and duration, an exposure mask is created by looking at each pixel in the final exposure mask, and finding the pixel in detector coordinates by de-applying aspect, and seeing that if the detector coordinates lie within the IPC geometry[2], add duration of aspect data to the image pixel.

---

[1] The unscreened images were published on CD-ROMs in March 1994 (McDowell et al. 1994).

[2] The IPC geometry is defined as all pixels $287 \leq x \leq 737$ and $288 \leq y \leq 738$ which are at least 15 pixels away from the rib centers of $x = 359.6$, $x = 656.8$, $y = 376.6$ and $y = 673.8$. All values are in PROS coordinates, the system used on the CD-ROM archive.

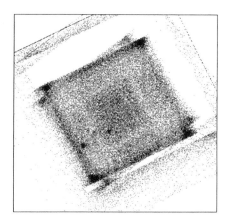 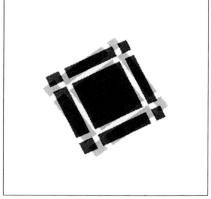

Figure 1. The *Einstein* unscreened IPC image for sequence I5803, along with its exposure mask created by the EINTOOLS task exp_make.

3. Background Maps

The main idea behind the background map algorithm is that the background for an individual image can be modeled as some combination of blank field ("deep survey") images and flat field ("bright Earth") images. The reason that the blank field image by itself is not a sufficient model is that each image (including the blank field) has a variable amount of particle induced and/or stray light background. Our algorithm accounts for this by including some (possibly negative!) amount of flat field data in the model background.

We assume that the blank field rate is constant over time, and thus the normalization we use for the deep survey image is based solely on the duration of the exposure. The normalization for the bright Earth image is set so that the total number of background counts (excluding sources) in the model is the same as those in the image, as determined by the following relation:

image counts = source counts + (deep survey counts + bright Earth counts)

Note that the background map algorithm does not explicitly take into account internal or particle induced background.

3.1. Background Map Algorithm

The algorithm used to create a background map is as follows: (1) calculate counts in the image due to sources[3] (SRC_CNTS); (2) calculate counts in deep survey and bright Earth maps[4] (DS_CNTS & BE_CNTS); then (3) for each set of aspect data and duration:

---

[3] See section 3.2.

[4] The counts are calculated using the bright edge region. See section 3.3.

1. Find number of photons in image which arrived during this set of aspect data (IM_CNTS)

2. Calculate deep survey factor (DSFAC):

$$\text{DSFAC} = \frac{\text{(aspect duration)}}{\text{(deep survey livetime)}}$$

3. Calculate bright Earth factor (BEFAC):

$$(\text{BEFAC}) \times (\text{BE\_CNTS}) = (\text{IM\_CNTS}) - (\text{DSFAC}) \times (\text{DS\_CNTS}) - \frac{\text{(aspect duration)}}{\text{(image livetime)}}(\text{SRC\_CNTS})$$

4. Apply BEFAC and DSFAC weights to bright Earth map and deep survey maps and sum these two images

5. Apply aspect to the summed image and add to background map

3.2. Source Counts Algorithm

The algorithm used to calculate the number of counts for each source in the image is; (1) calculate total counts in the source circle and in the background annulus [5] (TOT_CNTS & BG_CNTS); (2) calculate the area in the source circle and in the background annulus [6] (SRC_AREA & BG_AREA), and (3) compute the counts attributable to this source by the relation:

$$(\text{SRC\_CNTS}) = \text{CCC} \times \left((\text{TOT\_CNTS}) - (\text{BG\_CNTS})\frac{(\text{SRC\_AREA})}{(\text{BG\_AREA})}\right)$$

where CCC, the circle composite correction, is defined as the product

$$\text{CCC} = \text{(circle mirror scattering corr.)}\text{(circle point response corr.)}.$$

The user can generate a list of sources by using the PROS task LDETECT.

3.3. Masking Detector Hotspots

Our background map algorithm works best if the internal (particle induced) background is small, at a constant rate, and uniform over the field. In order to make this true, we must exclude the regions of very high internal background on the edges of the detector from the calculation of the normalization. This exclusion is performed with the bright edge filter on QPOE files and the bright edge region on the deep survey and bright Earth images[7].

---

[5] These counts are calculated using the bright edge filter. See section 3.3.

[6] All areas are calculated using the exposure mask.

[7] The bright edge filter and bright edge region describe identical areas. They differ because one is for filtering detector coordinates within QPOE files whereas the other is a region on an image.

Figure 2. The figure to the left is the broad band background map for the *Einstein* unscreened IPC image I5803, created by the EINTOOLS task bkfac_make. The figure to the right is the original Level One Processing background map created by the EINTOOLS task be_ds_rotate.

4. Eintools

The new PROS package EINTOOLS contains tasks for creating exposure masks and background maps for any *Einstein* IPC QPOE file. One can also create the original Level One Processing background maps using the bright Earth and deep survey factors from the original processing. The tasks allow the user to screen on any time filter, produce background maps for any (possibly nonstandard) PI range, and create exposure masks of any specified resolution.

This package will be available in the next major release of PROS. It is our hope that some of the data structures created for this package (such as a generic time-resolved aspect table) will be used in future programs generating exposure masks or background maps.

Acknowledgments. The PROS project is partially supported by NASA contracts NAS5-30934 (RSDC) and NAS8-30751 (Einstein).

References

Harnden, Jr., F. R., Fabricant, D. G., Harris, D. E., & Schwarz, J. 1984, Scientific Specification of the Data Analysis System for the Einstein Observatory (HEAO-2) Imaging Proportional Counter, SAO Special Report 393, (Cambridge, Smithsonian Astrophysical Observatory)

McDowell, J., Plummer, D., Prestwich, A., Manning, K., Van Stone, D., & Garcia, M. 1994, The Einstein Observatory IPC Unscreened Data Archive, CD-ROM Volumes 0–18, (Cambridge, Smithsonian Astrophysical Observatory)

The ASC Pipeline: Concept to Prototype

A. Mistry

TRW/AXAF Science Center, 60 Garden St., Cambridge, MA 02138

David Plummer, Robert Zacher

Smithsonian Astrophysical Observatory/AXAF Science Center, 60 Garden St., Cambridge, MA 02138

Abstract. Discusses the role of a pipeline in the *AXAF* Science Center Data System. Due to the complexity of the application, a prototyping effort was initiated to evaluate the risks involved with portability, open architecture, error handling/recovery, and distributed processing. Processes in the pipeline are initiated and monitored by a process control application developed using Perl. Currently, the process control application can attach IRAF, IDL, shell scripts, and various third party executables to create a number of data processing flows. An example process flow involves the acquisition of raw telemetry data from a data simulator to an image file. An error logging library was also developed to form the foundation for handling both minor and fatal errors.

1. Design Concept

The purpose of the *AXAF* Science Center Data System (ASC DS) is to provide the science community with useful *AXAF* data and the tools to process the data. The role of the pipeline is to provide the framework for creating data processing flows which produce the standard data products, and to provide an easy to use mechanism that enables scientists to create their own customized processing flows. Several basic items need to be incorporated in order to meet the needs of the ASC DS and the science community: the processing flows must be *programmable* to meet the needs of the user; the system must be *open* to external applications, either commercial or custom; the system must be *portable* to all platforms supported by the ASC DS; the system must be *reliable*, in that it must be fault tolerant and preserve data integrity; and the system must have *good performance*, in that it must perform its task in a reasonable amount of time. Finally, the system must be *easy to use*, in that it must provide a clear interface for the novice user as well as short-cuts for the experienced user.

2. Prototype Activity

The purpose of the prototyping activity was to determine if the design concept can be achieved before the project enters full scale design and development. After careful consideration of the goals that were set forth for the pipeline, a set of risk areas were defined. Those areas are: portability, error handling and recovery, open architecture, and distributed processing.

The prototype activity for the pipeline started in March and was completed by the end of July. During that time, a Process Control application was developed to allow us to assess the risk factors. The following section discusses how each of the risk areas were resolved.

2.1. Portability

A study of the portability issue was performed before any coding was done. From that research, two pieces of information were gathered: portability is simply the degree to which an application can be compiled and executed on a variety of different hardware platforms. As well, the language used for developing an application is irrelevant in determining its portability; provided that the application was developed using one of the popular development languages, and that no system-dependent routines were used.

The development language for the pipeline has not been selected yet. However, we are leaning towards a POSIX-complient system. POSIX is a group of standards focusing on hardware and software portability (Lewine 1994). The script language Perl was selected for the prototype because of its inherent portability and file/string manipulation capabilities (Wall & Schwartz 1992).

2.2. Error Handling and Recovery

The key to running the pipeline effectively and reliably will be its ability to detect and recover from errors. Through the prototyping effort, we learned that it is possible to trap the errors of an application regardless of what language it was developed under. The methodology is simple. When an application creates a child process, the parent will receive all error conditions that the child can not handle through `stderr` (Stevens 1992). Also, it is possible for the parent to trap the error condition before it affects the child. This scheme works since the Process Control is the parent of all processes that run in a pipeline. The recovery process is then simply a matter of querying the user for an action, or executing the proper error handling routine to correct the situation.

2.3. Open Architecture

The only way the Process Control application can be useful outside of the ASC environment and be flexible enough to adapt to changing technology is for the application to be open. The definition of an open system is:

Open System A system in which the specifications are made public (Lewine 1994).

Since users only need to know how to interface with the Process Control, and not how it works, we would like to add to that definition as follows:

Open System A system in which the *interface* specifications are made public.

An interface standard has not yet been developed, but work is in progress. The interface will be kept neat and simple. Once an interface standard has been published, all applications which follow that standard will operate with the Process Control, with no surprises.

2.4. Distributed Processing

The performance of a pipeline would be greatly increased if the processes were distributed over a number of machines (Jain 1991). The methodology is once again straightforward, in that a `daemon` process will be running on all machines that have been allocated for the Process Control. The Process Control will then assign a task in the pipeline to the available machines for execution. The data transfer from module to module will occur through sockets. In some cases the tasks themselves may be further distributed, but that responsibility will fall on the tasks themselves.

3. Conclusion

The purpose of any prototyping effort is to gather knowledge, to assess the risk factors, and to determine if a proposed design is possible. A fair amount of knowledge has been gained regarding open systems, portability, and distributed processing, through research of available literature and *trial and error*. Our knowledge base has grown to the point where we believe we can create a solid design and produce a reliable and easy to use product.

All of the above mentioned risk factors were assessed in the prototype. None of these factors seems to pose unreasonable risks. We learned several interesting lessons, which remain as concerns for the future:

Running IRAF in the Command Line mode. Pipes or sockets cannot be used in this case. If an IRAF task crashes, an error is reported via STDOUT, and no error code is returned to the shell. This makes it difficult for the process control to recover from a fatal error.

IDL. The number of IDL analysis tasks is limited to the number of site licenses. However remote procedure calls should solve the problem.

Perl. Perl lacks the ability to create complex data structures, which could make development of a complex system more challenging.

In summary the prototype proved that the original design concept is feasible.

References

Stevens, W. R. 1992, Advanced Programming in the UNIX Environment (Reading, Addison-Wesley)

Wall, L., Schwartz, R. 1992, Programming Perl (Sebastopol, O'Reilly & Associates)

Lewine, D. 1994, POSIX Programmer's Guide (Sebastopol, O'Reilly & Associates)

Jain, R. 1991, The Art of Computer Systems Performance Analysis (New York, Wiley)

Evolution of EUVE Guest Observer Data Reduction

E. C. Olson[1]

Center for EUV Astrophysics[2], University of California[3], 2150 N. Kittredge St., Berkeley, CA 94720

Abstract. The *Extreme Ultraviolet Explorer* (*EUVE*) Guest Observer (GO) Center is in its second year of data processing. The pipeline reduction system continues to evolve as additional requirements are added. Importantly, the original primary data products delivered to GOs over a year ago are correct. While the overall design of the reduction system has provided enough flexibility, many of its limitations are too strict. The applicability of this system to other event-oriented data reduction problems is discussed.

1. Introduction

The Extreme Ultraviolet Explorer (*EUVE*) Guest Observer (GO) Center is now in its second year of data processing. By using astronomical community standards like FITS and a common processing environment, IRAF, the Center was able to concentrate on software development and calibration analysis. The Center made full use of software developed at ST ScI and NOAO as well as CfA. The additional coordination required to enable this code reuse was well worth the time and effort. In an era of diminishing funding, these efforts must continue.

A large part of the success of the program is from selectively applying standard astronomical procedures and existing tools to the evolving EUVE data reduction problems. From its inception the GO Center expected to disseminate very complete raw data sets to GOs, calibration and reduction software, as well as the calibrated data products. Part of the motivation was that before launch, the operational characteristics of the telescope were unknown. Having maintained a stable reduction pipeline for over a year, the GO Center is considering streamlining (and reducing) the data products provided.

The aging core calibration tasks, written before launch in 1990, have had numerous improvements. These tasks are the result of various experiments in implementation: some were short lived, based on limited access to *EUVE* ground calibration data sets. Yet the initial design was general enough that they can support data sets and missions that were never considered in the design. Real *EUVE* data has been available since fall, 1992. By this time, all the major software components were completed and in final testing. By the spring of 1993

---

[1] http://www.cea.berkeley.edu/ericco/

[2] http://www.cea.berkeley.edu/

[3] http://www.berkeley.edu/

the software had been released a half dozen times. Under intense pressure and looming budget cuts, "fast-track" data sets were distributed to GOs for early *EUVE* data sets. By April, a functional reduction pipeline was in place. The fast-track data sets were immediately replaced with standard products and by summer, the data processing queue was eliminated. This effort was completed by the combined efforts of the entire GO Center staff. Since then additional functionality has been added to the software, the calibration data sets have been significantly improved, and the documentation has been written. Today, the GO Center is essentially providing the same raw (very complete) data products that it originally delivered, but with improved software and updated calibration data.

2. Background

The *EUVE* Guest Observer Center was created in the spring 1990 at the Center for EUV Astrophysics (CEA). The GO Center initially provided dedicated resources to the *EUVE* pointed spectroscopy program. The GO Center also provided *EUVE* spectrometer data and *EUVE*-specific analysis software. The GO Center is responsible for other activities including: science operations planning, GO proposal assessment and instrument calibration analysis.

The *EUVE* mission has two components after the initial in-orbit check out: a 6 month full-sky survey, and the pointed-mode Guest Observer phase. All *EUVE* data is archived at CEA on an optical disk jukebox. Data from the *EUVE* all-sky survey has been processed from raw telemetry into skymaps and catalogs at CEA.

Data processing for the pointed spectroscopy observations is handled by the GO Center. *EUVE* has seven photon counting detectors: three of the them produce spectral data. The band passes are 70–190 Å, 140–380 Å, and 280–760 Å for the short-, medium-, and long-wavelength detectors, respectively. *EUVE* photon event contain a detector position and time stamp.

EUVE generates approximately 100 GB of telemetry per year. With a typical observation lasting 1.0×10^5 s, about 150 GO observations are completed each year. Each observation corresponds to approximately 70 MB of night time data. It is expected that handling each observation will require several hundred MB of disk space and a workstation-class computer.

3. Software Design

The main design goals of the EGO Center software were to provide a complete data set in a timely fashion with calibration data and software to scientists to complete analysis on their proprietary observations. Providing *EUVE* telemetry was never discussed as possible data product since it is optimized for satellite communication, and data extraction was a complex task. To date, there are still parts of the *EUVE* telemetry stream that are not extracted due to a lack of resources and other issues.

When we decided to represent the telemetry in STSDAS Tables, the TABLES package was not yet available and IRAF did not have a layered package mechanism. Representing *EUVE* telemetry as several STSDAS Tables proved to be a relatively compact format, but the most important feature was the mere availability of the TABLES package from ST ScI. By providing *EUVE* data in this standard format, *EUVE* inherited all of the staff-years of programming effort that ST ScI put into the TABLES package (which continues to be upgraded). A recent enhancement to TABLES was the addition was ASCII table support, and we are looking forward to direct FITS BINTABLE support. The TABLES package also provided FITS compatibility: at the time, ASCII tables were part of the evolving FITS standard but binary tables were not.

At the same time, CfA and NOAO were coordinating development a new IRAF image format for handling event data called QPOE (Quick Position Oriented Event) for *ROSAT* data. Although QPOE was not developed for *EUVE*, we readily made use of this facility as well. In both cases, EUV software was developed without the benefit of a preexisting application interface. But by adhering to interface definitions, most of the integration was smooth.

The GO Center made no plans for reprocessing GO data products. By providing complete raw data sets, the only reason to reprocess would be to fix significant errors in these level zero data sets. Instead of the GO Center reprocessing the data, the GOs could reprocess their observations at any time since all the GO reduction software is provided in a portable IRAF environment. In fact, using IRAF has not eliminated portability problems. Some problems were due to the flexibility of the SunOS F77 compilers; others were due to system errors. Today, there are still a number of GOs who use less common platforms that have difficulty using the EUV package.

In addition to coordinating software development with other institutes, the GO software also had to coordinate development with internal groups at CEA. These components turned out to be the past ones completed. A key component was the system to access *EUVE* telemetry which significantly delayed testing of the GO software. However, by working closely with the operations group at CEA, this subsystem's performance has been excellent since its implementation. The benefits of cooperation and conforming to standards are huge, especially for smaller missions that are short-lived and low-cost.

Having made several courageous but necessary decisions about the software development environment and tools, the principal design issues could be addressed. First and foremost was the need for the calibration pipeline to be flexible and robust. A problem was that calibration pipeline was designed before calibration data were available. Based on the "end-to-end system" pipeline developed at CEA for analyzing *EUVE* sky survey data, the GO calibration pipeline tried to retain flexibility while providing efficient access to the observation data set. In the end, this calibration pipeline consisted of less then a dozen processing modules. Some of the modules are quite specific to *EUVE* data sets, but most are general purpose modules. These processing modules apply the calibration data to the photon events and produce QPOE images of the observation.

QPOE image files are the key element in processing *EUVE* data sets. They provide an excellent format for efficiently manipulating event-oriented data sets by selectively applying data filters that select subsets of the observation event data. In addition to filter, QPOE images can be used as if they are normal binned images by IRAF tasks that have no a priori knowledge of event oriented data. This enables all IRAF programs to be used with QPOE images.

The other key element for handling *EUVE* data sets are time-stamped tables. Each row in a time-stamped table contains a valid time stamp for the row. All *EUVE* engineering information are supplied in this format. Additionally, analysis programs provided by CEA also produce and manipulate time-stamped tables. By sharing this simple common format, these programs can all work together. Some of these tools are now being used on other missions. GOs can use engineering data, post-reduction analysis software, and independently developed software to filter the event data in QPOE files. It turned out that *EUVE* QPOE images could not be used directly as binned images, due to a number of calibration issues.

4. Conclusions

Developing software prior to the availability of calibration and data was not too problematic. In fact, it drove the requirement for a flexible solution, although a more general solution than the current implementation might not be desirable either. However, it does appear that there is a great deal of commonality in event-oriented data processing pipelines used by various missions. A survey of such missions would be in order before proceeding with the development of a multi-mission, event-oriented pipeline. CEA will collaborate with CfA on an event-oriented IRAF package that may provide the framework for such a multi-mission event pipeline, although that is not the focus of the collaboration. In retrospect, it is tempting to suggest that the data products and the processing pipeline could have been streamlined earlier. But in an environment where calibration data sets under go significant structural changes, and the instrument can have unexpected characteristics, it is wiser to plan for a worst-case scenario.

By creating a flexible reduction pipeline, CEA has been able to incrementally add supplemental processing modules. For example, modules for handling flat fielding errors and for handling imaging detectors have been added. The final version of the software will include tools for pointed imaging data analysis as well as sky survey data analysis. These new tools will be integrated with the existing tools that handle instrument distortion corrections, data selection, and calibrations.

The possibility of applying these techniques to the *AXAF* processing and other missions has been discussed recently. Although the general concepts may be applicable, it is clear that this specific implementation does not provide all the necessary features needed for multi-mission support. For example, the use of time stamped tables seems generally applicable. However, the "Comprehensive Event Pipeline" that applies calibration data sets to events is anything but comprehensive.

Acknowledgments. This research was supported by NASA contract NAS5–30180.

The OPUS Pipeline: A Partially Object-Oriented Pipeline System

J. Rose (CSC), R. Akella, S. Binegar, T. H. Choo, C. Heller-Boyer,
T. Hester, P. Hyde, R. Perrine (CSC), M. A. Rose, K. Steuerman

Space Telescope Science Institute, 3700 San Martin Dr., Baltimore, MD 21218

Abstract. The Post Observation Data Processing System (PODPS) for *Hubble Space Telescope* Science Operations Ground System was designed under the expectation that problems and bottlenecks would be resolved over time by bigger, faster computers. As a consequence, PODPS processes are highly coupled to other processes. Without touching the internal components of the system, the second generation system decouples the PODPS architecture, constructs a system with limited dependencies, and develops an architecture where global knowledge of the system's existence is unnecessary. Such a (partially) object-oriented system allows multiple processes to be distributed on multiple nodes over multiple paths.

The PODPS "pipeline" currently includes seven major processing steps organized in a linear sequence; each step builds on the previous step. Designed to run on a single minicomputer, the original architecture used VMS-specific interprocess communication facilities to control the operation of the system.

Today we are operating in a cluster environment with over 30 workstations and many gigabytes of disk space. As the volume of data and the amount of work done on these data increase, and as other reporting and monitoring requirements expand, it has become increasingly evident that the limitations imposed by the existing pipeline system are becoming a serious bottleneck.

We have proposed a solution to this bottleneck: opening up the architecture and taking advantage of the resources already available in the pipeline environment. Without touching the internal components of the system, the second generation pipeline at ST ScI decouples the PODPS architecture, constructs a system with limited dependencies, and develops an architecture where global knowledge of the system's existence is unnecessary. Using a simple blackboard architecture, such a (partially) object-oriented system allows multiple processes to be distributed on multiple nodes over multiple paths.

While the system is not being redesigned or rewritten in object-oriented languages, there is one major change to the system influenced by the dominant character of object-oriented systems: it is the objects themselves which know their own state, and that knowledge is what "controls" their processing. This simple understanding allows us to move from a procedural, linear, controlled, sequential processing system to one which is distributed, event-driven, parallel, extensible, and robust.

In the older system, the knowledge of an observation's state was maintained only by the controller process, which had exclusive access to a global memory area. The advantage of such an architecture of control is that race conditions and deadlocks are avoided by channeling access to the status information through a single point.

The disadvantages of this architecture are many. The worst aspect of this design for our purposes is the obvious control coupling: the controller "controls" and must know about all processes in the system in order to schedule activities. The result of the dependencies set up by this control coupling is that when any one of the system components fails, the entire system fails and must be restarted. Equally difficult is that the observation status information stored in memory is not persistent: if the controller exits, the memory of the status of observations in the pipeline is lost. While this architecture allows multiple copies of the system to be run independently on different machines (assuming separate paths), this is a trivial interpretation of distributed processing, and does not allow for multiple copies of selected processes focused on the same path. An effective solution to these problems is simply to eliminate the controller and to re-engineer the system around a simple blackboard architecture.

1. An Object-Oriented Blackboard Architecture

The blackboard model is both simple to understand and, in our case, simple to construct. The paradigm is that of a number of independent players with pieces of a jigsaw puzzle. One player might put a piece on the blackboard, and others will attempt to fit their pieces into the puzzle as connections become apparent. All the information necessary to complete the puzzle is posted on the blackboard, and seen by all players simultaneously. Additionally, the solution is achieved by independent players who need not communicate with each other, except through the blackboard.

Applying this model to the pipeline environment means that the status of each observation is "posted" to the blackboard. All processes can see the status of all observations at the same time. If a process notices that it has something to do (i.e., advance the status of a particular observation), then it does it. When complete, it reposts the revised status of that observation, and continues its search for something to do.

There are a variety of ways to implement the blackboard architecture: a shared global memory, a database relation, or a network of messages. But each of these solutions again requires a controller specifically to guard against the ambiguities of concurrency. Instead of selecting an architecture that requires an elaborate controller, we decided to use the concurrency controller already built into the operating system: the file system.

File systems constitute one of the central parts of any operating system, and have received a great deal of attention in their design to problems of concurrency. Especially in environments where multiple machines are clustered with common access to files on multiple devices, the operating system is required to deal explicitly with lockouts, exclusions, and race conditions. By using the file system we avoid duplicating a fairly complete, complex, and sophisticated piece of software with a proven ability to deal with concurrency problems.

Thus our "blackboard" is simply a directory on a commonly accessible disk. In a cluster of workstations and larger machines, if the protections are set appropriately, any process can "see" any file in the blackboard directory; the "posting" consists of creating or renaming a file in that directory.

When an observation enters the system the first pipeline process creates an "observation status file" in the blackboard directory. The name of that file contains the processing start time, its status (which pipeline steps have been completed), and the nine-character encoded name of the observation. The file itself is empty—all the information which the system requires is embedded in the name of the file. For example the following entries might be found on the OPUS blackboard:

```
19941105141608-CCCCCP_____.Y1GV0101T
19941105141628-CCCCCW_____.Y1GV0102T
19941105150000-CCW_____.U0CK0101T
```

The first indicates that the observation Y1GV0101T (which started pipeline processing on November 5, 1994 at 14:16:08) is currently being processed by the sixth component of the pipeline. The second observation, Y1GV0102T, has completed the fifth step and is waiting to be processed by the sixth step. Similarly, observation U0CK0101T has completed the second step and is waiting to be analyzed by the third step.

Each process in the system polls the blackboard directory for files with a particular status. For example the third processing step will poll for files with names like *-CCW*.* which indicates the first two steps have been completed and the observation is waiting for the third process. Upon finding a candidate, the process will attempt to rename the observation status file, replacing the "W" (waiting) with a "P" (processing). Because we can have multiple instances of this third process, this renaming can fail: another process might have renamed that observation status file already. This is not a problem, we just attempt to find another observation—better luck next time.

When processing for a step is complete, the observation status file is again renamed. If processing was successful, the "P" is changed to a "C" (complete), and the next process in the sequence of steps will eventually recognize this observation as a candidate. If processing is unsuccessful, the "P" is changed to an "E" (error), and the error collector process will notice a candidate for its handling. Once all processing, including archiving, is complete for an observation the status file can finally be erased (deleted) from the blackboard.

A polling environment raises a resource question: what percent of the CPU is consumed by the polling processes? While the answer to this question depends on the number of processes, the polling frequency, the nature of the CPU and the I/O bus, and other factors, it is still a bit of a red herring. When processes are busy converting, analyzing, calibrating, or plotting an observation, they are not polling. When there are no observations to process, that is, when the machine is relatively idle, there will be more polling going on as processes try to find something to do. In any case, on a VAX 4060 with twenty processes polling on a 5 second interval, less than five percent of the CPU is affected.

In implementing the blackboard pipeline we developed two categories of processes: those which have the polling loop implemented internally, and those which are run from an operating system shell which does the polling. The first category is for processes which have a significant amount of initialization to perform; those processes are designed to poll internally after that initialization is performed only once. Other processes, which have either a limited amount of initialization, or are treated as black-box executables, are handled as if they were simply part of a script.

Such a scripting shell clearly allows the pipeline to be extended in simple ways. Each process has an associated resource file which specifies the type of process (OPUS poller, IRAF task, DCL procedure, etc.), observation status triggers, as well as its success and exception triggers. To extend the pipeline to include another process, all that is required is the new resource file—no change to any existing code, other than a resource file, is necessary.

As implemented, there is nothing application-specific about either the problem or the solution. Any processing script which must be used on a large body of data sets can be turned into a blackboard distributed processing system. We have demonstrated this in a heterogeneous environment of ancient FORTRAN processes, modern IRAF tasks, and simple DCL procedures.

In order to facilitate monitoring and managing such a heterogeneous and distributed system, we have also developed a cluster of five Motif-based window managers to assist the operator:

the process manager to aid in monitoring what processes are doing (scanning a process blackboard);

the observation manager to assist in monitoring how observations are progressing, and whether there are any problem sets;

a disk monitor to continuously gauge the availability of resources;

a node monitor to graphically illustrate where activity is the heaviest; and

a network monitor to gauge traffic between processors and disks.

These Motif managers will be described more fully in a later paper.

Changing requirements are no longer seen as an exception in the software evolution process. We have learned that change is now part of the process, and we must build systems that are adaptable. Monolithic special-purpose systems have a proven track record of failure to adapt. A blackboard distributed processing system is, on the other hand, flexible, extensible, robust, and easy to build. By distributing the processing load across several workstations, the system can make better, if not optimal, use of an institution's resources. By using the concurrency protection built into the operating system, the blackboard architecture ensures a robust solution as well as a cost effective one.

References

Nii, H. P. 1989, in Blackboard Architectures and Applications, ed. V. Jagannathan, R. Dodhiawala, & L. Baum, (San Diego, Academic Press), p. xix

A Retrospective View of Miriad

R. J. Sault

Australia Telescope National Facility, CSIRO, P.O. Box 76, Epping, N.S.W., 2121, Australia

P. J. Teuben

Astronomy Department, University of Maryland, College Park, MD 20742

M. C. H. Wright

Astronomy Department, University of California, Berkeley, CA 94720

Abstract. Miriad is a radio interferometry data-reduction package, designed for taking raw data through to the image analysis stage. The Miriad project, begun in 1988, is now middle-aged. With the wisdom of hindsight, we review design decisions and some of Miriad's characteristics.

1. A Brief History

Miriad aims at being a "full service" radio interferometry data-reduction package, taking raw telescope data through to image analysis and publication-quality image displays. The Miriad project was initiated by the Berkeley Illinois Maryland Association (BIMA). This followed much internal debate, and a meeting in 1988 February where a number of external experts (from the AIPS, GIPSY, and IRAF camps) were invited to express their opinions on BIMA's off-line software options. BIMA's decision was to develop two streams. The first stream, the Miriad project, was to build on the experience with the Illinois Werong and Berkeley RALINT packages, and to develop a package up to, but excluding, the image analysis stage. The second, image-analysis stream, eventually died, and Miriad was extended to cover this area.

Miriad was designed and developed jointly by groups at the three BIMA sites. Most of the infrastructure was written in Illinois by one person, with significant input from Berkeley. The first astronomically useful applications (FITS readers/writers, imaging and deconvolution tasks) appeared in 1988 October. A fourth site became involved when two of the Miriad group members moved to the Australia Telescope National Facility in 1990, and began extending Miriad to meet the needs of that institute's interferometer.

2. Project Management and Economics

Having the project spread over three or four sites has its difficulties, especially when group members are responsible to individual institutes and not to the project as a whole. In the early stages, there were regular phone conferences, quarterly face-to-face meetings, and a barrage of e-mail. There was no real project leader, with the project advancing by the good will of the group members (a true anarchy). The pragmatic solution to the occasional disagreements between group members (and the interests they serve) was nick-named "the free market economy" (a true capitalism). Each piece of code has an "owner" who must approve any change to it. If an owner cannot be persuaded, anyone is free to submit alternative code, and the different codes compete in the open market for users. Generally this occurs only occasionally, but it has led to multiple, functionally similar, tasks of significantly varying quality. This "free market" approach results in some inefficient use of resources which perhaps could be avoided with a "centralized economy".

3. User Interface and Documentation

User interfaces were a controversial issue in the early stages of the project— no interface suits every use. The approach adopted was to make the "front-end" user interface quite separate from number-crunching tasks. Apart from running tasks, the main functions of a front-end program are to provide task documentation and to help assemble task parameters. These parameters are passed to the number-crunching tasks simply as command-line arguments. This trivial interface made it easy to develop front-end interfaces to suit different tastes; early front-end programs included a SunView windowed environment, a VT100 menu system, and a "dumb terminal" interface similar to POPS. This design means that the front-end task can be completely bypassed, with users initiating tasks at the host command line. More experienced users do this often, particularly with simpler tasks. It is also the best way to develop batch scripts (we prefer to write those in the powerful shell that the host provides, rather than to try to reinvent it).

Using an idea adapted from PGPLOT, task documentation is stored as comments as a preamble in the source code. A tool extracts this "help file" and stores it in a directory ready for use by the front-end programs (whenever a task is recompiled, its help file is also updated). Help files for subroutines are treated in a similar fashion. Although help files are good for specific information, additional documentation giving a greater overview is needed. User and programmer guides appeared early in Miriad's life. Later a guide specifically aimed at Australia Telescope users was developed. Although they are continually updated, re-assignments within BIMA have meant that their guides are now out-of-date. With the popularity of the WWW, a html version of Miriad's documentation was developed (e.g., see *http://bima.astro.umd.edu/bima/miriad* or *http://www.atnf.csiro.au/ATNF/miriad*). This hypertext version is generated automatically from the source of the users guide and the help files.

4. File Format Issues

Compatibility with FITS is clearly an important goal. Although we considered the possibility of making FITS the "native" file format of Miriad (as it was in Werong), we concluded that pure FITS was too inflexible for this purpose. However, we adopted an image data-set structure which shadowed FITS reasonably closely (thus making a translation between the two straightforward). This was not possible with visibility data-sets—the general characteristics of the uv FITS format were seen as too restrictive for our needs. The visibility format adopted is based on the RALINT design of Wilson Hoffman. It has proven to be significantly more flexible than the FITS style of handling visibility data, and has the added advantage of a much cleaner interface for the programmer. We have found it easy to use this format with a number of data forms not considered in the original design. Unfortunately, it has the penalty that the current implementation can be slow to read.

Our initial concept of Miriad was of a package working in a shared disk environment of VAXes, Suns and Crays—machine-independence of the data was seen as very important. Miriad data are stored in a canonical format on disk, with the i/o system performing needed conversions during the i/o process (this conversion is invisible to the applications programmer). Although this has been very successful, because of declining interest in VAXes and Crays, machine independence has not been as critical as we had anticipated.

5. Language and Portability Issues

Miriad was designed to be portable, with VMS and several UNIX variants being the initial target systems. Since then, Miriad has been ported, with minimal effort, to many UNIX-based systems. To aid portability, and to reduce our own workload, we have gladly used public-domain software where appropriate. Miriad's plotting tasks are based on PGPLOT, whereas some of the numerical code is based on LINPACK.

Although all the i/o system and ancillary tools are written in C, most of Miriad is written in FORTRAN. We used FORTRAN because we felt that most astronomers would be happiest with it (and we wanted to attract programmers), and because vector machines were important to us (the best vectorizing compilers continue to be FORTRAN ones). As the i/o system is written in C, at some level FORTRAN has to call C routines. This language barrier is invariably system dependent (there are six schemes used in the systems to which Miriad has been ported). We developed a tool which takes a system-independent interface description, and which produces a thin layer of system-dependent code that mates the FORTRAN and C parts of Miriad. Although we have always had some misgivings about a mixed-language system, this approach has worked reasonably well.

6. Visualization

Visualization and image display is one of Miriad's failings. In the initial stages, this area was split off to a sub-group to develop. Their plans were comprehensive, and their development lagged behind the rest of the project. As a stop-gap, a simple interim set of routines was adopted. Eventually, the comprehensive plan failed and that sub-group disbanded. Somewhat later, a second attempt was made. Although this did progress further, it also effectively failed, and this second sub-group also disbanded.

Meantime, the "interim" routines were slowly upgraded, and made to work under the X Windowing system, but they are still basic. These shortcomings have been somewhat alleviated by the development of a set of image display tasks using PGPLOT. Although a good plotting package, PGPLOT is not intended to be an image display package, and so its model of a display device is limiting.

7. Miriad's On-Line Component

One novel aspect of Miriad is that it has been integrated into the on-line system of the Hat Creek interferometer. The on-line system generally uses the Miriad user interface and documentation system. Thus, to some extent, the user interface remains the same from driving the telescope to producing publication quality output. The on-line system also writes the raw data directly in the Miriad uv data format (this requirement was one of the reasons why uv FITS was unsatisfactory).

8. A Niche Package

Miriad is continuing to be developed. It has proven to be a good and flexible environment for writing specialized applications, as well as for developing new algorithms of greater applicability, particularly to teach students. A natural question to ask is "why reinvent the wheel?" Certainly at the original planning meeting there was a strong voice (both from NRAO and some BIMA representatives) for using AIPS to solve BIMA's reduction problems. AIPS, however, did not satisfy a number of the criteria that BIMA felt were essential. We had the choice between a major development in AIPS, or a major development with a fresh system. A fresh start, a system more flexible than AIPS, and a more programmer-friendly environment were probably the deciding factors (politics may have also played a part).

In hindsight, it was the correct decision. We had a useful system comparatively quickly, and have been able to extend it with new algorithms and techniques at a good rate. The overhead of programming in AIPS would have, at best, slowed our software development. At worst, it would have completely dissuaded us from implementing many new applications.

Faced with AIPS++, what is the future of Miriad? Part of the success of Miriad is that it is not a huge package—it has been able to adapt and concentrate on specialized areas, and in a timely fashion. Miriad does not try to address the data-reduction needs of the entire radio-astronomy community, so the overheads are much less than those for AIPS++. We do not believe that mega-packages will, or should, swamp the small and mid-size packages; there will always be a place for these. At the same time Miriad is now showing some grey hairs—perhaps Miriad++ is needed.

Acknowledgments. We thank the many people who have helped make the package successful, in particular (in historical order) Wilson Hoffman, Brian Sutin, Lee Mundy, Neil Killeen, Jim Morgan, Bart Wakker, and Mark Stupar.

The IDL Astronomy User's Library

W. B. Landsman

Hughes STX Co., Code 681, NASA/GSFC, Greenbelt, MD 20771

Abstract. IDL (Interactive Data Language) is a commercial plotting, image processing, and programming language that is widely used in astronomy. Although IDL is a powerful language for spectral and image processing, it does not contain applications specific to astronomy. Therefore, in 1990 I created the IDL Astronomy User's Library (AUL), which is a collection of astronomy-related procedures written in the IDL language, available via anonymous FTP from *idlastro.gsfc.nasa.gov*.

I summarize recent developments in the IDL language of particular interest to astronomers. I then mention some additions to the IDL Astronomy Library since the previous report of Landsman (1993). Finally, I critically examine possible drawbacks to the use of IDL for astronomical data analysis.

1. Introduction

IDL[1] (Interactive Data Language) is a commercial plotting, image processing, programming, and graphical user interface (GUI) development language. It is a language designed to allow much more rapid scientific data analysis than is possible using FORTRAN or C. Features of IDL that promote this "hands-on" approach to data analysis include an interpreted, vectorized compiler, an expressive syntax, and numerous built-in spectral and image processing functions.

IDL is widely used in astronomy, especially in the analysis of space-based data, and in the fields of solar and planetary astronomy. Recent examples of astronomical software packages based on IDL include Bloch et al. (1993), Brekke (1993), McGlynn et al. (1993), Ewing et al. (1993), and Hall et al. (1994). In this paper I summarize recent additions to the native IDL language and to the IDL Astronomy User's Library, and consider possible drawbacks to the use of IDL for astronomical data analysis.

2. Enhancements to IDL

IDL has undergone substantial evolution since its initial release in 1982. Recent enhancements to IDL of particular interest to astronomers include the following:

---

[1] Research Systems Inc., Boulder CO

- IDL is now available for Microsoft Windows, Macintosh OS, most Unix workstations, and Vax/Alpha VMS. Little or no programming is required to transport IDL software between these different platforms.

- Many of the routines from the book "Numerical Recipes" (Press et al. 1992) are now licensed and incorporated into native IDL. The ease of use of the popular "Numerical Recipes" routines is significantly enhanced within the interpreted IDL environment, as the routines can be tested at the command level, and the results immediately printed or plotted.

- Native IDL now supports numerous disk file formats including CDF, HDF, netCDF, TIFF, GIF, and JPEG. (Complete FITS and STSDAS I/O support is available from the IDL Astronomy Library.) Note that because IDL operates in virtual memory, most steps of a procedure can be written without any particular disk format in mind. One simply needs an initial procedure call to convert a disk file into IDL variables, and a final procedure call at the end of the session to write variables back to disk.

- The GUI (widget) development tools have been significantly enhanced. For example, it is now possible to design a GUI without writing any code. The IDL widgets require a significantly higher level of programming sophistication than the traditional IDL command-line programming. Nevertheless, the widget tools are sufficiently simple that astronomers can develop their own personalized GUI tools, with only a modest investment of programming time.

- IDL is now sold by distributors in Japan and throughout Europe. About one-third of the users of the IDL Astronomy Library come from outside the United States.

3. Recent Additions to the IDL Astronomy User's Library

IDL is a general software package, used in such fields as remote sensing and medical imaging, in addition to astronomy. As such, it does not contain any procedures specific to astronomy. In 1990, I created the IDL Astronomy User's Library (IAUL), which is a collection of astronomy-related procedures written in the IDL language, available via anonymous FTP from idlastro.gsfc.nasa.gov (Landsman 1993). Astronomy-related IDL software contributed from the community is checked for appropriateness, accuracy, and programming standards. Thus, the site is intermediate between an unmoderated bulletin board and a unified data analysis package. The IAUL does not contain any instrument specific software, although it does contain pointers to other anonymous FTP sites containing instrument specific IDL software. (Anonymous FTP sites exist containing IDL reduction software for the *IUE*, *HST*/GHRS, *COBE*, *SOHO*, and *ROSAT* instruments.)

An important addition to the IAUL is the support for the 25 astronomical coordinate systems discussed by Greisen & Calabretta (1995). The coordinate conversion software was written by Rick Balsano and the more complicated transformations were verified by Imannuel Freedman. Additional procedures exist that recognize three ways that the world coordinates may be stored in

FITS keywords: (1) the original FITS/AIPS system with a reference pixel, rotation, CROTA2, and pixel scale CDELTi, (2) the IRAF/ST ScI system with a reference pixel and a coordinate description matrix CDi_j, and (3) the Greisen & Calabretta proposal which includes both a pixel scale and a rotation and skew matrix. The software also recognizes headers from the ST ScI Digitized Sky Survey and will apply the appropriate nonlinear transformation to the pixel coordinates.

A recent contribution from Tom McGlynn is a generalized FITS reader that supports both variable-length binary tables, and random groups. An especially important feature of this software is that binary table columns are directly mapped into the tags of IDL structure arrays. The full IDL data analysis capabilities (e.g., plotting, sorting, and subscripting) can then be applied to the structure variable.

Other recent additions to the IAUL include mathematics and statistics code to complement the intrinsic "Numerical Recipes" routines. Examples include code for principal components analysis (Murtagh & Heck 1987), Kolmogorov-Smirnov statistics, and cubic-spline smoothing.

4. Drawbacks of IDL?

Every language has its strengths and weaknesses. In this final section I examine five possible drawbacks a hypothetical astronomer might make to the use of IDL for his data analysis.

1. *IDL is an interpreted language, and thus slower than compiled languages such as* FORTRAN *or C.* This statement is technically true, but properly vectorized IDL code suffers little performance penalty compared to FORTRAN or C. (IDL code is usually most readable when it is properly vectorized, so that, in this case, "the good coincides with the beautiful.") In fact, IDL often outperforms pedestrian FORTRAN or C code, since the many built-in image and spectral processing functions of IDL are highly optimized. For the rare cases (such as generalized median filters) where the code cannot be vectorized, and where performance speed is crucial, IDL offers several different ways to link to a FORTRAN or C program, including a fast dynamic link to the executable.

2. *The use of virtual memory in IDL is not suitable for very large images.* While the virtual memory capabilities of workstations have dramatically improved in the past few years, it is also true that the increase in the size of CCDs has been just as dramatic. Very large ($\gtrsim$ 16 MB) images can still be processed with IDL, but the array must be processed in pieces, so that the simplicity of the IDL code is then lost.

3. *Professor X cannot use my IDL software since she does not have an IDL license.* This objection is valid, although Professor X can probably decipher the algorithm, since the IDL syntax was designed for clarity rather than brevity. In addition, IDL supports remote procedure calls (RPC), so that a licensed client can support a server without an IDL license.

4. *I cannot find a galaxy surface photometry package in IDL.* Most astronomical software is not written in IDL, and the software in the IDL Astronomy

Library is an order of magnitude less complete than that in the large public domain systems such as IRAF or MIDAS. For this reason, IDL is unlikely to be an astronomer's sole working environment, although it is likely to be his preferred environment for developing new code.

5. *I am not going to purchase an IDL license when there is public domain software available.* The purchase of a commercial license can be worthwhile if it saves programming time and costs, and significantly enhances one's data analysis capabilities. Some IDL features are duplicated by various types of public domain software. IDL may or may not be a worthwhile investment, depending on an astronomer's needs, skills, and budget.

Acknowledgments. The IDL Astronomy Library is funded under NASA grant NAS5-32583 to Hughes STX.

References

Bloch, J. J., Smith, B. W., & Edwards, B. C. 1993, in Astronomical Data Analysis Software and Systems II, ASP Conf. Ser., Vol. 52, eds. R. J. Hanisch, R. J. V. Brissenden, & J. Barnes (San Francisco, ASP), p. 243

Brekke, P. 1993, ApJS, 87, 443

Ewing, J. A., Isaacman, R., Gales, J. M., Chintala, S., Kryszak-Servin, P., & Galuk, K. G. 1993, in Astronomical Data Analysis Software and Systems II, ASP Conf. Ser., Vol. 52, eds. R. J. Hanisch, R. J. V. Brissenden, & J. Barnes (San Francisco, ASP), p. 367

Greisen, E. W., & Calabretta, M. 1995, this volume, p. 233

Hall, J. C., Fulton, E. E., Huenemoerder, D. P., Welty, A. D., & Neff, J. E. 1994, PASP, 106, 315

Landsman, W. B. 1993, in Astronomical Data Analysis Software and Systems II, ASP Conf. Ser., Vol. 52, eds. R. J. Hanisch, R. J. V. Brissenden, & J. Barnes (San Francisco, ASP), p. 246

McGlynn, T. A., White, N. E., & Scollick, K. 1993, in Astronomical Data Analysis Software and Systems III, ASP Conf. Ser., Vol. 61, eds. D. R. Crabtree, R. J. Hanisch, & J. Barnes (San Francisco, ASP), p. 34

Murtagh, F., & Heck A. 1987, Multivariate Data Analysis (Dordrecht, Reidel)

Press, W. H., Teukolsky, S. A., Vetterling, W. T., & Flannery, B. P. 1992, Numerical Recipes 2nd edition (New York, Cambridge University Press)

Part 11. Data Modeling and Analysis

Spectroscopic reduction and analysis programs at the DAO

G. Hill

Dominion Astrophysical Observatory, Herzberg Institute of Astronomy, National Research Council of Canada, 5071 West Saanich Road, Victoria BC V8X 4M6, Canada

Abstract. In this paper I outline the FORTRAN programs available at the DAO to process and synthesize spectra, loosely described under the generic name REDUCE. The viewpoint is that of a research scientist who develops and uses his own software, and who writes collaboratively for colleagues. Much of the software has been available on the DAO's VAX machines since the early 80s and is slowly being converted to SUNs, although recent developments involve the use of DEC Alphas. The rationale behind the various reduction and analytic software is outlined, as well as the way the programs are controlled, e.g., menu, questions and answers, keywords and values, or the cursor. Some particularly useful tools are mentioned, such as, interpolation, optimization, and error analysis.

1. Introduction

Since the mid-70s, I have been developing software for the reduction of digital spectra. Initially, the software was developed to process digitized scans of photographic spectra to measure radial velocities (RV) using techniques akin to those employed by Grant measuring engines, or the DAO equivalent: ARCTURUS. This involved line-by-line measurements (VELMEAS, Hill et al. 1982a), which was inadequate when blended profiles were encountered. Considering that binary stars were the major source of data, this was not much of an advance, but it did serve to give me experience with digital data and interactive graphics. When the *IUE* satellite flew, I spent a lot of time writing software to plot the data, and, later, to measure the lines by fitting Gaussian, Lorentzian, or rotational profiles to the data—*sans* reseau marks. This software, now called VLINE (Hill et al. 1982b), was combined with REDUCE (Hill et al. 1982c) to provide the basis for the next development. The need to measure blended line strengths spurred me (reluctantly) to modify REDUCE to include scans of the calibration wedges, in order to convert density to intensity. This led to the re-discovery of the "Baker density" (Baker 1925; de Vaucouleurs 1968), which provides a reliable way to generate intensity from the characteristic curve.

Early in the 80s, McLean (1981a,b) wrote a cross-correlation package for the velocity analysis of binary stars, and used data from the DAO to demonstrate the method. He applied it to the most difficult of systems—the W UMa stars—and showed that good velocities could result from the method. This was a breakthrough, and really marked the advent of "modern" binary-star velocity measurements. In 1981 I wrote software (VCROSS) (Hill 1982a) that also used the technique, following a visit from McLean's advisor, Ron Hilditch. It is now the heart of most of the velocity measures that I make.

As the software began to be developed I started exporting it, and continue to do so. Its use was originally on a collaborative basis but I have now abandoned that. But initial development of any new software is still on a collaborative basis. In this paper I outline the software I have been responsible for developing at the DAO. A previous summary is given by Hill & Fisher (1986a). Note that throughout the paper, program names are given in bold face.

2. Spectroscopic Data

2.1. Observing Context

In this section it is worth describing the data. The software was originally written for high dispersion spectroscopy, i.e., plates that were 8 in long taken, at 2.5–6.5 Å mm^{-1}, and (under-)sampled every 7–8 μm. Later, the software was modified to handle low dispersion spectra—(80 Å mm^{-1}). Both of these extremes produced problems: the former because the size of the data sets (typically 25,000 by 4) was too small by a factor of 3–4. and the latter because of the difficulty in automatically predicting the position of an arc line in a crowded spectrum accurately enough for it to be measured automatically. The latter problem was harder to solve. The storage problem required the switch to the VAX. In the mid-1980s, when the Reticon became available, it was easy to convert the software to deal with it, particularly when the detector was linear.

The thrust of the observing programs has been twofold: (1) to measure velocities of stars, and (2) to measure the line strengths, or equivalent widths (EW) of spectral lines. In the former case the measurements are quite routine for single stars or galaxies, since there are rarely any complications. A simple fit to the peak of the cross-correlation-function (CCF) will generally suffice, though a red-shifted emission-line galaxy spectrum might pose problems of interpretation. For binary stars the spectra are more complex, and a successful measurement demands a composite CCF or a series of spectral lines, depending on what you are doing. It is an exacting task to extract this information, even with a high signal-to-noise (S/N) CCF or line profile. The critical problem is matching line-shapes to the fitted function. When I first began the measurement of highly blended profiles, typified by W UMa systems, I fitted Gaussian profiles to the CCFs. They gave good mass ratios, but the systemic velocities from the two orbits taken independently almost always differed. Once realistic shapes were used to measure the CCFs, however, the systemic velocities became consistent and the precision of the velocity measures improved. It then became possible to measure masses to the 1–3% accuracy that one desires for the study of stellar evolution (Andersen 1991). I refer the reader to some recent papers on V566 Oph (Hill et al. 1989b) and V599 Aql (Hill & Khalesseh 1991).

These developments drove me toward more realistic fitting functions, mostly defined by the data themselves and not, as before, simply chosen by virtue of being analytic. This approach parallels the point-spread-function (PSF) concept used in direct imaging. The same problem is encountered in measuring line profiles. Here the observations have not been smoothed and added, like that of a CCF, and S/N is usually a problem. Again, without a realistic profile shape, an EW measurement may not be very good. The ability to measure

EWs with excellent precision for blended spectra has been enhanced by the work of Bagnuolo & Geis (1991), who developed a useful algorithm to separate out the component spectra. The tomographic method they employ derives from medicine, and has proven most useful, as shown by Hill & Khalesseh (1993) for V1425 Cyg, Hill & Holmgren (1994) for Y Cyg, and Hill et al. (1994) for CC Cas. Another method, developed by Simon & Sturm (1994), promises to be even better.

Another spur to software development was the availability of data with high S/N (~2000:1). Naturally, EW measurements are easier, but such high quality data need to be matched to stellar atmosphere models to derive directly the effective temperature (T_e), surface gravity (log g), rotational velocity $V \sin i$, abundance (log Fe/H), and microturbulence (ξ). To do, this the Model atmospheres of Kurucz, calculated over a suitable grid (every 500 K from 7500–11,500 K, 0.5 dex steps in log g from 3.0–4.5, unequal steps of 0.3–0.5 dex in log Fe/H from -1 to +1, and at 0, 1, 2, and 4 km^{-1} for ξ) are required. The size of this grid, when coupled with a requirement for spectral coverage of at least 500 Å (sampled at 0.01 Å), produces some difficulties in memory size. At high dispersion with the Reticon, we typically sample 1870 points and cover only 70 Å. Any analysis would require something larger than that to enable Hγ to be examined. (Note that in an A-star the hydrogen lines might extend over 100 Å and still not reach the continuum). From these figures, we need about 260 MB of memory to handle just 20 Å of spectrum. We are not there yet; our current need, without including log Fe/H as an unknown, is 50 MB.

2.2. Type of Observation

While the CCD and Reticon have simplified spectroscopy, they have also produced the potential for inaccuracy, since wavelength calibrations are made at different times. Unlike a photographic plate, where all the information needed to analyze it is carried on it (arcs, star, clear plate and calibration data), the digital detectors generate sequential files which must be tracked. It is, of course, a problem for the observer and not one intrinsic to the detector. With non-coudé systems, arcs should be taken before and after each stellar exposure. For coudé systems, where the system is stable, it is not necessary to get arcs with the In addition we need bias frames (CCD), darks perhaps, and certainly flat-fields. An echelle observation creates a particular difficulty, as well as an opportunity, but we do not have such a spectrograph at the DAO (thank God!). Shortridge's FIGARO software, however, is particularly effective in processing echelle spectra.

3. Comments on Software

My philosophy has largely been defined by my experience. As someone who started programming in FORTRAN in the early 1960s I have had many years to develop bad habits as well as to gain a good knowledge of the language. This, coupled with a structured approach and excellent software editors, has meant that a new program can be developed remarkably quickly for each new application. With the simplifications that came to us from the VAXs acquired throughout the early 1980s, I discovered that monolithic programs were far too inefficient. Generalizing problems seemed to get me into more trouble than it was worth. For that reason I have adopted the philosophy of writing independent programs to do different tasks. This philosophy covers graphics, as well as the application itself. Initially I wrote a "catch-all" plotting package, where one or two calls would define the whole graph, but I quickly abandoned it as being awkward and unwieldy. Now each application has its own graphics. This has simplified program development a great deal, since one does not need a manual to revise the graphics. Smaller programs are easier to understand and modify.

I have concluded that the software must be interactive, driven by the cursor and keywords—questions and answers must be minimized. It took a while before I realized that a system of keywords and associated values, tasks, etc., would be very useful. I developed my own decoding software to enable the strings to be interpreted (see Hill & Fisher 1986a,b). It really made program operation much easier. One could measure a velocity in about 15 minutes using ARCTURUS, so the computer had to give answers quickly. The programs must be easy to use, i.e., they all must have the same "look" or "feel." With the system of keywords and graphics, with mostly identical cursor commands, I have largely achieved this goal. All the software has manuals. But converting from printed (dedicated word-processor) to on-line manuals has been particularly burdensome.

Many of the repetitive tasks (measuring arcs, linearization, measuring velocities of single stars) are organized to work from a file of file names which can be processed in sequence, either automatically or manually. My favorite automatic feature is the playback mode, which stems from the robot welding on the Japanese assembly lines. Hard copy is a factor here since I have always felt it necessary to be able to retrace the reduction steps, particularly during the analysis process. For the automatic processing hard copy is therefore integral to the process.

Graphics have always been a problem. Initially the software was developed for a Tektronix 4012 terminal (768 by 1024) and its hard-copy unit. From Tektronix's **PLOT 10**, higher level graphics routines were written to facilitate program development (**TGRAPHLIB**, Hill 1986). With the growth of laser printers and PostScript, this software became inadequate, so I converted to **Lick Mongo** (Allen & Pogge 1991). This, too, created difficulties in exporting my software because **Lick Mongo** is not widely available. I am currently converting to **PGPLOT** (Pearson 1989), which is simpler to implement off-site.

The hardware platform also has an effect on things. The swing from VAXs to SUNs leaves software developers caught between two stools since one's clients are not all shifting at once. Thus I have been stuck with trying to maintain two systems (even the switch from VAXes to Alphas is not routine). Fortunately some simplifications have occurred along the way: the general adoption of a single hardware platform, FITS (Wells et al. 1981), and later, the large software packages IRAF and FIGARO (Shortridge 1987). But for reductions as specialized and as difficult as found in the binary star arena, "canned," generalized software cannot do the job.

4. Reduction of Spectra

In this section I deal with all the steps leading up to measuring a digital spectrum; analysis will be covered in the next section. Prior to that, the way that I do book-keeping is noted.

4.1. File Naming

Typically, a datafile name is based on the detector, telescope, year, and sequence number of the observation. A CCD image (number 1780) taken with the 1.22 m telescope in 1994 would have a file name C122941780. In addition to this, I follow the convention defined by Gulliver (1976) and Irwin (1978); this simple scheme facilitates record-keeping since each type of file or each reduction process simply alters the prefix. The arcs, clear plates, etc., are therefore uniquely identified with the stellar observations. The exact processing path depends on whether one has photographic or CCD (or Reticon) data. For photographic spectroscopy, the data are produced by a PDS machine which scans both arcs, the stellar spectrum, and a predetermined stretch of clear plate. The calibration is also scanned at this time. Such a series of scans yields the files listed in Table 1. As noted in the table, some of these steps are omitted for the digital detector.

Table 1.   File-naming convention for plate or file 122941780

Raw spectroscopic data		Processed data	
Calibration	T122941780*	Linearized in λ	W122941780
Clear plate	L122941780	Converted to intensity	I122941780*
Arc	F122941780	Normalize to continuum	R122941780
Stellar	S122941780	Converted to log λ	U122941780[†]

*Necessary only for photographic spectroscopy
[†]Generally unnecessary

4.2. FITS Files

The original DAO disk FITS file system was written by Poeckert in 1981, and has seen many incarnations since then. Each processing step that a spectrum undergoes is logged in the header to show what has happened to the data. This, along with the above file-naming sequence, allows the user to know the status of their reductions in an instant.

4.3. Photographic Reductions

These are routine since the scans all have related names.

4.4. Reticon Observations

In dealing with Reticon data, we try to emulate the photographic procedure. Given a long string of observations, the program RET72 allows one to produce a similar set of files. The data are plotted and examined, the flats are identified for a given night, and an average defined. Given the starting and ending file numbers, the data are normalized by the flats. RET72 automatically identifies arcs straddling stellar spectra and renames them, prefixing them with an "F" under the name of each stellar spectrum. The result for each stellar file is an "F-file" comprising the straddling files abutted end to end (2 dimensions). Incidentally, the arcs are automatically identified by a skewness value $\gg 1000$, but the algorithm will fail for emission-line spectra. The whole process can be done manually.

A similar scheme is planned to convert the CCD data that I am now acquiring at the DAO to one-dimensional (1-D) form. When using data from the AAT or WHT, I use the FIGARO software to produce the 1-D files that feed the reduction software. The arcs are measured by REDUCE (Hill et al. 1982a), an interactive program based on the method uses to measure arcs with ARCTURUS. The heart of the scheme is a standard plate—the prediction of a position x in μm—for a given λ, based on the spectrograph parameters. In essence, it is a differential method for finding the correction of each line from a predicted position. One must first identify one line with a given pixel (hardly an onerous task); that value is contained in the F-file which RET72 produces. Once a spectrum has been measured, the keystrokes are recorded and later measures are done in a playback mode. The same arcs, which might also straddle other stellar observations, are not re-measured but are updated according to the velocity correction to the sun.

Once the arcs are measured, the spectra are linearized in λ. The conversion is based on an interpolation using INTEP (Hill 1982b), which uses cubic splines to interpolate a reasonable curve *through* the data. It is an extraordinarily effective workhorse in all the reduction and analysis programs.

5. Required Measurements

As noted in Section 2.1 the types of measures we need to make are:

RV: Done either line-by-line, or all together. With binary stars, the measures may be very difficult;

λ: These are also very difficult, and are generally only made for single stars;

EW and $V \sin i$: The measures of line strengths and rotational velocities are also line-by-line and very difficult. The spectra often have low S/N;

High dispersion, high S/N: For high dispersion work we are looking at the physical parameters of single stars, which is a much easier, but CPU-intensive task. To date we have only worked on high S/N data ($\sim$2500:1) and small 20 Å stretches of spectrum. The software is ready to deal with spectra of B–A stars to measure T_e, $\log g$, $V \sin i$, $\log$ Fe/H and ξ.

6. Summary of Analysis Software

6.1. Prerequisites

All of these applications require wavelength linearized spectra, normalized to the continuum. The format of choice is FITS, although I occasionally write conversion software to provide the desired FITS format (the conversion from ASCII already exists). In this section I outline the programs in regular use at the DAO.

6.2. Reduction and Measurement of Spectra

Processing of Reticon observations. RET72 allows the user to plot raw data, average the flat field observations, normalize and assemble files in the desired forms of arc (F-files, 2-D) and stellar (S) files. It produces FITS files that can be measured for arc, linearized, and then rectified to the continuum by REDUCE. FIGARO performs the same function when we are dealing with CCD results from the WHT or AAT. RET72 is routinely used at the telescope, where it enables the observer to monitor the night's work very closely. Monitoring has been done off-site from Manitoba.

Radial velocity. VLINE measures up to 12 lines simultaneously. It requires a file of line identifications, and will fit either Gaussian, Lorentzian, rotational, or digital (PSF) profiles to selected non-contiguous pieces of the data.

VCROSS measures RV by cross-correlating selected pieces of spectrum with a reference spectrum or template. The reference spectrum may be real or synthetic (see Hill et al. 1993). It can fit analytic or digital profiles to the CCF, produce PSF digital profiles for use, create artificial binaries for testing, and combine inhomogeneous (data sampled at different wavelength intervals) FITS files. The $\ln \lambda$ conversion is done automatically. The last two features are transparent to the user. It can run from a list of file names interactively or automatically (only recommended for single stars).

VLINESUM measures RV by summing lines identified from a line list. It simulates an RV scanner, and is useful for working on visual binaries where the separations are small, and where the blended profiles will be further masked by the smoothing effect of a cross-correlation.

Wavelength, EW. VLINE measures up to 12 lines simultaneously, and will fit either Gaussian, Lorentzian, rotational, or digital (PSF) profiles to selected non-contiguous pieces of the data yielding EW, FWHM, and $V \sin i$. LINEMEAS is a less sophisticated version of VLINE, and only measures single (unblended) lines. It fits various profiles and is fast to use.

6.3. Analysis and Modeling

Binary stars. TOMOGRAPHY separates a series of composite spectra into two components. Given a series of spectra measured for RV (both components), it will deconvolve them into separate components. When cross-correlated with a reference spectrum, the resultant CCF provides the CCF PSF-profiles with which to re-measure the material, and hence to repeat the process.

Given files of RV and $JD_\odot$, RVORBIT calculates the spectroscopic orbit for any combination of parameters. It handles either single or double-lined spectroscopic binaries using either Sterne's (1941) method or the method of Lehmann-Filhés (1894). Errors are calculated by the usual error-matrix analysis and by Monte-Carlo or bootstrap statistics.

Spectral synthesis. ROTATION generates spectra from large files of synthetic intensity spectra derived from Kurucz's atmosphere models. The assembly of these spectra into a data-base is ongoing. Models are based on Roche geometry and Collin's (1963) formulation. Integration is based on the Gauss-Legendre scheme used by Collins and by Hill (1979) in LIGHT2 . The synthesized spectra are compared to the observations and then automatically solved for T_e, $\log g$,

$V \sin i$, log Fe/H and ξ. In certain cases (stars rotating at $\geq 50\%$ of breakup speed, with $i \ll 10°$), the inclination can be determined. It will be become extraordinarily useful for deriving physical parameters of "normal" stars, since it will happily interpolate automatically in four unknowns.

VSUN calculates the velocity correction to the sun ($RV_\odot$), as well as the heliocentric correction and Julian date. I usually use it to log the observations as I make them, and certainly use it at the end of the observing run. The file it produces, when combined with those generated from earlier observing, provides the data-base for **PHASES** (see below).

EPHEMERIS provides planning software for binary star observing. Given the site, date, UT range needed, and the ephemerides for the binary stars, the program outputs a table of phases for each night. It is possible to limit the phases (e.g., ranges around quadrature or eclipse), and to restrict observations above a given airmass (χ). It is also possible to pre-select a list of stars with varying phase constraints. This requires a constraint file of hour angle (HA) as a function of declination (not the same for all telescopes, even at the same latitude). The tables can be listed star-by-star over the nights desired or all stars on each night. Auxiliary tables giving HA, χ, and $RV_\odot$ may also be printed.

PHASES calculates the phases from the spectroscopic data-base produced by **VSUN**, and sorts and summarizes the data. Histograms of acquired phase data for each star help in the planning of an observing run.

Related tools and useful aids. Given a file of light-curve data (magnitude vs. phase or JD) **LIGHT2** will solve for any combination of parameters, e.g., relative radii and polar temperatures of either or both stars, and i. It can also provide line profiles for W UMa systems for use with **VCROSS**. It uses modified black-body intensities and linear limb darkening derived from theoretical models. It can be modified to use the intensity data **ROTATION** uses, once all the atmospheres have been calculated. It is described by Hill (1979, 1985) and by Hill & Rucinski (1993).

CURFIT is Bevington's (1969) venerable optimizing program—still remarkably potent. The algorithm has proven to be wonderfully flexible, and I use it throughout the software described above. It has been modified to enable the user to freely vary the parameters. All non-linear optimizing schemes have their warts when it comes to local or global minima, but, if used judiciously **CURFIT** does the job for these applications. In **ROTATION** we buttress the method by doing an extensive search through parameter space to verify the derived minima. In **LIGHT2** I use differing starting points. **CURFIT** is used to measure the arc lines in **REDUCE** and to perform all of the fitting in **VLINE** and **VCROSS**.

INTEP is an interpolation program based on cubic splines (Tsipouras & Cormier 1973) which has proven to be remarkably useful. It is stable, and fits a "reasonable" curve *through* data without the enhanced wiggles one sees in high order polynomials. The software for this, with some examples, is given by Hill (1982b). It has proven to be very reliable for the fitting of continua, unlike least-squares spline fits which do not go through the data. It is used for the re-binning of data, e.g., in the linearizing process, and within **VCROSS** when it is used to homogenize data prior to calculating the CCF.

7. Error Analysis

The calculation of errors is often the bane of any analysis, particularly when the parameters are correlated. In spectroscopic orbits, the classic problem involves the derivation of the longitude of periastron and the time of periastron passage. In addition CURFIT, as described by Bevington, has a bug in its error matrix analysis. I find that fits to CCFs often produce unrealistically low errors, even though the *relative* values derived from a list of measures are likely to be accurate. To get alternative estimates of the errors I have used Monte-Carlo and bootstrap statistics (e.g., Hill et al. 1989b).

Monte-Carlo methods. These are easy to implement—simply take the derived parameters, generate data for each time or phase, add the observational noise, and solve again. Make 100–1000 solutions, and then calculate the errors in each parameter.

Bootstrap statistics. These are equally easy to use, although it becomes impractical for CPU-bound analyses, such as solving a light curve or using ROTATION to calculate i for Vega (Gulliver et al. 1994). It would be interesting to see whether the DEC Alphas could do the job. The algorithm is straightforward: if you have n data points, take a random selection of n points from your data and analyze them. Repeat 100–1000 times, and sort the parameters in order. The values at the 16-th and 84-th percentile give twice the error.

Acknowledgments. I am indebted to my long-time friend and colleagues Drs. Ron Hilditch and Austin Gulliver for their support and help over the years. Without this close collaboration the software would never have been written.

References

Allen, S., & Pogge, R. 1991, Lick Mongo Manual (Santa Cruz, Univ. of California)
Andersen, J. 1991, A&AR, 3, 91
Baker, A. E. 1925, Proc. RAS Edinburgh, 45, 166
Bevington, P. R. 1969, Data Reduction and Error Analysis for the Physical Sciences (New York, McGraw-Hill)
Bagnuolo, W. G., & Geis, D. R. 1991, ApJ, 376, 266
Collins, G. W. 1963, ApJ, 138, 1134
de Vaucouleurs, G. 1968, AO, 7, 1513
Gulliver, A. F. 1976, Ph.D. Thesis, University of Toronto
Gulliver, A. F., Hill, G., & Adelman, S. J. 1994, ApJ, 429, L81
Hill, G. 1979, Publ. Dominion Astrophysical Observatory, 15, 297
Hill, G. 1982a, Publ. Dominion Astrophysical Observatory, 16, 59
Hill, G. 1982b, Publ. Dominion Astrophysical Observatory, 16, 67
Hill, G. 1985, LIGHT2 User Manual
Hill, G. 1986, TGRAPHLIB User Manual (Dominion Astrophysical Observatory)
Hill, G., Adelman, S. J., & Gulliver, A. F. 1993, PASP, 105, 748

Hill, G., & Fisher, W. A. 1986a, Publ. Dominion Astrophysical Observatory, 16, 159

Hill, G., & Fisher, W. A. 1986b, The Command System User Manual, DAO

Hill, G., Fisher, W. A., & Holmgren, D. 1989a, A&A, 211, 81

Hill, G., Fisher, W. A., & Holmgren, D. 1989b, A&A, 218, 152

Hill, G., Hilditch, R. W., Aikman, G. C. L., & Khalesseh, B. 1994, A&A, 282, 455

Hill, G., & Holmgren, D. 1994, A&A, in press

Hill, G., & Khalesseh, B. 1991, A&A, 245, 517

Hill, G., & Khalesseh, B. 1993, A&A, 276, 57

Hill, G., Ramsden, D., Fisher, W. A., & Morris, S. C. 1982a, Publ. Dominion Astrophysical Observatory 16, 11

Hill, G., & Rucinski, S. M. 1993, in Light Curve Modeling of Eclipsing Binary Stars, ed. E. F. Milone (New York, Springer-Verlag)

Hill, G., Poeckert, R., & Fisher, W. A. 1982b, Publ. Dominion Astrophysical Observatory, 16, 27

Hill, G., Poeckert, R., & Fisher, W. A. 1982c, Publ. Dominion Astrophysical Observatory, 16, 43

Irwin, A. 1978, Ph.D. Thesis, University of Toronto

Lehmann-Filhés, R. 1894, AN, 136, 17

McLean, B. J. 1981a, Ph.D. Thesis, University of St Andrews.

McLean, B. J. 1981b, MNRAS 195, 931

Pearson, T. J. 1989, PGPLOT User Manual (Pasadena, California Institute of Technology)

Shortridge, K. 1987, FIGARO Users Guide (Chilton, Rutherford Appleton Laboratory)

Simon, K. P., & Sturm, E. 1994, A&A, 281, 286

Tsipouras, J., & Cormier, R. V. 1973, Airforce Surveys in Geophysics, No. 272

Wells, D. C., Greisen, E. W., & Harten, R. H. 1981, A&AS, 44, 363

Interactive Fitting of EUVE Emission Line Spectra

M. Abbott

Center for EUV Astrophysics, 2150 N. Kittredge St., Berkeley, CA 94720-5030

Abstract. An interactive IRAF task for the analysis of *Extreme Ultraviolet Explorer* (*EUVE*) emission line spectra is presented. The line flux in extracted spectra may be measured by multiple techniques, including fitting individual lines. Blended lines may be fit simultaneously with the *EUVE* spectral PSF. To aid in identifying lines, the positions of known lines may be marked on the spectrum during the fitting process. Also, model *EUVE* spectra may be constructed and overlaid using plasma emissivities and source emission measures supplied by the user. The task is fully compatible with the calibration data set distributed by the *EUVE* Guest Observer (GO) Center and retrieves all needed instrumental parameters from that data set. It also makes use of the newly available IRAF support for windowing graphical user interfaces.

1. Rationale

Fitlines is an IRAF task currently under development at the Center for EUV Astrophysics. It was originally conceived of as a tool to aid in the wavelength calibration of the *EUVE* spectrometers, but may also prove useful as a spectral analysis tool. It was also a pilot project to investigate the development of tasks using the IRAF X11 support package.

The goal of **fitlines** is to provide an integrated environment in which to identify and measure the flux of emission lines in extracted *EUVE* spectra. The identification of lines is the most difficult part of the *EUVE* wavelength calibration effort. Observations of many different sources are used and most do not have previously published spectral analyses in the EUV regime.

Most of the capabilities of **fitlines** (plus a lot of additional ones) will be made available in the future in windows-based spectral reduction tasks already under development at NOAO. **Fitlines** is not intended as a general spectral reduction task, but rather seeks to complement such software by providing *EUVE*-specific functionality.

2. Implementation

Fitlines is implemented as an IRAF task and uses the newly released X11 IRAF support package for its user interface. It is written in SPP. Two separate major capabilities are present. The first is the ability to read in a table of line emissivities, display the tabulated lines on a plot of the spectrum, fold the emissivities for selected lines with a user-specified emission measure model for

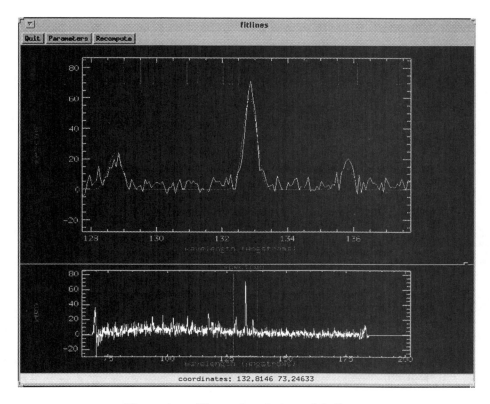

Figure 1. The main window of **fitlines**.

the source and the *EUVE* instrumental parameters to get a simulated spectrum, and overplot that on the observed spectrum. The goal is to help the user to decide which of many candidate IDs for a line are most plausible.

Figure 1 shows the main window of **fitlines**. The entire spectrum is plotted in the lower pane and a subsection is enlarged above. In the latter plot, the lines at the top mark the wavelengths of lines in the list of emissivities the user has supplied. Three of those lines have been selected and a model spectrum computed using them; it is overplotted on the actual spectrum.

Two more windows are illustrated in Figure 2. The lower window contains the user-specified emission measure of the source versus temperature. The user also supplies information about the distance to the source and the intervening interstellar absorption.

The upper window contains information about the list of plasma line emissivities. The three lines selected by the user are highlighted in the list and their emissivities versus temperature are plotted to the left. **Fitlines** attempts to simplify the identification procedure by forming aggregates of sets of closely-spaced lines that are not resolved with *EUVE*. Such aggregate lines are marked with an asterisk in the list in Figure 2 (all of them in this example). The user can display the components of any aggregate as desired; in the example, the user

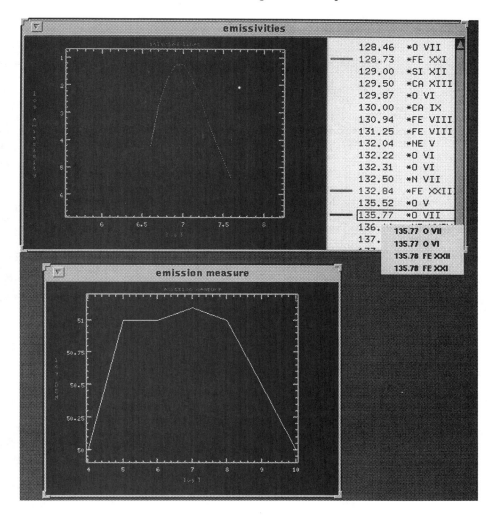

Figure 2. Display of line emissivities and the emission measure in the **fitlines** task.

has opened a pop-up window to see which lines make up the aggregate listed at a wavelength of 135.77 Å.

A second capability of **fitlines** is the measurement of line fluxes. Several common techniques are made available, including summing within regions, fitting functions, and fitting empirical line shapes. Where helpful, knowledge of the expected line width in an *EUVE* spectrum at any wavelength (determined from calibration data distributed by the *EUVE* Guest Observer Center) may be used to constrain fits.

Acknowledgments. This work was supported by NASA contract NAS5-30180.

On the Combination of Undersampled Multiframes

H.-M. Adorf

Space Telescope–European Coordinating Facility, European Southern Observatory, Karl-Schwarzschild-Str. 2, D-85748 Garching b. München

Abstract. Some cameras with digital detectors, such as the Wide Field and Planetary Camera 2 onboard the *Hubble Space Telescope*, undersample the joint point spread function of telescope and camera optics. In order to overcome undersampling, two or more frames of the same field may be obtained which are shifted with respect to each other by fractions of a pixel. During data analysis such multiframes must be "registered" and "combined". This contribution investigates a novel method for merging undersampled multiframes based on projections onto convex sets (POCS). The method does not require a point-spread function; it can cope with missing data, and may incorporate prior knowledge such as non-negativity and band-limit constraints.

1. Introduction

At present the Wide Field and Planetary Camera 2 (WFPC-2) is the primary science instrument on-board *HST*. Correcting *HST*'s spherical aberration, it displays a vastly improved image quality compared to WF/PC-1. However, it's CCDs coarse sampling cannot exploit *HST*'s high spatial resolving power. Obviously science programs such as photometry of dense star fields and the morphological classification of distant galaxies would benefit from a method allowing a proper combination of two or more "dithered" WFPC-2 frames, i.e., frames that are shifted with respect to each other by fractions of a pixel.

Resolution improvement by combining (potentially undersampled) multiframes has in the recent past been investigated by a number of researchers outside astronomy (for references see Adorf 1994). Within astronomy a combination method has been proposed by Lucy (1993 and references therein) and further developed by Hook & Lucy (1993 and references therein). Being a generalization of the well-known Richardson-Lucy restoration method, this algorithm requires knowledge of the point spread function (PSF).

An alternative combination algorithm is presented here which is based on the concept of projections onto convex sets (POCS, Sezan, & Stark 1983; Youla 1987 and references therein). The proposed algorithm is similar in spirit to the POCS-method for reconstructing irregularly sampled data series (Adorf 1993). The algorithm does not presume knowledge of the PSF; it can accommodate missing data, as well as bandpass and non-negativity constraints. When a non-negativity constraint is effective, the algorithm becomes non-linear.

2. The POCS-Based Combination Algorithm

Assuming that the relative shifts between the undersampled frames are known, the POCS-algorithm (for two undersampled frames) can be stated as follows:

1. guess initial high-resolution (hi-res) estimate
2. fractionally shift hi-res frame to register it with low-resolution (lo-res) frame #1 and replace pixel values in hi-res frame by those from lo-res frame #1
3. fractionally shift hi-res frame to register it with lo-res frame #2 and replace pixel values in hi-res frame by those from lo-res frame #2
4. apply bandpass constraint
5. optionally apply non-negativity constraint
6. if not converged, iterate starting from step 2

This reconstruction algorithm belongs to the class of POCS-algorithms since each of the steps 2 to 5 is a projection onto a convex set in the linear space of all images. The iteration provably converges if there is a common point in the intersection of all convex sets. The POCS-algorithm has been implemented in the Interactive Data Language (IDL) image processing package.

3. Registration

Using fine lock on guide stars, *HST* usually achieves a pointing precision of about 7 milliarcsec RMS, or better. However, for non-contiguous observations its absolute pointing accuracy on the detector (i.e., the difference between commanded and observed pointing) has been worse. Therefore, in practice, the registration parameters have to be estimated from the data—a non-trivial problem in the presence of undersampling.

Traditionally, registration parameters have been derived from position measurements of stars contained in the field. In the context of this work, the registration parameters were estimated using a cross-correlation technique: the asymmetric shape of the correlation function in the vicinity of its peak was exploited for locating the peak to subpixel accuracy. The correlation approach has the merit of not requiring the presence of point sources in the field. While this technique has passed a few simple tests, further investigations are necessary to prove that it is a viable general method for registering "dithered" WFPC-2 frames.

A remaining problem concerns the WFPC-2 CCD-chips, which have a *non-regular* pixel grid. Single rows with narrower pixels are interlaced with the normal rows. This pattern repeats itself. Thus there is no set of registration parameters globally valid across the field of view.

4. Application to WFPC-2 Data

The POCS combination algorithm was applied to public *HST* Early Release Observations of the distant, rich cluster of galaxies CL0939+4713 (*HST* program

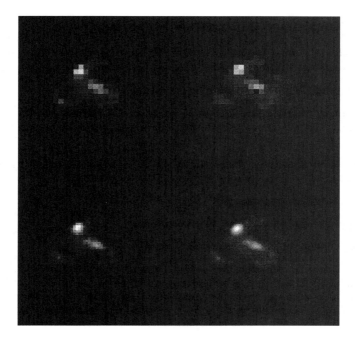

Figure 1. A "peculiar" galaxy in the distant, rich cluster CL0939+4713. **Top left:** part of a single coarsely sampled "subframe #1" at a scale of 0.1 arcsec/pixel; **top right:** part of the displaced subframe #2, shifted by $\Delta x = -0.13$ and $\Delta y = 0.31$ WFC pixels; **bottom left:** high-resolution output of the POCS-reconstruction algorithm; **bottom right:** same as before, but upsampled by a factor of four using sinc-interpolation. The combined high resolution frame reveals that the "peculiar" galaxy presumably consists of a pair of interacting galaxies.

number 5190). The data set has been analyzed by Dressler et al. (1994), who comment: "As remarkable as these images of systems at high redshift are, they point up the need for ... higher resolution... Substepping ... would dramatically increase the detail in these very high redshift systems ..."

The CL0939+4713 data set consists of 10 WFPC-2 exposures divided into two sets of 5 consecutive exposures. Each set was combined into a single frame using the STSDAS task "`crrej`", which simultaneously rejects cosmic ray hits. Subsequent work was carried out on chip #2 frames. The two undersampled frames in the working set still contained a large number of so-called "warm/hot pixels" at fixed CCD positions which were detected using a simple threshold algorithm. Data values at the location of warm/hot pixels were marked as "missing".

From the resulting two clean 800 × 800 pixel frames, two 128 × 128 pixel subframes centered on a peculiar galaxy were extracted (Figure 1a, b), and registered via cross-correlation. Fractional offsets of $\Delta x = -0.13$ and $\Delta y =$

+0.31 WFC pixels (i.e., $\Delta x = -13$ and $\Delta y = +31$ milliarcsec, respectively) were found.

The POCS combination algorithm was applied to these subframes in a staged approach, going through a low-resolution and a medium-resolution phase to the final high resolution. Altogether 30 iterations were executed. The resulting high-resolution frame (Figure 1c, d) nicely interpolates the missing data, and reveals fine spatial structure difficult, if not impossible, to gather from the two individual undersampled frames. The contrast in the high resolution image is also increased compared to the coarsely sampled frames.

5. Summary

A novel, iterative POCS (projections onto convex sets) based algorithm has been presented for combining two or more fractionally shifted and potentially undersampled observations of the same scene into a high-resolution image. The method does not require knowledge of the point spread function, and can accommodate missing data as well as bandpass non-negativity constraints. The algorithm can also be used for solely increasing the signal-to-noise ratio without resolution enhancement. The POCS-algorithm has been successfully used to generate high resolution images from *HST* WFPC-2 observations of the distant, faint galaxy cluster CL0939+4713, revealing morphological detail which cannot be easily discerned in the individual coarsely sampled observations.

References

Adorf, H.-M. 1993, in Proc. 5th ESO/ST-ECF Data Analysis Workshop, eds. P. Grosbøl et al. (Garching, ESO), p. 191

Adorf, H.-M. 1994, ST-ECF Newsl., 22, in press

Dressler, A., Oemler, J. Augustus, Sparks, W. B., & Lucas, R. A. 1995, preprint

Hook, R. N., Lucy, L. B. 1993, ST-ECF Newsl., 19, 6

Lucy, L. 1993, in Proc. Science with the Hubble Space Telescope, Chia Laguna, Sardinia, eds. P. Benvenuti and E. Schreier (Baltimore, Space Telescope Science Institute), p. 207

Sezan, M. I., & Stark, H. 1983, Appl. Opt., 22, 2781

Youla, D. C. 1987, in Image Recovery—Theory and Application, ed. H. Stark, (Orlando, Academic Press), p. 29

Interpolation of Irregularly Sampled Data Series—A Survey

H.-M. Adorf

Space Telescope–European Coordinating Facility, European Southern Observatory, Karl-Schwarzschild-Str. 2, D-85748 Garching b. München, Germany

Abstract. Many astronomical observations, including spectra and time series, consist of irregularly sampled data series, the analysis of which is more complicated than that of regularly spaced data sets. Therefore a viable strategy consists of resampling a given irregularly sampled data series onto a regular grid, in order to use conventional tools for further analysis. Resampling always requires some form of interpolation, which permits the construction of an underlying continuous function representing the discrete data. This contribution surveys the methods used in astronomy for the interpolation of irregularly sampled one-dimensional data series.

> There is virtually no literature
> on which method does best
> with which kind of problem.
> — Julian H. Krolik 1992

1. From Irregular to Regular Sampling

Many astronomical observations consist of irregularly sampled data series, e.g., spectra from spectrographs with non-linear wavelength scales, time series of light curves or radial velocity observations. It is well known that the analysis of irregularly spaced data sets is more complicated than that of regularly spaced ones. Therefore, as a promising strategy, one might consider resampling a given irregularly sampled data series onto a regular grid, in order to use conventional tools for further analysis.

Resampling always requires some form of interpolation or, in the presence of noise, *estimation*, which effectively allows the construction of an "underlying" continuous function representing the discrete data. The goal is to use an interpolation/estimation method which preserves the relevant information as much as possible. Therefore the assumptions on which the interpolation/estimation method rests must not preclude (or predetermine) the results of the subsequent analysis.

Constructing a continuous representation of an irregularly sampled data set may have a virtue of its own, since structure present in the data is often more clearly displayed by a continuous curve than by the scattered, clumped original data points, and may therefore more readily be picked up by the eye.

2. Interpolation Methods in Current Use

Numerous interpolation/estimation schemes exist, many of which generalise to the case of irregular sampling. Methods may be distinguished according to whether they are conceptually simple or complex, linear or non-linear, direct or iterative, and exact or approximate. Furthermore they may accommodate different statistical weights, respect a potentially existing non-negativity constraint, or allow an estimate of the interpolation error.

2.1. Direct Fourier Transform

This method (Scargle 1989) is applicable to data with arbitrary sampling. It allows to explicitly compute individual Fourier coefficients. The basic formulae are generalized to the case of unequal statistical weights. The method is equivalent to least-squares estimation of a single-frequency harmonic. By computing the Fourier coefficients at all required discrete frequencies and then Fourier-transforming back, a kind of interpolation can be obtained.

2.2. Compound Fourier Transform

The compound Fourier transform (FT) method (Meisel 1979) has been developed for gapped data. It uses a stack of individual discrete FTs. The exposition of the method is quite convoluted, and (therefore?) the method seems to be rarely used.

2.3. Matrix Inversion

Kuhn (1982) considers the irregular-to-regular resampling problem, and discusses two methods for solving the resulting matrix equation. One method is apparently applicable to jittered sampling, where the interpolation problem can be treated as a perturbation problem. Small off-diagonal terms in the interpolation matrix permit an iterative solution of the matrix equation. The other, direct method is applicable to regular sampling with missing data, and exploits the fact that in this case the interpolation matrix is circulant. Swan (1982) discusses the interpolation problem as an inverse problem in the presence of noise. The irregular sampling geometry makes the interpolation matrix ill-conditioned, leading to noise-amplifications, when straightforwardly applied to the data. Swan discusses two regularization strategies.

2.4. Least-Squares Estimation

This method, also known as minimum variance estimation, is a linear method applicable to arbitrary sampling patterns. For a signal with additive Gaussian noise it is a maximum likelihood estimation method. The LS-method permits the estimation of the probable interpolation error, and in its generalized formulation allows the inclusion of statistical weights. Barning (1963) devised an iterative scheme for the simultaneous LS-fitting of multiple sinusoids to data without weights. Ferraz-Mello (1981) advanced a formula for a weighted LS-fit (linear regression) of a single sinusoid model to zero-mean data. Gram-Schmidt orthogonalization is used to decouple the two sine and cosine terms of which the model is composed. He also considers the case when more than one harmonic is present in the data. Ferraz-Mello recommends an iterative scheme of subtracting previously fitted individual components ("pre-whitening"). Rybicki & Press

(1992) present a general formulation of the LS-estimation method for arbitrary sampling. An interpolation error estimate is provided.

2.5. Pre-Whitening and CLEAN Deconvolution

This method is applicable to a regularly sampled data series with missing data. Gray & Desikachary (1973), a year before the publication of CLEAN, employ an iterative scheme in the Fourier domain, where the Fourier transformed window pattern is identified in the direct Fourier transform of the data and subtracted. The required corresponding phase is estimated from the data. In a similar fashion Roberts et al. (1987) use a generalized, complex version of the CLEAN deconvolution method, to deconvolve the direct ("dirty") Fourier transform of the data, using the window function as "beam". CLEAN differs from Gray & Desikachary's method by using a smaller "gain" factor for the subtractions.

2.6. Autoregressive Maximum-Entropy Interpolation

This method by Fahlman & Ulrych (1982), applicable to a regularly sampled data series with missing data segments ("gaps"), attempts to fill the data gaps. Gap-filling starts with estimating an initial autoregressive (AR) model of order less than the length of the shortest known data segment. A maximum Burg-entropy algorithm is used to predict unknown data values. A new AR model is estimated from the entire data series, with known and predicted values, and iteration proceeds until convergence. The original method was improved by Brown & Christensen-Dalsgaard (1990) via inclusion of a bandpass-filtering preprocessing step. Despite the wide use of maximum entropy methods elsewhere in astronomy (particularly aperture synthesis imagery), ME interpolation seems not to have caught on widely (cf. Krolik 1992).

2.7. Polynomial Interpolation

This method (Groth 1975) is applicable to data with arbitrary sampling. A set of orthonormal polynomials is constructed over the set of irregular sampling locations. The interpolation is obtained as a linear superposition of these base polynomials. Statistical weights may be incorporated.

2.8. Trigonometric Polynomial Interpolation

This method is applicable to arbitrarily sampled band-limited, periodic (deterministic) data. It is based on the fact that a band-limited, periodic function can be exactly reconstructed from a sufficient number of irregularly spaced samples, using an explicit closed-form formula based on trigonometric polynomials (Adorf 1993). The method, originally devised by Cauchy in 1841, constructs a set of orthogonal basis functions over the set of irregular sampling locations. The method's linearity permits an estimate of the interpolation error. Interpolation with trigonometric polynomials may become an invaluable tool for high-fidelity interpolation of high signal-to-noise data.

2.9. POCS Interpolation

The POCS (projection onto convex sets) based interpolation method (Adorf 1993) is a conceptually simple, iterative scheme, directly applicable to regularly sampled observations with missing data, where the underlying continuous signal has a known (or assumed) fundamental period. The POCS method attempts to fill the data gaps by constructing a bandpass filtered version of the data series. The procedure iterates two "projection" operators: the first "completes" the original data series by filling gaps with arbitrary values, then applies a bandpass-filter; the second re-substitutes known data values. The POCS process provably converges, if the intersection of the convex sets is non-empty.

References

Adorf, H.-M. 1993, in Proc. 5th ESO/ST-ECF Data Analysis Workshop, eds. P. Grosbøl et al. (Garching, ESO), p. 191

Barning, F. J. M. 1963, Bull. Astron. Inst. Netherlands, 17, 22

Brown, T. M., & Christensen-Dalsgaard, J. 1990, ApJ, 349, 667

Fahlman, G. G., & Ulrych, T. J. 1982, MNRAS, 199, 53

Ferraz-Mello, S. 1981, AJ, 86, 619

Gray, D. F., & Desikachary, K. 1973, ApJ, 181, 523

Groth, E. J. 1975, ApJS, 29, 443

Krolik, J. H. 1992, in Statistical Challenges in Modern Astronomy, eds. E. D. Feigelson & G. J. Babu (New York, Springer-Verlag), p. 349

Kuhn, J. R. 1982, AJ, 87, 196

Meisel, D. D. 1979, AJ, 84, 116

Roberts, D. H., Lehár, J., & Dreher, J. W. 1987, AJ, 93, 968

Rybicki, G. B., & Press, W. H. 1992, ApJ, 398, 169

Scargle, J. D. 1989, ApJ, 343, 874

Swan, P. R. 1982, AJ, 87, 1608

Astronomical Data Analysis Software and Systems IV
ASP Conference Series, Vol. 77, 1995
R. A. Shaw, H. E. Payne, and J. J. E. Hayes, eds.

Calculating the Position and Velocity Components of HST

T. B. Ake

Astronomy Programs, Computer Sciences Corporation, Code 681/CSC, Goddard Space Flight Center, Greenbelt, MD 20771

Abstract. The *HST* uses an onboard model of its orbit to perform in real-time various control functions relating to spacecraft operations and observation support. The flight software uses the equation of center to solve Kepler's equation for two bodies. New coefficients for the ephemeris are updated every other day, and are archived as FITS keywords for each *HST* observation data set. Using these coefficients, an observer can perform a variety of calibration and characterization calculations relating to the orbital position and velocity of the telescope. We present here the methodology for computing the *HST* state vectors using this information.

1. Introduction

Many precise measurements with the *Hubble Space Telescope* (*HST*) require knowledge of its position and velocity during the observations. Orbital parallax, velocity aberration, Doppler shift, and light-travel time can all be significant when converting *HST* observations to the geocentric system, and from there to the barycenter of the solar system. In addition, the quality of observations are affected by the near-earth environment in which *HST* operates. An observer would be wise to understand such effects as scattered earth light, radiation background, and geomagnetically-induced motion on the data.

One way to determine the motion of *HST* during an observation is through the definitive orbit files that are archived at the Space Telescope Science Institute (ST ScI). Every other day the Flight Dynamics Facility (FDF) at Goddard Space Flight Center computes the position and velocity of the *HST* for the previous two days based on ranging measurements of the spacecraft. This information is forwarded to the ST ScI in the form of a list of *HST* state vectors for each minute of time, with each record giving the J2000 rectangular components of the position, in km, and velocity, in km s^{-1}, in the geocentric inertial coordinate system. The observer must find and extract the appropriate file(s) from the archive and interpolate the *HST* state vector data to the relevant times of the observations.

An easier method is to use the onboard ephemeris parameters that are provided with the data sets themselves. When the definitive orbit file is generated by the FDF, a set of ephemeris coefficients is created. The *HST* flight software uses these coefficients for various spacecraft control functions. During pipeline processing at the ST ScI, these are archived in the *.shh (non-astrometry) or *.dbm (astrometry) header file for each observation. We summarize using this information to compute the position and velocity of the *HST*.

Table 1. Onboard Ephemeris Model Parameters

FITS Keyword	FITS Description	Symbol
EPCHTIME	epoch time of parameters (secs since 1/1/85)	τ
MEANANOM	mean anomaly (radians)	M_0
FDMEANAN	1st derivative coef for mean anomly (revs/sec)	$\dot{M}$
SDMEANAN	2nd deriv coef for mean anomaly (revs/sec/sec)	$\ddot{M}$
ECCENTRY	eccentricity	e
SEMILREC	semi-latus rectum (meters)	$a(1-e^2)$
RASCASCN	right ascension of ascending node (revolutions)	Ω_0
RCASCNRV	rt chge right ascension ascend node (revs/sec)	$\dot{\Omega}$
ARGPERIG	argument of perigee (revolutions)	ω_0
RCARGPER	rate change of argument of perigee (revs/sec)	$\dot{\omega}$
COSINCLI	cosine of inclination	$\cos i$
SINEINCL	sine of inclination	$\sin i$
CIRVELOC	circular orbit linear velocity (meters/second)	V_C
TIMEFFEC	time parameters took effect (secs since 1/1/85)	–

2. Computations

The *HST* travels in a nearly circular orbit, with an altitude of about 600 km and velocity of 7.5 km s^{-1}. The orbital model used onboard is based on a simple two-body system, with perturbations to certain Keplerian elements due to the proximity of the earth. In Table 1 we list the relevant FITS keywords and descriptions that can be found in the header files, as well as the symbols used in the equations below.

The steps to compute the geocentric, rectangular coordinates for the *HST* using the parameters in Table 1 are as follows. First, the observer should verify that the correct onboard ephemeris has been archived with the data. The TIMEFFEC keyword specifies the beginning time at which the parameters are valid and its value should be within 2–3 days before the start of the observations.

For a time of interest, t, calculate the mean anomaly, M, from the initial position, M_0, at the epoch time of the parameters, τ,

$$M = M_0 + 2\pi \left[\dot{M}(t-\tau) + \frac{1}{2}\ddot{M}(t-\tau)^2 \right]. \quad (1)$$

Compute the true anomaly, ν, using the equation of center to solve Kepler's equation. This is typically expressed as a series in $\sin nM$ and e. For small e, terms only up to e^3 are needed (e.g., Smart 1965, equation V-85),

$$\nu = M + (2e - \frac{e^3}{4})\sin M + \frac{5}{4}e^2 \sin 2M + \frac{13}{12}e^3 \sin 3M.$$

Since trigonometric functions are computationally expensive, *HST* uses a different form of this equation involving only $\sin M$ and $\cos M$. Collecting like terms of e^n and expanding $\sin nM$, one can show

$$\nu = M + \sin M(2e + 3e^3 \cos^2 M - \frac{4}{3}e^3 \sin^2 M + \frac{5}{2}e^2 \cos M). \tag{2}$$

Once the true anomaly is known, then the distance from the center of the earth, r, is

$$r = \frac{a(1-e^2)}{1+e\cos\nu} \quad \text{(in meters)}. \tag{3}$$

The main perturbation on the orbital elements due to the non-spherical mass distribution of the earth is the regression of the ascending node, Ω, and the progression of perigee, ω, (Wertz 1978). The instantaneous values at t are

$$\Omega = 2\pi \left[\Omega_0 + \dot{\Omega}(t-\tau)\right] \quad \text{and} \tag{4}$$

$$\omega = 2\pi \left[\omega_0 + \dot{\omega}(t-\tau)\right]. \tag{5}$$

The geocentric *HST* position, in J2000 rectangular coordinates in meters, is then

$$\begin{aligned} x &= r\left[\cos\Omega\cos(\omega+\nu) - \cos i \sin\Omega\sin(\omega+\nu)\right], \\ y &= r\left[\sin\Omega\cos(\omega+\nu) + \cos i \cos\Omega\sin(\omega+\nu)\right], \\ z &= r\sin i \sin(\omega+\nu). \end{aligned} \tag{6}$$

The corresponding equations for radial velocity can be derived by differentiating those for position. Starting with x,

$$\dot{x} = \frac{x}{r}\dot{r} - r(2\pi\dot{\omega} + \dot{\nu})\left[\cos\Omega\sin(\omega+\nu) + \cos i \sin\Omega\cos(\omega+\nu)\right] - 2\pi\dot{\Omega}y.$$

To eliminate the $\dot{r}$ and $\dot{\nu}$ terms, we use a well-known property of elliptical orbits that the velocity can be represented as the vector sum of two constant velocities (Smart 1965). These are the velocity, μ/h, perpendicular to the radius vector and, $\mu e/h$, perpendicular to the semimajor axis, where $\mu = GM_\oplus$ and $h = r^2\dot{\nu} = \mu a(1-e^2)$. Designating $V_C = \mu/h$ as the circular velocity, we have

$$\begin{aligned} r\dot{\nu} &= V_C(1 + e\cos\nu) \quad \text{and} \\ \dot{r} &= eV_C \sin\nu. \end{aligned}$$

The rectangular velocities can now be determined from the onboard ephemeris parameters. Defining the auxiliary variables

$$\begin{aligned} a_0 &= \tfrac{1}{r}eV_C \sin\nu \quad \text{and} \\ a_1 &= V_C(1 + e\cos\nu) + 2\pi\dot{\omega}r, \end{aligned}$$

we have

$$\begin{aligned} \dot{x} &= a_0 x - a_1\left[\cos\Omega\sin(\omega+\nu) + \cos i \sin\Omega\cos(\omega+\nu)\right] - 2\pi\dot{\Omega}y, \\ \dot{y} &= a_0 y - a_1\left[\sin\Omega\sin(\omega+\nu) - \cos i \cos\Omega\cos(\omega+\nu)\right] - 2\pi\dot{\Omega}x, \\ \dot{z} &= a_0 z + a_1 \sin i \cos(\omega+\nu). \end{aligned} \tag{7}$$

Equations (1)–(7) yield the geocentric state vectors for *HST*.

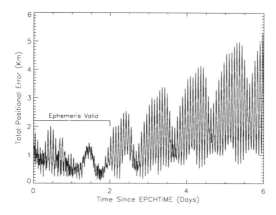

Figure 1. Typical positional errors from an on-board ephemeris.

3. Comparisons with Definitive Orbit Data

We can compare the results from the ephemeris model directly with the definitive orbit data. In Figure 1 we show the total error in position for several days during the first week of 1994 April. In this example the error is below 2 km for the two days during which the ephemeris is active. After this time, the error slowly increases, reaching about 5 km a week outside the two-day range of the model. The error in velocity was found to be only $0.01\,\mathrm{km\,s^{-1}}$ over the whole period investigated. Comparisons for other weeks indicates that the positional error can be as high as 4 km, but the velocity error is always very small, since the orbit is nearly circular.

4. Conclusions

Results using the onboard ephemeris are accurate enough for most needs. A positional error of 4 km translates to an uncertainty of 1 mas at 5.5 AU from the earth, so parallax errors are small for all but nearby passing asteroids and comets. The geographic position of *HST* can be determined to better than $0°03$, much more accurately than needed to compute effects due to the orbital environment. The error in velocity is much smaller than can be measured with the *HST* instruments. We conclude that using the onboard ephemeris eliminates the need to import definitive orbit data. This exemplifies the value of providing users with self-documenting data sets so that further analyses can be performed without resorting to additional outside information.

References

Smart, W. M. 1965, Spherical Astronomy (Cambridge, Cambridge University Press)
Wertz, J. R. 1978, in Spacecraft Attitude Determination and Control (Dordrecht, Reidel), p. 65

Robust Data Analysis Methods for Spectroscopy

P. Ballester

European Southern Observatory, Karl-Schwarzschild-Str. 2, D-85748 Garching, Germany

Abstract. This paper describes various methods for the analysis of spectroscopic data, particularly as they relate to wavelength calibration of echelle-format spectra. Methods that are robust to outlier values are highlighted.

1. Introduction

Classical statistical techniques usually lack the ability to reject outliers which appear in real data, resulting from measurement errors, mixed distributions, noise, etc. Different methods offer inherent robust properties in the presence of contamination. The application of robust methods to the domain of spectral data analysis is reviewed in this paper. A short introduction to robust regression by Least-Median of Squares is presented in Section 2. This algorithm is applied to the determination of echelle dispersion relations. Section 3 is an excerpt from a recently published article (Ballester 1994) related to the Hough transform and its application to echelle order detection and automated arc line identification. Wavelet transform can be useful in complementing robust techniques for multiresolution analysis, and Section 4 presents an application of the Mexican hat transform to the detection of spectral features.

2. Robust Regression

2.1. LS versus LMS

The sensitivity of least-squares (LS) regression to outliers is a traditional problem in data analysis, and as an alternative Rousseeuw (1987) introduced the Least Median of Squares (LMS) which consists of minimizing the term:

$$\text{med}\left\{(y_i - \alpha x_i - \beta)^2\right\}, i = 1, 2, .., N \tag{1}$$

This minimization cannot be obtained analytically and therefore requires repeated evaluations for different subsamples of size p drawn from the n observations. A complete trial would require $m = C_n^p$ evaluations which would rapidly become impracticable for large n and p values. Rousseeuw & Leroy (1987) determined the minimum number m of subsamples required to obtain a given probability α of drawing at least one subsample containing only good observations from a sample containing a fraction ϵ of outliers. By requiring α to be sufficiently close to 1 (e.g 0.95 or 0.99), m can be determined for given values of p and ϵ: $\alpha = 1 - (1 - (1 - \epsilon)^p)^m$

The smallest fraction ϵ of contamination that can cause the estimator to deviate arbitrarily from the estimate performed on the uncontaminated sample is called the breakdown point of the estimator. In the case of the LS estimator, only one outlier is sufficient to make the fitted parameters deviate arbitrarily from the expected value, therefore the breakdown point of LS is 0%. The LMS-based Reweighted Least Square (RLS) corresponds geometrically to finding the narrowest strip covering half of the observations. The breakdown point of the RLS is 50%. Most the other robust estimators do not attain a breakdown point of 30% (Rousseeuw & Leroy 1987). In the following sections of this paper, the applications have been realized using the PROGRESS algorithm available from the Statlib statistical library (e-mail: *statlib@lib.stat.cmu.edu*).

2.2. Iterative Dispersion Relation Determination

Wavelength calibration using an arc lamp exposure is a basic step in spectral calibration, usually requiring a careful analysis of errors and selection of lines, especially in echelle spectroscopy where arc spectra containing several hundreds of lines are commonly found. Several problems occur when using standard polynomial regression:

- LS regression assumes no error on the independent variables, whereas errors can occur both on the wavelength, due to line blending, misidentifications and quantization, and on the position, due to pixelation and centering errors.

- The lack of robustness to contamination will cause any outlier or misidentified line to affect the complete solution and introduce unnecessary errors into the residuals.

- In order to minimize the number of initial identifications, low-order relations are usually involved at the beginning of the calibration process (e.g., the echelle relation), introducing model errors.

Robust regression is an adapted method to take care of these problems. By combining Hough transform based automated arc line identification and robust regression, it is possible to perform an automated calibration, providing as information the central wavelength and the average dispersion in a single order. This first estimate is refined by Hough transform cross-matching and extended to the complete spectrum using the echelle relation $(m\lambda = f(x))$. However the accuracy of this relation is limited by several factors: in particular, optical misalignments occurring between echelle grating, cross-disperser, and detector. In the first iteration the unknown rotation angle is introduced into the error model of the echelle relation. Lines are then identified by a robust regression based iterative loop.

2.3. Results

Figure 1 shows the calibration of a Th-Ar exposure taken with the EMMI spectrograph at La Silla observatory. The wavelength range covered is 380–940 nm over 24 orders. The combination of arc line identification by Hough transform and robust regression allows to perform the calibration with the only information of central wavelength (679 nm ± 50 pixels), average dispersion (0.03 nm pixel^{-1} ± 30%) and absolute order number (24) for the relative order 18. The method

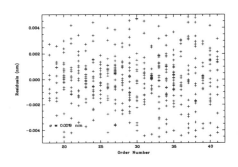

Figure 1. Th-Ar exposure obtained with the EMMI spectrograph at La Silla observatory and figure of residuals

requires no instrument dependent knowledge and allows for a relatively large inaccuracy of the initial solution. The figure of residuals shows a final accuracy of about 1/15 pixel rms. The Th-Ar line catalog was provided by H. Hensberge and corrected for line blending (De Cuyper & Hensberge 1995).

3. Hough Transform

A description of the Hough transform and its astrophysical applications can be found in Ballester (1994) and Ragazonni & Barbieri (1994). For the application of the HT to the detection of echelle orders, we use a representation of the HT assuming no preliminary segmentation of the image, since the sought-for features are brighter than the background. Accumulating a function of the intensity makes it possible to detect the orders by order of brightness.

The identification of arc lines consists of associating a list of line positions in pixel space with a list of reference wavelengths. The principle of the method is to perform all possible associations in a pixel-wavelength space. A three-dimensional HT allows us to detect, within a given range of central wavelength and average dispersion, the non-linear dispersion relation maximizing the number of associations. In the general case, the maximum is searched for in a cube of the Hough space, providing the parameters $(\lambda_c, \alpha, \beta)$ of the dispersion relation: $\lambda = \lambda_c + \alpha x(1 + \beta x)$. Usually the simplification $\beta = 0$ makes it possible to use a two-dimensional HT.

4. Wavelet Transform

The wavelet transform has been widely described (e.g., Starck, Murtagh, & Bijaoui 1995) and consists of the convolution product of a function with an analyzing wavelet. By choosing a wavelet which is the second derivative of a smoothing function, the wavelet coefficients become proportional to the second derivative of the smoothed signal. The Mexican hat transform involves the wavelet: $\psi(x) = (1 - x^2) \exp(-x^2/2)$ which is the second derivative of a Gaus-

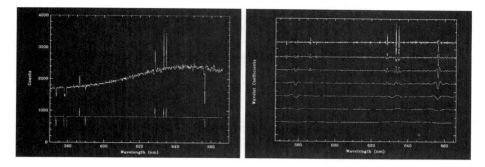

Figure 2. Spectrum of a standard star and associated binary mask (after scaling and translation), and dyadic Mexican hat transform of the above spectrum decomposed on 7 scales. The coefficients of the sharp positive features are concentrated at small scales. Coefficients of the absorption lines are maximized at different scales, depending on their spatial extension.

sian. Since the continuum of a spectrum varies smoothly, its second derivative will show increased values at the position of sharper spectral features. Figure 2 shows the Mexican hat transform of a spectrum presenting emission features, as well as absorption lines, of different widths. The wavelet coefficients become maximum (in absolute value) at different scales depending on the spatial extension of the features. The Mexican hat transform can be used to generate a multiresolution mask of the spectral features. The transformation applied to the wavelet coefficients at the different scales includes segmentation, generation of windows, and multiresolution recombination using a coarse-to-fine approach. After segmentation at each scale, the coefficients are compared between successive scales. The process starts from the largest scale and a feature will be retained at the next scale if its associated wavelet coefficients are larger. After recombination the mask is binarized and provides a schematic representation of the spectrum, providing the positions of the continuum and of the features.

Acknowledgments. I would like to thank P. Grosbøl, H. Hensberge, F. Murtagh, and J. L. Starck for helpful discussions.

References

Ballester P. 1994, A&A, 1011
De Cuyper J. P., & Hensberge H. 1995, this volume, p. 476
Ragazzoni R., & Barbieri C. 1994, PASP, 106, 683
Rousseeuw P. J., & Leroy A. M. 1987, Robust Regression and Outlier Detection (New York, Wiley)
Starck J. L., Murtagh, F., & Bijaoui, A. 1995, this volume, p. 279

Application of the Linear Quadtree to Astronomical Databases

P. Barrett
Universities Space Research Association & Laboratory for High Energy Astrophysics, NASA/Goddard SFC, Greenbelt, MD 20771

Abstract. Quadtrees have a wide range of applications, from graphics to image processing to spatial information systems. The use of linear quadtrees to represent spatial information has been widely used in geography, but rarely in astronomy. With the advent of the Guide Star Catalog and other large astronomical source lists, an efficient method of storing and accessing such spatial data is necessary. We show that encoding astronomical coordinates as a linear quadtree, instead of right ascension and declination as is typically done, can provide significant improvements in efficiency when accessing sources near a given spatial direction. We also discuss how the linear quadtree can aid in the correlation of source positions from different astronomical catalogs and how it can be applied to relational databases.

1. Introduction

This paper is a brief introduction to methods of storing multidimensional point data as it applies to astronomy. Such a vast amount of information cannot possibly be covered in a short paper; we have only tried to present the highlights to stir your interest. This study has been motivated by the publication of the Guide Star Catalog. The amount of data (nearly 800 MB) in this catalog can create problems with the storing and accessing of this data, if it resides in a (relational) database. The preliminary results of this study are intended to find solutions to such problems, so that large astronomical catalogs can be easily and efficiently handled by modern databases. A possible solution is the use of hierarchical data structures which provide for efficient methods of searching and storing of the data. Such problems are not new to the sciences. Geographers have had to deal with these problems since the initial development of Geographic Information Systems (GIS) and have therefore made important advances in the storage and access of vast amounts of point data. The information in this paper comes from such sources.

2. Data Structures for Multi-dimensional Point-Data

A common query of astronomical databases is the *range query* (i.e., region search), which requests all records from a database whose specified attributes are within given ranges. Techniques to search this type of query can be divided into two categories: those that organize data to be stored and those that organize

Data Structures	Search Operations	2-D Operations
non-hierarchical:		
sequential list	$O(N \cdot k)$	~ 200000
inverted list	$O(N^{1-1/k})$	~ 1000
fixed grid	$O(2^k \cdot F)$	~ 4
hierarchical:		
point quadtree	$O(k \cdot N^{1-1/k})$	~ 1000
k-d tree	$O(k \cdot N^{1-1/k})$	~ 1000
range tree	$O(log_2^k N + F)$	~ 400
MX quadtree	$O(2^d + F)$	~ 45
PR quadtree	$O(2^d + F)$	~ 45
bit-interleaving	$O(k \cdot log_2 N + F)$	~ 40

Table 1. A list of data structures and the typical number of operations required for a search. N and k are the number of records and search attributes of the list, respectively. For range queries, F is the number of records found. Column three gives the typical number of operations required for finding a single query ($F = 1$) of two attributes ($k = 2$, e.g., RA and Dec) on a list containing $N = 10^6$ records.

the embedding space from which the data are drawn. The binary search tree is an example of the former since boundaries of different regions in the search space are determined by the data being stored. Address computation methods such as radix searching are examples of the latter, since region boundaries are chosen from among locations that are fixed regardless of the content of the file. In a formal sense, the distinction is between *trees* and *tries*, respectively.

Each method has its advantages and disadvantages; some are more suitable to in-core data, while others, the bucket methods, ensure efficient access to disk data. Table 1 provides a list of these methods which are divided into hierarchical and non-hierarchical methods. For each method, column 2 gives an expression for the average number of operations for a search query, while column 3 gives the order-of-magnitude estimate of a 2-dimensional search for a file containing 10^6 records. It is evident that asymptotic (i.e., $\propto logN$) searches are superior to those that are not.

3. Two Proposed Methods of Astronomical Point-Data

In this section a brief description is given of two possible methods for storing astronomical records based on their coordinates. The first uses bit-interleaving of right ascension and declination, while the other encodes the coordinates as Gray codes. Both methods have the property that neighboring points also tend to be neighboring records. This property of the file should improve the efficiency of range searches. These two methods of encoding data are in effect space-filling curves.

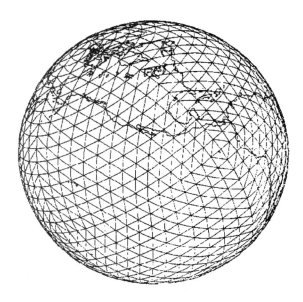

Figure 1. A quaternary triangular mesh (QTM) for a globe. The level 4 system of triangles viewed orthogonally from over the Pacific Ocean.

3.1. Bit-Interleaving

This first method of storing astronomical data, bit-interleaving, is the easiest to understand and to implement. A possible disadvantage is that it requires the data to be discrete. For present astronomical use, we believe that this property has minimal impact. Our proposal is as follows:

Right ascension and declination have ranges from 0° to 360° and −90° to +90°, respectively. Normally these coordinates are represented as real numbers, though they can also be represented in a discrete space with minimal impact on the data as long as the discrete space has adequate resolution. Right ascension and declination encoded as 32-bit integers have a resolution of about 1 milli-arcsecond which is completely adequate for almost all current astronomical catalogs. Once the coordinates are made discrete, then by a simple algorithm the bits of the two 32-bit integers are interleaved to form a single 64-bit integer or code.

3.2. Quaternary Triangular Mesh

The second proposed method is the Quaternary Triangular Mesh (QTM; see Figure 1). This method, as for the previous one, is based on the regular tessellation of space which is an infinitely repeatable pattern of a regular polygon (2-dimensional figure) or polyhedron (3-dimensional figure). The first method, bit-interleaving, uses the square as its regular polygon, whereas the QTM method uses the triangle. The QTM method has the properties that: (1) basic units are of almost equal size, (2) basic units are of almost equal shape, and (3) a set of units for which true adjacencies for neighboring elements are preserved.

The method first divides the globe into eight triangular regions whose vertices are at the north and south poles and at 0°, 90°, 180°, and 270° of the equator. This initial octahedron is better for fitting the requirements of polar symmetry and for mapping vertices along the equatorial plane. Figure 1 shows the level 4 system of triangles viewed orthogonally from over the Pacific Ocean.

Two other important properties of these methods are that they reduce k-dimensional space to a 1-dimensional space, and they are bucket methods. The advantage of encoding the data into a 1-dimensional space is that the resulting structure can be readily balanced, since balancing techniques for 1-dimensional data are well known (e.g., AVL trees). It has been shown that whether a data structure is balanced or not has a critical impact on the performance of operations such as search and update. The advantage of bucket methods is that data can be easily stored on disk. If the amount of data is too large to be stored in a single file, it can be divided into several more manageable files with little affect on the performance of searches and updates.

4. Conclusions

We present here two possible methods of storing astronomical point data or coordinates. They have some important properties, including a hierarchical structure for nearly unlimited resolution, and they reduce a k-dimensional space to a 1-dimensional space, which permits the construction of *balanced* binary trees and hence efficient data structures and searches. Finally, neighboring points typically have neighboring records for efficient range searches, and the ability to use bucket methods for storing the data easily and efficiently on disk.

References

Dutton, G. 1989, in Proc. Ninth International Symposium on Computered-Assisted Cartography (Falls Church, ASPRS/ACSM), p. 462

Goodchild, M., & Shiren, Y. 1989, NCGIA Tech. Paper (Santa Barbara, Univ. of California), p. 89

Laurini, R., & Thompson, D. 1992, in APIC Series: Fundamentals of Spatial Information Systems (San Diego, Academic), p. 133

Samet, H. 1989, The Design and Analysis of Spatial Data Structures (Reading, Addison-Wesley), p. 43

Wavelength Calibration at Moderately High Resolution

J.-P. De Cuyper and H. Hensberge

Koninklijke Sterrenwacht van België, Ringlaan 3, B-1180 Brussel, België

Abstract. Unrecognized blending in Thorium-Argon (Th-Ar) lamp spectra is a main contributor to calibration uncertainties. We present improved dispersion-dependent Th-Ar wavelengths for the features seen with moderate dispersion instruments.

1. Background

The calibration of astrophysical spectra in wavelength based on calibration lamp data suffers from different kinds of uncertainties. Besides inaccuracies in the observational phase, common data reduction practice adds "pixelation" errors, due to the fact that the centering method does not use the correct PSF. These model mismatch errors depend, for a given feature, on the sub-pixel location of the line center (David & Verschueren, private communication). In addition, there are dispersion-dependent errors, due to the fact that the laboratory input wavelengths refer to a pure line, while the features observed at astrophysical dispersions are generally blended to some extent. This aspect is the subject of our paper, with an application to Th-Ar lamps. Lastly, there is a lack of robustness by using a calibration relation with too many parameters with respect to the physics involved: echelle orders are often fitted independently, as well as sequences of spectra obtained under similar conditions (Hensberge & Verschueren 1989; Hensberge et al. 1995).

2. Anticipated Accuracy

Following T. M. Brown (1990), the centering accuracy for critically sampled Gaussian lines in the photon-noise limit is 0.01 pixel for lines with 3400 electrons detected in their central pixel (after extraction in cross-order direction). Usually as there are enough calibration lines available, the systematic effects on the centering and on the input data should not exceed this level. Therefore, even fairly weak blends cannot be included as calibrators without correcting their laboratory wavelength (Hensberge & Verschueren 1990). Figure 1 shows six lines, five of which are discretizations of a Gaussian PSF with $\sigma = 0.8$. One of them is actually a blend with a line having 10% of the intensity of the primary and situated one pixel to the right of the primary. The blend induces an error of 0.07 pixel compared to the wavelength of the primary line. The heavily blended lines commonly eliminated by observers correspond to substantially larger errors.

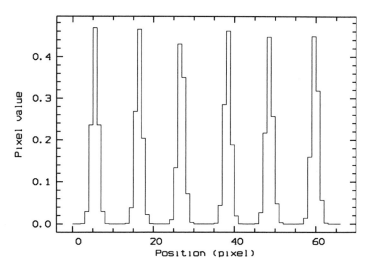

Figure 1. Shows five discretizations of a Gaussian line indistinguishable from a weak blend feature

3. Selected Wavelength Calibration Lines

Unblended lines are quite scarce at moderate dispersion. However, accurate calibration points can be provided by the many weakly blended lines if their input wavelengths are corrected for blending. Computations were done at four pixel scales between $\lambda/100,000$ and lines. The laboratory spectra were broadened to the desired resolution in order to determine the dispersion-dependent input wavelength of the features. The uncertainties present are due to several factors, such as the relative line intensities, the line width, the exact centering algorithm etc., and the details of the computations are discussed in De Cuyper & Hensberge (1995).

The laboratory data were taken from Palmer et al. (1983) for Th and of Minnhagen (1973) and Norlèn (1973) for Ar. The Ar to Th line ratios used are compatible with the lamps in use at ESO The spectrum was synthesized over a seven pixel wide interval for 20 different discretizations (shifted in steps of 0.05 pixel). The centering algorithm applied fits a Gaussian + constant. Improved wavelengths were derived for the features showing small (< 0.05 pixel) and stable (RMS over all discretizations < 0.015 pixel) corrections.

4. Results and Availability

We have composed input line lists at four pixel scales for direct use, containing a selection of commonly detectable, useful lines and their corrected wavelengths. In addition, a list containing the complete information and an algorithm permitting interpolation to other pixel sizes is provided. (For information see the

MIDAS-KSB-ORB Home Page[1].) P. Ballester is kindly acknowledged for making these lists publicly available through the ESO facilities (see the ESO Home Page[2]).

Table 1 summarizes the acceptance statistics. Figure 2 gives an example of the useful calibrators in an order of a CASPEC spectrum. When aiming for the highest possible accuracy, it is advisable to use more stringent criteria than those set in the computations. Moreover, this costs only a few of the selected lines.

Table 1. Statistics of useful calibrators for two wavelength intervals ($\lesssim$550 nm), including all lines and those measurable with sufficient accuracy.

Pixel Scale	All Lab. lines		Calibrators	
	Blue	Red	Blue	Red
0	7861	4046	1759	976
$\lambda/100\,000$	3470	2901	1159	824
$\lambda/50\,000$	1528	2069	736	699
$\lambda/33\,333$	848	1522	525	595
$\lambda/25\,000$	537	1149	393	506

5. Conclusions and Applications

Residuals of observed line positions with respect to calibration fits are in general still dominated by erroneous assumed laboratory wavelengths even after removal of the visually apparent blends or the apparent outliers. The introduction of dispersion-dependent wavelengths for the many weakly blended features at moderately high dispersion permits one to get the residuals considerably lower, to a few hundredths of a pixel. The wavelength corrections were obtained independently from the calibration fitting procedure, in contrast to clipping algorithms applied on residuals relative to such a relation. As a consequence, we can provide more realistic input wavelengths for Th-Ar lamps from below 300 to over 1000 nm in the dispersion range of interest.

Better input data not only lead directly to more precise calibration coefficients, but also provide the opportunity to derive a more appropriate mathematical representation for the calibration relation. Adequate results can then be obtained by few-parameter fits, which in the case of echelle spectroscopy cannot be represented as bivariate polynomials. The order dependence needs to be expressed as a ratio of two linear functions of the order number (Hensberge & Verschueren 1989).

---

[1] http://midas.oma.be/midas-ksb-orb-homepage.html

[2] http://server.hq.eso.org/eso-homepage.html

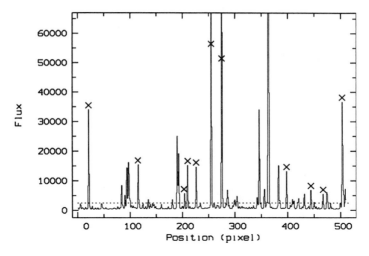

Figure 2. Shows an order (417–423 nm) from a CASPEC spectrum ($\lambda/33,333$). The useful calibrators lying above the indicated level (– – line) are marked with an x sign.

Acknowledgments. This research work was carried out in the framework of the project 'Service Centers and Research Networks' initiated and financed by the Belgian Federal Scientific Services (DWTC/SSTC) under contract number SC/005. The analysis profited from experience with calibration data taken during several runs at the European Southern Observatory; with the echelle spectrographs ECHELEC at the 1.5 m telescope, and CASPEC at the 3.6 m telescope. We acknowledge the observers A. G. A. Brown, E. J. de Geus, R. S. Lepoole and, in particular, W. Verschueren who was, in addition, deeply involved in the data reduction. H. Van Diest is thanked for computer assistance.

References

Brown, T. M. 1990, in CCD's in Astronomy, ASP Conf. Ser., Vol. 8, ed. G. H. Jacoby (San Francisco, ASP), p. 335
De Cuyper, J.-P., & Hensberge, H. 1995, A&A, in preparation
Hensberge, H., & Verschueren, W. 1989, ESO Messenger, 58, 51
Hensberge, H., & Verschueren, W. 1990, in Errors, Bias and Uncertainties in Astronomy, eds. F. Murtagh & C. Jaschek (Cambridge, Cambridge Univ. Press), p. 335
Hensberge, H., Verschueren, W., & De Cuyper, J.-P. 1995, A&A, in preparation
Minnhagen, L. 1973, J. Opt. Soc. Amer., 63, 1185
Norlèn G. 1973, Physica Scripta, 8, 249
Palmer, B. A., & Engleman, R. Jr. 1983, in Atlas of the Thorium Spectrum, ed. H. Sinoradzky (Los Alamos, Los Alamos National Laboratory)

An Approach for Obtaining Polarization Information from COBE-DMR

P. B. Keegstra[1], C. L. Bennett[2], G. F. Smoot[3]

Abstract. NASA's *Cosmic Background Explorer* (*COBE*)[4] Differential Microwave Radiometers (DMR) mapped the anisotropy of the Cosmic Microwave Background. The instrument is sensitive to linear polarization. The extraction of polarization information, however, is challenging, both algorithmically and computationally. Specifically, the model functions used must be defined with care, and the issue of degrees of freedom in the map that are not determined by the differential DMR data must be addressed. Managing the size of the calculation is also important.

1. DMR and Polarization

The DMR instrument measures differences in signal between pairs of antennae 60° apart on the sky. The 53 and 90 GHz receivers are sensitive to linear polarization. The microwave signal measured by each antenna of a pair has its E-vector lying radially outward from the spacecraft spin axis.

Because the DMR instrument is differential and every pixel is multiply sampled, sky maps are produced by solving a coupled system of linear equations (the normal equations) to minimize a chi-squared. This procedure can be extended to account for the additional degrees of freedom represented by polarization. The signal observed by a radiometer through an antenna pointed at a particular pixel varies with the orientation angle α as $S(\alpha) = I + Q\sin(2\alpha) + U\cos(2\alpha)$. I, Q, and U are Stokes parameters representing the state of linear polarization, and α is the angle between a vector in the plane of the radiometers and a fiducial vector on the sky for this pixel. Since each pixel is observed with a range of values of α, recovery of polarization information is possible. We define our fiducial vector to point from the center of each pixel towards the north pole. (Our usage differs by a factor of two from Born & Wolf (1975) so that we can identify the sky map of I with the DMR sky map made ignoring polarization.)

---

[1] Hughes STX Corporation

[2] Laboratory for Astronomy and Solar Physics, NASA/Goddard Space Flight Center

[3] Lawrence Berkeley Laboratory, University of California, Berkeley

[4] The National Aeronautics and Space Administration/Goddard Space Flight Center (NASA/GSFC) is responsible for the design, development, and operation of the *Cosmic Background Explorer* (*COBE*). Scientific guidance is provided by the *COBE* Science Working Group. GSFC is also responsible for the development of the analysis software and for the production of the mission data sets.

2. Defining the Model Functions

From the formula for the signal detected at each pixel, one can construct and minimize a chi-squared to generate the normal equations whose solution gives maps of I, Q, and U on the sky. These normal equations are analogous to the equations for a conventional DMR unpolarized sky map, except that each pixel now has three parameters (I, Q, and U) rather than a single temperature. The correlation of these parameters means that there are nine times as many elements in the normal equations. For example, the single element representing the plus pixel (pixel viewed by the antenna which contributes positively to the difference) by minus pixel off-diagonal contribution to the normal equations for a single observation becomes the following 3 by 3 submatrix:

$$\begin{array}{ccc} -1 & -\sin(2\alpha_-) & -\cos(2\alpha_-) \\ -\sin(2\alpha_+) & -\sin(2\alpha_+)\sin(2\alpha_-) & -\sin(2\alpha_+)\cos(2\alpha_-) \\ -\cos(2\alpha_+) & -\cos(2\alpha_+)\sin(2\alpha_-) & -\cos(2\alpha_+)\cos(2\alpha_-) \end{array}$$

where α_+ is the angle for the plus antenna, and α_- the same for the minus antenna.

The normal equations matrix is the sum, for all observations, of these contributions. The condition required to obtain a useful solution is that each time the same pair of pixels are observed, the same values of α must be used in evaluating the contribution to the normal equations. Note that the orientation of the spacecraft, and hence the angles α for the plus and minus antennae, are fixed by specifying the pointing of the plus and the minus antennae. Thus one way to satisfy the condition for obtaining a useful solution is by specifying that the angles α be calculated on the assumption that the antennae are pointing towards the centers of the pixels they are piercing.

The effect of this condition on the normal equations is to make the off-diagonal elements of the final summed matrix integral multiples of the above 3 by 3 matrix, where the multiplier is the number of times that the two pixels were observed together. Note that all the columns and rows of the above explicit 3 by 3 matrix are multiples of each other. This condition would not hold in the final summed matrix if not all α_+ and α_- were equal.

3. Accounting for the Null Eigenvectors

Since the DMR instrument is differential, it must follow that it cannot measure the absolute level of the sky temperature. Phrased another way, the average temperature of the pixels in a conventional DMR unpolarized skymap is not constrained by the DMR data. This is reflected in the algebraic form of the normal equations matrix, so that, if one were to diagonalize the matrix, one would find a single eigenvector whose eigenvalue was identically zero, and the associated eigenvector, viewed as a sky map, would have a constant value.

The added symmetry of the polarization normal equations means that in this case there are six null eigenvectors, the above mentioned map with constant I, and five additional maps with more complicated patterns. These maps are shown in the figures; since eigenvectors are dimensionless patterns, the grayscale value represents arbitrary units.

Care must be taken to preserve the symmetry which gives rise to these null eigenvectors, since this represents a genuine ambiguity in the experiment.

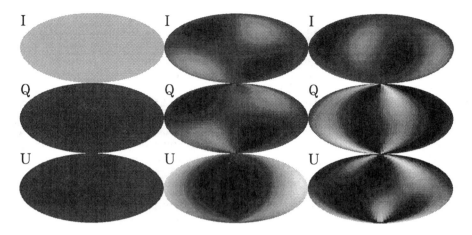

Figure 1. First, second, and third null eigenvectors. Each vector consists of projections of the full sky for each of I, Q, and U.

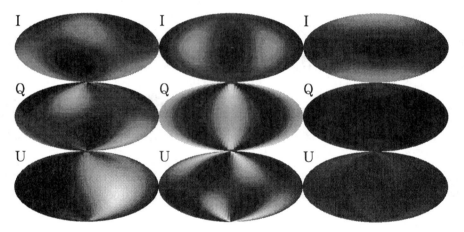

Figure 2. Fourth, fifth, and sixth null eigenvectors.

Forcibly breaking the symmetry by using different values of α for different observations of the same pair of pixels will not produce any new information, but rather will cause large amounts of these null eigenvectors to couple into the final output map.

The condition imposed by the iterative solution algorithm is approximately that the projection of the null eigenvectors on the I, Q, and U maps taken together, is zero. This can leave large contributions from the null eigenvectors in the Q and U maps, so long as these are offset in the I map. Since we can calculate the I map much more reliably without polarization, what we are most interested in from these calculations is maps of Q and U. Consequently, as our final step in producing polarized maps, we explicitly project out the contribution of the null eigenvectors over the subspace consisting of just the Q and U maps.

4. Conclusions

The 53 and 90 GHz radiometers of the *COBE* DMR instrument are sensitive to linear polarization. It is straightforward to define the matrix system whose solution gives best fit solutions for I, Q, and U skymaps. However, the fact that these equations require a 3 by 3 floating-point submatrix rather than a single integer for each off-diagonal pair means that the memory requirements for solving these systems at the standard DMR resolution, using 6144 pixels, are not achievable. Since the memory required scales as the number of pixels to the three-halves power, reducing the size of the calculation one step, to 1536 pixels, allows the calculation to be tractable. To minimize the effect of temperature gradients over these larger pixels, and also to reduce cross-talk into the Q and U maps, the best estimate of the sky temperature from a standard resolution map is subtracted from each data point.

The correct prescription for formulating the model functions associated with each observation has been obtained. The null eigenvectors, which describe particular sky patterns which cannot be detected by DMR due to its differential nature, have been characterized, and a prescription for producing Q and U maps minimizing the contribution from these eigenvectors has been implemented. This projection operation is straightforward, and can be applied rather easily to Monte Carlo simulations of models for the polarized sky, thus facilitating comparison between models and the mission data.

References

Bennett, C., et al. 1992, ApJ, 391, 466

Born, M., & Wolf, E. 1975, Principles of Optics, Fifth Edition (New York, Pergamon Press)

Smoot, G., et al. 1990, ApJ, 360, 685

Discrimination of Point-like Objects in Astronomical Images using Surface Curvature

A. Llebaria, P. Lamy, P. Malburet

Laboratoire d'Astronomie Spatiale du CNRS, BP 8, F-13376 Marseille Cedex 12

Abstract. A new method for the discrimination of point-like objects in astronomical images is presented. The method makes use of the surface curvature of the image, without any *a priori* knowledge of the shape of the point-like objects.

1. Introduction

The accurate detection and subsequent removal of point-like objects in astronomical images is a frequent problem. We mention, as examples, the case of stars pervading an extended objects, or cosmic ray impacts on CCD images. The problem is often complicated by various distortions affecting the point-like objects, such as, saturation, variable Point Spread Function (PSF) in the field-of-view, and irregular images which preclude the use of fitting methods. We propose a new approach to the problem based on the properties of surface curvature which is related to peaks, pits and flats in images and therefore contain the key information to discriminate point-like objects.

The intensity of an image can be viewed as a two-dimensional surface in a three-dimensional space with the two spatial coordinates forming two of the three dimensions and the intensity axis forming the third dimension. Surface curvature will describe how much this surface is bending in a given direction at a particular point; it involves the first and second partial derivatives with respect to the spatial coordinates and thus involves the spatial neighborhood of the point. It is readily understandable that surface curvature will turn out be a powerful means of discriminating objects in images.

2. Method

Practical experiments with the surface curvature of astronomical images have quickly revealed the superiority of dealing with the log of the intensity, $\log(B)$, rather than with the intensity itself. The justifications of this transformation are easily understood because, (1) the range of curvatures will be narrower on the $\log(B)$ images than on the intensity images, insuring a better detection of faint objects, and (2) point-like objects, such as stars may be viewed, at least in a first approximation, as having a Gaussian shape. The logarithmic transformation will generate similar paraboloids (i.e., having the same parameters) for which the maximum of curvature takes place at their tops. Practically, the logarithmic

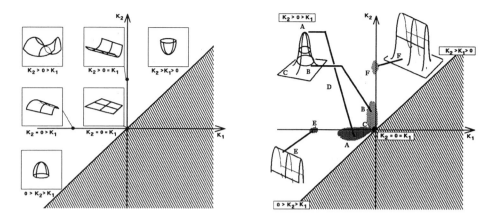

Figure 1. (Left side). Surface curvature visualized in the plane of the two principal curvatures k_1 and k_2. (Right side). A practical k_1, k_2 diagram for an astronomical image. The regions of interest are related to specific parts or curvatures of real objects

transformation of the star images will exhibit a coarse paraboloidal shape which can be easily localized from the maximum of curvature.

A detailed analysis of surface curvature can be found in Peet and Sahota (1985). This analysis shows that at each regular point of a surface we can easily determine k_1 and k_2, the local minimum and maximum curvatures, respectively. k_1 and k_2 are invariant under rigid motions of the surface. Two other curvatures are often defined; the Gaussian or total curvature $K = k_1 k_2$, and the mean curvature $M = (k_1 + k_2)/2$; but it turns out that the principal curvatures k_1 and k_2 are best suited to the detailed characterization required by our problem.

As curvature is a geometric property of continuous surfaces, we must apply a continuous model to the discrete array of 2-D data in order to deduce local derivatives, and from this the local curvatures. Even if more complex models can be used, a simplified facet model of second order (Halarick 1984), is good enough when a small set of points is involved in the estimation. For the 3×3 patch window the practical calculation of k_1 and k_2 is extremely simple. The surface $z(x, y) = \log B$ is approximated by a quadric $z(x, y) = a_{00} + a_{10}x + a_{01}y + a_{20}x^2 + a_{02}y^2 + a_{11}xy$ and the coefficients are determined using the least squares fit to the real surface on a 3×3 neighborhood of the pixel under consideration.

It is quite convenient to discuss the surface curvature in the (k_1, k_2) space as illustrated in Figure 1. Using the above convention, $k_2 \geq k_1$, and only the corresponding half plane is therefore allowed. Also, according to the above definitions, if a curve on the surface is concave up, the surface curvature in that direction is positive and *vice-versa*. The allowed half-plane of Figure 1 (left side) is divided in three regions by the k_1 and k_2 axis. The two axis themselves correspond to parabolic points, with ridges on the k_1 axis ($k_1 < k_2 = 0$), and valleys on the k_2 axis ($k_2 > 0 = k_1$). The second quadrant $k_2 > 0 > k_1$, is the region of the saddle points. The elliptic points are located in the remaining two half-quadrants, distinguished by their concavity: pits for $k_2 > k_1 > 0$, and peaks for $k_1 < k_2 < 0$. Obviously, a plane is at the origin $k_1 = k_2 = 0$.

3. Discrimination of Point-Like Objects on the (k_1, k_2) Diagram

Let us now consider how the above theoretical considerations apply to real astronomical images. Figure 1 (right side) is a sketch of the (k_1, k_2) diagram based on a CCD image which exhibits an extended source; the tail of comet P/Halley, a rich star field, and various defects and artifacts. The features of this (k_1, k_2) diagram are sufficiently general to localize the various domains of interest and to establish the appropriate criteria for the discrimination.

The main feature of the diagram is a "butterfly" pattern composed of two wings A and B and a central core C. Both wings pertain to stars, wing A to their top portions and wing B to their bottom portions. These latter parts are fairly asymmetric saddle points leaning to valleys, and explain the closeness of the wing B to the k_2 axis. In fact, the orientation of the wing, as given for instance by the slope k_2/k_1 of their symmetry axis, is a very good estimate of the anisotropy of the surface curvature. For instance, elongated (trailed) star images will induce a rotation of wing A. The central core C obviously correspond to the flat part of the image; essentially the extended object and the background. Finally, two additional features are conspicuous in the diagram, the blobs E and F respectively localized on the k_1 and k_2 axis. They correspond to elongated structures in the original image which involve both ridges (top parts) and valleys (bottom parts). These structures are typically bad columns in the CCD and saturated star images which spill off along the column direction.

It is now a very simple matter to establish criteria for discriminating objects in the original image. These criteria are written in terms on conditions on k_1 and k_2 or on a combination of the two. From the initial image we define a mask (a binary image) defining the subset of pixels in the original image which belongs to the "flat" regions. The complementary subset includes all invalid points from stars, saturations, etc.

4. Restitution of the Background

The purpose of the next and final phase is to replace patches of invalid pixels in the original intensity image by a local interpolation, the so-called pyramidal interpolation, using a non-linear multiresolution method which resembles the Haar wavelet analysis and synthesis.

The procedure starts by shrinking the original $N \times M$ pixels image to a $N/2 \times M/2$ image. Each pixel in the latter corresponds to an area of 2×2 pixels in the former. Their value is a mean of valid pixels in the 2×2 related area. If 3 or 4 pixels in this area are invalid, the mean is classed as invalid. After this first pass, we obtain an image of means and a corresponding binary image or mask of valid or invalid means. Repeating this procedure builds another series of images (means and mask) of lower resolution. The image of means is called the analysis image.

The procedure is repeated l times until either the extinction of patches of invalid pixels is obtained, or N_{l+1} or M_{l+1} are too small (that is equal to or less than 4 pixels on a side). If some final pixels are invalid, they are replaced by values estimated from a global adjustment of a quadratic surface over the remaining valid pixels. After this, all pixels are valid at the lowest resolution level and the synthesis process can start.

The synthesis is done by (1) expanding the synthesis image built in the previous $l - 1st$ step by 2×2, (2) substitute invalid pixels in the analysis image at this step l with the corresponding pixels from the above expanded image, and take this image (with substituted pixels) as the new synthesis image at this step. The procedure stops when the initial resolution is recovered. The method reconstitutes, in each patch of invalid pixels, the low resolution estimate achieved with valid pixel neighbors.

This procedure needs a minimal percentage of valid pixels ($\sim 50\%$) to obtain significant results. When the number of valid pixels is very low, the threshold to obtain a significant mean from a 2×2 area in the analysis step can be reduced to one valid pixel.

5. Conclusion

We have presented a new method for the discrimination of point-like objects using surface curvature. The characterization by the curvature offers a very general and flexible method which can be adapted to any specific problem without any necessary *a priori* knowledge of the shape of the point-like objects. A few trials are basically what is needed to determine the threshold for discrimination. Another potential application is the removal of cosmic rays events from single exposure CCD images (when the exposure has been duplicated, the preferred approach is obviously the comparison of the two images). These impact events appear as very sharp peaks which may be unambiguously discriminated by their curvature.

References

Lamy, P. L., Pederson, H., & Vio, R., 1987, A&A, 187, 661

Peet, F. G., & Sahorta, T. S. 1985, IEEE Trans. Pattern Anal. Machine Intelligence, Vol. PAMI-7, 734

Halarick 1984, IEEE Trans. Pattern Anal. Machine Intelligence, Vol. PAMI-6, 58

Automated Globular Cluster Photometry with DASHA

O. M. Smirnov and A. P. Ipatov

Institute of Astronomy of the Russian Academy of Sciences / Isaac Newton Institute, Moscow Branch, 48 Pyatnitskaya Str., Moscow 109017 Russia

Abstract. We describe a tool in the PCIPS environment for finishing the job of globular cluster photometry started by DAOPHOT II: starting with the instrumental magnitudes derived by DAOPHOT II, DASHA can automatically produce final color-magnitude diagrams.

1. Introduction

Photometric observations of globular clusters usually involve enormous amounts of data. A typical data set can consist of several dozen CCD frames produced with two or more different filters and at different exposures, with anything from a few hundred to a few thousand stars present in each frame. Powerful software tools are required if the reductions are to be accomplished at the cost of CPU hours rather than man-hours. The process of data reduction involves two phases:

1. Photometry of individual CCD frames, giving as results several dozen per-frame lists of instrumental magnitude measurements.

2. Calibrating magnitudes, cross-identifying objects between frames and collecting their measurements. The results are magnitudes (averaged across all the frames) in every available filter.

The first phase can be performed using one of several PSF-fitting stellar photometry packages currently available to the community (PSF fitting is required since the frames are usually too crowded for simple aperture photometry). Of these perhaps, DAOPHOT II (Stetson 1991) is by far the most comprehensive, allowing near-automatic reductions of large batches of data. We are unaware of any packages that provide a complete reduction path for the second phase, which is why we developed DASHA. DASHA can automate the whole process of globular cluster photometry, from instrumental magnitude lists produced by DAOPHOT II, to final color-magnitude diagrams. For example, having reduced 45 CCD frames (3 filters/15 frames each) with DAOPHOT II, you can cross-identify all the stars, calibrate them, produce a table of mean magnitudes in each filter, and obtain color-magnitude diagrams, all within one or two hours.

2. Design Goals

Several design goals were involved in the development of DASHA:

2.1. Ease of Use

DASHA is implemented in the PCIPS environment (Smirnov & Piskunov 1994), and takes full advantage of its graphical user interface (GUI). In addition, DASHA constantly displays various plots that allow a user to keep track of exactly what's happening to the data.

2.2. Data Format Compatibility with DAOPHOT II

DAOPHOT II was selected as a base for DASHA, for two reasons: First, a full version of DAOPHOT II, PCDAOPHOT II, has been implemented under PCIPS. Therefore, both phases of the reduction process can be performed in one environment; and second, DAOPHOT II handles crowded fields very well. Globular cluster observations are usually very crowded.

2.3. Flexibility

Functionally, DASHA consists of several modules. Using them in different combinations allows the user to accommodate variations in the reduction process.

3. Components of DASHA

DASHA is oriented towards star lists produced by DAOPHOT II as source data. Each star list is an ASCII table with one line per star. The final output lists contain columns for coordinates, instrumental magnitude, magnitude error, goodness-of-fit statistics, etc. The same format (with some extensions) is used throughout DASHA for intermediate and final results.

DASHA contains several modules:

StarVis is a tool for visualizing star lists. It can produce plots and histograms of any columns in the star list, and allows the user to filter a list by selecting regions on the plot using the mouse. StarVis is used throughout the reduction process to accomplish different tasks. For example, the source star lists produced by DAOPHOT II can be filtered using magnitude vs.. magnitude error and magnitude vs. goodness-of-fit plots to discard bad measurements. StarVis is also useful for reviewing intermediate results, as well as producing final color-magnitude diagrams.

Calibrate produces a simulated star field by plotting the stars of a list on an image, and allows the user to select calibration standards and specify their photometric magnitudes.

Cross-identify and merge observations (XID) merges star lists for different frames of one filter. XID can automatically cross-identify the stars across all the frames, compute weighted averages for the magnitudes, and write the results into a single merged list. In the event that some lists can not be calibrated (i.e., they originate from long-exposure frames where only faint stars are measured by DAOPHOT II, the brighter stars being saturated, and thus do not contain any photometric standards), XID can also automatically select secondary calibration standards from other, already calibrated, lists, and perform cross-calibration.

Cross-identify and compile filter table (XFilt) compiles the merged lists for all available filters into a single filter table, containing mean magnitudes for each star as measured and calibrated in each filter. In order to do this, XFilt automatically cross-identifies stars between merged lists.

Create color table (ColTab) converts a filter table into color tables. The user can specify which columns to place into the color table (i.e., if the filter table contains measurements in B, V, R, and I, ColTab can make tables of $BVRI$, $B-V$, $V-R$, etc.) A color table can be considered the final output of DASHA, since it contains magnitudes and color indices for every star observed. Using StarVis, the user can produce a color-magnitude diagram based on the color table.

4. Automatic Cross-Identification of Objects

A feature of DASHA (or more specifically, the XID and XFilt modules) that is crucial to its productivity is the ability to automatically cross-identify objects between frames. Frames are not always aligned in position, so to perform a cross-identification, DASHA must first find the positional offset between them. DASHA finds the offset by automatically selecting a set of positional standards, i.e., objects that seem to be present in both frames. Once it has the standards, it can compute the mean offset; and once it has the mean offset, the rest is trivial.

Clearly, the actual selection of standards is not so trivial. DASHA approaches the problem much like a human would: it compares the two frames to see which patterns of stars seem similar. More specifically, it starts out by selecting at random a bunch of two-star patterns on one frame, and tries to find similar patterns on the second frame (each pair of corresponding patterns is called a cluster.) Next, it attempts to grow each cluster by adding a star at random to the pattern on the first frame. If the second frame also has a star at the expected position, both stars are added to the respective patterns. When DASHA repeats this step over and over, clusters with correctly cross-identified patterns tend to grow, while misidentified ones don't. DASHA drops those clusters that do not grow for a specific time

Once the clusters contain enough stars, DASHA performs a few more idle loops to let any remaining misidentifications perish, and then takes the remaining stars as positional standards. Typically, it finds over a hundred standards in several seconds.

5. First Practical Results

Using DASHA, we obtained BVI photometry of the globular cluster NGC 5927. Unique observational data was kindly provided by Dr. G. Alcaino (Isaac Newton Institute, Chile), and consisted of 17 CCD frames in the B filter, 17 in I, and 13 in V (telescope: ESO's 2.2 m at La Silla). The most interesting diagram, B vs. $B-I$, is included here (Figure 1a); Figure 1b is a histogram of the distribution of B magnitudes. The final color table contains over 5000 stars, of which over

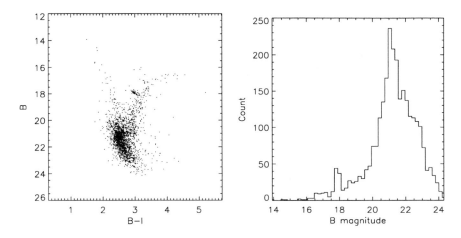

Figure 1. (a) Color-magnitude diagram, (b) B magnitude distribution for NGC 5927

2000 are measured in all three filters. The whole process, starting with AllStar's results, and ending with color-magnitude diagrams, took little over an hour.

$BVRI$ photometry of three other globular clusters, NGC362, NGC5286, and NGC7099, is in progress at the time of writing.

6. Feedback and Further Information

We welcome any questions and comments about DASHA. Please contact Oleg Smirnov by e-mail at *oms@inasan.rssi.ru*. DASHA and the PCIPS image processing system are commercial products. Please contact Oleg Smirnov for details.

Acknowledgments. We would like to thank ST ScI for the financial support that made this presentation possible.

References

Smirnov, O. M., & Piskunov, N. E. 1994, in Astronomical Data Analysis Software and Systems III, ASP Conf. Ser., Vol. 61, eds. D. R. Crabtree, R. J. Hanisch, & J. Barnes (San Francisco, ASP), p. 245

Stetson, P. B. 1992, in Astronomical Data Analysis Software and Systems I, ASP Conf. Ser., Vol. 25, eds. D. M. Worrall, C. Biemesderfer, & J. Barnes (San Francisco, ASP), p. 297

MACSQIID: A package for the reduction of data from the SQIID Infrared Camera

J. W. MacKenty

Space Telescope Science Institute, 3700 San Martin Drive, Baltimore MD 21218

Abstract. The SQIID camera was developed at NOAO's Kitt Peak National Observatory to provide multi-band infrared imaging. It is well suited for galaxy surface photometry as it has a uniform, stable flat field and a large field of view. We have developed a software package to facilitate the reduction of data from this instrument for this purpose. This package is appropriate for use with observations where the telescope is moved a small fraction of the field of view between exposures. This package uses the stability of the camera to construct flat field calibrations over several nights. The flat field and image summation algorithms avoid the use of the median and instead rely upon a combination of source detection in the $J + H + K$ summed frames, sigma clipping, and iteration. The sky illumination pattern is determined from the residuals between observations of different targets on the sky.

1. Introduction

We have developed a software package to facilitate the reduction of data from the NOAO SQIID (Simultaneous Quad-color Infrared Imaging Device) camera (Ellis et al. 1992). Our approach emphasizes the design of a software package targeted towards a specific instrument and a limited set of observational projects. Rather than build software for the general problem of reducing infrared images from a large range of instruments, this package is intended only for use with the SQIID camera. Therefore, it incorporates, and to some extent conceals from the user, considerable knowledge about the camera. Further, while the SQIID camera can be used to map large regions of the sky by constructing mosaics, this package is appropriate only for use with observations where the telescope is moved a limited fraction of the field of view between exposures and the entire set of exposures of a target possesses some degree of overlap.

The SQIID camera employs four 256×256 pixel PtSi array detectors which, although their quantum efficiency is rather modest (6.6% at J to 3.4% at K), have excellent cosmetic qualities, good linearity, outstanding temporal stability, and little sensitivity to the effects of previous exposures. SQIID incorporates dichroic beam splitters and re-imaging optics to simultaneously view the same field in the J (1.2 μm), H (1.6 μm), K (2.2 μm), and L (3.5 μm) passbands. Due to its very low sensitivity, the $L-$band channel is not incorporated in the MACSQIID software package. SQIID has a scale of $1''\!.36$ pixel^{-1} and a field of view of $5'\!.5$ on the KPNO 1.3 m telescope, and $0''\!.43$ pixel^{-1} and $1'\!.8$ on the KPNO 4 m telescope.

2. A Methodology for the Calibration of Infrared Images

The goal of the MACSQIID package is to produce images from a set of observations which are free from the various instrumental and atmospheric effects. An observing run consists of many (typically 5 to 20) exposures (60 to 300 s each) of a series of targets. We will refer to each set of exposures as a "dataset". Within each dataset the telescope was offset a distance less than the SQIID's field of view and larger than the extent of the target(s). For larger targets, a few exposures of equal duration are obtained of a nearby, blank area on the sky.

2.1. Basic Calibration Equation

In each SQIID exposure, we detect the flux distribution of the target field of view plus the sky and instrumental backgrounds. The detected image, $I_{(x,y,t)}$, is related to the original scene, $O_{x,y}$, by:

$$O_{(x,y)} = [(I_{(x,y,t)} - Dark_{(x,y)}) \times Flat_{(x,y)}] - Background_{(t)} - Sky_{(x,y)}.$$

The individual pixels in each image have unique sensitivities ("flat field") and signal internal to the instrument ("dark current"). The dark current is directly measured by obtaining many exposures with a cold dark slide inserted behind the instrument's input aperture. The MACSQIID package makes the following assumptions. First, that the dark current and flat field are constant over a significant number of datasets (in practice they are stable over a week long observing run). Second, that the sky background can be modeled separately as a uniform level which varies rapidly with time ("background") and as an illumination pattern which is constant within a given dataset ("sky"). It is assumed that the "sky" pattern is mainly the result of large scale illumination effects (e.g., moonlight) and of the telescope structure, therefore it is taken to be constant with the small motions of the telescope which occur within a dataset.

2.2. Iterative Approach

The basic approach to the solution of the calibration problem is successive iterations which improve the quality of the flat, sky, and background calibrations. This process is implemented in two loops. The outer loop operates on each individual dataset. It calibrates (via the Basic Calibration Equation) the image from each exposure, then shifts and averages the exposures for each target. Pixels flagged as "bad" in the dark or flat field are excluded from this average. The averaged images from three passbands are then mapped to a common coordinate system (usually the K band's due to its lower S/N) and summed. This summed image is then scanned to locate "sources" based on a photon statistics model of the expected sky noise. Finally, all pixels in the original individual images which correspond to sources in the summed image are flagged.

The inner loop simultaneously utilizes all observations from one night (or even an entire observing run). Working on one passband at a time, the background for each exposure, the average flat field, and the sky for each dataset (i.e., target) is determined. To avoid the biases inherent in median filtering, a combination of the exclusion of pixels flagged as containing sources and iterative sigma clipping is used for each part of this process. The sky is determined from the residual of the spatially smoothed average of the exposures of each target. This gives the sky a S/N comparable to that achieved in the flat field (typically a few hundred samples per pixel) and is justified by the recognition that the sky illumination pattern is out of focus (i.e., does not contain significant high spatial frequency structure). The inner loop is executed a user determined number of times (typically 3 to 5) in order to achieve convergence for most of the pixels. Pixels with poor solutions may be excluded from further use in the outer loop.

The outer loop is repeated until the image quality is satisfactory. This usually requires 3 or 4 iterations. The first pass of the outer loop is provided with a simple "bootstrap" flat field solution constructed from a median of the images in each dataset. The user must select which exposures are to be included in the inner loop (e.g., those without bright sources and sky exposures) and which to use to determine the sky illumination pattern for each dataset.

3. Implementation

We start with many J, H, and K images in IRAF files. IRAF tasks are used to make appropriate dark calibration frames. Also, the relative offsets, rotations, and scale changes between the J, H, and K passbands are determined with the IRAF geometry tasks. The MACSQIID package then provides four tasks:

BUILD creates a "dsf" file containing all images, calibration files, and miscellaneous data for a dataset. It is driven by a simple ASCII script file. The use of a script file for control rather than an interactive interface permits painless re-execution should it become necessary to start over. Non-linearities in the detectors at high signal levels are corrected for at this stage.

PROCESS provides an interactive task to calibrate and examine images. This implements the "outer loop" portion of the calibration process defined in § 2.2. The user is presented with a simple menu of one key commands. These support the basic calibration (including an initial bootstrap calibration using a median based flat field generated on the fly), geometric mapping between J, H, and K channels, the alignment between exposures (star marking and cross correlation of source pixels), and source detection and flagging. Considerable statistical information is displayed during the execution of the various calibration steps. The user has the option to exclude individual exposures from the combined image. This task also provides for the interactive examination of calibration data and raw, calibrated, and flagged images using a named pipe connection to SAOIMAGE.

CALIB updates the background, flat field, and sky calibrations in a large number of dataset simultaneously. This implements the "inner loop" of the calibration process defined in § 2.2. This batch program is driven by an ASCII script file. It provides copious reporting and statistics to assess the quality of its solutions.

EXTRACT produces IRAF format images from the internal format "dsf" file. This is generally used at the completion of the calibration process to allow further display and analysis within the IRAF system.

The MACSQIID package uses several programming methods which reflect the intention to produce a specific rather than a general purpose software package. The package embeds considerable knowledge about the SQIID instrument. This is mostly contained in "include" files as defined parameters. These serve to hide details of SQIID from the user (e.g., the noise model "knows" about the system gain factor). The code uses a large data structure (implemented within FORTRAN common blocks) which places all of the images in system memory. This supports fast interactive performance to encourage the user to explore the data and experiment with alternative calibrations. This data structure is common to each task and directly maps onto the "dsf" file format.

Crucial to the goal of keeping the coding effort limited in scope was the availability and use of existing interfaces to the external environments. The IRAF "imfort" interface was used to access IRAF format images at the start and end phases of the calibration process. The "sao-iis"[1] package by Jim Wright (formerly at CFHT) provides simple and efficient means of displaying image data and interacting with the SAOIMAGE display tool.

4. Conclusions

This package provides a means of achieving near-optimal (i.e., background or photon noise limited) calibration of data from the SQIID camera. The use of object detection and iterative sigma clipping results in good flat fields at all spatial scales. The separation of the sky background into a time variable background component and a target (dataset) variable illumination pattern permits the recovery of the calibration accuracy inherent in the flat fields. This depends, in part, on the recognition that the sky illumination pattern is intrinsically out of focus in the detector plane.

This package was designed to encourage the user to interactively examine datasets and to understand the quality of the calibration(s) achieved while placing the computationally intensive steps into batch programs. It simplifies the data management task by grouping all of the observations and calibration files for a dataset into a single large file of a custom design yet provides IRAF format files for further analysis subsequent to the calibration process. The code development effort was kept to reasonable levels with the use of existing interface packages ("imfort" and "sao-iis").

Acknowledgments. The assistance of Mike Merrill and Ian Gatley at the NOAO in understanding the SQIID camera and its data is gratefully acknowledged. Gina Jones' skill and patience in the testing of this package is also appreciated. Copies of this package are available from the author.

References

Ellis, T. et al. 1992, Proc. SPIE, 1765, 94

---

[1] ftp://ftp.cfht.hawaii.edu/pub/sao-iis

EMSAO: Radial Velocities from Emission Lines in Spectra

D. J. Mink, W. F. Wyatt

Harvard-Smithsonian Center for Astrophysics, 60 Garden St., Cambridge, MA 02138

Abstract. Many extragalactic objects for which radial velocities are desired display strong emission lines. For large surveys, the interactive determination of line centers and calculation of redshifts is too slow. The EMSAO task of the RVSAO IRAF package automatically finds emission lines, fits them with single or multiple Gaussians, and combines the determined redshifts weighting them by the uncertainty in the line centers. In addition to being used in the CfA and other redshift surveys, EMSAO has been used to study sky line shifts in 15 years of observations to characterize the uncertainty in instrumental measurements.

1. Introduction

When SAO moved its redshift program from minicomputers to workstations, it was decided that NOAO IRAF, a standard reduction system, be used, rather than the previous custom Forth system (Tonry & Wyatt 1988). At the time, no specialized radial velocity software existed inside IRAF, so the RVSAO package was developed. While the redshifts of many spectra can be computed by cross-correlating them against templates, that process is more complicated for spectra exhibiting strong emission lines. For large surveys, the interactive determination of line centers and calculation of redshifts by a program such as IRAF's SPLOT, is simply too slow. EMSAO, a companion to the cross-correlation task XCSAO (Kurtz et al. 1992), was written to find emission lines automatically, compute redshifts for each identified line, and combine them into a single radial velocity. The results may be graphically displayed or printed. The graphic cursor may be used to change fit and display parameters.

2. How EMSAO Works

EMSAO takes a list of 1-dimensional wavelength-calibrated spectra in either IRAF one-dimensional or MULTISPEC format. Dispersion information is read from the world coordinate system keywords in the image headers. After reading the spectrum, an initial velocity guess is used to search for lines from an input line list. That velocity guess may come from an input parameter, a velocity in the image header, or an identification of one line in the spectrum. This last method uses a table of lines and the wavelength range over which each one should be the strongest line. This table can be modified by the user to match a given dataset. The search for the rest of the emission lines is carried out in a similar

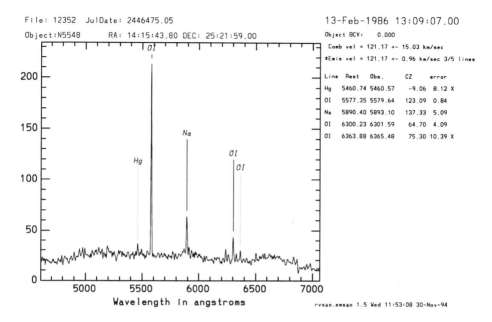

Figure 1. This EMSAO Summary page shows line centers and fit information for a night sky spectrum

way, driven by a different table with smaller wavelength tolerances. After the lines are identified, Gaussian profiles are fit to more exactly determine the line centers. A third table identifies combined lines, such as the N1–H–N2 triplet at 6548 Å, 6563 Å, and 6584 Å, and multiple Gaussians are fit simultaneously.

After redshift velocities are determined for each of the lines, an error-weighted mean is computed, omitting the velocities from those lines whose fits do not meet certain criteria. If good cross-correlation and emission line velocities exist for a given object, a combined velocity is also computed. The results may be graphically displayed with emission and/or absorption lines (from a fourth table) labeled as in Figure 1, logged in any of several formats to a file or text terminal, and/or written into the spectrum's header. Figure 2 shows the most verbose tabulation of EMSAO results, with wavelength center and velocity information for each line. If the results are displayed graphically, individual identified lines may be added to or subtracted from the emission line velocity, the spectrum may be edited, and several other conditions of the fit may be changed.

3. Testing

To test EMSAO adequately, a lot of emission line spectra were needed. The Smithsonian Astrophysical Observatory has archived 27,000 galaxy and stellar spectra from its Z-Machine spectrograph. Since a sky spectrum is filed with each object spectrum, there exist 27,000 easily-available spectra of night sky emission lines. As these mercury, oxygen, and sodium lines are airglow lines

```
rvsao.emsao 1.5 NOAO/IRAF V2.10.3BETA mink@cfa165 Mon 18:09:02 28-Nov-94

12352 Object: N5548      RA: 14:15:43.80 Dec: 25:21:59.0
Observed 13-Feb-1986 13:09:07.00 = JD 2446475.0480 BCV:     0.00
Combined vel =    145.01 +-    15.66 km/sec, z= 0.0005
Correlation vel =   145.01 +-     4.50 km/sec, z= 0.0005 R=   40.1
Emission vel =    121.17 +-     0.96 km/sec, z= 0.0004 for 3/5 lines


Line  Rest lam   Obs. lam    Pixel       z         vel      dvel      eqw      wt
 Hg    5460.74    5460.57    813.09   -0.000      -9.06     8.12     2.86    0.000 X
 OI    5577.35    5579.64    916.48    0.0004    123.09     0.84    66.60    0.936
 Na    5890.40    5893.10   1178.47    0.0005    137.33     5.09    21.10    0.025
 OI    6300.23    6301.59   1499.71    0.0002     64.70     4.09     9.93    0.039
 OI    6363.88    6365.48   1548.07    0.0003     75.30    10.39     5.01    0.000 X
```

Figure 2. This is a tabulation, with report_mode=1, of the EMSAO results displayed graphically in Figure 1.

```
00001  2443575.69726    21   0   2804 44   3.5  2804  44 E 1 1     21.38      0.00      0.00
00002  2445219.24201    13   0   1828 34   4.8  1828  34 E 1 1     13.27      0.00      0.00
00003  2443575.79326   -37   0      0  0   0.0     0   0 E 1 1    -37.00      0.00      0.00
. . .
27169  2449342.03237    -2   1  35477 62   2.1    -2  62 E 3 3      4.85    -17.34    -32.84
27170  2449336.76664   -28   0   4069 45   2.9   -28  45 E 3 3    -27.92    -28.44    -38.31
27171  2449300.78227   -17   0   5465 26   6.2   -17  26 E 3 3     -9.00    -30.71    -41.82
```

Figure 3. A portion of the tabulated results for 27,171 night sky spectra produced by EMSAO with report_mode=3.

from the upper atmosphere or emission from artificial sources such as street lights, they should exhibit a negligible Doppler line shift.

An IRAF program was already being used to convert Z-Machine spectra from SAO's internal archive format to IRAF .imh and .pix files (Mink & Wyatt 1992). A substitute line list was set up with all of the bright night sky lines. Since the mercury and sodium line strengths varied significantly, they were dropped in favor of the more stable oxygen lines. The 6300 Å [O I] line can show significant variations in intensity within a single night, but that is a story for another paper. The screen output of EMSAO for an individual night sky spectrum is shown in Figure 1. An IRAF CL script was written to run through the entire archive, or portions thereof, with a single command.

The results were tabulated in a file shown in Figure 3, where the columns are reduced file number, Julian Date of observation, emission line velocity and error, cross-correlation velocity (from a galaxy template) with error and R-value, combination velocity and error, quality flag, number of lines found, number of lines fit, and velocities in $\text{km}\,\text{s}^{-1}$ for [O I] lines at 5577.35 Å, 6300.23 Å, and 6363.88 Å.

Due to the distribution of lines in the calibration lamp spectrum, the position of the [O I] line at 6300 Å was most certain. Figure 4 shows the velocity shift computed from the change in the position of the center of that emission line over the 15-year lifetime of the Z-Machine spectrograph on the 1.5 m Tillinghast Reflector. Each vertical grouping is one month's dark-time run, with

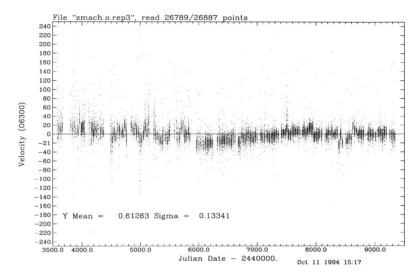

Figure 4. Velocity shift in 6300 Å OI night sky line over 15 years as observed from Mt. Hopkins, Arizona

larger gaps usually indicating summer telescope shutdown. It is obvious that some runs had a large scatter in sky "velocities," but for the most part, the sky "velocity" distribution is within the 63 km s^{-1} that a one-pixel shift in the emission line would cause.

4. Access to Software

RVSAO can be obtained via anonymous FTP from *cfa-ftp.harvard.edu* in the *pub/iraf* directory. On-line documentation with examples is available on the World Wide Web at the RVSAO home page[1].

References

Kurtz, M. J., Mink, D. J., Wyatt, W. F., Fabricant, D. G., Torres, G., Kriss, G., & Tonry, J. L. 1992, in Astronomical Data Analysis Software and Systems I, ASP Conf. Ser., Vol. 25, eds. D. M. Worrall, C. Biemesderfer, & J. Barnes (San Francisco, ASP), p. 432

Mink, D. J., & Wyatt, W. F. 1992, in Astronomical Data Analysis Software and Systems I, ASP Conf. Ser., Vol. 25, eds. D. M. Worrall, C. Biemesderfer, & J. Barnes (San Francisco, ASP), p. 439

Tonry, J. L., & Wyatt, W. F. 1988, CFA Z-Machine Data Analysis Software (Cambridge, Smithsonian Astrophysical Observatory)

---

[1] http://tdc-www.harvard.edu/iraf/rvsao/rvsao.html

A Technique for Determining Proper Motions from Schmidt Plate Scans at ST ScI

R. L. Williamson II
Space Telescope Science Institute, 3700 San Martin Drive, Baltimore, MD 21218

D. J. MacConnell
Computer Sciences Corporation, Space Telescope Science Institute, 3700 San Martin Drive, Baltimore, MD 21218

W. J. Roberts
Baltimore, MD 21210

Abstract. In this paper we describe a program of determining proper motions from digitized scans of 1950's-era Palomar Sky Survey (POSS) Schmidt Plates and 1980's-era Palomar Oschin Schmidt 'Quick V' plates that were taken for the *Hubble Space Telescope* (*HST*) Guide Star Catalogue.

1. Introduction

Using two PDS 2020G microdensitometers, scanning the early 1950's-epoch POSS E plates north of $\delta = -18°$ was completed at the Space Telescope Science Institute about three years ago. These scans, together with scans of the Palomar Oschin Schmidt "Quick V" plates taken for the *HST* Guide Star Catalogue provided a baseline of about 30 years from which proper motions may be determined.

2. The Method

Extractions from the scan database are typically done over small regions, no larger than 14′ on a side, so that distortions are minimized. An in-house software program generates an inventory of all of the objects in an extracted field for each epoch, and outputs a file that contains positions (in both linear units and equatorial coordinates), rough magnitudes, and a "star" or "non-star" classification for each object. Objects that are classified as stars in both frames are matched using a nearest-neighbor algorithm in both spatial directions. The program eliminates those with residuals $\geq 2.25\sigma$ as potential proper motion stars, but retains in the solution those with proper motions, μ, not exceeding $0\rlap{.}''035\,\mathrm{yr}^{-1}$. The number of reference stars remaining is typically 25 to 30. After matches are made and accepted, we then determine a solution for an affine transformation from one frame to the other using the least-squares, singular value decomposition routines from Press et al. (1992). With our material, we

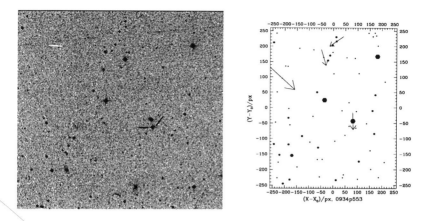

Figure 1. POSS Image of region around PG0923+533 and resulting plot from program.

have the most confidence in results for stars with visual magnitudes in the range $11 \leq V \leq 18$, and with proper motions in the range $0\rlap.{''}05\,\mathrm{yr}^{-1} \leq \mu \leq 0\rlap.{''}25\,\mathrm{yr}^{-1}$. We work in either of two ways which we call "Selected-Target" and "Survey" modes.

3. Selected-Target Mode

In this mode, the position of a specific star is centered in an extraction window for each epoch, and the motion is determined for that star. We have been obtaining proper motions of: candidate nearby stars for H. Jahreiß, verifying that some having motions up to $0\rlap.{''}3\,\mathrm{yr}^{-1}$ which are not listed in the Luyten or Giclas catalogues; hot DAs from the Palomar-Green (PG) Survey for J. Liebert; sdO stars in the PG survey for R. Saffer; faint, high-latitude carbon stars; and low-mass, post-AGB stars for M. Parthasarathy. Other collaborations are pending.

Figure 1 shows a print (*left*) of the area around one of our single targets, PG0923+533, and the resulting plot (*right*) from the analysis of the field for proper motions. Arrows indicate the magnitude and direction of the proper motion for the stars. We determined a proper motion for this target of $\mu = 0\rlap.{''}047\,\mathrm{yr}^{-1}$ and $\theta = 182°$.

4. Survey Mode

We also carried out a search over the full 6° field of POSS Area 321 near the North Galactic Pole using a 50% overlapping sub-plate technique. Our epoch difference is 31.92 yr compared with 17.04 yr for Luyten's neighboring area 322 and 12.98 yr for area 478 (Luyten 1973). This method is vulnerable to cosmetic defects, especially for faint objects, and the accuracy of the star/non-star classifier.

Of 123 Luyten proper motion stars in the common area, we find 44 certain and two possible correspondences; the faintest Luyten R mag of these is 19.8.

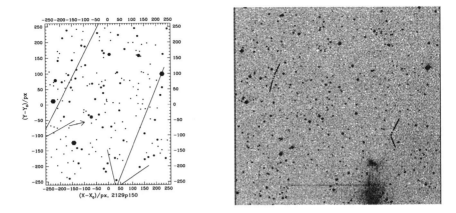

Figure 2. Plot from the program of the area around target PG2129+150 and the POSS Image of the region.

Of 77 Luyten stars not found, 24 are too faint [R(Luy) $\geq$ 19.9] to appear on the "Quick V" plate and two are too bright (8th mag). Of the 51 remaining, 34 have $\mu \leq 0''.06 \, \text{yr}^{-1}$—i.e., perhaps too small for us to detect. Of the remaining 17, six are near our faint or bright limit. The largest undetected Luyten proper motion is $0''.18 \, \text{yr}^{-1}$ for a R=18.9 star. This and other cases of missing known proper motion stars, as well as spurious proper motions, may be attributable to the star/non-star classifier being too conservative and hence eliminating real stars from one or both frames.

Such a case is illustrated in Figure 2 which is a plot showing the stars matched in the field of PG2129+150; the target star, marked with a "t" in the figure, show no motion, but two others, separated by about 9', seem to be a common proper motion pair with $\mu = 0''.35 \, \text{yr}^{-1}$ and $\theta = 156°$. However, upon inspecting the actual photograph, also in Figure 2, we see that two stars of the right separation and position angle are present in each position. For these spurious motions to be computed, it must be that the star/non-star classifier classified the northern member of each pair as a "star" and the southern members as "non-stars" in the early-epoch plate output file, and then reversed the classifications in the late-epoch plate.

Acknowledgments. We are pleased to acknowledge the support of the Space Telescope Science Institute and its Catalogues and Surveys Branch, and in particular, B. Lasker and B. McLean.

References

Press, W. H., Flannery, B. P., Teukolsky, S. A., & Vettering, W. T. 1992, Numerical Recipes in FORTRAN: The Art of Scientific Computing (Cambridge, Cambridge Univ. Press)

Towards a General Definition for Spectroscopic Resolution

A. W. Jones, J. Bland-Hawthorn

Anglo-Australian Observatory, P.O. Box 296, Epping, NSW 2121 Australia

P. L. Shopbell

Dept. of Space Physics & Astronomy, Rice University, P.O. Box 1892, Houston, TX 77251

Abstract. Judged by their instrumental profiles, spectrometers fall into two basic classes—Lorentzian and Gaussian—with many other line profile functions ($sinc^m$ functions, Voigt functions, Airy functions, etc.) falling into one of these two categories in some limit. We demonstrate that the Rayleigh, Sparrow, and Houston resolution criteria are of limited use compared to the "equivalent width" of the line profile.

1. Introduction

Modern day spectrographs ultimately rely on the interference of a finite number of beams that traverse different optical paths to form a signal (Bell 1972). The spectrometer disperses the incoming light into a finite number of wavelength (energy) intervals, where the size of the resolution element ($\delta\lambda$) is set by the bandwidth limit imposed by the dispersing element. Different dispersive techniques produce a variety of instrumental profiles. A long-slit spectrometer in the diffraction limit produces a $sinc^2$ wavelength response, a property shared with acousto-optic filters. In practice, optical and mechanical defects within either device tend to make the instrumental response more Gaussian in form. The response of the Fourier Transform Spectrometer is fundamentally the sinc function, although this response is commonly apodized to produce a profile with better side-lobe behavior. An internally reflecting cavity (e.g., Fabry-Perot filter) generates an instrumental response given by the periodic Airy function. In the limit of high finesse (the periodic interval $\Delta\lambda$ divided by the line FWHM $\delta\lambda$), the Airy function reduces to the Lorentzian function.

2. Resolution Criteria

The resolution element $\delta\lambda$, or more formally the *spectral purity*, is the smallest measurable wavelength difference at a given wavelength λ. In the case of rectangular and triangular functions, the (average) instrumental width is unambiguous; for more complex functions, a characteristic width can be more difficult to define.

	$f(x)$	Area	Sparrow
G	$\exp(-\ln 16 x^2/\delta x^2)$	$\left(\dfrac{\delta x}{2}\right)\sqrt{\dfrac{\pi}{\ln 2}}$	$\dfrac{\delta x}{\sqrt{2\ln 2}}$
L	$\dfrac{1}{1+(2x/\delta x)^2}$	$\left(\dfrac{\delta x}{2}\right)\pi$	$\dfrac{\delta x}{\sqrt{3}}$
A	$\dfrac{1}{1+\alpha\sin^2(\pi x/\Delta x)}$	$\dfrac{\Delta x}{\sqrt{1+\alpha}}$	$\lim\limits_{\alpha\to\infty}\dfrac{\delta x}{\sqrt{3}}$
S	$\operatorname{sinc}^2\left(\dfrac{2.7831 x}{\delta x}\right)$	$\left(\dfrac{\pi}{2.7831}\right)\delta x$	$0.9364\,\delta x$
V_1	$\dfrac{a}{\pi}\displaystyle\int_{-\infty}^{+\infty}\dfrac{\exp(-y^2)}{\left(x\sqrt{\ln 16}/\delta x - y\right)^2 + a^2}dy$	$\lim\limits_{a\to 0}\left(\dfrac{\delta x}{2}\right)\sqrt{\dfrac{\pi}{\ln 2}}$	$\lim\limits_{a\to 0}\dfrac{\delta x}{\sqrt{2\ln 2}}$
V_2	$\dfrac{a^2}{\sqrt{\pi}}\displaystyle\int_{-\infty}^{+\infty}\dfrac{\exp(-y^2)}{\left(2x/\delta x - y\right)^2 + a^2}dy$	$\lim\limits_{a\to\infty}\left(\dfrac{\delta x}{2}\right)\dfrac{\sqrt{\pi}}{a}$	$\lim\limits_{a\to\infty}\dfrac{\delta x}{\sqrt{3}}$

Table 1. Resolution criteria for common spectral line functions in terms of the FWHM δx (see Figure 1). The Airy function is periodic over the interval Δx.

Rayleigh criterion. Lord Rayleigh (1879) first derived the resolved distance of two identical, diffraction-limited point sources with the aid of Bessel functions. This separation arises when the peak of one Bessel function falls on the first zero point of the other function. The often quoted resolution criterion, $1.220 f\lambda/L$, where f is the focal ratio of the imaging system, was only intended for use in this context. Thus, we do not investigate this criterion further.

Houston criterion. The usual metric in astronomy is to adopt the "full width at half maximum" (FWHM) as a suitable definition of spectral purity $\delta\lambda$. Houston (1926) ventured that this property can be used to define the natural separation of two identical lines which are resolved from each other.

Sparrow criterion. Sparrow (1916) suggested a clever alternative, which depends on the property of the summed line profiles. As we move the lines closer together from far apart, a minimum develops. Sparrow suggested that a natural definition for resolution results at the line separation where the saddle point first develops (i.e., the gradient at the peak of the summed profile is zero). More formally, for an instrumental response given by $f(x)$, then two sources are resolved at a separation of σ_L (Sparrow limit) when both of the following conditions are satisfied:

$$\frac{d}{dx}\left[f(x) + f(x+\sigma_L)\right] = 0,$$

$$\frac{d^2}{dx^2}\left[f(x) + f(x+\sigma_L)\right] = 0.$$

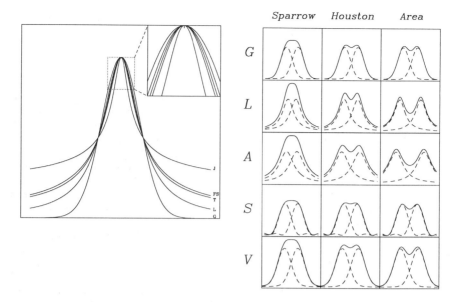

Figure 1. (a) Five functions with the same FWHM that demonstrate a gradual trend in the ratio of core power to wing power: (G) Gaussian, (L) Lorentzian, (T) $x^{-1} \tanh x$, (FS) Fraser-Suzuki function, (J) the most extreme case to date discovered by AWJ. All of these functions have continuous higher derivatives. In order for the function to remain everywhere continuous, the neighborhood of the peak becomes narrower (see inset) to compensate for the more extreme curvature near the FWHM. (b) Three different possible criteria for the resolution of two identical instrumental functions: (G) Gaussian, (L) Lorentzian, (A) Airy, (S) sinc^2, (V) Voigt.

3. Logical Dilemma

We illustrate a *reductio ad absurdum* arising from conventional definitions of spectral resolution, with the aid of a series of functions that are everywhere continuous and have the same FWHM, but have varying core to wing ratios (see Figure 1(a)). Long-slit spectrometers usually have rather Gaussian (G) profiles, whereas Fabry-Perot interferometers approach Lorentzian (L) at high finesse. The extreme wing-to-core ratio function (J) is not unlike the *spatial* response function of the *Hubble Space Telescope* prior to installation of the COSTAR optics. The Houston criterion implies that all of the profiles have equal resolving ability. In contrast, the Sparrow criterion would lead one to believe that narrow cores and large wings have better resolution capabilities (see Table 1 and Figure 1(b)). *This is clearly not physical.* A resolved core does not guarantee that two profiles are clearly separated, since the wing contribution remains unresolved. Intuitively, one way to see this is in terms of the shot noise constraint. The uncertainty in finding the line centroid depends inversely on the signal under the profile. As we increase the wing power, for a constant line flux, the signal in the core decreases dramatically.

4. Comparison of Resolution Criteria

In Table 1, we compare five different functional forms which arise in observational astronomy. Each of the functions has been expressed in terms of its FWHM δx; the Voigt function is expressed in two different ways to emphasize its Lorentzian and Gaussian behavior in the different limits. The various cases are illustrated in Figure 1(b). The figure shows that the Sparrow criterion is the least stringent, followed by the Houston criterion. The equivalent width (area) criterion is least forgiving to the Lorentzian (and therefore Airy) function. At first glance, it would appear that the Houston criterion is sufficient to resolve two lines. However, to avoid the logical dilemma described in the previous section, we should say that the lines are *properly resolved* by the area criterion.

We propose that the equivalent width (the area of the line profile divided by the peak height) is a better measure of the spectral purity of an instrumental function. This is more physical for a number of reasons: (1) the equivalent width, not the FWHM, constitutes the average width of the profile, (2) the equivalent width is more representative of the total signal under the line profile, (3) the equivalent width is a better discriminant of the wing behavior of a line profile, and (4) the equivalent width has an entirely general definition for an arbitrary positive function. For Gaussian spectrometers (e.g., long-slit devices), the correction amounts to no more than 6%. For Lorentzian spectrometers (e.g., Fabry-Perot filters), this leads to a large correction factor (50–60%) to the more standard use of the line profile FWHM.

Acknowledgments. AWJ acknowledges a studentship (austral winter, 1994) at the Anglo-Australian Observatory.

References

Bell, R. J. 1972, Introductory Fourier Transform Spectroscopy (New York, Academic Press)

Fraser R. D. B., & Suzuki, E. 1969, Anal. Chem., 41, 37

Houston, W. V. 1926, ApJ, 64, 81

Rayleigh, Lord 1879, Phil. Mag., 8, 261

Sparrow, C. M. 1916, ApJ, 44, 76

ively unaffected for dark current 0.11 counts pix^{-1} ksec^{-1}.-->

A Test for Weak Cosmic Ray Events on CCD Exposures

J. Bland-Hawthorn

Anglo-Australian Observatory, P.O. Box 296, Epping, N.S.W. 2121 Australia

S. Serjeant

Department. of Astrophysics, University of Oxford, Oxford OX1 3RH, UK

P. L. Shopbell

Department. of Space Physics & Astronomy, Rice University, P.O. Box 1892, Houston, TX 77251

Abstract. It is notoriously difficult to identify weak cosmic ray events in long exposure observations. The majority of "de-glitch" programs have little difficulty in identifying the bright events. The faint events—either individual or "splatter" around bright events—are rather more problematic. We propose a simple test to evaluate the effectiveness of a de-glitch program in removing weak events. The method is to compare the probability distribution function (PDF) of cosmic ray events between an observation and a dark frame matched in exposure time. We illustrate the basic idea using dark exposures from a Tek 1024×1024 CCD with varying exposure lengths and read-out times.

1. Observations

We have obtained dark frames using the Tek 1024×1024 CCD at the AAT 3.9 m with varying exposure lengths (15, 30, 60, and 120 min) and read-out times (FAST, SLOW, XTRASLOW). The histogram of each frame shows the contribution from the bias, read and dark noise. A millisecond exposure was used to remove the bias and read noise contribution to each histogram. The additional contribution from the dark noise is well calibrated at 0.11 counts pix^{-1} ksec^{-1}. It is assumed that the remaining events are related to cosmic rays. Only half the CCD frame was used because long exposures revealed a weak intensity gradient in the dark response on the other half.

2. Method

We define $n(E)\,dE$ to be the number of cosmic ray events with energies (expressed in counts) in the range $(E, E + dE)$. The cumulative distribution is

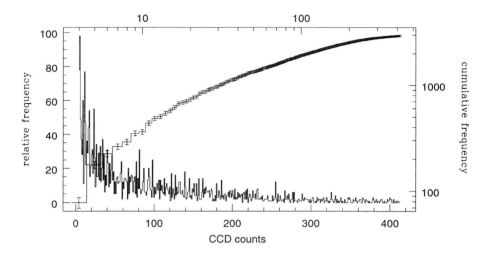

Figure 1. Histogram of cosmic ray events in a two hour dark frame. The monotonic curve is the cumulative histogram of these events. The error bars are Poissonian and not independent.

then

$$P(E) = \int_0^E n(E)\, dE \tag{1}$$

When $n(E) \propto E^\beta$, the slope of the plot $\log P$ vs $\log E$ is proportional to $\beta + 1$. In Figure 1, the noisy histogram is the bias/dark/read noise subtracted dark frame. The monotonic curve is a plot of $\log P$ vs. $\log E$ which is found to be rather well defined and reproducible over the different exposures. We propose that de-glitch programs should compute the PDF in this way for both the data and a matched dark frame (same exposure, same read-out time). The PDF for the data is determined from all events identified by the de-glitch algorithm. Since the energetic events are easier to find, the bright end of both PDFs will be well matched. In Figure 2, where the de-glitch PDF turns over at low energy—presumably but not necessarily at an energy greater than or equal to the turnover in the dark PDF—gives some idea as to how effective the algorithm has been in removing the weaker events.

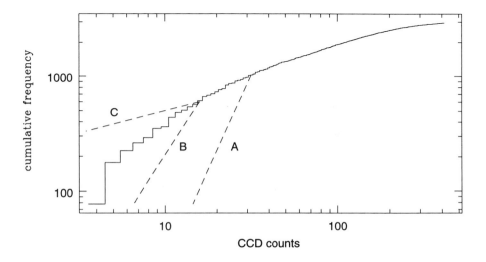

Figure 2. The cumulative histogram from Figure 1 with which to compare the performance of a de-glitch algorithm. Three cases are illustrated: algorithm A is too conservative, algorithm B more reliable, and algorithm C has mistaken real data with faint cosmic rays.

Automated Spectral Reduction in the IRAF Fabry-Perot Package

P. L. Shopbell

Rice University, Department of Space Physics, Houston, TX 77251-1892

J. Bland-Hawthorn

AAO, P.O. Box 296, Epping, NSW 2121, AUSTRALIA

Abstract. As introduced at ADASS I and II (Bland-Hawthorn, Shopbell, & Cecil 1992), a Fabry-Perot analysis package for IRAF[1] is under development as a joint effort of ourselves and the IRAF group. In this paper, we describe an important component of the Fabry-Perot package, the **fpplot** task for spectral plotting and fitting. While this task has many similarities with the familiar **splot** and **specplot** tasks in the **onedspec** package, **fpplot** has been optimized and extended specifically for use with imaging Fabry-Perot data. The task provides for the display and analysis of grids of spectra, including functions for binning, scaling, masking, and overplotting spectra. The most important features of **fpplot** use the IRAF `nlfit` and `inlfit` nonlinear fitting libraries to perform both interactive and background fitting of Fabry-Perot spectra. Automated techniques are essential for quantifying the thousands of spectra in a Fabry-Perot data cube for velocity and photometric studies. An example is given from current work involving the starburst galaxy M82.

1. Motivation

Since their inception, the complexity of imaging Fabry-Perot interferometers has made the reduction of their data a daunting task. Large data sets and complex instrumental profiles have caused the majority of Fabry-Perot observations to be interpreted simply as velocity maps, ignoring the enormous amount of photometric information also present. A primary goal of the IRAF Fabry-Perot package is to enable the *photometric* reduction of imaging Fabry-Perot data.

Once the characteristic Airy instrumental profile of the Fabry-Perot etalon is removed from the data, by a process we call "phase calibration" (Shopbell, Bland-Hawthorn, & Cecil 1992), data visualization can take two forms:

1. A sequence of spatial monochromatic images, each separated by a small increment in wavelength ($\sim 1\,\text{Å}$).

---

[1] IRAF is distributed by the National Optical Astronomy Observatories, which is operated by the Association of Universities for Research in Astronomy, Inc. (AURA) under cooperative agreement with the National Science Foundation.

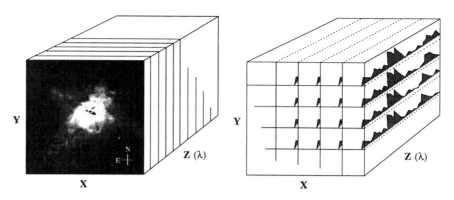

Figure 1. Two forms of imaging Fabry-Perot data visualization. The left cube depicts a series of monochromatic spatial frames sampled regularly in wavelength. The right cube depicts a grid of spectra sampled regularly in the spatial dimensions.

2. A spatial grid of spectra, sampled at the pixel resolution of the CCD ($\sim 0\rlap{.}''5$).

These forms are illustrated in Figure 1.

Early stages of the reduction of Fabry-Perot data, including cosmetic cleaning, flat-fielding, and alignment, employ the first visualization model. The later stages of reduction, such as phase correction, sky subtraction, and spectral fitting, employ primarily the second visualization model. While there are clearly tools available for the manipulation and display of images and image mosaics, there is currently a lack of useful visualization tools for application to the spectral domain of Fabry-Perot data. **fpplot** has been designed to assist the user in the advanced analysis stages of imaging Fabry-Perot data.

2. Features

The **fpplot** task in the IRAF Fabry-Perot package has many features found in the **splot** and **specplot** tasks in the **onedspec** and **twodspec** packages, as well as many options added specifically for Fabry-Perot data analysis. First there is the display of spectra. The displayed spectral grid merely represents a "window" onto the full spectral cube. The limits of the view may therefore be shifted and zoomed to view large-scale trends or details of individual spectra. Next, the spectra may be arbitrarily binned in the spatial dimensions, allowing the user to view large spatial variations. Binning in the spectral dimension is also possible. Additionally, one can scale the spectra. The spectra may be scaled to a variety of limits, including the extremes of the entire data set or each spectrum's extremes (i.e., autoscaling). In addition to the above, many overplotting options are provided for the comparison of data with other data or models. Additional Fabry-Perot spectral cubes, spectral fits, and image contour maps may be overplotted. Also included is the sophisticated spectral fitting module **fpplot** which allows for both interactive and background fitting of spectra with multiple-component Gaussian functions. The interactive form is very

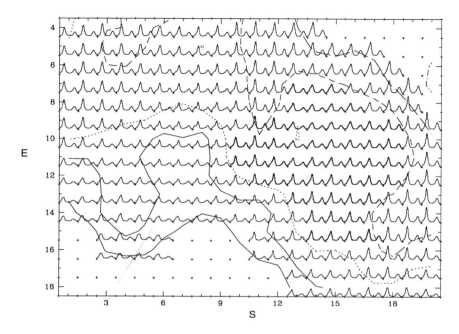

Figure 2. A spectral grid from Fabry-Perot data of the starburst galaxy M82 (Hα/[N II]), illustrating an overlaid contour map, two irregular areas of masked spectra, and a large area of spectra fitted automatically with a four-component Gaussian model.

similar to that provided by the **splot** task, including point rejection, residual plotting, etc. The automatic form uses spectra that have already been fitted to determine initial conditions for fitting additional spectra. Currently Gaussian functions and a linear continuum are used; Lorentzian and Voigt line fitting, as well as non-linear continuum fitting are under development. Lastly, to enable efficient automated fitting, full support is provided for spatial masking, via the IRAF PLIO routines (Tody 1988). Using masks, the user can remove bad pixels, sky regions, etc. from the fitting process. Figure 2 demonstrates several of these features using Fabry-Perot data from a central region of the starburst galaxy M82 (Shopbell 1995).

3. Fitting

The most significant capability of the **fpplot** task is that of fitting spectra. A small (512 × 512) CCD, assuming a useful data coverage of 50%, yields over 125,000 distinct Fabry-Perot spectra. If the instrument characteristics are well understood, these spectra can be fit to yield not only radial velocities, but emission line fluxes and dispersions as well. However, such a study clearly requires an automated means of fitting the emission line spectra.

The fitting portions of **fpplot** are modeled after those found in the **splot** task (in particular, those activated by the 'd' and 'k' keys). Single Gaussian functions can be fit; multiple Gaussians can be deblended. As with **splot**, **fpplot** employs IRAF's nlfit and inlfit nonlinear least squares fitting routines (Davis 1991) to fit the Gaussian profiles and, optionally, a linear continuum.

Major additions to **fpplot** enable it to fit large numbers of spectra in an almost entirely automated fashion. The user need only fit a few "characteristic" spectra interactively. **fpplot** will then propagate these fits across the desired region, using the fits of spatially nearby spectra as initial guesses for the current spectrum's fit. This propagation method is similar to that used in the FIGARO **longslit** package (Wilkins & Axon 1992). Because Fabry-Perot spectra are typically of low spectral resolution and encompass small wavelength ranges, this method of fitting works especially well. The typical problems of incomplete spectral coverage and inadequate continuum are still present however, as well as difficulties arising from multiple spectral components and rapidly varying spatial features. The use of an iterative procedure involving interactive verification of selective fits appears to solve these problems adequately.

Acknowledgments. The authors would like to thank the National Optical Astronomy Observatories, and the IRAF group in particular, for their continued support of this project. Partial support of P.L.S. has been provided by the Sigma Xi Grants-In-Aid of Research program and the Texas Space Grant Consortium.

References

Bland-Hawthorn, J., Shopbell, P. L., & Cecil, G. 1992, in Astronomical Data Analysis Software and Systems I, ASP Conf. Ser., Vol. 25, eds. D. M. Worrall, C. Biemesderfer, & J. Barnes (San Francisco, ASP), p. 393

Davis, L. 1991, NLFIT/INLFIT README files, IRAF v2.10 distribution (Tucson, NOAO)

Shopbell, P. L., Bland-Hawthorn, J., & Cecil, G. 1992, in Astronomical Data Analysis Software and Systems I, ASP Conf. Ser., Vol. 25, eds. D. M. Worrall, C. Biemesderfer, & J. Barnes (San Francisco, ASP), p. 442

Shopbell, P. L. 1995, Ph.D. Thesis, Rice University

Tody, D. 1988, PLIO README files, IRAF v2.10 distribution (Tucson, NOAO)

Wilkins, T. N., & Axon, D. J. 1992, in Astronomical Data Analysis Software and Systems I, ASP Conf. Ser., Vol. 25, eds. D. M. Worrall, C. Biemesderfer, & J. Barnes (San Francisco, ASP), p. 427

Robust Estimation and Image Combining

C. Y. Zhang

Space Telescope Science Institute, 3700 San Martin Dr., Baltimore, MD 21218

Abstract. A task GCOMBINE in the STSDAS package, for combining images, is briefly reviewed. With a stack of input images, one can identify bad pixels by comparing the deviation of a pixel from the mean with a user-specified threshold times the clipping sigma. It is crucial that estimation of the mean and sigma be robust against unidentified outliers, in order to reject cosmic ray events effectively. A variety of weighting schemes and means for storing noise characteristics are discussed. Information in neighboring pixels can also be used in a global way to identify cosmic rays.

1. Introduction

In images taken with CCD cameras, cosmic ray (CR) events become a common and annoying problem. Space-based CCD cameras, such as the Wide Field/Planetary Camera (WFPC) of the *Hubble Space Telescope* (*HST*), are especially subject to a large flux of CR events. The problem becomes even more prominent for the new WFPC2. In these cases, it is recommended that long science observations be broken into at least two exposures (CR-SPLIT) to ensure that CR events can later be identified and removed (see, e.g., Wide Field and Planetary Camera 2 Instrument Handbook, 1994).

2. Overview of GCOMBINE

The task GCOMBINE, in STSDAS, is designed mainly for *HST* images in GEIS format, but is not restricted to them. A parameter, *groups*, allows a user to select a range of groups to combine. A combined image is either a weighted average or median of a set of input images with bad pixels, including CR events, removed. As in the IRAF task IMCOMBINE, bad pixels can be rejected using masking, threshold levels, and various rejection algorithms.

A list of input error images can be used for the algorithm, *errclip* to reject bad pixels and to compute the pixel-wise weight when the parameter, *weight*, is set to **pixelwise**.

There are a variety of rejection algorithms from which to choose for rejecting bad pixels after possible masking and thresholding operations. These are the **minmax, ccdclip, rsigclip, avsigclip**, and the **errclip** algorithms. In the following, we will discuss some of these algorithms in more detail.

3. Scaling and Weighting Schemes

In order to combine images with different exposures, one must first bring them to a common level. For the i'th image, let us denote D_i as the original pixel value, Z_i the zero level, $\langle Z \rangle$ the average of Z_i, S_i the divisive scaling factor, $\langle S \rangle$ the average of S_i, and d_i the scaled pixel value, we write

$$d_i = (D_i - Z_i + \langle Z \rangle) * \langle S \rangle / S_i . \tag{1}$$

This will bring the images to the mean sky level of $\langle Z \rangle$ and common exposure level of $\langle S \rangle$. In the code, Equation (1) is translated into

$$d_i = D_i / s_i - z_i , \tag{2}$$

where $s_i = S_i/\langle S \rangle$ and $z_i = (Z_i - \langle Z \rangle)/s_i$. The normalized scaling factors s_i and z_i are stored in an output log file.

When the combined image is defined as a weighted average, it is useful to distinguish between uniform and pixelwise weighting schemes. By pixelwise weighting, we mean that the weight is calculated on a pixel-by-pixel basis. It requires an input "error map" associated with each input image, and the weight factor at a given pixel is then calculated as a reciprocal of the square of the error at that pixel. By uniform weighting, we mean that the weight is a constant for all pixels in an input image. In the case of uniform weighting, one has a variety of choices, such as, the exposure time, the mean, median, or mode of the input image. In a relatively flat image, one may choose a reciprocal of an averaged variance as a uniform weight.

4. Robust Estimates of Means and Clipping Sigmas - RSIGCLIP

After a mean is computed, one computes a deviation of the pixel from the mean and then compares it with a "clipping sigma". If the deviation exceeds a specified threshold, κ, times the clipping sigma, then the pixel under test is regarded as being bad. After the bad pixel is rejected, one iterates the process, but now the mean and the clipping sigma are re-calculated, excluding the bad pixels already rejected in the previous iterations. This iteration terminates when no more pixels are rejected or the retained number of pixels is fewer than the user-specified minimum number of retained pixels.

It is obvious that if there are outliers to be identified, the simple mean of all measurements not excluding the unidentified outliers will be significantly biased by often too large values of the outliers. The effects of the unidentified outliers on the estimate of the clipping sigma could be even more severe, since the deviation is being squared in computing the sigma. We discuss the robust sigma clipping algorithm, **rsigclip**, in detail.

4.1. Effects of Poisson Scaling

Assuming the Poisson noise dominates over the read noise, and, for simplicity, ignoring the effects of the zero flux level offset, Equation (2) is reduced to $d_i = D_i/s_i$. The sigma in the scaled image becomes, $\sigma_i^s = \sigma_i^u/s_i$, where σ_i^s and σ_i^u are the sigmas corresponding to the scaled and unscaled data. For the Poisson distribution, $(\sigma_i^u)^2 = \langle D_i \rangle/g$, where g is the gain. The optimal mean of the unscaled data is given by $\langle D_i \rangle = \langle d_i \rangle s_i$, where $\langle d_i \rangle$ is the proper mean of the scaled data. We thus have, $(\sigma_i^s)^2 = \langle d_i \rangle/(s_i g)$. Therefore, even if the gain is unknown, the relative variance, v_i, in the scaled data is given by $\langle d_i \rangle/s_i$. Taking the weights $w_i = 1/v_i = s_i/\langle d_i \rangle$, the average sample sigma is given by

$$\langle \sigma \rangle^2 = \frac{\sum w_i (d_i - \langle d_i \rangle)^2}{\sum w_i} \frac{n}{n-1}. \tag{3}$$

Noting that $w_i = s_i/\langle d_i \rangle$, the estimated absolute value of the sigma in the scaled data is then given (see, e.g., Bevington 1969) by

$$(\sigma_i^s)^2 = \frac{\xi_i}{n-1} \sum \frac{(d_i - \langle d_i \rangle)^2}{\xi_i}, \tag{4}$$

where the Poisson scaling factor is given by $\xi_i = \langle d_i \rangle/s_i$.

4.2. Robust Estimates of the Mean and Sigma

It is emphasized that no matter how accurate Equation (4) is when the data follow exactly the Poisson distributions, the presence of unidentified outliers, which *do not* follow the same Poisson statistics at all, will inevitably make the estimates of Equation (4) unrealistic. It is where the "robust" estimation of the mean and sigma should come into play.

The method for estimating the mean used in **rsigclip** is robust. We exclude the maximum and minimum values before computing the mean, because they are most vulnerable to being bad. The effect is to assign a zero weight to the two extremal points, while assigning normal weights to the rest of data. This estimation of the so-called trimmed mean, used in the GCOMBINE and IMCOMBINE tasks, is an estimate of the mean robust against unidentified outliers.

To minimize influence of unidentified outliers on the clipping sigma, Equation (3) is modified. We set a "normal" range of the data values to exclude the most deviant points on both sides from the mean. For points that are within the range, the normal weights, $s_i/\langle d_i \rangle$, are used, as in Equation (3). Much smaller weights are assigned to the points outside of the range, farther from the mean: a weight of s_i/d_i is used for points on the high side of the mean, and a weight of $0.001 s_i/\langle d_i \rangle$ is applied to points on the low side of the mean. An outlier, if present, will get the negligible weight and therefore have minimal influence on the clipping σ. This scheme works fine for identifying CR events, even in the case of only a few images to combine.

5. CCDCLIP and ERRCLIP Algorithms

If the noise in the i'th unscaled image can be obtained from the CCD noise model, using the error propagation principle and the relation that $\langle D_i \rangle = \langle d_i \rangle s_i$, we find that the optimal clipping sigma in the scaled image should be

$$(\sigma_i^s)^2 = \left(\frac{1}{s_i^2}\right) \left\{ \frac{\sigma_{rd}^2}{g^2} + \frac{\langle d_i \rangle s_i}{g} + s^2 \langle d_i \rangle^2 s_i^2 \right\}, \tag{5}$$

where σ_{rd} is the readout noise in number of electrons, g is the gain factor in units of e^-/DN, and s is the fractional "sensitivity noise."

In the code, the trimmed mean or median is used. Substituting the trimmed mean or median into Equation (5) gives estimation of the clipping sigma. This clipping sigma does not involve in any computations of the square of the deviations, is thus robust against the unidentified outliers. If the error maps exist for input images, one does not have to compute the clipping sigma, because the sigma is available from the input error maps for each individual pixels. The testing shows that *errclip* works well for WFPC images. if one makes error maps using, e.g., the **imcalc** task based on the CCD noise model. In doing so, it is important that the error provided in the input error maps should simultaneously be scaled as the input images.

6. Information in Neighboring Pixels

In the above discussions, only the information along a stack of input images for a given output pixel is used. It is conceivable that information in neighboring pixels, such as patterns of various astronomical objects, can be used to distinguish CR events. In the case of, say, only two images to combine, it becomes mandatory to use as much of the information in the image plane as possible. In the GCOMBINE task the information contained in the neighboring pixels is globally taken into account for the case of two images to combine.

It is conceivable that the pixel values of the two images for the same target, through the same filter, must be closely correlated with each other. If one plots pixel values of one image versus those of the other, it is seen that more than 99.5% pixels are located in between the two envelope curves, where the pixel values of one image equal those of the other plus and minus three sigma. The outliers are almost exclusively located outside the enveloped region. The locations of outliers relative to the rest of neighboring "normal" pixels show in a global way that it is possible to identify outliers based on the direct difference of the pixel values of the two images at a given pixel of the combined image. If the difference exceeds the user-specified κ times of the sigma, the larger one is rejected as long as the very negative pixels are excluded first. The clipping sigma, in this case, can only be obtained with either the CCD noise model or the input error map. Tests show that this method is very effective in removing cosmic rays.

Acknowledgments. I am grateful to F. Valdes, R. White, and K. Ratnatunga for very stimulating discussions.

References

Bevington, P. R. 1969, Data Reduction and Error Analysis for the Physical Sciences (New York, McGraw-Hill)

Burrows, C. J., ed. 1994, Wide Field and Planetary Camera 2 Instrument Handbook (Baltimore, Space Telescope Science Institute)

ADASS '94 - A Summary And A Look To The Future

G. H. Jacoby

National Optical Astronomy Observatories[1], *P.O. Box 26732, Tucson, AZ 85719*

Abstract. Papers presented at this meeting are classified according to 12 steps in the astronomical research process. In considering future software needs, I present five projects to enhance the astronomer's research effectiveness, and note that large telescope projects and information databases will also challenge software developers. Obvious trends emerging in astronomical software are noted, and the need for two more are predicted.

1. Applicability of ADASS Software to Astronomy

The ADASS conferences are intended to foster discussions between astronomers and software developers so that the computational needs of researchers are understood and met. Are the software developers, in fact, addressing the needs of the astronomical community?

I estimated the applicability of the papers presented at this conference by counting the number of papers in each of 12 categories identified as steps in an observational research project. Table 1 lists these steps along with the paper count. Some papers provide software in more than one area; others do not fall into any of these classes, but still provide critical support (e.g., seven papers on FITS, an essential tool usually taken for granted).

The plurality of papers fall into the reduction and analysis step, as expected from the original focus of the ADASS conferences (and suggested by the conference title). There is, however, a healthy diversity outside this theme. In particular, there is an impressive number of papers in the crucial modeling step, which represents "science extraction" from the data. Additionally, there is a large force dealing with data distribution and access, a step that has grown in importance now that large databases are available.

No one seems to be dealing with judgment and subjective issues by using software (e.g., identifying important scientific problems, judging proposal merit, writing papers). It is probably wise to avoid these until artificial intelligence software matures further. Thus, the ADASS conferences now transcend reduction software, and are not at all IRAF users meetings as some originally feared.

---

[1]National Optical Astronomy Observatories, operated by the Association of Universities for Research in Astronomy (AURA) under a cooperative agreement with the National Science Foundation.

Table 1. Applications Where Software Helps Astronomers.

	Task	Paper Count
1	Identify outstanding problem	
2	Literature/catalog/database search	17
3	Prepare/submit telescope proposal	6
4	Technical/scientific proposal evaluations	2
5	Schedule telescopes	2
6	Instrument/telescope control	11
7	On-line/quick-look data verification	3
8	Reduction & analysis	39
9	Modeling/statistics/visualization	16
10	Write paper/prepare graphics	2
11	Disseminate information & data (see #2)	
12	Communicate with colleagues	3
	Items 11 & 12 lead back to item 1	

2. Software For 1999

I see two primary drivers for software needs five years hence. On this timescale, it is the very big projects which get our attention because they require 3–5 years of development. Additionally, recurring and growing problems demand solutions if we wish to avoid being consumed by them.

2.1. Large Telescopes

The biggest projects are the new telescopes. My view is biased toward ground-based optical work, but astronomers working in other regimes will have similar concerns. The large advanced technology telescopes expected by the year 2000 present special challenges to software developers. Adaptive optics telescopes will require very complex, highly distributed computer systems to achieve the high spatial resolution they promise, and to control their exotic instrumentation; spatial resolution drives detectors toward tiny pixels in huge arrays. To run the telescopes and instruments, and to diagnose failures rapidly, essential software includes clear and carefully engineered GUIs. To quickly estimate data quality, fast and insightful visualization techniques are needed. The vast data arrays will tax networks, I/O systems, and storage capabilities, so compression techniques will be crucial. Perhaps lossy compression will be used most of the time rather than rarely, as is current practice.

2.2. Large Databases

Large databases are being built (e.g., Sloan digital sky survey, *HST* archives, NOAO archives) which already are changing the way astronomers do research. Using telescopes may become a minor part of observational astronomy, but first, access to archives must be easy and fast. Software will be needed to browse the catalogs and to download data "samples" (i.e., representative subsets) to verify that the catalog delivers the advertised information. Again, the need for compression techniques and the acceptance of lossy data may be necessary to live within network bandwidths and on-line storage limits. Also, the data quality must be documented and attached to the database entries (e.g., was it photometric, were images $< 0\rlap{.}''2$, what were the filter characteristics?).

2.3. Large Journals

A comment I hear often is "I missed your paper—I don't have time to read the journals anymore." If this trend continues, research will become pointless. One extreme view is to eliminate paper journals, for which no one has shelf space anyway. The move to electronic publishing provides a solution: rather than subscribing to a few journals, astronomers can subscribe to selected topics in all the journals. For example, one could subscribe to all papers dealing with lithium abundances. When a paper is "published" on lithium in any journal, an e-mail message is sent to the subscriber indicating that the paper can be downloaded. The embarrassment of missing a critical paper in one's own specialty is avoided, and the astronomer is protected from information overload. A concern is that scientists will be channeled into narrow disciplines without serendipitously reading interesting material in other fields.

One journal tool is nearly here: full text on-line searches. We already have abstract searches, but these place researchers at the mercy of authors who may not realize the value of all aspects of a paper.

2.4. Education

Another frequent complaint is that researchers spend an increasing fraction of their time writing grant proposals. With success rates dropping, funding is a major time sink. This becomes a vicious circle; rather than learning about nature, scientists are learning about funding.

Despite the public's keen interest in astronomy (e.g., the Comet SL9/Jupiter encounter), money continues to be a problem. While public excitement and support for astronomy may not correlate perfectly, there is some relation. Perhaps, if we keep new discoveries interesting and frequent by tightly integrating the research environment with education and the media, the proposal writing cycle can be broken. To do so requires that reduction, analysis, and modeling software be smarter so that scientific results can be released within days rather than years. We also need visualization tools; not everyone has a graphic artist to turn a scientific result into a picture for non-specialists.

2.5. Experts on the Desk

Most of what scientists do is a sequence of operations they've done before. Can they teach software to perform those repetitive steps so they can concentrate on the subjective aspects of interpreting results? Expert software in photometry, spectroscopy, and statistics, for example, could alleviate a lot of the tedium. Can a photometrist program be given 15 images of a field to find all variable stars having periods of 5 to 60 days to build a Cepheid finding expert? Can a spectroscopist program be taught to classify stars and nebulae and derive abundances? Can a statistician program be taught to answer questions like: "Are these two distributions different, what is the confidence level that these lithium detections are real, what are the errors on Cepheid period determinations?"

3. Themes

After four ADASS meetings, clear software trends are emerging.

3.1. Clear Trends

Graphical User Interfaces (GUIs) – I prefer command line interfaces because they can be built into scripts, allow type ahead, and require little screen area and few computer resources. Nevertheless, most people prefer GUIs because they provide easy navigation through increasingly complex software packages. But, if a task doesn't need a GUI, don't build one just because you can.

World Wide Web (WWW) – "The Web" has become a major resource, growing from a curiosity to a serious research tool in just two years.

Tbytes – Disk space used to be measured in MB. Now we talk about GB. The standard is becoming TB.

Object Oriented Systems (OO) – OO programming concepts seem esoteric, but there are advantages to thinking this way. Improved performance and high level programmability are driving databases and programming tools to use OO.

FITS – FITS usage has been a trend for over 10 years. FITS continues to develop (7 papers) and to generate tremendous interest among software folks. Astronomy is fortunate to have a highly viable data standard; congratulations to the originators of FITS (Wells, Greisen, & Harten 1981)!

IDL – IDL grew from the field of astronomy to become a viable commercial product. IDL's future appears healthy thanks to its use in more lucrative markets (e.g., medicine and earth sciences) which feed software back to astronomy.

IRAF – After $\sim$ 10 years in the community, and with development for various projects (e.g., *HST*, PROS, *EUVE*, and NOAO), IRAF has become a basic research tool in astronomy.

3.2. Future Trends

It is difficult to identify a trend which has not yet begun, but let me suggest that astronomy needs the following two.

Education software – The public deserves to know as much about the universe as we do and in a timely manner. Electronic picture books (see the paper by Brown p. 3) can improve information turnaround, but we need hands-on experiments, too.

Error Propagation – The failure to track errors properly from detected photons to final answers has plagued subfields of astronomy and created controversies lasting decades. Analysis systems need to help astronomers with the subtle details of forming valid statistics and errors.

4. Conclusion

More and more, astronomers rely on software in every aspect of their jobs. ADASS conferences provide programmers the opportunity to present new software that helps astronomers in their daily work, and astronomers are encouraged to attend the ADASS meetings to become more effective researchers.

References

Brown, R. 1995, this volume, p. 3
Wells, D. C., Greisen, E. W., & Harten, R. H. 1981, A&AS 44, 363

Author Index

Abbott, M., **453**
Abbott, T. M. C., **341**
Accomazzi, A., 28, 32, **36**
Adorf, H.-M., 52, **456**, **460**
Ake, T. B., **464**
Akella, R., 429
Akritas, M., 311
Albrecht, M. A., 58, 221
Alexander, D., 331
Allan, D. J., **199**
Allen, J. S., 176
Allen, R. J., 191
Andernach, H., 48
Angelini, L., 117, 219
Ansari, S. G., **155**
Antia, B., 44
Aparicio, A., 323
Appleton, P. N., 289
Asson, D. J., 65

Baade, D., 221
Baffa, C., 25
Bailey, J. A., 113
Ballester, P., **468**
Barrett, P., **472**
Barry, K., 207
Basart, J. P., **289**
Baum, S. A., 158
Bennett, C. L., 480
Bhatnagar, A., **109**
Bijaoui, A., 279
Binegar, S., 429
Blackburn, J. K., **367**
Bland-Hawthorn, J., 503, **507**, 510
Bloch, J., **203**
Borne, K. D., **158**
Bridger, A., 379
Brown, R. A., **3**
Brummell, N., **15**
Brunner, R. J., **169**
Busetta, M., 383
Bushouse, H. A., **345**
Busko, I., **315**
Butcher, J. A., 211

Calabretta, M., 233
Chavan, A. M., **58**

Chen, J., **406**, 410
Chen, K., 44
Cherkauer, K., 272
Choo, T. H., 429
Christian, C., 44
Clayton, C. A., **371**
Connolly, A. J., 169
Conroy, M., **207**, 406
Corbet, R. H. D., **211**
Corcoran, M., **219**
Craig, N., 44
Cresitello-Dittmar, M., 176
Curtis, P., 99

Dal Fiume, D., **387**
Daly, P. N., **375**, **379**
Davis, L. E., 297
de Carvalho, R. R., **272**
De Cuyper, J.-P., **476**
DePonte, J., 353, 406, **410**
Djorgovski, S. G., 272
Duesterhaus, M., 62

Ebert, R., 84
Egret, D., 52, 84
Eichhorn, G., **28**, 32, 36
Eisenhamer, J. D., 121
Erdwurm, W., 185

Farrell, T. J., **113**, **391**
Farris, A., **191**
Favata, F., **383**
Fayyad, U., 272
Feigelson, E., **311**
Fini, L., **40**
Fitzpatrick, M. J., **225**
Folk, M., 229
Freeman, M., 357
Frontera, F., 387
Fruchter, A., 158

Gaetz, T., 357
Gallart, C., 323
Garcia, M., 418
Gavryusev, V., **25**
George, I., 219
Giani, E., 25
Giommi, P., **117**

523

Gooch, R., **144**
Gothoskar, P., 253
Grant, C. S., 28, **32**, 36, 48
Gray, A., 272
Greisen, E. W., **233**
Grosbøl, P., 221
Gruendl, R. A., 335
Gulati, R. K., **253**
Gupta, R., 253

Hamabe, M., 173
Hanisch, R. J., 301
Harris, D. E., **48**
He, L. X., 289
Heck, A., 52
Heller-Boyer, C., 429
Hensberge, H., 476
Hester, T., 429
Hill, G., **443**
Hillberg, B., 361
Hook, R. N., **293**
Horaguchi, T., 173
Hughes, J. P., 357
Hulbert, S. J., **121**
Hyde, P., 429

Ichikawa, S., **173**
Ipatov, A. P., 488
Ishee, J., 3
Isherwood, P., 383

Jackson, R. E., **52, 55**
Jacoby, G. H., **518**
Jennings, D. G., **229**
Jerius, D., **357**
Jones, A. W., **503**
Jordan, J. M., **176**

Kearns, K., **331**
Keegstra, P. B., **480**
Khobragade, S., 253
Kleiner, S. C., **136**
Koekemoer, A., 52
Krist, J., **349**
Kurtz, M., 28, 32, 36

Lallo, C., 3
Lammers, U., 383
Lamy, P., 484
Landsman, W. B., **437**
Larkin, C., 211

Levay, Z. G., 121
Lewis, J. W., **327**
Llebaria, A., **484**
Long, K. S, 158
Louys, M., 268
Lucy, L. B., 293
Lupton, R. H., 169

Macconnell, D. J., 500
MacKenty, J. W., **492**
Malburet, P., 484
Malkov, O. Yu., **182, 257**
Mandel, E., **125**
Manning, K. R., **353**, 410
Mathieu, J. J., 383
McDonald, K., **44**
McDowell, J. C., 207, 418
McGlynn, T. A., 219
Micol, A., 155
Mink, D. J., **496**
Mistry, A., **422**
Mo, J., **301**
Morgan, J. A., **129**
Morrison, J. E., **179**
Mueller, Th., 345
Mukai, K., 219
Murray, S. S., 28, 32, 36
Murtagh, F., 52, **260, 264**, 268, 279

Nguyen, D., **361**
Nicastro, L., 387
Nichols, D., 99
Nicinski, T., 237
Nousek, J. A., 211

Olson, E., **425**
Orlandini, M., 387
Osborne, J. P., 211

Page, C. G., **215**
Parmar, A. N., 383
Pasian, F., **68**
Pásztor, L., **319, 323**
Paterson, M. J., 379
Payne, H. E., **65**
Pedelty, J. A., 289
Pence, W. D., 219, 229, **245**
Peng, W., **237**
Percival, J. W., **140**
Peron, M., **221**
Perrine, R., 429

Phillips, A. C., **297**
Pickup, D. A., 379
Pintar, J. A., 185
Piskunov, N. E., 133
Plummer, D., 422
Podgorski, W., 357
Pollizzi, J. A. III, **162**
Primini, F., 331, 353, 406

Ramaiyer, K., 169
Rasmussen, B. F., **72**
Regan, M. W., **335**
Richmond, A., **62**
Richon, J., **166**
Riegler, G., **8**
Roberts, B., 44
Roberts, J. W., 500
Roden, J., 272
Rosa, M. R., 345
ROSAT/ASCA/XTE Development Team, 241
Rose, J., **429**
Rose, M. A., 429
Rosenberger, J., 311
Rots, A., 219

Sault, R. J., **433**
Schlegel, E., 62
Schmidt, D., 410, **414**
Seaman, R., **247**
Serjeant, S., 507
Shaw, R. A., 121
Shopbell, P. L., 503, 507, **510**
Shortridge, K., 113, 391
Simon, R., 207
Smale, A., 62
Smareglia, R., 68
Smirnov, O. M., **133**, 182, 257, **488**
Smoot, G. F., 480
Starck, J.-L., 260, **268**, **279**
Steuerman, K., 429
Strom, K. M., **76**
Stroozas, B. A., 44
Szalay, A. S., 169

Terrett, D. L., **395**
Teuben, P. J., **398**, 433
Theiler, J., 203
Tody, D., **89**, 125
Toomre, J., 15
Tóth, L. V., 319

Travisano, J., **80**
Trifoglio, M., 387
Trueblood, M., **185**

Vacanti, G., 383
Van Buren, D., **84**, **99**
Van Stone, D., 410, **418**
Vílchez, J. M., 323

Wang, Z., **402**
Warmels, R. H., 341
Weir, N., 272
Wells, D., 52, **148**
White, N. E., 62, 117
Williamson, R. L. II, **500**
Wright, M. C. H., 433
Wu, N., **305**
Wyatt, W. F., 496

Yom, S., 62
Yoshida, M., 173
Yoshida, S., 173

Zacher, R., 422
Zeilinger, W., 260
Zhang, C.-Y., **514**

Index

2dF, 113, 391
3-dimensional visualization, 15, 144

abstracts, 28, 36
ADAM, 375, 395
ADS, *see* Astrophysics Data System
Advanced Visual Systems (AVS), 148
ALEXIS, 203
ALICE, 379
analysis
 clustering, 272
 data, 109, 121, 191, 311, 367, 387, 433, 443, 507, 510
 multiresolution, 260, 268, 279
 software, 480
 spectroscopy, 133
 time series, 460
applications software
 DAOPHOT II, 133, 488
 DASHA, 133, 488
 EMSAO, 496
 GUIDARES, 182
 IDL, 109, 182, 422, 437
 IRAF, 89, 121, 207, 225, 305, 402, 410, 414, 418, 422, 496
 LaTeX, 58, 65, 76
 MIDAS, 221, 341
 Miriad, 129, 433
 NCSA Mosaic, 68, 76, 80, 182
 NEMO, 398
 PCIPS, 133, 488
 PROS, 207, 406, 410, 414
 Starlink, 371, 375
 StarView, 80, 158
 STSDAS, 402
 Sybase, 72
 Tiny Tim, 349
 WIP, 129
archival research, 221
archives, 28, 166, 191
 distributed, 40, 55, 68
 EUVE, 44
 HST, 80, 158, 162
 IUE publications, 155
 MOKA, 173
 NOAO, 247
 SAX, 387
artificial intelligence, 257, 429
ASC, 207, 422

associations, 323
astrometry, 179, 500
Astrophysics Data System (ADS), 32, 36
AstroWeb, 52
AXAF, 207, 357, 361, 422

background
 cosmic microwave, 480
BATSE, 176
Bayesian modeling, 315
BIMA, 129
blackboard model, 429

C++, 162
calibration, 221, 453
 wavelength, 468, 476
catalogs, 28, 32, 48, 55, 76, 80, 327, 472
 Guide Star, 182, 257, 500
 HST, 158
 IRAS PSC, 264
CCD
 data analysis, 341, 456
 missing data, 456
 photometry, 133, 484, 488
 testing, 341
centering algorithms, 476
CEPPAD, 109
CGS4, 375, 379
checksum, 247
cirrus, infrared, 289
classification, 264, 272
clients, 62, 89
clustering analysis, 272
collaborative tools, 99
color-magnitude diagrams, 488
Common Lisp Object System (CLOS), 199
compression, image, 268, 279
Compton observatory, 176
coordinates, 233, 472
 world, 437
cosmic microwave background, 480
cosmic rays, 507, 514
COSSC, 176
cross-identification of objects, 488
cyberspace, 99

DADS, 158

527

DAO, 443
DAOPHOT II, 133, 488
DASHA, 133, 488
data, 241
    acquisition, 25, 387
    analysis, 109, 121, 191, 311, 367, 387, 429, 433, 443, 507, 510
    classification, 221
    dictionary, 32
    exchange, 237
    formats, 225, 229
        FITS, 129, 191, 203, 207, 211, 219, 229, 233, 237, 241, 245, 247, 367, 383, 437
        HDF, 229
        netCDF, 229
        QPOE, 44, 207, 215, 353, 406, 418, 425
    interchange, 398
    model, 191, 199, 207
    modeling, 315
    organization, 221
    quick-look, 68
    retrieval, 166
    structures, 191, 207
    two-dimensional, 472
    verification, 247
databases, 48, 58, 72, 99, 253, 387, 472
    image, 84
    large, 185
    mirroring, 185
    object-oriented, 169
deconvolution, 279, 349
detectors, 341, 507
digital libraries, 28
Digitized Sky Survey, 179
distributed
    archives, 40, 55, 68
    processing, 422
    software, 25
DRAMA, 113, 391

e-mail interface, 58, 166
echelle spectroscopy, 133, 468, 476
education, 3
EGRET, 176
Einstein observatory
    IPC, 418
    On-Line Service, 48
EINTOOLS package, 418

electronic
    picturebooks, 3
    publication, 76, 129, 155
emission lines, 297, 496
EMSAO, 496
ephemeris, 464
error
    detection, 247
    handling, 422
    maps, 514
    systematic, 476
ESO, 341
EUVE, 44, 327, 425, 453
event list, 215
ExInEd, 3

FITS data format, 129, 191, 207, 211, 219, 229, 233, 241, 245, 247, 367, 383
    keywords, 203
    tables, 237, 437
FITSIO, 367
FORTRAN, 245, 367, 375, 433, 437
FTOOLS, 367
fuzzy sets, 319, 323

galaxies, 496
geometry, 148
Ginga, 211
GONG, 185
Gopher, 52
graphical user interface (GUI), 89, 109, 121, 125, 133, 136, 140, 158, 173, 395, 437, 453
graphics, 129
guest observer, 44, 425
GUIDARES, 182
Guide Star Catalog (GSC), 182, 257, 500

HDF data format, 229
HEASARC, 62
helioseismology, 185
Hough transform, 468
HRI, *ROSAT*, 353
Hubble Space Telescope (*HST*), 80, 162, 305
    archive, 158
    FOS, 345
    position and velocity, 464
    PSFs, 349

Index 529

WF/PC, 301
WF/PC2, 456
Hypertext Markup Language (HTML), 76

IDL, 109, 182, 422, 437
image
    analysis, 99
    combining, 514
    compression, 268, 279
    conversion, 225
    databases, 84
    deconvolution, 279, 349
    display, 125
    processing, 133, 279, 289, 297, 315, 335, 456, 476, 484, 507, 514
    registration, 297
    restoration, 279, 293, 301, 305
        insufficient sampling, 456
        projection onto convex sets, 456
imaging, infrared, 335, 492
imfort, 492
indexing, 52, 55
information
    retrieval, 28
    systems, 36
infrared, 25
    cirrus, 289
    data, 84
    imaging, 335, 492
inheritance, 391
instrumentation, 113, 140, 387, 391, 496, 510
inter-process communication, 125, 357
interferometry, 433, 510
Internet services
    Einstein On-Line Service, 48
    Astrophysics Data System, 32, 36
    AstroVR, 99
    AstroWeb, 52
    digital libraries, 28
    e-mail, 58, 166
    electronic publication, 76, 129, 155
    Gopher, 52
    news, 52
    WAIS, 32, 36, 52

    WWW, 28, 32, 36, 40, 44, 52, 58, 62, 68, 72, 76, 80, 84, 162, 176, 182
interpolation, 460
interpreted languages, 437
inverse problems, 460
IPC, *Einstein observatory*, 418
IRAF, 89, 121, 207, 225, 305, 402, 410, 414, 418, 422, 425, 496
IRAS
    images, 289
    Point Source Catalog, 264
IRCAM3, 379
irregular sampling, 460
IUE, 253
    publication archive, 155

La Silla Observatory, 341
languages
    C++, 162
    FORTRAN, 245, 367, 375, 433, 437
    Hypertext Markup Language (HTML), 76
    IDL, 109, 182, 422, 437
    interpreted, 437
    MOO, 99
LaTeX, 58, 65, 76
lattice models, 319
LBT, 40
legacy systems, 429
libraries
    digital, 28
    subroutines, *see* subroutine libraries
linear quadtrees, 472
literature searching, 28

M82, 510
mass storage, 15
massively-parallel machines, 15
mathematical morphology, 289
Maximum Entropy method, 215, 301, 305
Message Passing Interface (MPI), 361
methods, mathematical
    centering, 476
    clustering, 272
    deconvolution, 279, 349
    Hough transform, 468
    interpolation, 460

Maximum Entropy, 215, 301, 305
multiresolution, 260, 268, 279
numerical, 398, 510
pyramidal median transform, 268
Richardson-Lucy, 301
statistical, 253, 311, 331, 507
wavelet transform, 279, 468
MIDAS, 221, 341
Miriad, 129, 433
mirroring, database, 185
missions, space
    ALEXIS, 203
    AXAF, 207, 357, 361, 422
    Compton observatory, 176
    Einstein observatory, 48, 418
    EUVE, 44, 327, 453
    Ginga, 211
    Hubble Space Telescope, 80, 158, 162, 301, 305, 345, 349, 456, 464
    IRAS, 264, 289
    IUE, 155, 253
    POLAR, 109
    ROSAT, 117, 215, 353, 406
    SAX, 383, 387
    SWAS, 136
    XTE, 62, 241
modeling
    Bayesian, 315
    data, 315
    scattered light, 345
models
    blackboard, 429
    data, 191, 199, 207
    lattice, 319
    noise, 514
    selection, 319, 323
    spatial, 319, 323
    spectra, 453
    structural, 148
MOKA, 173
morphology, mathematical, 289
Mosaic, *see* NCSA Mosaic
mosaicing, 335
motion blur, 301
multi-processing, 357
multiresolution analysis, 260, 268, 279
multiuser software, 99

NASA, Astrophysics Program, 8

NCSA Mosaic, 68, 76, 80, 182
NEMO, 398
netCDF data format, 229
neural networks, 253, 264
news, 52
NGC 6822, 323
noise
    cleaning, 484
    models, 514
null eigenvectors, 480
numerical methods, 398, 510

object
    classification, 264
    cross-identification, 488
    detection, 260, 268, 279
object-oriented
    data modeling, 191
    databases, 169
    programming, 391
Occam, 379
on-line services, 48, 84, 99
open
    architecture, 207, 414, 422
    systems, 25
optical media, 162
optimization, 305

Palomar Sky Survey, 500
parallel processing, 357
parameter estimation, 331
PCA, 437
PCIPS, 133, 488
Perl, 65, 422
PGPLOT, 129
photometry, 44, 293, 335, 507
    CCD, 133, 484, 488
    surface, 492
photon-event list, 215
picturebooks, electronic, 3
pipeline, 422, 429
point
    patterns, 323
    processes, 319
    response function (PRF), 353, 406
    spread function (PSF), 293, 301, 349
    spread functions (PSF) matching, 297
POLAR, 109

polarization, 480
porting software, *see* software migration
position, 464
posix, 422
power spectrum, 460
procmail, 65
programming
    object-oriented, 391
    real-time, 391
    tools, 245
    visual, 148
project management, 40
proper motions, 500
proposal handling, 58, 62, 65
PROS, 207, 406, 410, 414
publication
    archive, 155
    electronic, 76, 129
pyramidal median transform, 268

QE, 406
QPOE data format, 44, 207, 215, 353, 406, 418, 425
quadtrees, linear, 472
quick-look data, 68

radial velocity, 496
radio telescope, 402
radiometer, 480
real-time programming, 113, 391
redshift, 496
regionalized variables, 319
regions, 414
regression, robust, 468
Remote Proposal Submission (RPS), 62
Richardson-Lucy method, 301
robust regression, 468
robustness, 514
ROSAT, 117, 215, 353, 406

sampling, 315, 460
sao-iis, 492
satellites, *see* missions, space
SAX, 383, 387
scattered light, 345
SCCA, 311
scheduling, 136
Schmidt plates, 500
searching, 55

biases, 327
literature, 28
servers, 62, 89
signal estimation, 315
simulation, 357
    source, 353
sky surveys, 215, 272, 327, 500
software
    analysis, 99, 109, 121, 133, 367, 387, 429, 433, 443
    applications, *see* applications software
    architecture, 429
    development, 62
    distributed, 25
    distribution, 371
    migration, 371, 375
    multiuser, 99
    portability, 245, 422
    systems, 410
sources
    classification, 257, 272
    detection, 327, 484
    detection in event lists, 215
    M82, 510
    NGC 6822, 323
    simulation, 353, 361
    variability, 117
space missions, *see* missions, space
spaceball, 144
spatial
    models, 319, 323
    region descriptors, 414
    statistics, 319, 323
    structure, 323
spectral
    analysis, 133
    classification, 253
spectroscopy, 44, 121, 253, 443, 453, 496, 510
    echelle, 133, 468, 476
    resolution, 503
    submillimeter, 402
SQIID camera, 335, 492
SQL, 32, 68
srcinten, 406
standards, 241
Starlink, 371, 375
stars, 253
    stellar dynamics, 398

stellar interiors, 185
Sun, 185
variability, 297
StarView, 80, 158
statistics
  methods, 253, 311, 331, 507
  spatial, 319, 323
structural models, 148
STSDAS tables, 402
subroutine libraries
  FITSIO, 367
  PGPLOT, 129
surface
  curvature, 484
  photometry, 492
surveys
  Digitized Sky, 179
  EUVE, 44
  Palomar Sky, 500
  Palomar Sky II, 272
  ROSAT, 215
Sybase, 72
systematic errors, 476
systems
  architecture, 84
  information, 36, 40
  legacy, 429
  open, 25
  software, 410

tapes, 241
Tcl/Tk, 58, 89, 113, 395
telescope
  control, 25
  design, 148
  radio, 402
thorium lamp, 476
time series analysis, 460
Tiny Tim, 349
tools
  collaborative, 99
  programming, 245
  scheduling, 136
training, 371
transform
  Hough, 468
  pyramidal median, 268
  wavelet, 279, 468
transputers, 379
turbulent convection, 15

UKIRT, 379
UNIX, 357, 371, 375
user interface management system (UIMS), 89
user interfaces, 68, 72, 129, 162
  e-mail, 58, 166
  graphical, 89, 109, 121, 125, 133, 136, 140, 158, 173, 395, 437, 453

variability, 117
  stars, 297
velocity, 464
visual programming, 148
visualization, 15, 144, 148, 433
VMS, 375
volume rendering, 15

wavelength calibration, 468, 476
wavelet transform, 279, 468
WDB, 72
Wide Area Information Server (WAIS), 32, 36, 52
widget, 25, 89, 99, 109, 125, 140, 387, 437
WIP, 129
WIYN, 140
world coordinates, 437
World Wide Web (WWW), 28, 32, 36, 40, 44, 52, 58, 62, 68, 72, 76, 80, 84, 162, 176, 182

X Window System, 125
X-ray, 117, 207, 211, 383, 387
XLIB, 89
Xt, 89
XTE, 62, 241

Colophon

These Proceedings were prepared as a single LaTeX document, using a style based on the PASP conference style. The 122 independent LaTeX documents containing the contributed manuscripts were automatically processed into a form suitable for incorporation into a skeleton document, along with the front and back matter. The table of contents, indices, and cross-references between contributions were done in pretty much standard LaTeX fashion. The list of participants was provided by *ferberts associates*.

The decision to create a single large document was based on the hope that it would facilitate the production of the electronic version of these proceedings. It also made it easy to incorporate late papers into an almost-final version. A UNIX Makefile helped keep all of the pieces up-to-date.

Pages were typeset in 10 point Computer Modern fonts on pages approximately the same size as those in the final product. Tomas Rokicki's *dvips* program was used to scale the output up to 11 point for the final proofs, which were printed at ST ScI on an HP LaserJet 4M, using 600 dpi Computer Modern fonts constructed especially for this printer. After being reduced by the publisher, the final result should be true 10 point Computer Modern fonts at an effective resolution of 660 dpi.

All 120 manuscripts were received electronically, either via anonymous ftp or e-mail. Of these 119 contained usable LaTeX markup.

There are 119 figures, for which we received 130 Encapsulated PostScript (EPS) files and only eight figures on paper. We have clearly reached the point where EPS can be considered a near-universal standard. We scanned the eight printed figures (and the conference photograph), so that all of the figures would be available in the electronic version. All of the submitted EPS files were printable, although some were too big (in bytes) to be useful. In particular, some images were sent to us as color images, essentially making them 3 times bigger than necessary. For these files we created images at a lower resolution and saved them as black and white PostScript. Many figures had erroneous BoundingBox comments, forcing authors to resort to the `plotfiddle` macro. Since this makes the pages unscalable, we fixed all of the BoundingBox comments and replaced all of the `plotfiddle` commands—this was easier than it sounds, and should have been explained to authors in the first place. We formatted one of the 24 tables into an EPS file so that it could be included in landscape orientation (page 157). The conference photograph, with its overlay and identifier list, was done the same way.